Elliptically Contoured Models in Statistics and Portfolio Theory

Arjun K. Gupta • Tamas Varga • Taras Bodnar

Elliptically Contoured Models in Statistics and Portfolio Theory

Second Edition

Springer

Arjun K. Gupta
Department of Mathematics and Statistics
Bowling Green State University
OH, USA

Tamas Varga
Damjanich, Budapest
Hungary

Taras Bodnar
Department of Mathematics
Humboldt-University of Berlin
Berlin, Germany

ISBN 978-1-4939-5328-8 ISBN 978-1-4614-8154-6 (eBook)
DOI 10.1007/978-1-4614-8154-6
Springer New York Heidelberg Dordrecht London

Softcover reprint of the hardcover 2nd edition 2013

Printed on acid-free paper

Springer is part of Springer Science+Business Media (www.springer.com)

Dedicated to the memory of my parents the late Smt. Leela and Sh. Amarnath Gupta.

(AKG)

Dedicated to my children Terézia, Julianna, and Kristóf.

(TV)

Dedicated to my wife Olha and to my children Bohdan and Anna-Yaroslava.

(TB)

Preface

In multivariate statistical analysis, elliptical distributions have recently provided an alternative to the normal model. Most of the work, however, is spread out in journals throughout the world and is not easily accessible to the investigators. Fang, Kotz, and Ng presented a systematic study of multivariate elliptical distributions; however, they did not discuss the matrix variate case. Fang and Zhang have summarized the results of generalized multivariate analysis which include vector as well as the matrix variate distributions. On the other hand, Fang and Anderson collected research papers on matrix variate elliptical distributions, many of them published for the first time in English. They published very rich material on the topic, but the results are given in paper form which does not provide a unified treatment of the theory. Therefore, it seemed appropriate to collect the most important results on the theory of matrix variate elliptically contoured distributions available in the literature and organize them in a unified manner that can serve as an introduction to the subject.

The book will be useful for researchers, teachers, and graduate students in statistics and related fields whose interests involve multivariate statistical analysis and its application into portfolio theory. Parts of this book were presented by Arjun K. Gupta as a one semester course at Bowling Green State University. Knowledge of matrix algebra and statistics at the level of Anderson is assumed. However, Chap. 1 summarizes some results of matrix algebra. This chapter also contains a brief review of the literature and a list of mathematical symbols used in the book.

Chapter 2 gives the basic properties of the matrix variate elliptically contoured distributions, such as the probability density function and expected values. It also presents one of the most important tools of the theory of elliptical distributions, the stochastic representation.

The probability density function and expected values are investigated in detail in Chap. 3.

Chapter 4 focuses on elliptically contoured distributions that can be represented as mixtures of normal distributions.

The distributions of functions of random matrices with elliptically contoured distributions are discussed in Chap. 5. Special attention is given to quadratic forms.

Characterization results are given in Chap. 6.

The next three chapters are devoted to statistical inference. Chapter 7 focuses on estimation results, whereas Chap. 8 is concerned with hypothesis testing problems. Inference for linear models is studied in Chap. 9.

Chapter 10 deals with the application of the elliptically contoured distributions for modeling financial data. We present distributional properties of the estimated main characteristics of optimal portfolios, like variance and expected return assuming that the asset returns are elliptically contoured distributed. The joint distributions of the estimated parameters of the efficient frontier are derived as well as we provide exact inference procedures for the corresponding population values. We also study the distributional properties of the estimated weights of the global minimum variance portfolio in detail.

In Chap. 11, we consider a further extension of matrix variate elliptically contoured distributions that allows us to model the asymmetry in data. Here, first the multivariate skew normal distribution is presented and its matrix generalization is discussed. We also study the main properties of this distribution, like moments, the density function, and the moment-generating function. Next, the skew t-distribution is introduced as well as the general class of matrix variate skew elliptically contoured distributions. Moreover, we present the distributional properties of quadratic forms in skew elliptical distributions and discuss the inference procedures. An application into portfolio theory is discussed as well. Finally, an up-to-date bibliography has been provided, along with author and subject indexes. The materials in the first nine chapters are from the book Elliptically Contoured Models in Statistics by the first two authors. The material in Chaps. 10 and 11 is taken from the papers of the authors. Permission of their publishers Kluwer Academic Publishers (http://www.wkap.com), Japan Statistical Society (http://www.jss.gr.jp), Springer (http://www.springer.com), and Taylor and Francis (http://www.tandfonline.com/) is gratefully acknowledged.

We would like to thank the Department of Mathematics and Statistics, Bowling Green State University, and the Department of Mathematics, Humboldt University of Berlin, for supporting our endeavor and for providing the necessary facilities to accomplish the task. The first author is thankful to the Biostatistics Department, University of Michigan, for providing him the opportunity to organize the material in its final form. Thanks are also due to Professors D. K. Nagar, M. Siotani, J. Tang, and N. Nguyen for many helpful discussions. He would also like to acknowledge his wife, Meera, and his children, Alka, Mita, and Nisha, for their support throughout the writing of the book. The second author is thankful to his mother Edit for her support in the early stages of the work on this book. The third author acknowledges the support of the Department of Statistics, European University Viadrina and the German Research Foundation (DFG) via the Research Unit 1735 "Structural Inference in Statistics: Adaptation and Efficiency". Thanks are also due to Professors W. Schmid and Y. Yelejko. He is also greatly thankful to his wife Olha and to his children Bohdan and Anna-Yaroslava for providing considerable help during the preparation of the book.

Finally our sincere thanks to Marc Strauss, Springer, for his help at every stage of the completion of this book.

Bowling Green, USA — Arjun K. Gupta
Budapest, Hungary — Tamas Varga
Berlin, Germany — Taras Bodnar

Contents

Part IV Applications

Acronyms

We denote matrices by capital bold letters, vectors by small bold letters and scalars by small letters. We use the same notation for a random variable and its values. Also the following notations will be used in the sequel.

$\mathbb{R}^p$	the p-dimensional real space
$\mathscr{B}(\mathbb{R}^p)$	the Borel sets in $\mathbb{R}^p$
S_p	the unit sphere in $\mathbb{R}^p$
$\mathbb{R}^+$	the set of positive real numbers
$\mathbb{R}_0^+$	the set of nonnegative real numbers
$\chi_A(x)$	the indicator function of A, that is $\chi_A(x) = 1$ if $x \in A$ and $\chi_A(x) = 0$ if $x \notin A$
$\chi(x \geq t)$	the same as $\chi_{[t,\infty)}(x)$ (t is a real number)
$\mathbf{A} \in \mathbb{R}^{p \times n}$	$\mathbf{A}$ is a $p \times n$ real matrix
a_{ij}	the (i, j)th element of matrix $\mathbf{A}$
$\mathbf{A}'$	transpose of $\mathbf{A}$
$rk(\mathbf{A})$	rank of $\mathbf{A}$
$\mathbf{A} > \mathbf{0}$	the square matrix $\mathbf{A}$ is positive definite (see also Sect. 1.2)
$\mathbf{A} \geq \mathbf{0}$	the square matrix $\mathbf{A}$ is positive semidefinite (see also Sect. 1.2)
$\lvert\mathbf{A}\rvert$	determinant of the square matrix $\mathbf{A}$
$tr(\mathbf{A})$	trace of the square matrix $\mathbf{A}$
$etr(\mathbf{A})$	$\exp(tr(\mathbf{A}))$ if $\mathbf{A}$ is a square matrix
$\lVert\mathbf{A}\rVert$	norm of $\mathbf{A}$ defined by $\lVert\mathbf{A}\rVert = \sqrt{tr(\mathbf{A}'\mathbf{A})}$
$\mathbf{A}^{-1}$	inverse of $\mathbf{A}$
$\mathbf{A}^-$	generalized inverse of $\mathbf{A}$, that is $\mathbf{A}\mathbf{A}^-\mathbf{A} = \mathbf{A}$ (see also Sect. 1.2)
$\mathbf{A}^{1/2}$	let the spectral decomposition of $\mathbf{A} \geq 0$ be $\mathbf{G}\mathbf{D}\mathbf{G}'$, and define $\mathbf{A}^{1/2} = \mathbf{G}\mathbf{D}^{1/2}\mathbf{G}'$ (see also Sect. 1.2)
$O(p)$	the set of $p \times p$ dimensional orthogonal matrices
$\mathbf{I}_p$	the $p \times p$ dimensional identity matrix
$\mathbf{e}_p$	the p-dimensional vector whose elements are 1's; that is, $\mathbf{e}_p = (1, 1, \ldots, 1)'$ real matrix
$\mathbf{A} \otimes \mathbf{B}$	Kronecker product of the matrices $\mathbf{A}$ and $\mathbf{B}$ (see also Sect. 1.2)

$\mathbf{A} > \mathbf{B}$	$\mathbf{A} - \mathbf{B}$ is positive definite
$\mathbf{A} \geq \mathbf{B}$	$\mathbf{A} - \mathbf{B}$ is positive semidefinite
$vec(\mathbf{A})$	the vector $\begin{pmatrix} \mathbf{a}_1 \\ \mathbf{a}_2 \\ \vdots \\ \mathbf{a}_n \end{pmatrix}$ where $\mathbf{a}_i$ denotes the ith column of $p \times n$ matrix $\mathbf{A}$, $i = 1, 2, \ldots, n$
$J(\mathbf{X} \to f(\mathbf{X}))$	the Jacobian of the matrix transformation f
$\mathbf{X} \sim \mathscr{D}$	the random matrix $\mathbf{X}$ is distributed according to the distribution $\mathscr{D}$
$\mathbf{X} \approx \mathbf{Y}$	the random matrices $\mathbf{X}$ and $\mathbf{Y}$ are identically distributed
$Cov(\mathbf{X})$	covarinace matrix of the random matrix $\mathbf{X}$; that is $Cov(\mathbf{X}) = Cov(vec(\mathbf{X}'))$
$\phi_{\mathbf{X}}(\mathbf{T})$	the characteristic function of the random matrix $\mathbf{X}$ at $\mathbf{T}$; that is $E(etr(i\mathbf{T}'\mathbf{X}))$, $\mathbf{X}, \mathbf{T} \in I\!R^{p \times n}$

For a review of Jacobians, see Press (1972) and Siotani, Hayakawa and Fujikoshi (1985). We also use the following notations for some well known probability distributions.

UNIVARIATE DISTRIBUTIONS:

$N(\mu, \sigma^2)$ normal distribution; its probability density function is

$$f(x) = \frac{1}{\sqrt{2\pi}\sigma} \exp\left\{ -\frac{(x-\mu)^2}{2\sigma^2} \right\},$$

where $\mu \in I\!R$, $\sigma \in I\!R^+$, and $x \in I\!R$

$B(a, b)$ beta distribution; its probability density function is

$$f(x) = \frac{1}{\beta(a,b)} x^{a-1}(1-x)^{b-1},$$

where $a > 0$, $b > 0$, $\beta(a,b) = \frac{\Gamma(a)\Gamma(b)}{\Gamma(a+b)}$, and $0 < x < 1$

t_n Student's t-distribution; its probability density function is

$$f(x) = \frac{\Gamma\left(\frac{n+1}{2}\right)}{\sqrt{n\pi}\Gamma\left(\frac{n}{2}\right)} \left(1 + \frac{x^2}{n}\right)^{-\frac{n+1}{2}},$$

where $n > 0$, and $x \in I\!R$

χ^2_n chi-square distribution; its probability density function is

$$f(x) = \frac{1}{2^{\frac{n}{2}}\Gamma\left(\frac{n}{2}\right)} x^{\frac{n}{2}-1} \exp\left\{-\frac{x}{2}\right\},$$

where $n > 0$, and $x \geq 0$

χ_n chi distribution; its probability density function is

$$f(x) = \frac{1}{2^{\frac{n}{2}-1}\Gamma\left(\frac{n}{2}\right)} x^{n-1} \exp\left\{-\frac{x^2}{2}\right\},$$

where $n > 0$, and $x \geq 0$

$F_{n,m}$ F distribution; its probability density function is

$$f(x) = \frac{\Gamma\left(\frac{n+m}{2}\right)}{\Gamma\left(\frac{n}{2}\right)\Gamma\left(\frac{m}{2}\right)} \left(\frac{n}{m}\right)^{\frac{n}{2}} \frac{x^{\frac{n}{2}-1}}{\left(1+\frac{n}{m}x\right)^{\frac{n+m}{2}}},$$

where $n, m = 1, 2, \ldots,$, and $x > 0$

$U_{p,m,n}$ U distribution, which is the same as the distribution of $\prod_{i=1}^{p} v_i$; where v_i's are independent and

$$v_i \sim B\left(\frac{n+1-i}{2}, \frac{m}{2}\right)$$

For the U distribution, see Anderson (2003), pp. 307–314.

MULTIVARIATE DISTRIBUTIONS:

$N_p(\mu, \Sigma)$ multivariate normal distribution; its characteristic function is

$$\phi_{\mathbf{x}}(\mathbf{t}) = \exp\left\{i\mathbf{t}'\mu + \frac{1}{2}\mathbf{t}'\boldsymbol{\Sigma}\mathbf{t}\right\},$$

where $\mathbf{x}, \mathbf{t}, \mu \in I\!R^p$, $\Sigma \in I\!R^{p\times p}$, and $\boldsymbol{\Sigma} \geq \mathbf{0}$

$D(m_1, \ldots, m_p; m_{p+1})$ Dirichlet distribution; its probability density function is

$$f(\mathbf{x}) = \frac{\Gamma\left(\sum_{i=1}^{p+1} m_i\right)}{\prod_{i=1}^{p+1}\Gamma(m_i)} \prod_{i=1}^{p} x_i^{m_i-1} \left(1 - \sum_{i=1}^{p} x_i\right)^{m_{p+1}-1},$$

where $\mathbf{x} = (x_1, x_2, \ldots, x_p)' \in I\!R^p$, $0 < \sum_{i=1}^{p} x_i < 1$, and $m_i > 0$, $i = 1, 2, \ldots, p$

$SMT_\nu(\alpha)$ multivariate skew t-distribution; its probability density function is

$$f_\nu(\mathbf{y}) = 2 f_{T_\nu}(\mathbf{y}) F_{T_{\nu+p}}\left(\frac{\alpha'\mathbf{y}}{(\nu + \mathbf{y}'\Sigma^{-1}\mathbf{y})^{\frac{1}{2}}}\sqrt{\nu+p}\right),$$

where $f_{T_k}(\cdot)$ and $F_{T_k}(\cdot)$ denote the probability density function and the cumulative distribution function of central t-distribution with k degrees of freedom, respectively; $\mathbf{y} \in I\!R^p$, $\nu > 0$, $\alpha \in I\!R^p$, $\Sigma \in I\!R^{p\times p}$, and $\Sigma > \mathbf{0}$.

$SMC(\alpha)$ multivariate skew Cauchy distribution; its probability density function is

$$f(\mathbf{y}) = \frac{2\Gamma\left(\frac{p+1}{2}\right)}{(\pi)^{\frac{p+1}{2}}}\left(1 + \sum_1^p y_j^2\right)^{-\frac{p+1}{2}} F_{T_{p+1}}\left(\frac{\alpha'\mathbf{y}\sqrt{p+1}}{\left(1 + \sum_1^p y_j^2\right)^{\frac{1}{2}}}\right),$$

where $F_{T_k}(\cdot)$ denotes the cumulative distribution function of central t-distribution with k degrees of freedom; $\mathbf{y} \in I\!R^p$, $\alpha \in I\!R^p$, $\Sigma \in I\!R^{p\times p}$, and $\Sigma > \mathbf{0}$.

$CSN_{p,q}(\mu, \Sigma, \mathbf{D}, \nu, \Delta)$ closed skew normal distribution; its probability density function is

$$g_{p,q}(\mathbf{y}) = C\phi_p(\mathbf{y}; \mu, \Sigma)\Phi_q[\mathbf{D}(\mathbf{y} - \mu); \nu, \Delta],$$

with

$$C^{-1} = \Phi_q[\mathbf{0}; \nu, \Delta + \mathbf{D}\Sigma\mathbf{D}'] \tag{1}$$

where $\phi_l(\mathbf{x}; \mu, \Sigma)$ and $\Phi_l(\mathbf{x}; \mu, \Sigma)$ denote the probability density function and the cumulative distribution function of the l-dimensional normal distribution with mean vector μ and covariance matrix Σ, respectively; $\mathbf{y} \in I\!R^p$, $p, q \geq 1$, $\mu \in I\!R^p$, $\nu \in I\!R^q$, $\mathbf{D} \in I\!R^{q\times p}$, $\Sigma \in I\!R^{p\times p}$, $\Sigma > \mathbf{0}$, $\Delta \in I\!R^{q\times q}$, and $\Delta > \mathbf{0}$.

MATRIX VARIATE DISTRIBUTIONS:

$N_{p,n}(\mathbf{M}, \Sigma \otimes \Phi)$ matrix variate normal distribution; its characteristic function is

$$\phi_{\mathbf{X}}(\mathbf{T}) = etr\left\{i\mathbf{T}'\mathbf{M} + \frac{1}{2}\mathbf{T}'\boldsymbol{\Sigma}\mathbf{T}\boldsymbol{\Phi}\right\},$$

where $\mathbf{M}, \mathbf{X}, \mathbf{T} \in I\!R^{p\times n}$, $\boldsymbol{\Sigma} \in I\!R^{p\times p}$, $\boldsymbol{\Sigma} \geq \mathbf{0}$, $\boldsymbol{\Phi} \in I\!R^{n\times n}$, and $\boldsymbol{\Phi} \geq \mathbf{0}$

$W_p(\boldsymbol{\Sigma}, n)$ Wishart distribution; its probability density function is

$$f(\mathbf{X}) = \frac{|\mathbf{X}|^{\frac{n-p-1}{2}} etr\left\{-\frac{1}{2}\boldsymbol{\Sigma}^{-1}\mathbf{X}\right\}}{2^{\frac{np}{2}}|\boldsymbol{\Sigma}|^{\frac{n}{2}}\Gamma_p\left(\frac{n}{2}\right)},$$

where $\mathbf{X} \in I\!R^{p\times p}$, $\mathbf{X} > \mathbf{0}$, $\boldsymbol{\Sigma} \in I\!R^{p\times p}$, $\boldsymbol{\Sigma} > \mathbf{0}$, p, n are integers, $n \geq p$, and

$$\Gamma_p(t) = \pi^{\frac{p(p-1)}{4}} \prod_{i=1}^{p} \Gamma\left(t - \frac{i-1}{2}\right)$$

$B_p^I(a, b)$ matrix variate beta distribution of type I; its probability density function is

$$f(\mathbf{X}) = \frac{|\mathbf{X}|^{a-\frac{p+1}{2}}|\mathbf{I}_p - \mathbf{X}|^{b-\frac{p+1}{2}}}{\beta_p(a, b)},$$

where $a > \frac{p-1}{2}$, $b > \frac{p-1}{2}$, $\beta_p(a, b) = \frac{\Gamma_p(a)\Gamma_p(b)}{\Gamma_p(a+b)}$, $\mathbf{X} \in I\!R^{p\times p}$, and $\mathbf{0} < \mathbf{X} < \mathbf{I}_p$

$B_p^{II}(a, b)$ matrix variate beta distribution of type II; its probability density function is

$$f(\mathbf{X}) = \frac{|\mathbf{X}|^{a-\frac{p+1}{2}}|\mathbf{I}_p + \mathbf{X}|^{-(a+b)}}{\beta_p(a, b)},$$

where $a > \frac{p-1}{2}$, $b > \frac{p-1}{2}$, $\mathbf{X} \in I\!R^{p\times p}$, and $\mathbf{X} > \mathbf{0}$

$T_{p,n}(m, \mathbf{M}, \boldsymbol{\Sigma}, \boldsymbol{\Phi})$ matrix variate T distribution; its probability density function is

$$f(\mathbf{X}) = \frac{\pi^{-\frac{np}{2}}\Gamma_p\left(\frac{n+m+p-1}{2}\right)}{\Gamma_p\left(\frac{m+p-1}{2}\right)|\boldsymbol{\Sigma}|^{\frac{n}{2}}|\boldsymbol{\Phi}|^{\frac{p}{2}}}|\mathbf{I}_p + \boldsymbol{\Phi}^{-1}(\mathbf{X}-\mathbf{M})\boldsymbol{\Sigma}^{-1}(\mathbf{X}-\mathbf{M})'|^{-\frac{n+m+p-1}{2}},$$

where $m > 0$, $\mathbf{M}, \mathbf{X}, \mathbf{T} \in I\!R^{p\times n}$, $\boldsymbol{\Sigma} \in I\!R^{p\times p}$, $\boldsymbol{\Sigma} > \mathbf{0}$, $\boldsymbol{\Phi} \in I\!R^{n\times n}$, and $\boldsymbol{\Phi} > \mathbf{0}$

$E_{p,n}(\mathbf{M}, \Sigma \otimes \Phi, \psi)$ matrix variate elliptically contoured distribution; its characteristic function

$$\phi_{\mathbf{X}}(\mathbf{T}) = etr(i\mathbf{T}'\mathbf{M})\psi(tr(\mathbf{T}'\Sigma\mathbf{T}\Phi)),$$

where $\mathbf{T} : p \times n$, $\mathbf{M} : p \times n$, $\Sigma : p \times p$, $\Phi : n \times n$, $\Sigma \geq \mathbf{0}$, $\Phi \geq \mathbf{0}$, and $\psi : [0, \infty) \to \mathbb{R}$

For further discussion of $B_p^I(a,b)$, $B_p^{II}(a,b)$, see Olkin and Rubin (1964) and Javier and Gupta (1985b), and for results on $T_{p,n}(m, \mathbf{M}, \Sigma, \Phi)$, see Dickey (1967).

Part I
Preliminaries

Chapter 1
Preliminaries

1.1 Introduction and Literature Review

Matrix variate distributions have been studied by statisticians for a long time. The first results on this topic were published by Hsu and Wishart. These distributions proved to be useful in statistical inference. For example, the Wishart distribution is essential when studying the sample covariance matrix in the multivariate normal theory. Random matrices can also be used to describe repeated measurements on multivariate variables. In this case, the assumption of the independence of the observations, a commonly used condition in statistical analysis, is often not feasible. When analyzing data sets like these, the matrix variate elliptically contoured distributions can be used to describe the dependence structure of the data. This is a rich class of distributions containing the matrix variate normal, contaminated normal, Cauchy and Student's t-distributions. The fact that the distributions in this class possess certain properties, similar to those of the normal distribution, makes them especially useful. For example, many testing procedures developed for the normal theory to test various hypotheses can be used for this class of distributions, too.

Matrix variate elliptically contoured distributions represent an extension of the concept of elliptical distributions from the vector to the matrix case. Important distribution results on vector variate elliptical distributions were derived by Kelker (1970), Chu (1973), Dawid (1977) and Cambanis, Huang, and Simons (1981). Quadratic forms in elliptical distributions were studied by Cacoullos and Koutras (1984), Fang and Wu (1984), Anderson and Fang (1987), and Smith (1989). Problems related to moments were considered by Berkane and Bentler (1986a). Characterization results were given by Kingman (1972), Khatri and Mukerjee (1987), and Berkane and Bentler (1986b). Kariya (1981), Kuritsyn (1986), Anderson, Fang, and Hsu (1986), Jajuga (1987), Cellier, Fourdrinier, and Robert (1989) and Grübel and Rocke (1989) focused on inference problems. Asymptotic results were obtained by Browne (1984), Hayakawa (1987), Khatri (1988) and Mitchell (1989). Special aspects of elliptical distributions were discussed

A.K. Gupta et al., *Elliptically Contoured Models in Statistics and Portfolio Theory*,
DOI 10.1007/978-1-4614-8154-6_1, © Springer Science+Business Media New York 2013

by Khatri (1980), Sampson (1983), Mitchell and Krzanowski (1985), Cacoullos and Koutras (1985), Khattree and Peddada (1987), and Cléroux and Ducharme (1989). Krishnaiah and Lin (1986) introduced the concept of complex elliptical distributions. Some of the early results in elliptical distributions were summarized in Muirhead (1982) and Johnson (1987). More extensive reviews of papers on this topic were provided by Chmielewski (1981), and Bentler and Berkane (1985). The most recent summary of distribution results was given by Fang, Kotz, and Ng (1990).

Some of the papers mentioned above also contain results on matrix variate elliptically contoured distributions; for example, Anderson and Fang (1987), and Anderson, Fang, and Hsu (1986). Other papers, like Chmielewski (1980), Richards (1984), Khatri (1987), and Sutradhar and Ali (1989), are also concerned with matrix variate elliptical distributions. Fang and Anderson (1990) is a collection of papers on matrix variate elliptical distributions. Many of these papers were originally published in Chinese journals and this is their first publication in English. Fang and Zhang (1990) have provided an excellent account of spherical and related distributions.

The purpose of the present book is to provide a unified treatment of the theory of matrix variate elliptically contoured distributions, to present the most important results on the topic published in various papers and books, to give their proofs and show how these results can be applied in portfolio theory.

1.2 Some Results from Matrix Algebra

In this section, we give some results from matrix algebra which are used in the subsequent chapters. Except for the results on the generalized inverse, we do not prove the theorems since they can be found in any book of linear algebra (e.g. Magnus and Neudecker, 1988). Other books, like Anderson (2003) and Muirhead (1982), discuss these results in the appendices of their books.

Definition 1.1. Let $\mathbf{A}$ be a $p \times p$ matrix. Then, $\mathbf{A}$ is called

(i) Symmetric if $\mathbf{A}' = \mathbf{A}$.
(ii) Idempotent if $\mathbf{A}^2 = \mathbf{A}$.
(iii) Nonsingular if $|\mathbf{A}| \neq 0$.
(iv) Orthogonal if $\mathbf{A}\mathbf{A}' = \mathbf{A}'\mathbf{A} = \mathbf{I}_p$.
(v) Positive semidefinite and this is denoted by $\mathbf{A} \geq \mathbf{0}$ if $\mathbf{A}$ is symmetric and for every p-dimensional vector $\mathbf{v}$, $\mathbf{v}'\mathbf{A}\mathbf{v} \geq 0$.
(vi) Positive definite and this is denoted by $\mathbf{A} > \mathbf{0}$ if $\mathbf{A}$ is symmetric and for every p-dimensional nonzero vector $\mathbf{v}$, $\mathbf{v}'\mathbf{A}\mathbf{v} > 0$.
(vii) Permutation matrix if in each row and each column of $\mathbf{A}$ exactly one element is 1 and all the others are 0.
(viii) Signed permutation matrix if in each row and each column of $\mathbf{A}$ exactly one element is 1 or -1 and all the others are 0.

Theorem 1.1. *Let* $\mathbf{A}$ *be* $p \times p$, *and* $\mathbf{B}$ *be* $q \times p$. *Then, we have the following results.*

(i) If $\mathbf{A} > \mathbf{0}$, *then* $\mathbf{A}^{-1} > \mathbf{0}$.
(ii) If $\mathbf{A} \geq \mathbf{0}$, *then* $\mathbf{BAB}' \geq \mathbf{0}$.
(iii) If $q \leq p$, $\mathbf{A} > \mathbf{0}$ *and* $rk(\mathbf{B}) = q$, *then* $\mathbf{BAB}' > \mathbf{0}$.

Definition 1.2. Let $\mathbf{A}$ be a $p \times p$ matrix. Then, the roots (with multiplicity) of the equation

$$|\mathbf{A} - \lambda I_p| = 0$$

are called the characteristic roots of $\mathbf{A}$.

Theorem 1.2. *Let* $\mathbf{A}$ *be a* $p \times p$ *matrix and* $\lambda_1, \lambda_2, \ldots, \lambda_p$ *its characteristic roots. Then,*

(i) $|\mathbf{A}| = \prod_{i=1}^{p} \lambda_i$.
(ii) $tr(\mathbf{A}) = \sum_{i=1}^{p} \lambda_i$.
(iii) $rk(\mathbf{A}) =$ *the number of nonzero characteristic roots.*
(iv) $\mathbf{A}$ *is nonsingular if and only if the characteristic roots are nonzero.*
(v) Further, if we assume that $\mathbf{A}$ *is symmetric, then the characteristic roots of* $\mathbf{A}$ *are real.*
(vi) $\mathbf{A}$ *is positive semidefinite if and only if the characteristic roots of* $\mathbf{A}$ *are nonnegative.*
(vii) $\mathbf{A}$ *is positive definite if and only if the characteristic roots of* $\mathbf{A}$ *are positive.*

The next theorem gives results on the rank of matrices.

Theorem 1.3. *(i) Let* $\mathbf{A}$ *be a* $p \times q$ *matrix. Then,* $rk(\mathbf{A}) \leq \min(p,q)$ *and* $rk(\mathbf{A}) = rk(\mathbf{A}') = rk(\mathbf{AA}') = rk(\mathbf{A}'\mathbf{A})$. *If* $p = q$, *then* $rk(\mathbf{A}) = p$ *if and only if* $\mathbf{A}$ *is nonsingular.*
(ii) Let $\mathbf{A}$ *and* $\mathbf{B}$ *be* $p \times q$ *matrices. Then,* $rk(\mathbf{A}+\mathbf{B}) \leq rk(\mathbf{A}) + rk(\mathbf{B})$.
(iii) Let $\mathbf{A}$ *be a* $p \times q$, $\mathbf{B}$ *a* $q \times r$ *matrix. Then,* $rk(\mathbf{AB}) \leq \min(rk(\mathbf{A}), rk(\mathbf{B}))$. *If* $p = q$ *and* $\mathbf{A}$ *is nonsingular then* $rk(\mathbf{AB}) = rk(\mathbf{B})$.

Definition 1.3. Let $\mathbf{A}$ be a $p \times q$ matrix. If $rk(\mathbf{A}) = min(p,q)$, then $\mathbf{A}$ is called a full rank matrix.

In the following theorem we list some of the properties of the trace function.

Theorem 1.4. *(i) Let* $\mathbf{A}$ *be a* $p \times p$ *matrix. Then,* $tr(\mathbf{A}) = tr(\mathbf{A}')$, *and* $tr(c\mathbf{A}) = ctr(\mathbf{A})$ *where* c *is a scalar.*
(ii) Let $\mathbf{A}$ *and* $\mathbf{B}$ *be* $p \times p$ *matrices. Then,* $tr(\mathbf{A}+\mathbf{B}) = tr(\mathbf{A}) + tr(\mathbf{B})$.
(iii) Let $\mathbf{A}$ *be a* $p \times q$, $\mathbf{B}$ *a* $q \times p$. *Then,* $tr(\mathbf{AB}) = tr(\mathbf{BA})$.

Theorem 1.5. *Let the $p \times p$ matrix $\mathbf{A}$ be defined by*

$$a_{ij} = \begin{cases} x, \text{ if } i = j \\ y, \text{ if } i \neq j \end{cases}.$$

Then, $|\mathbf{A}| = (x-y)^{p-1}(x+(p-1)y)$.

Now we give some matrix factorization theorems.

Theorem 1.6. *(Singular value decomposition of a matrix)*
Let $\mathbf{A}$ be a $p \times q$ matrix with $p \geq q$. Then, there exist a $p \times p$ orthogonal matrix $\mathbf{G}$, a $q \times q$ orthogonal matrix $\mathbf{H}$ and a $q \times q$ positive semidefinite diagonal matrix $\mathbf{D}$ such that

$$\mathbf{A} = \mathbf{G} \begin{pmatrix} \mathbf{D} \\ \mathbf{0} \end{pmatrix} \mathbf{H},$$

where $\mathbf{0}$ denotes the $(p-q) \times q$ zero matrix. Moreover, $rk(\mathbf{D}) = rk(\mathbf{A})$.

Theorem 1.7. *(Spectral decomposition of a symmetric matrix)*
Let $\mathbf{A}$ be a $p \times p$ symmetric matrix. Then, there exist a $p \times p$ orthogonal matrix $\mathbf{G}$ and a $p \times p$ diagonal matrix $\mathbf{D}$ such that

$$\mathbf{A} = \mathbf{G}\mathbf{D}\mathbf{G}'. \tag{1.1}$$

Moreover, if $\mathbf{A}$ is of the form (1.1) then the diagonal elements of $\mathbf{D}$ are the characteristic roots of $\mathbf{A}$.

Definition 1.4. Let $\mathbf{A}$ be a $p \times p$ positive semidefinite matrix with spectral decomposition $\mathbf{A} = \mathbf{G}\mathbf{D}\mathbf{G}'$. Let $\mathbf{D}^{1/2}$ be the diagonal matrix whose elements are the square roots of the elements of $\mathbf{D}$. Then we define $\mathbf{A}^{1/2}$ as $\mathbf{A}^{1/2} = \mathbf{G}\mathbf{D}^{1/2}\mathbf{G}'$.

Theorem 1.8. *Let $\mathbf{A}$ and $\mathbf{B}$ be $p \times p$ matrices. Assume $\mathbf{A}$ is positive definite and $\mathbf{B}$ is positive semidefinite. Then, there exist a $p \times p$ nonsingular matrix $\mathbf{C}$ and a $p \times p$ diagonal matrix $\mathbf{D}$ such that*

$$\mathbf{A} = \mathbf{C}\mathbf{C}' \quad \text{and} \quad \mathbf{B} = \mathbf{C}\mathbf{D}\mathbf{C}'. \tag{1.2}$$

Moreover, if $\mathbf{A}$ and $\mathbf{B}$ are of the form (1.2), then the diagonal elements of $\mathbf{D}$ are the roots of the equation $|\mathbf{B} - \lambda\mathbf{A}| = 0$.

Theorem 1.9. *Let $\mathbf{A}$ be a $p \times q$ matrix with $rk(\mathbf{A}) = q$. Then there exist a $p \times p$ orthogonal matrix $\mathbf{G}$ and a $q \times q$ positive definite matrix $\mathbf{B}$ such that*

$$\mathbf{A} = \mathbf{G} \begin{pmatrix} \mathbf{I}_q \\ \mathbf{0} \end{pmatrix} \mathbf{B},$$

where $\mathbf{0}$ denotes the $(p-q) \times q$ zero matrix.

Theorem 1.10. *(The rank factorization of a square matrix)*
Let $\mathbf{A}$ *be a* $p \times p$ *matrix with* $rk(\mathbf{A}) = q$. *Then, there exists a* $p \times q$ *matrix* $\mathbf{B}$ *of rank* q *such that* $\mathbf{A} = \mathbf{B}\mathbf{B}'$.

Theorem 1.11. *(Vinograd's Theorem)*
Assume $\mathbf{A}$ *be a* $p \times q$, *and* $\mathbf{B}$ *is a* $p \times r$ *matrix, where* $q \leq r$. *Then,* $\mathbf{A}\mathbf{A}' = \mathbf{B}\mathbf{B}'$ *if and only if there exists a* $q \times r$ *matrix* $\mathbf{H}$ *with* $\mathbf{H}\mathbf{H}' = \mathbf{I}_q$ *such that* $\mathbf{B} = \mathbf{A}\mathbf{H}$.

Theorem 1.12. *Let* $\mathbf{A}$ *be a* $p \times p$, *symmetric idempotent matrix of rank* q. *Then, there exists a* $p \times p$ *orthogonal matrix* $\mathbf{G}$ *such that*

$$\mathbf{A} = \mathbf{G}\begin{pmatrix} \mathbf{I}_q & \mathbf{0} \\ & \mathbf{0}\ \mathbf{0} \end{pmatrix}\mathbf{G}',$$

where the $\mathbf{0}$*'s denote zero matrices of appropriate dimensions.*

Theorem 1.13. *Let* $\mathbf{A}_1, \mathbf{A}_2, \ldots, \mathbf{A}_n$ *be* $p \times p$, *symmetric, idempotent matrices. Then, there exists a* $p \times p$ *orthogonal matrix such that* $\mathbf{G}'\mathbf{A}_i\mathbf{G}$ *is diagonal for every* $1 \leq i \leq n$ *if and only if* $\mathbf{A}_i\mathbf{A}_j = \mathbf{A}_j\mathbf{A}_i$ *for every* $1 \leq i, j \leq n$.

Theorem 1.14. *Let* $\mathbf{A}_1, \mathbf{A}_2, \ldots, \mathbf{A}_n$ *be* $p \times p$, *symmetric, idempotent matrices. Then, there exists a* $p \times p$ *orthogonal matrix* $\mathbf{G}$ *such that*

$$\mathbf{G}'\mathbf{A}_1\mathbf{G} = \begin{pmatrix} \mathbf{I}_{r_1} & \mathbf{0} \\ \mathbf{0} & \mathbf{0} \end{pmatrix}, \mathbf{G}'\mathbf{A}_2\mathbf{G} = \begin{pmatrix} \mathbf{0} & \mathbf{0} & \mathbf{0} \\ \mathbf{0} & \mathbf{I}_{r_2} & \mathbf{0} \\ \mathbf{0} & \mathbf{0} & \mathbf{0} \end{pmatrix},$$

$$\ldots, \mathbf{G}'\mathbf{A}_n\mathbf{G} = \begin{pmatrix} \mathbf{0} & \mathbf{0} & \mathbf{0} \\ \mathbf{0} & \mathbf{I}_{r_n} & \mathbf{0} \\ \mathbf{0} & \mathbf{0} & \mathbf{0} \end{pmatrix},$$

where $r_i = rk(\mathbf{A}_i)$, $i = 1, \ldots, n$, *if and only if* $\mathbf{A}_i\mathbf{A}_j = \mathbf{0}$ *for every* $i \neq j$.

Next, we give some results for the Kronecker product, also called direct product of matrices.

Definition 1.5. Let $\mathbf{A} = (a_{ij})$ be a $p \times q$, $\mathbf{B}$ an $r \times s$ matrix. Then the Kronecker product of $\mathbf{A}$ and $\mathbf{B}$, denoted by $\mathbf{A} \otimes \mathbf{B}$, is the $(pr) \times (qs)$ matrix defined by

$$\mathbf{A} \otimes \mathbf{B} = \begin{pmatrix} a_{11}\mathbf{B} & a_{12}\mathbf{B} & \ldots & a_{1q}\mathbf{B} \\ a_{21}\mathbf{B} & a_{22}\mathbf{B} & \ldots & a_{2q}\mathbf{B} \\ \vdots & \vdots & \ldots & \vdots \\ a_{p1}\mathbf{B} & a_{p2}\mathbf{B} & \ldots & a_{pq}\mathbf{B} \end{pmatrix}.$$

Theorem 1.15. *(i) If* c *and* d *are scalars, then* $(c\mathbf{A}) \otimes (d\mathbf{B}) = cd(\mathbf{A} \otimes \mathbf{B})$.
(ii) If $\mathbf{A}$ *and* $\mathbf{B}$ *are of equal dimension, then*

$$(\mathbf{A}+\mathbf{B}) \otimes \mathbf{C} = (\mathbf{A} \otimes \mathbf{C}) + (\mathbf{A} \otimes \mathbf{C}), \quad \textit{and} \quad \mathbf{C} \otimes (\mathbf{A}+\mathbf{B}) = (\mathbf{C} \otimes \mathbf{A}) + (\mathbf{C} \otimes \mathbf{B}).$$

(iii) $(\mathbf{A} \otimes \mathbf{B}) \otimes \mathbf{C} = \mathbf{A} \otimes (\mathbf{B} \otimes \mathbf{C})$.
(iv) $(\mathbf{A} \otimes \mathbf{B})' = \mathbf{A}' \otimes \mathbf{B}'$.
(v) *If* $\mathbf{A}$ *and* $\mathbf{B}$ *are square matrices then*

$$tr(\mathbf{A} \otimes \mathbf{B}) = tr(\mathbf{A})tr(\mathbf{B}).$$

(vi) *If* $\mathbf{A}$ *is* $p \times q$, $\mathbf{B}$ *is* $r \times s$, $\mathbf{C}$ *is* $q \times n$, *and* $\mathbf{D}$ *is* $s \times v$, *then*

$$(\mathbf{A} \otimes \mathbf{B})(\mathbf{C} \otimes \mathbf{D}) = (\mathbf{AC}) \otimes (\mathbf{BD}).$$

(vii) *If* $\mathbf{A}$ *and* $\mathbf{B}$ *are nonsingular matrices, then* $\mathbf{A} \otimes \mathbf{B}$ *is also nonsingular and*

$$(\mathbf{A} \otimes \mathbf{B})^{-1} = \mathbf{A}^{-1} \otimes \mathbf{B}^{-1}.$$

(viii) *If* $\mathbf{A}$ *and* $\mathbf{B}$ *are orthogonal matrices, then* $\mathbf{A} \otimes \mathbf{B}$ *is also orthogonal.*
(ix) *If* $\mathbf{A}$ *and* $\mathbf{B}$ *are positive semidefinite matrices, then* $\mathbf{A} \otimes \mathbf{B}$ *is also positive semidefinite.*
(x) *If* $\mathbf{A}$ *and* $\mathbf{B}$ *are positive definite matrices, then* $\mathbf{A} \otimes \mathbf{B}$ *is also positive definite.*
(xi) *If* $\mathbf{A}$ *is* $p \times p$, $\mathbf{B}$ *is* $q \times q$ *matrix, then* $|\mathbf{A} \otimes \mathbf{B}| = |\mathbf{A}|^q |\mathbf{B}|^p$.
(xii) *If* $\mathbf{A}$ *is* $p \times p$, $\mathbf{B}$ *is* $q \times q$, $\lambda_1, \lambda_2, \ldots, \lambda_p$ *are the characteristic roots of* $\mathbf{A}$, *and* $\mu_1, \mu_2, \ldots, \mu_q$ *are the characteristic roots of* $\mathbf{B}$, *then* $\lambda_i \mu_j$, $i = 1, \ldots, p$, $j = 1, \ldots, q$ *are the characteristic roots of* $\mathbf{A} \otimes \mathbf{B}$.

Theorem 1.16. *Let* $\mathbf{A}_1$ *and* $\mathbf{A}_2$ *be* $p \times q$, $\mathbf{B}_1$ *and* $\mathbf{B}_2$ *be* $r \times s$ *nonzero matrices. Then,* $\mathbf{A}_1 \otimes \mathbf{B}_1 = \mathbf{A}_2 \otimes \mathbf{B}_2$, *if and only if there exists a nonzero real number* c *such that* $\mathbf{A}_2 = c\mathbf{A}_1$ *and* $\mathbf{B}_2 = \frac{1}{c}\mathbf{B}_1$.

Definition 1.6. Let $\mathbf{X}$ be a $p \times n$ matrix and denote the columns of $\mathbf{X}$ by $\mathbf{x}_1, \mathbf{x}_2, \ldots, \mathbf{x}_n$. Then $vec(\mathbf{X}) = \begin{pmatrix} \mathbf{x}_1 \\ \mathbf{x}_2 \\ \vdots \\ \mathbf{x}_n \end{pmatrix}$.

Theorem 1.17. *(i) Let* $\mathbf{X}$ *be a* $p \times n$, $\mathbf{A}$ $q \times p$, *and* $\mathbf{B}$ $n \times m$ *matrices. Then,*

$$vec((\mathbf{AXB})') = (\mathbf{A} \otimes \mathbf{B}')vec(\mathbf{X}').$$

(ii) Let $\mathbf{X}$ *and* $\mathbf{Y}$ *be* $p \times n$, $\mathbf{A}$ $p \times p$, *and* $\mathbf{B}$ $n \times n$ *matrices. Then*

$$tr(\mathbf{X}'\mathbf{AYB}) = (vec(\mathbf{X}'))'(\mathbf{A} \otimes \mathbf{B}')vec(\mathbf{Y}').$$

(iii) Let $\mathbf{X}$ *and* $\mathbf{Y}$ *be* $p \times n$ *dimensional matrices. Then*

$$tr(\mathbf{X}'\mathbf{Y}) = (vec(\mathbf{X}'))'vec(\mathbf{Y}').$$

For a more extensive study of the Kronecker product, see Graybill (1969).

Now we give some results on the generalized inverse of a matrix. Since this concept is not so widely used in statistical publications, we prove the theorems in this part of the chapter. For more on generalized inverse, see Rao and Mitra (1971).

Definition 1.7. Let $\mathbf{A}$ be a $p \times q$ matrix. If there exists a $q \times p$ matrix $\mathbf{B}$ such that $\mathbf{ABA} = \mathbf{A}$, then $\mathbf{B}$ is called a generalized inverse of $\mathbf{A}$ and is denoted by $\mathbf{A}^-$.

It follows from the definition that $\mathbf{A}^-$ is not necessarily unique. For example, for any real number a, $(1, a)$ is a generalized inverse of $\begin{pmatrix} 1 \\ 0 \end{pmatrix}$. However, if $\mathbf{A}$ is a nonsingular square matrix, then $\mathbf{A}^-$ is unique as the following theorem shows.

Theorem 1.18. *Let $\mathbf{A}$ be a $p \times p$ nonsingular matrix. Then, $\mathbf{A}^{-1}$ is the one and only one generalized inverse of $\mathbf{A}$.*

PROOF: The matrix $\mathbf{A}^{-1}$ is a generalized inverse of $\mathbf{A}$ since $\mathbf{A}\mathbf{A}^{-1}\mathbf{A} = \mathbf{I}_p\mathbf{A} = \mathbf{A}$. On the other hand, from $\mathbf{A}\mathbf{A}^-\mathbf{A} = \mathbf{A}$ we get $\mathbf{A}^{-1}\mathbf{A}\mathbf{A}^-\mathbf{A}\mathbf{A}^{-1} = \mathbf{A}^{-1}\mathbf{A}\mathbf{A}^{-1}$. Hence, $\mathbf{A}^- = \mathbf{A}^{-1}$. ■

Next, we show that every matrix has a generalized inverse.

Theorem 1.19. *Let $\mathbf{A}$ be a $p \times q$ matrix. Then $\mathbf{A}$ has a generalized inverse.*

PROOF: First, we prove that the theorem is true for diagonal matrices. Let $\mathbf{D}$ be $n \times n$ diagonal and define the $n \times n$ diagonal matrix $\mathbf{B}$ by

$$b_{ii} = \begin{cases} \frac{1}{d_{ii}}, & \text{if } d_{ii} \neq 0 \\ 0, & \text{if } d_{ii} = 0 \end{cases}.$$

Then, $\mathbf{DBD} = \mathbf{D}$. Hence $\mathbf{B}$, is a generalized inverse of $\mathbf{D}$.

Next, assume that $\mathbf{A}$ is a $p \times q$ matrix with $p \geq q$. Using Theorem 1.6, we can find a $p \times p$ orthogonal matrix $\mathbf{G}$, a $q \times q$ orthogonal matrix $\mathbf{H}$, and a positive semidefinite diagonal matrix $\mathbf{D}$ such that

$$\mathbf{A} = \mathbf{G}\begin{pmatrix} \mathbf{D} \\ \mathbf{0} \end{pmatrix}\mathbf{H}.$$

We already know that $\mathbf{D}$ has a generalized inverse. Define

$$\mathbf{B} = \mathbf{H}'\left(\mathbf{D}^-, \mathbf{0}\right)\mathbf{G}'.$$

Then, we obtain

$$\mathbf{ABA} = \mathbf{G}\begin{pmatrix} \mathbf{D} \\ \mathbf{0} \end{pmatrix}\mathbf{HH}'\left(\mathbf{D}^-, \mathbf{0}\right)\mathbf{G}'\mathbf{G}\begin{pmatrix} \mathbf{D} \\ \mathbf{0} \end{pmatrix}\mathbf{H} = \mathbf{G}\begin{pmatrix} \mathbf{D} \\ \mathbf{0} \end{pmatrix}\mathbf{H} = \mathbf{A}.$$

So $\mathbf{B}$ is a generalized inverse of $\mathbf{A}$. If $\mathbf{A}$ is $p \times q$ dimensional with $p < q$, then $\mathbf{A}'$ has a generalized inverse $\mathbf{B}$. So $\mathbf{A}'\mathbf{B}\mathbf{A}' = \mathbf{A}'$. Therefore, $\mathbf{A}\mathbf{B}'\mathbf{A} = \mathbf{A}$. Hence, $\mathbf{B}'$ is a generalized inverse of $\mathbf{A}$. ■

We know that if $\mathbf{A}$ is a nonsingular square matrix, then $(\mathbf{A}^{-1})' = (\mathbf{A}')^{-1}$. The generalized inverse has the same property as the next theorem shows.

Theorem 1.20. *Let* $\mathbf{A}$ *be a* $p \times q$ *matrix. Then,* $(\mathbf{A}^{-})' = (\mathbf{A}')^{-}$*; that is* $\mathbf{B}$ *is a generalized inverse of* $\mathbf{A}$*, if and only if* $\mathbf{B}'$ *is a generalized inverse of* $\mathbf{A}'$*.*

PROOF: First, assume $\mathbf{B}$ is a generalized inverse of $\mathbf{A}$. Then, $\mathbf{ABA} = \mathbf{A}$. Hence, $\mathbf{A}'\mathbf{B}'\mathbf{A}' = \mathbf{A}'$. So $\mathbf{B}'$ is a generalized inverse of $\mathbf{A}'$.

On the other hand, assume $\mathbf{B}'$ is a generalized inverse of $\mathbf{A}'$. Then, $\mathbf{A}'\mathbf{B}'\mathbf{A}' = \mathbf{A}'$. Hence, $\mathbf{ABA} = \mathbf{A}$ and therefore $\mathbf{B}$ is a generalized inverse of $\mathbf{A}$. ■

For nonsingular square matrices of equal dimension, we have $(\mathbf{AB})^{-1} = \mathbf{B}^{-1}\mathbf{A}^{-1}$. However, for the generalized inverse, $(\mathbf{AB})^{-} = \mathbf{B}^{-}\mathbf{A}^{-}$ does not always hold. For example, consider $\mathbf{A} = (1,0)$ and $\mathbf{B} = \begin{pmatrix} 1 \\ 0 \end{pmatrix}$. Then, for any real numbers a and b, $\begin{pmatrix} 1 \\ a \end{pmatrix}$ is a generalized inverse of $\mathbf{A}$ and $(1,b)$ is a generalized inverse of $\mathbf{B}$. However, $\mathbf{B}^{-}\mathbf{A}^{-} = 1 + ab$ is a generalized inverse of $\mathbf{AB} = 1$ only if $a = 0$ or $b = 0$. In special cases, however, $(\mathbf{AB})^{-} = \mathbf{B}^{-}\mathbf{A}^{-}$ is true as the next theorem shows.

Theorem 1.21. *Let* $\mathbf{A}$ *be a* $p \times q$*,* $\mathbf{B}$ *a* $p \times p$*, and* $\mathbf{C}$ *a* $q \times q$ *matrix. Assume* $\mathbf{B}$ *and* $\mathbf{C}$ *are nonsingular. Then,* $(\mathbf{BAC})^{-} = \mathbf{C}^{-1}\mathbf{A}^{-}\mathbf{B}^{-1}$.

PROOF: The matrix $\mathbf{F}$ is a generalized inverse of $(\mathbf{BAC})^{-}$ iff $\mathbf{BACFBAC} = \mathbf{BAC}$. This is equivalent to $\mathbf{ACFBA} = \mathbf{A}$; that is $\mathbf{CFB} = \mathbf{A}^{-}$, or $\mathbf{F} = \mathbf{C}^{-1}\mathbf{A}^{-}\mathbf{B}^{-1}$. ■

Theorem 1.22. *Let* $\mathbf{A}$ *be a* $p \times q$ *matrix with* $rk(\mathbf{A}) = q$*. Then,* $\mathbf{A}^{-}\mathbf{A} = \mathbf{I}_q$.

PROOF: It follows from Theorem 1.6, that there exists a $p \times p$ orthogonal matrix $\mathbf{G}$, a $q \times q$ orthogonal matrix $\mathbf{H}$, and a $q \times q$ positive definite diagonal matrix $\mathbf{D}$ such that

$$\mathbf{A} = \mathbf{G}\begin{pmatrix} \mathbf{D} \\ \mathbf{0} \end{pmatrix}\mathbf{H}.$$

Then, $\mathbf{AA}^{-}\mathbf{A} = \mathbf{A}$ can be written as

$$\mathbf{G}\begin{pmatrix} \mathbf{D} \\ \mathbf{0} \end{pmatrix}\mathbf{HA}^{-}\mathbf{A} = \mathbf{G}\begin{pmatrix} \mathbf{D} \\ \mathbf{0} \end{pmatrix}\mathbf{H}.$$

Premultiplying the last equation by $\mathbf{G}'$, we obtain

$$\begin{pmatrix} \mathbf{D} \\ \mathbf{0} \end{pmatrix}\mathbf{HA}^{-}\mathbf{A} = \begin{pmatrix} \mathbf{D} \\ \mathbf{0} \end{pmatrix}\mathbf{H}.$$

Hence,

$$\begin{pmatrix} \mathbf{DHA^-A} \\ \mathbf{0} \end{pmatrix} = \begin{pmatrix} \mathbf{DH} \\ \mathbf{0} \end{pmatrix}.$$

and consequently $\mathbf{DHA^-A} = \mathbf{DH}$. Now, $\mathbf{D}$ and $\mathbf{H}$ are $q \times q$ nonsingular matrices, so we get $\mathbf{A^-A} = \mathbf{I}_q$. ■

Theorem 1.23. *Let* $\mathbf{A}$ *be a* $p \times p$ *matrix of rank* q *and let* $\mathbf{A} = \mathbf{BB}'$ *be a rank factorization of* $\mathbf{A}$ *as it is defined in Theorem 1.10. Then,* $\mathbf{B^-B} = \mathbf{I}_q$ *and* $\mathbf{B^-AB^{-\prime}} = \mathbf{I}_q$*. Moreover,* $\mathbf{B^{-\prime}B^-}$ *is a generalized inverse of* $\mathbf{A}$.

PROOF: Since $\mathbf{B}$ is $p \times q$ dimensional and $rk(\mathbf{B}) = q$, from Theorem 1.22 we get $\mathbf{B^-B} = \mathbf{I}_q$ and

$$\mathbf{B^-AB^{-\prime}} = \mathbf{B^-BB'B^{-\prime}} = \mathbf{B^-B(B^-B)'} = \mathbf{I}_q\mathbf{I}_q = \mathbf{I}_q .$$

We also have

$$\mathbf{AB^{-\prime}B^-A} = \mathbf{BB'B^{-\prime}B^-BB'} = \mathbf{B(B^-B)'(B^-B)B'} = \mathbf{BI}_q\mathbf{I}_q\mathbf{B'} = \mathbf{BB'} = \mathbf{A}.$$ ■

1.3 A Functional Equation

We close this chapter with a result from the theory of functional equations that will prove to be useful in the derivation of many theorems about elliptically contoured distributions. The theorem gives the solution of a variant of Hamel's equation (or Cauchy's equation).

Theorem 1.24. *Let* f *be a real function defined on* $\mathbb{R}_0^+$ *the set of nonnegative numbers. Assume that* f *is bounded in each finite interval and satisfies the equation*

$$f(x+y) = f(x)f(y) \quad \textit{for all} \quad x,y \in \mathbb{R}_0^+ .$$

Then either $f(x) = 0$ *for all* x *or* $f(x) = e^{ax}$ *where* $a \in \mathbb{R}$.

PROOF: See Feller (1957), p. 413. ■

Corollary 1.1. *Let* f *be a bounded, not identically zero function defined on* $\mathbb{R}_0^+$*. If* f *satisfies the equation*

$$f(x+y) = f(x)f(y) \quad \textit{for all} \quad x,y \in \mathbb{R}_0^+$$

then $f(x) = e^{-kx}$ *where* $k \geq 0$.

Part II
Definition and Distributional Properties

Chapter 2
Basic Properties

2.1 Definition

In the literature, several definitions of elliptically contoured distributions can be found, e.g. see Anderson and Fang (1982b), Fang and Chen (1984), and Sutradhar and Ali (1989). We will use the following definition given in Gupta and Varga (1994b).

Definition 2.1. Let $\mathbf{X}$ be a random matrix of dimensions $p \times n$. Then, $\mathbf{X}$ is said to have a matrix variate elliptically contoured (m.e.c.) distribution if its characteristic function has the form

$$\phi_{\mathbf{X}}(\mathbf{T}) = etr(i\mathbf{T}'\mathbf{M})\psi(tr(\mathbf{T}'\Sigma\mathbf{T}\Phi))$$

with $\mathbf{T} : p \times n$, $\mathbf{M} : p \times n$, $\Sigma : p \times p$, $\Phi : n \times n$, $\Sigma \geq \mathbf{0}$, $\Phi \geq \mathbf{0}$, and $\psi : [0, \infty) \to \mathbb{R}$.

This distribution will be denoted by $E_{p,n}(\mathbf{M}, \Sigma \otimes \Phi, \psi)$.

Remark 2.1. If in Definition 2.1 $n = 1$, we say that $\mathbf{X}$ has a vector variate elliptically contoured distribution. It is also called multivariate elliptical distribution. Then the characteristic function of $\mathbf{X}$ takes on the form

$$\phi_{\mathbf{x}}(\mathbf{t}) = exp(i\mathbf{t}'\mathbf{m})\psi(\mathbf{t}'\Sigma\mathbf{t}),$$

where $\mathbf{t}$ and $\mathbf{m}$ are p-dimensional vectors. This definition was given by many authors, e.g. Kelker (1970), Cambanis, Huang and Simons (1981) and Anderson and Fang (1987). In this case, in the notation $E_{p,n}(\mathbf{M}, \Sigma \otimes \Phi, \psi)$, the index n can be dropped; that is, $E_p(\mathbf{m}, \Sigma, \psi)$ will denote the distribution $E_{p,1}(\mathbf{m}, \Sigma, \psi)$.

Remark 2.2. It follows from Definition 2.1 that $|\psi(t)| \leq 1$ for $t \in \mathbb{R}_0^+$.

The following theorem shows the relationship between matrix variate and vector variate elliptically contoured distributions.

A.K. Gupta et al., *Elliptically Contoured Models in Statistics and Portfolio Theory*, DOI 10.1007/978-1-4614-8154-6_2, © Springer Science+Business Media New York 2013

Theorem 2.1. *Let* $\mathbf{X}$ *be a* $p \times n$ *random matrix and* $\mathbf{x} = vec(\mathbf{X}')$. *Then,* $\mathbf{X} \sim E_{p,n}(\mathbf{M}, \Sigma \otimes \Phi, \psi)$ *if and only if* $\mathbf{x} \sim E_{pn}(vec(\mathbf{M}'), \Sigma \otimes \Phi, \psi)$.

PROOF: Note that $\mathbf{X} \sim E_{p,n}(\mathbf{M}, \Sigma \otimes \Phi, \psi)$ iff

$$\phi_{\mathbf{X}}(\mathbf{T}) = etr(i\mathbf{T}'\mathbf{M})\psi(tr(\mathbf{T}'\Sigma\mathbf{T}\Phi)) \tag{2.1}$$

On the other hand, $\mathbf{x} \sim E_{pn}(vec(\mathbf{M}'), \Sigma \otimes \Phi, \psi)$ iff

$$\phi_{\mathbf{x}}(\mathbf{t}) = exp(i\mathbf{t}'vec(\mathbf{M}'))\psi(\mathbf{t}'(\Sigma \otimes \Phi)\mathbf{t}).$$

Let $\mathbf{t} = vec(\mathbf{T}')$. Then

$$\phi_{\mathbf{x}}(\mathbf{t}) = exp(i(vec(\mathbf{T}'))'vec(\mathbf{M}'))\psi((vec(\mathbf{T}'))'(\Sigma \otimes \Phi)vec(\mathbf{T}')). \tag{2.2}$$

Now, using Theorem 1.17, we can write

$$(vec(\mathbf{T}'))'vec(\mathbf{M}') = tr(\mathbf{T}'\mathbf{M}) \tag{2.3}$$

and

$$(vec(\mathbf{T}'))'(\Sigma \otimes \Phi)vec(\mathbf{T}') = tr(\mathbf{T}'\Sigma\mathbf{T}\Phi). \tag{2.4}$$

From (2.1), (2.2), (2.3), and (2.4) it follows that $\phi_{\mathbf{X}}(\mathbf{T}) = \phi_{\mathbf{x}}(vec(\mathbf{T}'))$. This completes the proof. ■

The next theorem shows that linear functions of a random matrix with m.e.c. distribution have elliptically contoured distributions also.

Theorem 2.2. *Let* $\mathbf{X} \sim E_{p,n}(\mathbf{M}, \Sigma \otimes \Phi, \psi)$. *Assume* $\mathbf{C} : q \times m$, $\mathbf{A} : q \times p$, *and* $\mathbf{B} : n \times m$ *are constant matrices. Then,*

$$\mathbf{AXB} + \mathbf{C} \sim E_{q,m}(\mathbf{AMB} + \mathbf{C}, (\mathbf{A}\Sigma\mathbf{A}') \otimes (\mathbf{B}'\Phi\mathbf{B}), \psi).$$

PROOF: The characteristic function of $\mathbf{Y} = \mathbf{AXB} + \mathbf{C}$ can be written as

$$\begin{aligned}
\phi_{\mathbf{Y}}(\mathbf{T}) &= E(etr(i\mathbf{T}'\mathbf{Y})) \\
&= E(etr(i\mathbf{T}'(\mathbf{AXB} + \mathbf{C}))) \\
&= E(etr(i\mathbf{T}'\mathbf{AXB}))etr(i\mathbf{T}'\mathbf{C}) \\
&= E(etr(i\mathbf{BT}'\mathbf{AX}))etr(i\mathbf{T}'\mathbf{C}) \\
&= \phi_{\mathbf{X}}(\mathbf{A}'\mathbf{TB}')etr(i\mathbf{T}'\mathbf{C}) \\
&= etr(i\mathbf{BT}'\mathbf{AM})\psi(tr(\mathbf{BT}'\mathbf{A}\Sigma\mathbf{A}'\mathbf{TB}'\Phi))etr(i\mathbf{T}'\mathbf{C}) \\
&= etr(i\mathbf{T}'(\mathbf{AMB} + \mathbf{C}))\psi(tr(\mathbf{T}'(\mathbf{A}\Sigma\mathbf{A}')\mathbf{T}(\mathbf{B}'\Phi\mathbf{B}))).
\end{aligned}$$

This is the characteristic function of $E_{q,m}(\mathbf{AMB} + \mathbf{C}, (\mathbf{A}\Sigma\mathbf{A}') \otimes (\mathbf{B}'\Phi\mathbf{B}), \psi)$. ■

Corollary 2.1. *Let* $\mathbf{X} \sim E_{p,n}(\mathbf{M}, \Sigma \otimes \Phi, \psi)$, *and let* $\Sigma = \mathbf{AA}'$ *and* $\Phi = \mathbf{BB}'$ *be rank factorizations of* Σ *and* Φ. *That is,* $\mathbf{A}$ *is* $p \times p_1$ *and* $\mathbf{B}$ *is* $n \times n_1$ *matrix, where* $p_1 = rk(\Sigma)$, $n_1 = rk(\Phi)$. *Then,*

$$\mathbf{A}^{-}(\mathbf{X}-\mathbf{M})\mathbf{B}'^{-} \sim E_{p_1,n_1}(\mathbf{0}, \mathbf{I}_{p_1} \otimes \mathbf{I}_{n_1}, \psi).$$

Conversely, if $\mathbf{Y} \sim E_{p_1,n_1}(\mathbf{0}, \mathbf{I}_{p_1} \otimes \mathbf{I}_{n_1}, \psi)$, *then*

$$\mathbf{AYB}' + \mathbf{M} \sim E_{p,n}(\mathbf{M}, \Sigma \otimes \Phi, \psi).$$

with $\Sigma = \mathbf{AA}'$ *and* $\Phi = \mathbf{BB}'$.

PROOF: Let $\mathbf{X} \sim E_{p,n}(\mathbf{M}, \Sigma \otimes \Phi, \psi)$, $\Sigma = \mathbf{AA}'$ and $\Phi = \mathbf{BB}'$ be the rank factorizations of Σ and Φ. Then, it follows from Theorem 2.2 that

$$\mathbf{A}^{-}(\mathbf{X}-\mathbf{M})\mathbf{B}'^{-} \sim E_{p_1,n_1}(\mathbf{0}, (\mathbf{A}^{-}\Sigma\mathbf{A}'^{-}) \otimes (\mathbf{B}^{-}\Phi\mathbf{B}'^{-}), \psi).$$

Using Theorem 1.23, we get $\mathbf{A}^{-}\Sigma\mathbf{A}'^{-} = \mathbf{I}_{p_1}$ and $\mathbf{B}^{-}\Phi\mathbf{B}'^{-} = \mathbf{I}_{n_1}$, which completes the proof of the first part of the theorem. The second part follows directly from Theorem 2.2. ■

If $\mathbf{x} \sim E_p(\mathbf{0}, \mathbf{I}_p, \psi)$, then it follows from Theorem 2.2 that $\mathbf{Gx} \sim E_p(\mathbf{0}, \mathbf{I}_p, \psi)$ for every $\mathbf{G} \in O(p)$. This gives rise to the following definition.

Definition 2.2. The distribution $E_p(\mathbf{0}, \mathbf{I}_p, \psi)$ is called spherical distribution.

A consequence of the definition of the m.e.c. distribution is that if $\mathbf{X}$ has m.e.c. distribution, then $\mathbf{X}'$ also has m.e.c. distribution. This is shown in the following theorem.

Theorem 2.3. *Let* $\mathbf{X} \sim E_{p,n}(\mathbf{M}, \Sigma \otimes \Phi, \psi)$. *Then,* $\mathbf{X}' \sim E_{n,p}(\mathbf{M}', \Phi \otimes \Sigma, \psi)$.

PROOF: We have

$$\begin{aligned}
\phi_{\mathbf{X}'}(\mathbf{T}) &= E(etr(i\mathbf{T}'\mathbf{X}')) \\
&= E(etr(i\mathbf{XT})) \\
&= E(etr(i\mathbf{TX})) \\
&= etr(i\mathbf{TM})\psi(tr(\mathbf{T}\Sigma\mathbf{T}'\Phi)) \\
&= etr(i\mathbf{T}'\mathbf{M}')\psi(tr(\mathbf{T}'\Phi\mathbf{T}\Sigma)).
\end{aligned}$$

This is the characteristic function of $\mathbf{X}' \sim E_{n,p}(\mathbf{M}', \Phi \otimes \Sigma, \psi)$. ■

The question arises whether the parameters in the definition of a m.e.c. distribution are uniquely defined. The answer is they are not. To see this assume that a, b, and c are positive constants such that $c = ab$, $\Sigma_2 = a\Sigma_1$, $\Phi_2 = b\Phi_1$,

$\psi_2(z) = \psi_1\left(\frac{1}{c}z\right)$. Then, $E_{p,n}(\mathbf{M},\Sigma_1 \otimes \Phi_1,\psi_1)$ and $E_{p,n}(\mathbf{M},\Sigma_2 \otimes \Phi_2,\psi_2)$ define the same m.e.c. distribution. However, this is the only way that two formulae define the same m.e.c. distribution as shown in the following theorem.

Theorem 2.4. *Let* $\mathbf{X} \sim E_{p,n}(\mathbf{M}_1,\Sigma_1 \otimes \Phi_1,\psi_1)$ *and at the same time* $\mathbf{X} \sim E_{p,n}(\mathbf{M}_2, \Sigma_2 \otimes \Phi_2,\psi_2)$. *If* $\mathbf{X}$ *is nondegenerate, then there exist positive constants a, b, and c such that* $c = ab$, *and* $\mathrm{M}_2 = \mathrm{M}_1$, $\Sigma_2 = a\Sigma_1$, $\Phi_2 = b\Phi_1$, *and* $\psi_2(z) = \psi_1\left(\frac{1}{c}z\right)$.

PROOF: The proof follows the lines of Cambanis, Huang and Simons (1981). First of all, note that the distribution of $\mathbf{X}$ is symmetric about $\mathbf{M}_1$ as well as about $\mathbf{M}_2$. Therefore, $\mathbf{M}_1 = \mathbf{M}_2$ must hold. Let $\mathbf{M} = \mathbf{M}_1$. Let us introduce the following notations

$$\Sigma_l = {}_l\sigma_{ij},\ i,j = 1,\ldots,p; \quad l = 1,2,$$
$$\Phi_l = {}_l\phi_{ij}, i,j = 1,\ldots,n; \quad l = 1,2.$$

Let $\mathbf{k}(a)$ denote the p-dimensional vector whose ath element is 1 and all the others are 0 and $\mathbf{l}(b)$ denote the n-dimensional vector whose bth element is 1 and all the others are 0. Since $\mathbf{X}$ is nondegenerate, it must have an element $x_{i_0 j_0}$ which is nondegenerate. Since $x_{i_0 j_0} = \mathbf{k}'(i_0)\mathbf{X}\mathbf{l}(j_0)$, from Theorem 2.2 we get

$$x_{i_0 j_0} \sim E_1(m_{i_0 j_0},\ {}_1\sigma_{i_0 i_0}\ {}_1\phi_{j_0 j_0},\psi_1)$$

and

$$x_{i_0 j_0} \sim E_1(m_{i_0 j_0},\ {}_2\sigma_{i_0 i_0}\ {}_2\phi_{j_0 j_0},\psi_2).$$

Therefore, the characteristic function of $x_{i_0 j_0} - m_{i_0 j_0}$ is

$$\begin{aligned}\phi(t) &= \psi_1(t^2\ {}_1\sigma_{i_0 i_0}\ {}_1\phi_{j_0 j_0}) \\ &= \psi_2(t^2\ {}_2\sigma_{i_0 i_0}\ {}_2\phi_{j_0 j_0})\end{aligned} \tag{2.5}$$

with $t \in \mathbb{R}$.

Since, ${}_l\sigma_{i_0 i_0}$ and ${}_l\phi_{j_0 j_0}$ $(l = 1,2)$ are diagonal elements of positive semidefinite matrices, they cannot be negative, and since $x_{i_0 j_0}$ is nondegenerate, they cannot be zero either. So, we can define

$$c = \frac{{}_2\sigma_{i_0 i_0}\,{}_2\phi_{j_0 j_0}}{{}_1\sigma_{i_0 i_0}\,{}_1\phi_{j_0 j_0}}.$$

Then, $c > 0$ and $\psi_2(z) = \psi_1\left(\frac{1}{c}z\right)$ for $z \in [0,\infty)$.

We claim that $\Sigma_2 \otimes \Phi_2 = c(\Sigma_1 \otimes \Phi_1)$. Suppose this is not the case. Then, there exists $\mathbf{t} \in \mathbb{R}^{pn}$ such that $\mathbf{t}'(\Sigma_2 \otimes \Phi_2)\mathbf{t} \neq c\mathbf{t}'(\Sigma_1 \otimes \Phi_1)\mathbf{t}$. From Theorem 1.17, it follows that there exists $\mathbf{T}_0 \in \mathbb{R}^{p\times n}$ such that $tr(\mathbf{T}_0'\Sigma_2\mathbf{T}_0\Phi_2) \neq ctr(\mathbf{T}_0'\Sigma_1\mathbf{T}_0\Phi_1)$.

Define $\mathbf{T} = u\mathbf{T}_0$, $u \in I\!R$. Then, the characteristic function of $\mathbf{X} - \mathbf{M}$ at $u\mathbf{T}_0$ is

$$\psi_1(utr(\mathbf{T}_0'\Sigma_1\mathbf{T}_0\Phi_1)) = \psi_2(uctr(\mathbf{T}_0'\Sigma_1\mathbf{T}_0\Phi_1)).$$

On the other hand, the characteristic function of $\mathbf{X} - \mathbf{M}$ at $u\mathbf{T}_0$ can be expressed as $\psi_2(utr(\mathbf{T}_0'\Sigma_2\mathbf{T}_0\Phi_2))$. So

$$\psi_2(uctr(\mathbf{T}_0'\Sigma_1\mathbf{T}_0\Phi_1)) = \psi_2(utr(\mathbf{T}_0'\Sigma_2\mathbf{T}_0\Phi_2)). \tag{2.6}$$

If $tr(\mathbf{T}_0'\Sigma_1\mathbf{T}_0\Phi_1) = 0$ or $tr(\mathbf{T}_0'\Sigma_2\mathbf{T}_0\Phi_2) = 0$, then from (2.6) we get that $\psi(u) = 0$ for every $u \in I\!R$. However, this is impossible since $\mathbf{X}$ is nondegenerate.

If $tr(\mathbf{T}_0'\Sigma_1\mathbf{T}_0\Phi_1) \neq 0$ and $tr(\mathbf{T}_0'\Sigma_1\mathbf{T}_0\Phi_1) \neq 0$, then define

$$d = c\frac{tr(\mathbf{T}_0'\Sigma_1\mathbf{T}_0\Phi_1)}{tr(\mathbf{T}_0'\Sigma_2\mathbf{T}_0\Phi_2)}.$$

Then, $d \neq 0$, $d \neq 1$, and from (2.6) we get $\psi_2(u) = \psi_2(du)$. By induction, we get

$$\psi_2(u) = \psi_2(d^n u) \quad \text{and} \quad \psi_2(u) = \psi_2\left(\left(\frac{1}{d}\right)^n u\right), \; n = 1, 2, \ldots.$$

Now either $d^n \to 0$ or $\left(\frac{1}{d}\right)^n \to 0$, and from the continuity of the characteristic function and the fact that $\psi_2(0) = 1$ it follows that $\psi_2(u) = 0$ for every $u \in I\!R$. However, this is impossible. So, we must have $\Sigma_2 \otimes \Phi_2 = c(\Sigma_1 \otimes \Phi_1)$. From Theorem 1.16 it follows that there exist $a > 0$ and $b > 0$ such that $\Sigma_2 = a\Sigma_1$, $\Phi_2 = b\Phi_1$, and $ab = c$. This completes the proof. ■

An important subclass of the class of the m.e.c. distributions is the class of matrix variate normal distributions.

Definition 2.3. The $p \times n$ random matrix $\mathbf{X}$ is said to have a matrix variate normal distribution if its characteristic function has the form

$$\phi_{\mathbf{X}}(\mathbf{T}) = etr(i\mathbf{T}'\mathbf{M})etr\left(-\frac{1}{2}\mathbf{T}'\Sigma\mathbf{T}\Phi\right),$$

with $\mathbf{T} : p \times n$, $\mathbf{M} : p \times n$, $\Sigma : p \times p$, $\Phi : n \times n$, $\Sigma \geq \mathbf{0}$, $\Phi \geq \mathbf{0}$. This distribution is denoted by $N_{p,n}(\mathbf{M}, \Sigma \otimes \Phi)$.

The next theorem shows that the matrix variate normal distribution can be used to represent samples taken from multivariate normal distributions.

Theorem 2.5. *Let* $\mathbf{X} \sim N_{p,n}(\mu\mathbf{e}_n', \Sigma \otimes \mathbf{I}_n)$, *where* $\mu \in I\!R^p$. *Let* $\mathbf{x}_1, \mathbf{x}_2, \ldots, \mathbf{x}_n$ *be the columns of* $\mathbf{X}$. *Then,* $\mathbf{x}_1, \mathbf{x}_2, \ldots, \mathbf{x}_n$ *are independent identically distributed random vectors with common distribution* $N_p(\mu, \Sigma)$.

PROOF: Let $\mathbf{T} = (\mathbf{t}_1, \mathbf{t}_2, \ldots, \mathbf{t}_n)$ be $p \times n$ matrix. Then

$$\phi_{\mathbf{X}}(\mathbf{T}) = etr\left(i \begin{pmatrix} \mathbf{t}_1' \\ \mathbf{t}_2' \\ \vdots \\ \mathbf{t}_n' \end{pmatrix} (\mu, \mu, \ldots, \mu)\right) etr\left(-\frac{1}{2} \begin{pmatrix} \mathbf{t}_1' \\ \mathbf{t}_2' \\ \vdots \\ \mathbf{t}_n' \end{pmatrix} \Sigma (\mathbf{t}_1, \mathbf{t}_2, \ldots, \mathbf{t}_n)\right)$$

$$= exp\left(i \sum_{j=1}^{n} \mathbf{t}_j' \mu\right) exp\left(-\frac{1}{2} \sum_{j=1}^{n} \mathbf{t}_j' \Sigma \mathbf{t}_j\right) \neg$$

$$= \prod_{j=1}^{n} exp\left(i\mathbf{t}_j' \mu - \frac{1}{2} \mathbf{t}_j' \Sigma \mathbf{t}_j\right),$$

which shows that $\mathbf{x}_1, \mathbf{x}_2, \ldots, \mathbf{x}_n$ are independent, each with distribution $N_p(\mu, \Sigma)$. ■

2.2 Probability Density Function

If $\mathbf{X} \sim E_{p,n}(\mathbf{M}, \Sigma \otimes \Phi, \psi)$ defines an absolutely continuous elliptically contoured distribution, Σ and Φ must be positive definite. Assume this is not the case. For example, $\Sigma \geq \mathbf{0}$ but Σ is not positive definite. Then, from Theorem 1.7, it follows that $\Sigma = \mathbf{G}\mathbf{D}\mathbf{G}'$ where $\mathbf{G} \in O(n)$, and $\mathbf{D}$ is diagonal and $d_{11} = 0$. Let $\mathbf{Y} = \mathbf{G}'(\mathbf{X} - \mathbf{M})$. Then, $\mathbf{Y} \sim E_{p,n}(\mathbf{0}, \mathbf{D} \otimes \Phi, \psi)$, and the distribution of $\mathbf{Y}$ is also absolutely continuous. On the other hand, $y_{11} \sim E_1(0, 0, \psi)$ so y_{11} is degenerate. But the marginal of an absolutely continuous distribution cannot be degenerate. Hence, we get a contradiction. So, $\Sigma > \mathbf{0}$ and $\Phi > \mathbf{0}$ must hold when the m.e.c. distribution is absolutely continuous.

The probability density function (p.d.f.) of a m.e.c. distribution is of a special form as the following theorem shows.

Theorem 2.6. *Let $\mathbf{X}$ be a $p \times n$ dimensional random matrix whose distribution is absolutely continuous. Then, $\mathbf{X} \sim E_{p,n}(\mathbf{M}, \Sigma \otimes \Phi, \psi)$ if and only if the p.d.f. of $\mathbf{X}$ has the form*

$$f(\mathbf{X}) = |\Sigma|^{-\frac{n}{2}} |\Phi|^{-\frac{p}{2}} h(tr((\mathbf{X} - \mathbf{M})' \Sigma^{-1} (\mathbf{X} - \mathbf{M}) \Phi^{-1})), \tag{2.7}$$

where h and ψ determine each other for specified p and n.

PROOF: I. First, we prove that if $\mathbf{X} \sim E_{p,n}(\mathbf{M}, \Sigma \otimes \Phi, \psi)$ and $E_{p,n}(\mathbf{M}, \Sigma \otimes \Phi, \psi)$ is absolutely continuous, then the p.d.f. of $\mathbf{X}$ has the form (2.7).

Step 1. Assume that $\mathbf{M} = \mathbf{0}$ and $\Sigma \otimes \Phi = \mathbf{I}_{pn}$. Then, $\mathbf{X} \sim E_{p,n}(\mathbf{0}, \mathbf{I}_p \otimes \mathbf{I}_n, \psi)$. We want to show that the p.d.f. of $\mathbf{X}$ depends on $\mathbf{X}$ only through $tr(\mathbf{X}'\mathbf{X})$. Let $\mathbf{x} = vec(\mathbf{X}')$. From Theorem 2.1 we know that $\mathbf{x} \sim E_{pn}(\mathbf{0}, \mathbf{I}_{pn}, \psi)$. Let $\mathbf{H} \in O(pn)$, then, in view of Theorem 2.2,

$$\mathbf{Hx} \sim E_{pn}(\mathbf{0}, \mathbf{HH}', \psi) = E_{pn}(\mathbf{0}, \mathbf{I}_{pn}, \psi).$$

Thus, the distribution of $\mathbf{x}$ is invariant under orthogonal transformation. Therefore, using Theorem 1.11, we conclude that the p.d.f. of $\mathbf{x}$ depends on $\mathbf{x}$ only through $\mathbf{x}'\mathbf{x}$. Let us denote the p.d.f. of $\mathbf{x}$ by $f_1(\mathbf{x})$. We have $f_1(\mathbf{x}) = h(\mathbf{x}'\mathbf{x})$. Clearly, h only depends on p, n, and ψ. It follows from Theorem 1.17, that $\mathbf{x}'\mathbf{x} = tr(\mathbf{X}'\mathbf{X})$. Thus, denoting the p.d.f. of $\mathbf{X}$ by $f(\mathbf{X})$, we get $f(\mathbf{X}) = h(tr(\mathbf{X}'\mathbf{X}))$.

Step 2. Now, let $\mathbf{X} \sim E_{p,n}(\mathbf{M}, \Sigma \otimes \Phi, \psi)$. From Corollary 2.1, it follows that $\mathbf{Y} = \Sigma^{-\frac{1}{2}}(\mathbf{X}-\mathbf{M})\Phi^{-\frac{1}{2}} \sim E_{p,n}(\mathbf{0}, \mathbf{I}_p \otimes \mathbf{I}_n, \psi)$. Therefore, if $g(\mathbf{Y})$ is the p.d.f. of $\mathbf{Y}$, then $g(\mathbf{Y}) = h(tr(\mathbf{Y}'\mathbf{Y}))$. The Jacobian of the transformation $\mathbf{Y} \to \mathbf{X}$ is $|\Sigma^{-\frac{1}{2}}|^n|\Phi^{-\frac{1}{2}}|^p$. So the p.d.f. of $\mathbf{X}$ is

$$\begin{aligned} f(\mathbf{X}) &= h\left(tr\left(\Phi^{-\frac{1}{2}}(\mathbf{X}-\mathbf{M})'\Sigma^{-\frac{1}{2}}\Sigma^{-\frac{1}{2}}(\mathbf{X}-\mathbf{M})\Phi^{-\frac{1}{2}}\right)\right)|\Sigma|^{-\frac{n}{2}}|\Phi|^{-\frac{p}{2}} \\ &= |\Sigma|^{-\frac{n}{2}}|\Phi|^{-\frac{p}{2}}h\left(tr\left((\mathbf{X}-\mathbf{M})'\Sigma^{-1}(\mathbf{X}-\mathbf{M})\Phi^{-1}\right)\right). \end{aligned}$$

II. Next, we show that if a random matrix $\mathbf{X}$ has the p.d.f. of the form (2.7), then its distribution is elliptically contoured. That is, assume that the $p \times n$ random matrix $\mathbf{X}$ has the p.d.f.

$$f(\mathbf{X}) = |\Sigma|^{-\frac{n}{2}}|\Phi|^{-\frac{p}{2}}h\left(tr\left((\mathbf{X}-\mathbf{M})'\Sigma^{-1}(\mathbf{X}-\mathbf{M})\Phi^{-1}\right)\right),$$

then we want to show that $\mathbf{X} \sim E_{p,n}(\mathbf{M}, \Sigma \otimes \Phi, \psi)$. Let $\mathbf{Y} = \Sigma^{-\frac{1}{2}}(\mathbf{X}-\mathbf{M})\Phi^{-\frac{1}{2}}$. Then, the p.d.f. of $\mathbf{Y}$ is $g(\mathbf{Y}) = h(tr(\mathbf{Y}'\mathbf{Y}))$. Let $\mathbf{y} = vec(\mathbf{Y}')$. Then, the p.d.f. of $\mathbf{y}$ is $g_1(\mathbf{y}) = h(\mathbf{y}'\mathbf{y})$. The characteristic function of $\mathbf{y}$ is

$$\phi_{\mathbf{y}}(\mathbf{t}) = \int_{\mathbb{R}^{pn}} \exp(i\mathbf{t}'\mathbf{y})h(\mathbf{y}'\mathbf{y})d\mathbf{y},$$

where $\mathbf{t} \in \mathbb{R}^{pn}$.

Next, we prove that if $\mathbf{t}_1$ and $\mathbf{t}_2$ are vectors of dimension pn such that $\mathbf{t}_1'\mathbf{t}_1 = \mathbf{t}_2'\mathbf{t}_2$, then $\phi_{\mathbf{y}}(\mathbf{t}_1) = \phi_{\mathbf{y}}(\mathbf{t}_2)$. Using Theorem 1.11, we see that there exists $\mathbf{H} \in O(pn)$, such that $\mathbf{t}_1'\mathbf{H} = \mathbf{t}_2'$. Therefore,

$$\begin{aligned} \phi_{\mathbf{y}}(\mathbf{t}_2) &= \int_{\mathbb{R}^{pn}} \exp(i\mathbf{t}_2'\mathbf{y})h(\mathbf{y}'\mathbf{y})d\mathbf{y} \\ &= \int_{\mathbb{R}^{pn}} \exp(i\mathbf{t}_1'\mathbf{H}\mathbf{y})h(\mathbf{y}'\mathbf{y})d\mathbf{y}. \end{aligned}$$

Let $\mathbf{z} = \mathbf{H}\mathbf{y}$. The Jacobian of the transformation $\mathbf{y} \to \mathbf{z}$ is $|\mathbf{H}'|^{pn} = 1$. So

$$\begin{aligned} \int_{\mathbb{R}^{pn}} \exp(i\mathbf{t}_1'\mathbf{H}\mathbf{y})h(\mathbf{y}'\mathbf{y})d\mathbf{y} &= \int_{\mathbb{R}^{pn}} \exp(i\mathbf{t}_1'\mathbf{z})h(\mathbf{z}'\mathbf{H}\mathbf{H}'\mathbf{z})d\mathbf{z} \\ &= \int_{\mathbb{R}^{pn}} \exp(i\mathbf{t}_1'\mathbf{z})h(\mathbf{z}'\mathbf{z})d\mathbf{z} \\ &= \phi_{\mathbf{y}}(\mathbf{t}_1). \end{aligned}$$

This means that $\phi_{\mathbf{y}}(\mathbf{t}_1) = \phi_{\mathbf{y}}(\mathbf{t}_2)$. Therefore, $\phi_{\mathbf{y}}(\mathbf{t})$ is a function of $\mathbf{t}'\mathbf{t}$, which implies that $\phi_{\mathbf{Y}}(\mathbf{T})$ is a function of $tr(\mathbf{T}'\mathbf{T})$. Therefore, there exists a function ψ such that $\phi_{\mathbf{Y}}(\mathbf{T}) = \psi(tr(\mathbf{T}'\mathbf{T}))$. That is, $\mathbf{y} \sim E_{p,n}(\mathbf{0}, \mathbf{I}_p \otimes \mathbf{I}_n, \psi)$. Using Corollary 2.1, we get $\mathbf{X} \sim E_{p,n}(\mathbf{M}, \Sigma \otimes \Phi, \psi)$. ∎

Next we prove a lemma which will be useful for further study of m.e.c. distributions.

Lemma 2.1. *Let f be a function $f : A \times \mathbb{R}^p \to \mathbb{R}^q$, where A can be any set. Assume there exists a function $g : A \times \mathbb{R} \to \mathbb{R}^q$ such that $f(a,\mathbf{x}) = g(a,\mathbf{x}'\mathbf{x})$ for any $a \in A$ and $\mathbf{x} \in \mathbb{R}^p$. Then, we have*

$$\int_{\mathbb{R}^p} f(a,\mathbf{x})d\mathbf{x} = \frac{2\pi^{\frac{p}{2}}}{\Gamma\left(\frac{p}{2}\right)} \int_0^\infty r^{p-1} g(a,r^2)dr$$

for any $a \in A$.

PROOF: Let $\mathbf{x} = (x_1, x_2, \ldots, x_p)'$ and introduce the polar coordinates

$$\begin{aligned}
x_1 &= r sin\theta_1 sin\theta_2 \ldots sin\theta_{p-2} sin\theta_{p-1} \\
x_2 &= r sin\theta_1 sin\theta_2 \ldots sin\theta_{p-2} cos\theta_{p-1} \\
x_3 &= r sin\theta_1 sin\theta_2 \ldots cos\theta_{p-2} \\
&\vdots \\
x_{p-1} &= r sin\theta_1 cos\theta_2 \\
x_p &= r cos\theta_1 ,
\end{aligned}$$

where $r > 0$, $0 < \theta_i < \pi$, $i = 1, 2, \ldots, p-2$, and $0 < \theta_{p-1} < 2\pi$. Then, the Jacobian of the transformation $(x_1, x_2, \ldots, x_p) \to (r, \theta_1, \theta_2, \ldots, \theta_{p-1})$ is

$$r^{p-1} sin^{p-2}\theta_1 sin^{p-3}\theta_2 \ldots sin\theta_{p-2}.$$

We also have $\mathbf{x}'\mathbf{x} = r^2$. Thus,

$$\begin{aligned}
&\int_{\mathbf{x}\in\mathbb{R}^p} f(a,\mathbf{x})d\mathbf{x} = \int_{\mathbf{x}\in\mathbb{R}^p} g(a,\mathbf{x}'\mathbf{x})d\mathbf{x} \\
&= \int_0^\infty \int_0^\pi \int_0^\pi \cdots \int_0^{2\pi} g(a,r^2) r^{p-1} sin^{p-2}\theta_1 sin^{p-3}\theta_2 \ldots sin\theta_{p-2} d\theta_{p-1} \ldots d\theta_2 d\theta_1 dr \\
&= \frac{2\pi^{\frac{p}{2}}}{\Gamma\left(\frac{p}{2}\right)} \int_0^\infty r^{p-1} g(a,r^2)dr.
\end{aligned}$$

∎

The next theorem is due to Fang, Kotz, and Ng (1990).

Theorem 2.7. *Let* $g : I\!R_0^+ \to I\!R_0^+$ *be a measurable function. Then, there exists a constant c such that*

$$cg(tr(\mathbf{X}'\mathbf{X})), \quad \mathbf{X} \in I\!R^{p\times n}$$

is the p.d.f. of the $p \times n$ *random matrix* $\mathbf{X}$ *if and only if*

$$0 < \int_0^\infty r^{pn-1} g(r^2) dr < \infty.$$

Moreover, the relationship between g and c is given by

$$c = \frac{\Gamma\left(\frac{pn}{2}\right)}{2\pi^{\frac{pn}{2}} \int_0^\infty r^{np-1} g(r^2) dr}.$$

PROOF: By definition, $cg(tr(\mathbf{X}'\mathbf{X})), \mathbf{X} \in I\!R^{p\times n}$, is the p.d.f. of a $p \times n$ random matrix $\mathbf{X}$ iff $cg(\mathbf{y}'\mathbf{y}), \mathbf{y} \in I\!R^{pn}$ is the p.d.f. of a pn-dimensional random vector $\mathbf{y}$. On the other hand, $cg(\mathbf{y}'\mathbf{y}), \mathbf{y} \in I\!R^{pn}$ is the p.d.f. of a pn-dimensional random vector $\mathbf{y}$ iff

$$\int_{I\!R^{pn}} cg(\mathbf{y}'\mathbf{y}) d\mathbf{y} = 1.$$

From Lemma 2.1, we get

$$\int_{I\!R^{pn}} cg(\mathbf{y}'\mathbf{y}) d\mathbf{y} = c \frac{2\pi^{\frac{pn}{2}}}{\Gamma\left(\frac{pn}{2}\right)} \int_0^\infty r^{pn-1} g(r^2) dr.$$

Hence, we must have

$$0 \leq \int_0^\infty r^{pn-1} g(r^2) dr < \infty.$$

and

$$c = \frac{\Gamma\left(\frac{pn}{2}\right)}{2\pi^{\frac{pn}{2}} \int_0^\infty r^{np-1} g(r^2) dr}.$$

■

2.3 Marginal Distributions

Using Theorem 2.2, we can derive the marginal distributions of a m.e.c. distribution.

Theorem 2.8. *Let* $\mathbf{X} \sim E_{p,n}(\mathbf{M}, \Sigma \otimes \Phi, \psi)$*, and partition* $\mathbf{X}$*,* $\mathbf{M}$*, and* Σ *as*

$$\mathbf{X} = \begin{pmatrix} \mathbf{X}_1 \\ \mathbf{X}_2 \end{pmatrix}, \quad \mathbf{M} = \begin{pmatrix} \mathbf{M}_1 \\ \mathbf{M}_2 \end{pmatrix}, \quad \text{and} \quad \Sigma = \begin{pmatrix} \Sigma_{11} & \Sigma_{12} \\ \Sigma_{21} & \Sigma_{22} \end{pmatrix},$$

where $\mathbf{X}_1$ *is* $q \times n$, $\mathbf{M}_1$ *is* $q \times n$, *and* Σ_{11} *is* $q \times q$, $1 \leq q < p$. *Then,*

$$\mathbf{X}_1 \sim E_{q,n}(\mathbf{M}_1, \Sigma_{11} \otimes \Phi, \psi).$$

PROOF: Let $\mathbf{A} = (\mathbf{I}_q, \mathbf{0})$ be of dimensions $q \times p$. Then, $\mathbf{AX} = \mathbf{X}_1$, and from Theorem 2.2, we obtain $\mathbf{X}_1 \sim E_{q,n}\left((\mathbf{I}_q, \mathbf{0})\mathbf{M}, \left((\mathbf{I}_q, \mathbf{0})\Sigma \begin{pmatrix} \mathbf{I}_q \\ \mathbf{0} \end{pmatrix}\right) \otimes \Phi, \psi\right)$, i.e. $\mathbf{X}_1 \sim E_{q,n}(\mathbf{M}_1, \Sigma_{11} \otimes \Phi, \psi)$. ∎

If we partition $\mathbf{X}$ vertically, we obtain the following result.

Theorem 2.9. *Let* $\mathbf{X} \sim E_{p,n}(\mathbf{M}, \Sigma \otimes \Phi, \psi)$, *and partition* $\mathbf{X}$, $\mathbf{M}$, *and* Φ *as*

$$\mathbf{X} = (\mathbf{X}_1, \mathbf{X}_2), \quad \mathbf{M} = (\mathbf{M}_1, \mathbf{M}_2), \quad \text{and} \quad \Phi = \begin{pmatrix} \Phi_{11} & \Phi_{12} \\ \Phi_{21} & \Phi_{22} \end{pmatrix},$$

where $\mathbf{X}_1$ *is* $p \times m$, $\mathbf{M}_1$ *is* $p \times m$, *and* Φ_{11} *is* $m \times m$, $1 \leq m < n$. *Then,*

$$\mathbf{X}_1 \sim E_{p,m}(\mathbf{M}_1, \Sigma \otimes \Phi_{11}, \psi). \tag{2.8}$$

PROOF: From Theorem 2.3, it follows that

$$\mathbf{X}' = \begin{pmatrix} \mathbf{X}_1' \\ \mathbf{X}_2' \end{pmatrix} \sim E_{n,p}\left(\begin{pmatrix} \mathbf{M}_1' \\ \mathbf{M}_2' \end{pmatrix}, \begin{pmatrix} \Phi_{11} & \Phi_{12} \\ \Phi_{21} & \Phi_{22} \end{pmatrix} \otimes \Sigma, \psi\right)$$

Then (2.8) follows directly from Theorem 2.8. ∎

Theorem 2.10. *Let* $\mathbf{X} \sim E_{p,n}(\mathbf{M}, \Sigma \otimes \Phi, \psi)$, *then* $x_{ij} \sim E_1(m_{ij}, \sigma_{ii}\phi_{jj}, \psi)$.

PROOF: The result follows from Theorems 2.8 and 2.9. ∎

Remark 2.3. It follows from Theorems 2.8 and 2.9 that if $\mathbf{X} \sim E_{p,n}(\mathbf{M}, \Sigma \otimes \Phi, \psi)$ and $\mathbf{Y}$ is a $q \times m$ submatrix of $\mathbf{X}$, then $\mathbf{Y}$ also has m.e.c. distribution; $\mathbf{Y} \sim E_{q,m}(\mathbf{M}^*, \Sigma^* \otimes \Phi^*, \psi)$.

2.4 Expected Value and Covariance

In this section, the first two moments of a m.e.c. distribution will be derived. In Chap. 3, moments of higher orders will also be obtained.

Theorem 2.11. *Let* $\mathbf{X} \sim E_{p,n}(\mathbf{M}, \Sigma \otimes \Phi, \psi)$.

(a) If $\mathbf{X}$ *has finite first moment, then* $E(\mathbf{X}) = \mathbf{M}$.
(b) If $\mathbf{X}$ *has finite second moment, then* $Cov(\mathbf{X}) = c\Sigma \otimes \Phi$, *where* $c = -2\psi'(0)$.

PROOF: Step 1. First, let us assume $\mathbf{M} = \mathbf{0}$ and $\Sigma \otimes \Phi = \mathbf{I}_{pn}$. Then, $\mathbf{X} \sim E_{p,n}(\mathbf{0}, \mathbf{I}_p \otimes \mathbf{I}_n, \psi)$.

(a) In view of Theorem 2.2, we have

$$(-\mathbf{I}_p)\mathbf{X} \sim E_{p,n}(\mathbf{0}, \mathbf{I}_p \otimes \mathbf{I}_n, \psi).$$

Therefore, $E(\mathbf{X}) = E(-\mathbf{X})$, and $E(\mathbf{X}) = \mathbf{0}$.

(b) Let $\mathbf{x} = vec(\mathbf{X}')$. Then $\mathbf{x} \sim E_{pn}(\mathbf{0}, \mathbf{I}_{pn}, \psi)$. The characteristic function of $\mathbf{x}$ is $\phi_{\mathbf{x}}(\mathbf{t}) = \psi(\mathbf{t}'\mathbf{t})$, where $\mathbf{t} = (t_1, \ldots, t_{pn})'$. Then,

$$\frac{\partial \phi_{\mathbf{x}}(\mathbf{t})}{\partial t_i} = \frac{\partial \psi\left(\sum_{l=1}^{pn} t_l^2\right)}{\partial t_i} = 2t_i \psi'\left(\sum_{l=1}^{pn} t_l^2\right).$$

So,

$$\frac{\partial^2 \phi_{\mathbf{x}}(\mathbf{t})}{\partial t_i^2} = 2\psi'\left(\sum_{l=1}^{pn} t_l^2\right) + 4t_i^2 \psi''\left(\sum_{l=1}^{pn} t_l^2\right).$$

and if $i \neq j$, then

$$\frac{\partial^2 \phi_{\mathbf{x}}(\mathbf{t})}{\partial t_j \partial t_i} = 4t_i t_j \psi''\left(\sum_{l=1}^{pn} t_l^2\right).$$

Therefore,

$$\left.\frac{\partial^2 \phi_{\mathbf{x}}(\mathbf{t})}{\partial t_i^2}\right|_{\mathbf{t}=\mathbf{0}} = 2\psi'(0) \quad \text{and} \quad \left.\frac{\partial^2 \phi_{\mathbf{x}}(\mathbf{t})}{\partial t_i \partial t_j}\right|_{\mathbf{t}=\mathbf{0}} = 0 \quad \text{if} \quad i \neq j.$$

Thus, $Cov(\mathbf{x}) = -2\psi'(0)\mathbf{I}_{pn}$.

Step 2. Now, let $\mathbf{X} \sim E_{p,n}(\mathbf{M}, \Sigma \otimes \Phi, \psi)$. Let $\Sigma = \mathbf{A}\mathbf{A}'$ and $\Phi = \mathbf{B}\mathbf{B}'$ be the rank factorizations of Σ and Φ. Then, from Corollary 2.1, it follows that $\mathbf{Y} = \mathbf{A}^-(\mathbf{X} - \mathbf{M})\Phi'^- \sim E_{p_1,n_1}(\mathbf{0}, \mathbf{I}_{p_1} \otimes \mathbf{I}_{n_1}, \psi)$ and $\mathbf{X} = \mathbf{A}\mathbf{Y}\mathbf{B}'$. Using Step 1 ,we get the following results:

(a) $E(\mathbf{Y}) = \mathbf{0}$. Hence $E(\mathbf{X}) = \mathbf{A}\mathbf{0}\mathbf{B}' + \mathbf{M} = \mathbf{M}$.

(b) Let $\mathbf{x} = vec(\mathbf{X}')$, $\mathbf{y} = vec(\mathbf{Y}')$, and $\mu = vec(\mathbf{M}')$. Then $\mathbf{x} = (\mathbf{A} \otimes \mathbf{B})\mathbf{y} + \mu$, and $Cov(\mathbf{y}) = -2\psi'(0)\mathbf{I}_{pn}$ and so

$$\begin{aligned} Cov(\mathbf{x}) &= -2\psi'(0)(\mathbf{A} \otimes \mathbf{B})\mathbf{I}_{pn}(\mathbf{A}' \otimes \mathbf{B}') \\ &= -2\psi'(0)(\mathbf{A}\mathbf{A}') \otimes (\mathbf{B}\mathbf{B}') \\ &= -2\psi'(0)\Sigma \otimes \Phi. \end{aligned}$$

■

Corollary 2.2. *With the conditions of Theorem 2.11, the ith $(i = 1,\dots,p)$ column of the matrix* $\mathbf{X}$ *has the covariance matrix* $\phi_{ii}\Sigma$ *and the jth row* $(j = 1,\dots,n)$ *has the covariance matrix* $\sigma_{jj}\Phi$.

Corollary 2.3. *With the conditions of Theorem 2.11,*

$$Corr(x_{ij},x_{kl}) = \frac{\sigma_{ik}\phi_{jl}}{\sqrt{\sigma_{ii}\sigma_{kk}\phi_{jj}\phi_{ll}}},$$

that is, the correlations between two elements of the matrix $\mathbf{X}$*, depend only on* Σ *and* Φ *but not on* ψ.

PROOF: From Theorem 2.11, we get $Cov(x_{ij},x_{kl}) = c\sigma_{ik}\phi_{jl}$, $Var(x_{ij}) = c\sigma_{ii}\phi_{jj}$, and $Var(x_{kl}) = c\sigma_{kk}\phi_{ll}$, where $c = -2\psi'(0)$. Therefore

$$\begin{aligned} Corr(x_{ij},x_{kl}) &= \frac{c\sigma_{ik}\phi_{jl}}{\sqrt{c^2\sigma_{ii}\sigma_{kk}\phi_{jj}\phi_{ll}}} \\ &= \frac{\sigma_{ik}\phi_{jl}}{\sqrt{\sigma_{ii}\sigma_{kk}\phi_{jj}\phi_{ll}}}. \end{aligned}$$

■

2.5 Stochastic Representation

In Cambanis, Huang, and Simons (1981) the stochastic representation of vector variate elliptically contoured distribution was obtained using a result of Schoenberg (1938). This result was extended to m.e.c. distributions by Anderson and Fang (1982b). Shoenberg's result is given in the next theorem.

Theorem 2.12. *Let* ψ *be a real function* $\psi : [0,\infty) \to I\!R$. *Then,* $\psi(\mathbf{t}'\mathbf{t})$, $\mathbf{t} \in I\!R^k$ *is the characteristic function of a k-dimensional random variable* $\mathbf{x}$*, if and only if* $\psi(u) = \int_0^\infty \Omega_k(r^2u)dF(r)$, $u \geq 0$, *where F is a distribution function on* $[0,\infty)$ *and* $\Omega_k(\mathbf{t}'\mathbf{t})$, $\mathbf{t} \in I\!R^k$ *is the characteristic function of the k-dimensional random variable* $\mathbf{u}_k$ *which is uniformly distributed on the unit sphere in* $I\!R^k$. *Moreover,* $F(r)$ *is the distribution function of* $r = (\mathbf{x}'\mathbf{x})^{\frac{1}{2}}$.

PROOF: Let us denote the unit sphere in $I\!R^k$ by S_k:

$$S_k = \{\mathbf{x}|\mathbf{x} \in I\!R^k; \mathbf{x}'\mathbf{x} = 1\},$$

and let A_k be the surface area of S_k i.e.

$$A_k = \frac{2\pi^{\frac{k}{2}}}{\Gamma\left(\frac{k}{2}\right)}.$$

First, assume $\psi(u) = \int_0^\infty \Omega_k(r^2u)dF(r)$. Let r be a random variable with distribution function $F(r)$, and let $\mathbf{u}_k$ be independent of r and uniformly distributed on S_k. Define $\mathbf{x} = r\mathbf{u}_k$. Then, the characteristic function of $\mathbf{x}$ is

$$\begin{aligned}
\phi_{\mathbf{x}}(\mathbf{t}) &= E(exp(i\mathbf{t}'\mathbf{x})) \\
&= E(exp(i\mathbf{t}'r\mathbf{u}_k)) \\
&= E\{E(exp(i\mathbf{t}'r\mathbf{u}_k)|r)\} \\
&= \int_0^\infty E(exp(i\mathbf{t}'r\mathbf{u}_k)|r=y)dF(y) \\
&= \int_0^\infty \phi_{\mathbf{u}_k}(y\mathbf{t})dF(y) \\
&= \int_0^\infty \Omega_k(y^2\mathbf{t}'\mathbf{t})dF(y).
\end{aligned}$$

Therefore, $\psi(\mathbf{t}'\mathbf{t}) = \int_0^\infty \Omega_k(y^2\mathbf{t}'\mathbf{t})dF(y)$ is indeed the characteristic function of the k-dimensional random vector $\mathbf{x}$. Moreover,

$$F(y) = P(r \le y) = P((r^2)^{\frac{1}{2}} \le y) = P(((r\mathbf{u}_k)'(r\mathbf{u}_k))^{\frac{1}{2}} \le y) = P((\mathbf{x}'\mathbf{x})^{\frac{1}{2}} \le y).$$

Conversely, assume $\psi(\mathbf{t}'\mathbf{t})$ is the characteristic function of a k-dimensional random vector $\mathbf{x}$. Let $G(\mathbf{x})$ be the distribution function of $\mathbf{x}$. Let $d\omega_k(\mathbf{t})$ denote the integration on S_k. We have $\psi(u) = \psi(u\mathbf{t}'\mathbf{t})$ for $\mathbf{t}'\mathbf{t} = 1$, and therefore we can write

$$\begin{aligned}
\psi(u) &= \frac{1}{A_k}\int_{S_k} \psi(u\mathbf{t}'\mathbf{t})d\omega_k(\mathbf{t}) \\
&= \frac{1}{A_k}\int_{S_k} \phi_{\mathbf{x}}(\sqrt{u}\mathbf{t})d\omega_k(\mathbf{t}) \\
&= \frac{1}{A_k}\int_{S_k}\int_{\mathbb{R}^m} \exp(i\sqrt{u}\mathbf{t}'\mathbf{x})dG(\mathbf{x})d\omega_k(\mathbf{t}) \\
&= \int_{\mathbb{R}^m} \left(\frac{1}{A_k}\int_{S_k} \exp(i\sqrt{u}\mathbf{x}'\mathbf{t})d\omega_k(\mathbf{t})\right) dG(\mathbf{x}) \\
&= \int_{\mathbb{R}^m} \Omega_k\left((\sqrt{u}\mathbf{x})'(\sqrt{u}\mathbf{x})\right) dG(\mathbf{x}) \\
&= \int_{\mathbb{R}^m} \Omega_k(u\mathbf{x}'\mathbf{x})dG(\mathbf{x}) \\
&= \int_0^\infty \Omega_k(uy^2)dF(y),
\end{aligned}$$

where $F(y) = P((\mathbf{x}'\mathbf{x})^{\frac{1}{2}} \le y)$. ■

Now, we can derive the stochastic representation of a m.e.c. distribution.

Theorem 2.13. *Let* $\mathbf{X}$ *be a* $p \times n$ *random matrix. Let* $\mathbf{M}$ *be* $p \times n$, Σ *be* $p \times p$, *and* Φ *be* $n \times n$ *constant matrices,* $\Sigma \geq \mathbf{0}$, $\Phi \geq \mathbf{0}$, $rk(\Sigma) = p_1$, $rk(\Phi) = n_1$. *Then,*

$$\mathbf{X} \sim E_{p,n}(\mathbf{M}, \Sigma \otimes \Phi, \psi) \tag{2.9}$$

if and only if

$$\mathbf{X} \approx \mathbf{M} + r\mathbf{A}\mathbf{U}\mathbf{B}', \tag{2.10}$$

where $\mathbf{U}$ *is* $p_1 \times n_1$ *and* $vec(\mathbf{U}')$ *is uniformly distributed on* $S_{p_1 n_1}$, r *is a nonnegative random variable,* r *and* $\mathbf{U}$ *are independent,* $\Sigma = \mathbf{A}\mathbf{A}'$, *and* $\Phi = \mathbf{B}\mathbf{B}'$ *are rank factorizations of* Σ *and* Φ. *Moreover,* $\psi(u) = \int_0^\infty \Omega_{p_1 n_1}(r^2 u) dF(r)$, $u \geq 0$, *where* $\Omega_{p_1 n_1}(\mathbf{t}'\mathbf{t})$, $\mathbf{t} \in I\!R^{p_1 n_1}$ *denotes the characteristic function of* $vec(\mathbf{U}')$, *and* $F(r)$ *denotes the distribution function of* r. *The expression,* $\mathbf{M} + r\mathbf{A}\mathbf{U}\mathbf{B}'$, *is called the stochastic representation of* $\mathbf{X}$.

PROOF: First, assume $\mathbf{X} \sim E_{p,n}(\mathbf{M}, \Sigma \otimes \Phi, \psi)$. Then, it follows from Corollary 2.1, that $\mathbf{Y} = \mathbf{A}^-(\mathbf{X} - \mathbf{M})\mathbf{B}'^- \sim E_{p_1,n_1}(\mathbf{0}, \mathbf{I}_{p_1} \otimes \mathbf{I}_{n_1}, \psi)$. Thus,

$$\mathbf{y} = vec(\mathbf{Y}') \sim E_{p_1 n_1}(\mathbf{0}, \mathbf{I}_{p_1 n_1}, \psi).$$

So, $\psi(\mathbf{t}'\mathbf{t})$, $\mathbf{t} \in I\!R^{p_1 n_1}$ is a characteristic function and from Theorem 2.12, we get

$$\psi(u) = \int_0^\infty \Omega_{p_1 n_1}(y^2 u) dF(y), \quad u \geq 0,$$

which means that $\mathbf{y} \approx r\mathbf{u}$, where r is nonnegative with distribution function $F(y)$, $\mathbf{u}$ is uniformly distributed on $S_{p_1 n_1}$, and r and $\mathbf{u}$ are independent. Therefore, we can write $\mathbf{y} \approx r\mathbf{u}$, where $\mathbf{u} = vec(\mathbf{U}')$. Now, using Corollary 2.1 again, we get

$$\mathbf{X} \approx \mathbf{A}\mathbf{Y}\mathbf{B}' + \mathbf{M} \approx \mathbf{M} + r\mathbf{A}\mathbf{U}\mathbf{B}'.$$

Conversely, suppose $\mathbf{X} \approx \mathbf{M} + r\mathbf{A}\mathbf{U}\mathbf{B}'$. Let $\mathbf{u} = vec(\mathbf{U}')$. Define

$$\psi(u) = \int_0^\infty \Omega_{p_1 n_1}(y^2 u) dF(y),$$

where $F(y)$ is the distribution function of r, $u \geq 0$. Then, it follows from Theorem 2.12, that $\psi(\mathbf{t}'\mathbf{t})$, $\mathbf{t} \in I\!R^{p_1 n_1}$ is the characteristic function of $r\mathbf{u}$. So $r\mathbf{u} \sim E_{p_1 n_1}(\mathbf{0}, \mathbf{I}_{p_1 n_1}, \psi)$ and hence,

$$r\mathbf{U} \sim E_{p_1 n_1}(\mathbf{0}, \mathbf{I}_{p_1} \otimes \mathbf{I}_{n_1}, \psi).$$

Therefore,

$$\mathbf{X} \approx \mathbf{M} + r\mathbf{A}\mathbf{U}\mathbf{B}' \sim E_{p,n}(\mathbf{M}, (\mathbf{A}\mathbf{A}') \otimes (\mathbf{B}\mathbf{B}'), \psi) = E_{p,n}(\mathbf{M}, \Sigma \otimes \Phi, \psi). \qquad \blacksquare$$

It may be noted that the stochastic representation is not uniquely defined. We can only say the following.

Theorem 2.14. *$\mathbf{M}_1 + r_1\mathbf{A}_1\mathbf{U}\mathbf{B}_1'$ and $\mathbf{M}_2 + r_2\mathbf{A}_2\mathbf{U}\mathbf{B}_2'$, where $\mathbf{U}$ is $p_1 \times n_1$, are two stochastic representations of the same $p \times n$ dimensional nondegenerate m.e.c. distribution if and only if $\mathbf{M}_1 = \mathbf{M}_2$, and there exist $\mathbf{G} \in O(p_1)$, $\mathbf{H} \in O(n_1)$, and positive constants a, b, and c such that $ab = c$, $\mathbf{A}_2 = a\mathbf{A}_1\mathbf{G}$, $\mathbf{B}_2 = b\mathbf{B}_1\mathbf{H}$, and $r_2 = \frac{1}{c}r_1$.*

PROOF: The "if" part is trivial. Conversely, let $\mathbf{X} \approx \mathbf{M}_1 + r_1\mathbf{A}_1\mathbf{U}\mathbf{B}_1'$ and $\mathbf{X}_2 \approx \mathbf{M}_2 + r_2\mathbf{A}_2\mathbf{U}\mathbf{B}_2'$. Then

$$\mathbf{X} \sim E_{p,n}(\mathbf{M}_1, (\mathbf{A}_1\mathbf{A}_1') \otimes (\mathbf{B}_1\mathbf{B}_1'), \psi_1)$$

and

$$\mathbf{X} \sim E_{p,n}(\mathbf{M}_2, (\mathbf{A}_2\mathbf{A}_2') \otimes (\mathbf{B}_2\mathbf{B}_2'), \psi_2),$$

where $\psi_i(u) = \int_0^\infty \Omega_{p_1 n_1}(y^2 u)dF_i(y)$, and $F_i(y)$ denotes the distribution function of r_i, $i = 1, 2$.

It follows, from Theorem 2.4, that $\mathbf{M}_1 = \mathbf{M}_2$, and there exist $a^2 > 0$, $b^2 > 0$, and $c^2 > 0$ such that $a^2b^2 = c^2$, $\mathbf{A}_2\mathbf{A}_2' = a^2\mathbf{A}_1\mathbf{A}_1'$, $\mathbf{B}_2\mathbf{B}_2' = b^2\mathbf{B}_1\mathbf{B}_1'$, and $\psi_2(z) = \psi_1\left(\frac{1}{c^2}z\right)$. Now, from Theorem 1.11, it follows that there exist $\mathbf{G} \in O(p_1)$ and $\mathbf{H} \in O(n_1)$ such that $\mathbf{A}_2 = a\mathbf{A}_1\mathbf{G}$, and $\mathbf{B}_2 = b\mathbf{B}_1\mathbf{H}$. Since, $\psi_2(z) = \psi_1\left(\frac{1}{c^2}z\right)$, we have

$$\begin{aligned}
\psi_2(z) &= \int_0^\infty \Omega_{p_1 n_1}(y^2 z)dF_2(y) \\
&= \psi_1\left(\frac{z}{c^2}\right) \\
&= \int_0^\infty \Omega_{p_1 n_1}\left(y^2\frac{z}{c^2}\right)dF_1(y) \\
&= \int_0^\infty \Omega_{p_1 n_1}\left(\left(\frac{y}{c}\right)^2 z\right)dF_1(y) \\
&= \int_0^\infty \Omega_{p_1 n_1}(t^2 z)dF_1(ct).
\end{aligned}$$

Therefore $F_2(y) = F_1(cy)$, and

$$P(r_2 < y) = P(r_1 < cy) = P\left(\frac{r_1}{c} < y\right).$$

Hence, $r_2 = \frac{1}{c}r_1$. ■

Remark 2.4. It follows, from Theorem 2.13, that $\mathbf{U}$ does not depend on ψ. On the other hand, if p_1 and n_1 are fixed, ψ and r determine each other.

Remark 2.5. Let $E_{p,n}(\mathbf{0}, \Sigma \otimes \Phi, \psi)$ and $r\mathbf{AUB}'$ be the stochastic representation of $\mathbf{X}$. Then, $\mathbf{A}^-\mathbf{XB}'^- \approx r\mathbf{U}$, and $tr((\mathbf{A}^-\mathbf{XB}'^-)'(\mathbf{A}^-\mathbf{XB}'^-)) \approx tr(r^2\mathbf{U}'\mathbf{U})$. Now,

$$\begin{aligned} tr((\mathbf{A}^-\mathbf{XB}'^-)'(\mathbf{A}^-\mathbf{XB}'^-)) &= tr(\mathbf{B}^-\mathbf{X}'\mathbf{A}'^-\mathbf{A}^-\mathbf{XB}'^-) \\ &= tr(\mathbf{X}'\mathbf{A}'^-\mathbf{A}^-\mathbf{XB}'^-\mathbf{B}^-) \\ &= tr(\mathbf{X}'\Sigma^-\mathbf{X}\Phi^-) . \end{aligned}$$

Here we used $\mathbf{A}'^-\mathbf{A}^- = \Sigma^-$, which follows from Theorem 1.23. On the other hand, $tr(\mathbf{U}'\mathbf{U}) = 1$. Therefore, we get $r^2 \approx tr(\mathbf{X}'\Sigma^-\mathbf{X}\Phi^-)$.

If an elliptically contoured random matrix is nonzero with probability one, then the terms of the stochastic representation can be obtained explicitly. First we introduce the following definition.

Definition 2.4. Let $\mathbf{X}$ be a $p \times n$ matrix. Then its norm, denoted by $||\mathbf{X}||$, is defined as

$$||\mathbf{X}|| = \left(\sum_{i=1}^{p} \sum_{j=1}^{n} x_{ij}^2 \right)^{\frac{1}{2}} .$$

That is, $||\mathbf{X}|| = (tr(\mathbf{X}'\mathbf{X}))^{\frac{1}{2}}$, and if $n = 1$, then we have $||\mathbf{x}|| = (\mathbf{x}'\mathbf{x})^{\frac{1}{2}}$.

The proof of the following theorem is based on Muirhead (1982).

Theorem 2.15. *Let* $\mathbf{X} \sim E_{p,n}(\mathbf{0}, \mathbf{I}_p \otimes \mathbf{I}_n, \psi)$ *with* $P(\mathbf{X} = \mathbf{0}) = 0$. *Then,* $\mathbf{X} = ||\mathbf{X}|| \frac{\mathbf{X}}{||\mathbf{X}||}$, $P(||\mathbf{X}|| > 0) = 1$, $vec\left(\frac{\mathbf{X}'}{||\mathbf{X}||}\right)$ *is uniformly distributed on* S_{pn}, *and* $||\mathbf{X}||$ *and* $\frac{\mathbf{X}}{||\mathbf{X}||}$ *are independent. That is,* $||\mathbf{X}|| \frac{\mathbf{X}}{||\mathbf{X}||}$ *is the stochastic representation of* $\mathbf{X}$.

PROOF: Since $\mathbf{X} = \mathbf{0}$ iff $tr(\mathbf{X}'\mathbf{X}) = 0$, $P(||\mathbf{X}|| > 0) = 1$, follows so we can write $\mathbf{X} = ||\mathbf{X}|| \frac{\mathbf{X}}{||\mathbf{X}||}$. Define $\mathbf{x} = vec(\mathbf{X}')$. Then, $\mathbf{x} \sim E_{pn}(\mathbf{0}, \mathbf{I}_{pn}, \psi)$ and $||\mathbf{X}|| = ||\mathbf{x}||$. Hence, $\mathbf{x} = ||\mathbf{x}|| \frac{\mathbf{x}}{||\mathbf{x}||}$.

Let $T(\mathbf{x}) = \frac{\mathbf{x}}{||\mathbf{x}||}$, and $\mathbf{G} \in O(pn)$. Then, we get $\mathbf{Gx} \sim E_{pn}(\mathbf{0}, \mathbf{I}_{pn}, \psi)$, so $\mathbf{x} \approx \mathbf{Gx}$ and $T(\mathbf{x}) \approx T(\mathbf{Gx})$. On the other hand,

$$T(\mathbf{Gx}) = \frac{\mathbf{Gx}}{||\mathbf{Gx}||} = \frac{\mathbf{Gx}}{||\mathbf{x}||} = \mathbf{G}T(\mathbf{x}) .$$

Hence, $T(\mathbf{x}) \approx \mathbf{G}T(\mathbf{x})$. However, the uniform distribution is the only one on S_{pn} which is invariant under orthogonal transformation. So, $T(\mathbf{x})$ is uniformly distributed on S_{pn}.

Now, we define a measure μ on S_{pn}. Fix $B \subset I\!R_0^+$ Borel set. Let $A \subset S_{pn}$ be a Borel set. Then,

$$\mu(A) = P(T(\mathbf{x}) \in A | \|\mathbf{X}\| \in B).$$

Since $\mu(I\!R^{pn}) = 1$, μ is a probability measure on S_{pn}.

Let $\mathbf{G} \in O(pn)$. Then, $\mathbf{G}^{-1}\mathbf{x} \approx \mathbf{x}$, and we have

$$\begin{aligned}\mu(\mathbf{G}A) &= P(T(\mathbf{x}) \in \mathbf{G}A | \|\mathbf{x}\| \in B)\\ &= P(\mathbf{G}^{-1}T(\mathbf{x}) \in A | \|\mathbf{x}\| \in B)\\ &= P(T(\mathbf{G}^{-1}\mathbf{x}) \in A | \|\mathbf{x}\| \in B)\\ &= P(T(\mathbf{G}^{-1}\mathbf{x}) \in A | \|\mathbf{G}^{-1}\mathbf{x}\| \in B)\\ &= P(T(\mathbf{x}) \in A | \|\mathbf{x}\| \in B)\\ &= \mu(A).\end{aligned}$$

Thus, $\mu(A)$ is a probability measure on S_{pn}, invariant under orthogonal transformation, therefore, it must be the uniform distribution. That is, it coincides with the distribution of $T(\mathbf{x})$. So, $\mu(A) = P(T(\mathbf{x}) \in A)$, from which it follows that

$$P(T(\mathbf{x}) \in A | \|\mathbf{x}\| \in B) = P(T(\mathbf{x}) \in A).$$

Therefore, $T(\mathbf{x})$ and $\|\mathbf{x}\|$ are independently distributed. Returning to the matrix notation, the proof is completed. ■

Muirhead (1982) has given the derivation of the p.d.f. of r in the case when $\mathbf{x} \sim E_p(\mathbf{0}, \mathbf{I}_p, \psi)$ and $\mathbf{x}$ is absolutely continuous. Now for the elliptically contoured random matrices, the following theorem can be stated.

Theorem 2.16. *Let $\mathbf{X} \sim E_{p,n}(\mathbf{0}, \Sigma \otimes \Phi, \psi)$ and $r\mathbf{AUB}'$ be a stochastic representation of $\mathbf{X}$. Assume $\mathbf{X}$ is absolutely continuous and has the p.d.f.*

$$f(\mathbf{X}) = |\Sigma|^{-\frac{n}{2}} |\Phi|^{-\frac{p}{2}} h(tr(\mathbf{X}'\Sigma^{-1}\mathbf{X}\Phi^{-1})).$$

Then, r is also absolutely continuous and has the p.d.f.

$$g(r) = \frac{2\pi^{\frac{pn}{2}}}{\Gamma\left(\frac{pn}{2}\right)} r^{pn-1} h(r^2), \quad r \geq 0.$$

PROOF: Step 1. First we prove the theorem for $n = 1$. Then, $\mathbf{x} \sim E_p(\mathbf{0}, \Sigma, \psi)$ and so

$$\mathbf{y} = \mathbf{A}^{-1}\mathbf{x} \sim E_p(\mathbf{0}, \mathbf{I}_p, \psi).$$

Therefore $\mathbf{y}$ has the p.d.f. $h(\mathbf{y}'\mathbf{y})$. Let us introduce polar coordinates:

$$\begin{aligned}
y_1 &= r sin\theta_1 sin\theta_2 \dots sin\theta_{p-2} sin\theta_{p-1}\\
y_2 &= r sin\theta_1 sin\theta_2 \dots sin\theta_{p-2} cos\theta_{p-1}\\
y_3 &= r sin\theta_1 sin\theta_2 \dots cos\theta_{p-2}\\
&\vdots\\
y_{p-1} &= r sin\theta_1 cos\theta_2\\
y_p &= r cos\theta_1 ,
\end{aligned}$$

where $r > 0$, $0 < \theta_i < \pi$, $i = 1,2,\dots,p-2$, and $0 < \theta_{p-1} < 2\pi$. We want to express the p.d.f. of $\mathbf{y}$ in terms of $r, \theta_1, \dots, \theta_{p-1}$. The Jacobian of the transformation $(y_1, y_2, \dots, y_p) \to (r, \theta_1, \dots, \theta_{p-1})$ is $r^{p-1} sin^{p-2}\theta_1 sin^{p-3}\theta_2 \dots sin\theta_{p-2}$. On the other hand, $\mathbf{y}'\mathbf{y} = r^2$. Therefore, the p.d.f. of $(r, \theta_1, \dots, \theta_{p-1})$ is

$$h(r^2) r^{p-1} sin^{p-2}\theta_1 sin^{p-3}\theta_2 \dots sin\theta_{p-2}.$$

Consequently, the p.d.f. of r is

$$\begin{aligned}
g(r) &= r^{p-1} h(r^2) \int_0^{2\pi}\int_0^{\pi}\dots\int_0^{\pi} sin^{p-2}\theta_1 sin^{p-3}\theta_2 \dots sin\theta_{p-2} d\theta_1 d\theta_2 \dots d\theta_{p-2} d\theta_{p-1}\\
&= r^{p-1} h(r^2) \frac{2\pi^{\frac{p}{2}}}{\Gamma\left(\frac{p}{2}\right)}.
\end{aligned}$$

Step 2. Now let $\mathbf{X} \sim E_{p,n}(\mathbf{0}, \Sigma \otimes \Phi, \psi)$ and $\mathbf{X} \approx r\mathbf{AUB}'$. Define $\mathbf{x} = vec(\mathbf{X}')$, and $\mathbf{u} = vec(\mathbf{U}')$. Then, $\mathbf{x} \sim E_{pn}(\mathbf{0}, \Sigma \otimes \Phi, \psi)$, $\mathbf{x}$ has p.d.f. $\frac{1}{|\Sigma \otimes \Phi|} h(\mathbf{x}'(\Sigma \otimes \Phi)^{-1}\mathbf{x})$, and $\mathbf{x} \approx r(\mathbf{A} \otimes \mathbf{B})\mathbf{u}$. Using Step 1 we get the following as the p.d.f. of r,

$$g(r) = r^{pn-1} h(r^2) \frac{2\pi^{\frac{pn}{2}}}{\Gamma\left(\frac{pn}{2}\right)}.$$

■

The stochastic representation is a major tool in the study of m.e.c. distributions. It will often be used in further discussion.

Cambanis, Huang and Simons (1981), and Anderson and Fang (1987) derived the relationship between the stochastic representation of a multivariate elliptically contoured distribution and the stochastic representation of its marginals. This result is given in the next theorem.

Theorem 2.17. *Let* $\mathbf{X} \sim E_{p,n}(\mathbf{0}, \mathbf{I}_p \otimes \mathbf{I}_n, \psi)$ *with stochastic representation* $\mathbf{X} \approx r\mathbf{U}$. *Let* $\mathbf{X}$ *be partitioned into*

$$\mathbf{X} = \begin{pmatrix} \mathbf{X}_1 \\ \mathbf{X}_2 \\ \vdots \\ \mathbf{X}_m \end{pmatrix},$$

where $\mathbf{X}_i$ *is* $p_i \times n$ *matrix,* $i = 1, \ldots, m$. *Then,*

$$\begin{pmatrix} \mathbf{X}_1 \\ \mathbf{X}_2 \\ \vdots \\ \mathbf{X}_m \end{pmatrix} \approx \begin{pmatrix} rr_1\mathbf{U}_1 \\ rr_2\mathbf{U}_2 \\ \vdots \\ rr_m\mathbf{U}_m \end{pmatrix},$$

where r, $(r_1, r_2, \ldots, r_m)$, $\mathbf{U}_1, \mathbf{U}_2, \ldots, \mathbf{U}_m$ *are independent,* $r_i \geq 0$, $i = 1, \ldots, m$, $\sum_{i=1}^m r_i^2 = 1$,

$$(r_1^2, r_2^2, \ldots, r_{m-1}^2) \sim D\left(\frac{p_1 n}{2}, \frac{p_2 n}{2}, \ldots, \frac{p_{m-1} n}{2}; \frac{p_m n}{2}\right), \tag{2.11}$$

and $vec(\mathbf{U}_i')$ *is uniformly distributed on* $S_{p_i n}$, $i = 1, 2, \ldots, m$.

PROOF: Since $\mathbf{X} \approx r\mathbf{U}$, we have

$$\begin{pmatrix} \mathbf{X}_1 \\ \mathbf{X}_2 \\ \vdots \\ \mathbf{X}_m \end{pmatrix} \approx r\mathbf{U},$$

where r and $\mathbf{U}$ are independent. Thus it suffices to prove that

$$\mathbf{U} \approx \begin{pmatrix} r_1\mathbf{U}_1 \\ r_2\mathbf{U}_2 \\ \vdots \\ r_m\mathbf{U}_m \end{pmatrix}.$$

Note that $\mathbf{U}$ does not depend on ψ, so we can choose $\psi(z) = \exp\left(-\frac{z}{2}\right)$, which means $\mathbf{X} \sim N_{p,n}(\mathbf{0}, \mathbf{I}_p \otimes \mathbf{I}_n)$. It follows that $\mathbf{X}_i \sim N_{p_i,n}(\mathbf{0}, \mathbf{I}_{p_i} \otimes \mathbf{I}_n)$ and $\mathbf{X}_i$'s are mutually independent, $i = 1, \ldots, m$.

Now,

$$\mathbf{U} \approx \frac{\mathbf{X}}{\|\mathbf{X}\|} = \left(\frac{\mathbf{X}_1'}{\|\mathbf{X}\|}, \frac{\mathbf{X}_2'}{\|\mathbf{X}\|}, \ldots, \frac{\mathbf{X}_m'}{\|\mathbf{X}\|}\right)'.$$

From Theorem 2.15 it follows that $\mathbf{X}_i = \|\mathbf{X}_i\| \frac{\mathbf{X}_i}{\|\mathbf{X}_i\|}$, where $\|\mathbf{X}_i\|$ and $\frac{\mathbf{X}_i}{\|\mathbf{X}_i\|}$ are independent and $vec\left(\frac{\mathbf{X}_i'}{\|\mathbf{X}_i\|}\right) \approx \mathbf{u}_{p_i n}$ which is uniformly distributed on $S_{p_i n}$. Since, $\mathbf{X}_i$'s are independent, $\|\mathbf{X}_i\|$ and $\frac{\mathbf{X}_i}{\|\mathbf{X}_i\|}$ are mutually independent, $i = 1, 2, \ldots, m$. Therefore, we get

$$\mathbf{U} \approx \left(\frac{\|\mathbf{X}_1\|}{\|\mathbf{X}\|} \frac{\mathbf{X}_1'}{\|\mathbf{X}_1\|}, \frac{\|\mathbf{X}_2\|}{\|\mathbf{X}\|} \frac{\mathbf{X}_2'}{\|\mathbf{X}_2\|}, \ldots, \frac{\|\mathbf{X}_m\|}{\|\mathbf{X}\|} \frac{\mathbf{X}_m'}{\|\mathbf{X}_m\|} \right)'.$$

Define $r_i = \frac{\|\mathbf{X}_i\|}{\|\mathbf{X}\|}$, and $\mathbf{U}_i = \frac{\mathbf{X}_i}{\|\mathbf{X}_i\|}$, $i = 1, 2, \ldots, m$. Since $\|\mathbf{X}\| = \left(\sum_{i=1}^m \|\mathbf{X}_i\|^2\right)^{\frac{1}{2}}$, r_i's are functions of $\|\mathbf{X}_1\|, \|\mathbf{X}_2\|, \ldots, \|\mathbf{X}_m\|$. Hence, $(r_1, r_2, \ldots, r_m)$, $\mathbf{U}_1, \mathbf{U}_2, \ldots, \mathbf{U}_m$ are independent. Moreover, $\|\mathbf{X}_i\|^2 = tr(\mathbf{X}_i'\mathbf{X}_i) \sim \chi^2_{p_i n}$ and $\|\mathbf{X}_i\|^2$'s are independent. Now, it is known that

$$\left(\frac{\|\mathbf{X}_1\|^2}{\sum_{i=1}^m \|\mathbf{X}_i\|^2}, \frac{\|\mathbf{X}_2\|^2}{\sum_{i=1}^m \|\mathbf{X}_i\|^2}, \ldots, \frac{\|\mathbf{X}_m\|^2}{\sum_{i=1}^m \|\mathbf{X}_i\|^2} \right) \sim D\left(\frac{p_1 n}{2}, \frac{p_2 n}{2}, \ldots, \frac{p_{m-1} n}{2}; \frac{p_m n}{2} \right)$$

(see Johnson and Kotz, 1972). Consequently, $(r_1^2, r_2^2, \ldots, r_{m-1}^2)$ has the distribution (2.11). ■

Corollary 2.4. *Let* $\mathbf{X} \sim E_{p,n}(\mathbf{0}, \mathbf{I}_p \otimes \mathbf{I}_n, \psi)$ *with stochastic representation* $\mathbf{X} \approx r\mathbf{U}$. *Let* $\mathbf{X}$ *be partitioned into*

$$\mathbf{X} = \begin{pmatrix} \mathbf{X}_1 \\ \mathbf{X}_2 \end{pmatrix},$$

where $\mathbf{X}_1$ *is* $q \times n$ *matrix,* $1 \leq q < p$. *Then,* $\begin{pmatrix} \mathbf{X}_1 \\ \mathbf{X}_2 \end{pmatrix} \approx \begin{pmatrix} rr_1\mathbf{U}_1 \\ rr_2\mathbf{U}_2 \end{pmatrix}$, *where* r, (r_1, r_2), $\mathbf{U}_1, \mathbf{U}_2$ *are independent,* $r_i \geq 0$, $i = 1, 2$, $r_1^2 + r_2^2 = 1$, *and* $r_1^2 \sim B\left(\frac{qn}{2}, \frac{(p-q)n}{2}\right)$. *Also* $vec(\mathbf{U}_1')$ *is uniformly distributed on* S_{qn} *and* $vec(\mathbf{U}_2')$ *is uniformly distributed on* $S_{(p-q)n}$.

2.6 Conditional Distributions

First, we derive the conditional distribution for the vector variate elliptically contoured distribution. We will follow the lines of Cambanis, Huang, and Simons (1981). The following lemma will be needed in the proof.

Lemma 2.2. *Let* x *and* y *be one-dimensional nonnegative random variables. Assume that* y *is absolutely continuous with probability density function* $g(y)$. *Denote the distribution function of* x *by* $F(x)$. *Define* $z = xy$. *Then,* z *is absolutely continuous on* $\mathbb{R}^+$ *with p.d.f.*

$$h(z) = \int_0^\infty \frac{1}{x} g\left(\frac{z}{x}\right) dF(x). \tag{2.12}$$

If $F(0) = 0$, then z is absolutely continuous on $\mathbb{R}_0^+$, and if $F(0) > 0$, then z has an atom of size $F(0)$ at zero. Moreover, a conditional distribution of x given z is

$$P(x \le x_0 | z = z_0) = \begin{cases} \frac{1}{h(z_0)} \int_{(0,w]} \frac{1}{x_0} g\left(\frac{z_0}{x_0}\right) dF(x_0) & \text{if } x_0 \ge 0,\ z_0 > 0, \text{ and } h(z_0) \ne 0 \\ 1 & \text{if } x_0 \ge 0, \text{ and } z_0 = 0 \\ & \text{or } x_0 \ge 0,\ z_0 > 0 \text{ and } h(z_0) = 0 \\ 0 & \text{if } x_0 < 0. \end{cases} \tag{2.13}$$

PROOF:

$$\begin{aligned} P(0 < z \le z_0) &= P(0 < xy \le z_0) \\ &= \int_0^\infty P(0 < xy \le z_0 | x = x_0) dF(x_0) \\ &= \int_0^\infty P(0 < y \le \frac{z_0}{x_0}) dF(x_0) \\ &= \int_0^\infty \int_{\left(0, \frac{z_0}{x_0}\right]} g(y) dy dF(x_0). \end{aligned}$$

Let $t = x_0 y$. Then, $y = \frac{t}{x_0}$, $dy = \frac{1}{x_0} dt$ and

$$\begin{aligned} \int_0^\infty \int_{\left(0, \frac{z_0}{x_0}\right]} g(y) dy dF(x_0) &= \int_0^\infty \int_{(0,z_0]} \frac{1}{x_0} g\left(\frac{t}{x_0}\right) dt dF(x_0) \\ &= \int_{(0,z_0]} \int_0^\infty \frac{1}{x_0} g\left(\frac{t}{x_0}\right) dF(x_0) dt \end{aligned}$$

and this proves (2.12).

Since, y is absolutely continuous, $P(y = 0) = 0$. Hence,

$$P(\chi_{\{0\}}(z) = \chi_{\{0\}}(x)) = 1. \tag{2.14}$$

Therefore, if $F(0) = 0$, then $P(z = 0) = 0$, and so z is absolutely continuous on R_0^+. If $F(0) > 0$, then $P(z = 0) = F(0)$ and thus z has an atom of size $F(0)$ at zero.

Now, we prove (2.13). Since $x \ge 0$, we have $P(x \le x_0) = 0$ if $x_0 < 0$. Hence, $P(x \le x_0 | z) = 0$ if $x_0 < 0$. If $x_0 \ge 0$, we have to prove that the function $P(x \le x_0 | z)$ defined under (2.13) satisfies

$$\int_{[0,r]} P(x \le x_0 | z) dH(z) = P(x \le x_0, z \le r),$$

where $H(z)$ denotes the distribution function of z and $r \ge 0$. Now,

$$\begin{aligned}\int_{[0,r]} P(x \le x_0|z)dH(z) &= P(x \le x_0|z=0)H(0) + \int_{(0,r]} P(x \le x_0|z)dH(z)\\ &= H(0) + \int_{(0,r]} \frac{1}{h(z)}\left(\int_{(0,x_0]} \frac{1}{x} g\left(\frac{z}{x}\right) dF(x)\right) h(z)dz\\ &= H(0) + \int_{(0,x_0]} \int_{(0,r]} \frac{1}{x} g\left(\frac{z}{x}\right) dz dF(x).\end{aligned}$$

Let $u = \frac{z}{x}$. Then, $J(u \to z) = x$, and so

$$\int_{(0,r]} \frac{1}{x} g\left(\frac{z}{x}\right) dz = \int_{(0,\frac{r}{x}]} g(u)du = P\left(0 < y \le \frac{r}{x}\right).$$

Hence,

$$\begin{aligned}H(0) + \int_{(0,x_0]} \int_{(0,r]} \frac{1}{x} g\left(\frac{z}{x}\right) dz dF(x) &= H(0) + \int_{(0,x_0]} P\left(0 < y \le \frac{r}{x}\Big| x = x_0\right) dF(x_0)\\ &= H(0) + P\left(0 < x \le x_0, 0 < y \le \frac{r}{x}\right)\\ &= H(0) + P(0 < x \le x_0, 0 < xy \le r)\\ &= P(z=0) + P(0 < x \le x_0, 0 < z \le r)\\ &= P(x=0, z=0) + P(0 < x \le x_0, 0 < z \le r)\\ &= P(0 \le x \le x_0, 0 \le z \le r)\\ &= P(x \le x_0, z \le r),\end{aligned}$$

where we used (2.14). ■

Now, we obtain the conditional distribution for spherical distributions.

Theorem 2.18. *Let $\mathbf{x} \sim E_p(\mathbf{0}, \mathbf{I}_p, \psi)$ with stochastic representation $r\mathbf{u}$. Let us partition $\mathbf{x}$ as $\mathbf{x} = \begin{pmatrix} \mathbf{x}_1 \\ \mathbf{x}_2 \end{pmatrix}$, where $\mathbf{x}_1$ is q-dimensional $(1 \le q < p)$. Then, the conditional distributionœ of $\mathbf{x}_1$ given $\mathbf{x}_2$ is $(\mathbf{x}_1|\mathbf{x}_2) \sim E_q\left(\mathbf{0}, I_q, \psi_{\|\mathbf{x}_2\|^2}\right)$, and the stochastic representation of $(\mathbf{x}_1|\mathbf{x}_2)$ is $r_{\|\mathbf{x}_2\|^2}\mathbf{u}_1$, where $\mathbf{u}_1$ is q-dimensional. The distribution of $r_{\|\mathbf{x}_2\|^2}$ is given by*

$$a) \qquad P(r_{a^2} \le y) = \frac{\int_{(a,\sqrt{a^2+y^2}]} (w^2 - a^2)^{\frac{q}{2}-1} w^{-(p-2)} dF(w)}{\int_{(a,\infty)} (w^2 - a^2)^{\frac{q}{2}-1} w^{-(p-2)} dF(w)} \qquad (2.15)$$

for $y \geq 0$ if $a > 0$ and $F(a) < 1$,

$$b) \qquad P(r_{a^2} = 0) = 1 \quad \text{if} \quad a = 0 \quad \text{or} \quad F(a) = 1. \tag{2.16}$$

Here F denotes the distribution function of r.

PROOF: From Corollary 2.4, we have the representation

$$\begin{pmatrix} \mathbf{x}_1 \\ \mathbf{x}_2 \end{pmatrix} \approx \begin{pmatrix} rr_1\mathbf{u}_1 \\ rr_2\mathbf{u}_2 \end{pmatrix}.$$

Using the independence of r, r_1, $\mathbf{u}_1$ and $\mathbf{u}_2$, we get

$$\begin{aligned}(\mathbf{x}_1|\mathbf{x}_2) &\approx (rr_1\mathbf{u}_1|rr_2\mathbf{u}_2 = \mathbf{x}_2) \\ &= (rr1\mathbf{u}_1|r(1-r_1^2)^{\frac{1}{2}}\mathbf{u}_2 = \mathbf{x}_2) \\ &= (rr_1|r(1-r_1^2)^{\frac{1}{2}}\mathbf{u}_2 = \mathbf{x}_2)\mathbf{u}_1,\end{aligned}$$

and defining $r_0 = (rr_1|r(1-r_1^2)^{\frac{1}{2}}\mathbf{u}_2 = \mathbf{x}_2)$, we see that r and $\mathbf{u}_1$ are independent; therefore, $(\mathbf{x}_1|\mathbf{x}_2)$ has a spherical distribution.

Next, we show that

$$(rr_1|r(1-r_1^2)^{\frac{1}{2}}\mathbf{u}_2 = \mathbf{x}_2) \approx \left((r^2 - \|\mathbf{x}_2\|^2)^{\frac{1}{2}}|r(1-r_1^2)^{\frac{1}{2}} = \|\mathbf{x}_2\|\right).$$

If $r(1-r_1^2)^{\frac{1}{2}}\mathbf{u}_2 = \mathbf{x}_2$, then $\left(r(1-r_1^2)^{\frac{1}{2}}\mathbf{u}_2\right)' r(1-r_1^2)^{\frac{1}{2}}\mathbf{u}_2 = \|\mathbf{x}_2\|^2$ and therefore, $r^2(1-r_1^2) = \|\mathbf{x}_2\|^2$. Hence, we get $r^2 - r^2r_1^2 = \|\mathbf{x}_2\|^2$, thus $r^2r_1^2 = r^2 - \|\mathbf{x}_2\|^2$ and $rr_1 = (r^2 - \|\mathbf{x}_2\|^2)^{\frac{1}{2}}$. Therefore,

$$(rr_1|r(1-r_1^2)^{\frac{1}{2}}\mathbf{u}_2 = \mathbf{x}_2) \approx ((r^2 - \|\mathbf{x}_2\|^2)^{\frac{1}{2}}|r(1-r_1^2)^{\frac{1}{2}}\mathbf{u}_2 = \mathbf{x}_2).$$

If $\mathbf{x}_2 = \mathbf{0}$, then $\|\mathbf{x}_2\| = 0$, and using the fact that $\mathbf{u}_1 \neq \mathbf{0}$ we get

$$\begin{aligned}((r^2 - \|\mathbf{x}_2\|^2)^{\frac{1}{2}}|r(1-r_1^2)^{\frac{1}{2}}\mathbf{u}_2 = \mathbf{x}_2) &= ((r^2 - \|\mathbf{x}_2\|^2)^{\frac{1}{2}}|r(1-r_1^2)^{\frac{1}{2}} = 0) \\ &= ((r^2 - \|\mathbf{x}_2\|^2)^{\frac{1}{2}}|r(1-r_1^2)^{\frac{1}{2}} = \|\mathbf{x}_2\|).\end{aligned}$$

If $\mathbf{x}_2 \neq \mathbf{0}$, then we can write

$$\begin{aligned}&((r^2 - \|\mathbf{x}_2\|^2)^{\frac{1}{2}}|r(1-r_1^2)^{\frac{1}{2}}\mathbf{u}_2 = \mathbf{x}_2) \\ &= \left((r^2 - \|\mathbf{x}_2\|^2)^{\frac{1}{2}}|r(1-r_1^2)^{\frac{1}{2}}\mathbf{u}_2 = \|\mathbf{x}_2\|\frac{\mathbf{x}_2}{\|\mathbf{x}_2\|}\right)\end{aligned}$$

$$= \left((r^2 - \|\mathbf{x}_2\|^2)^{\frac{1}{2}} | r(1-r_1^2)^{\frac{1}{2}} = \|\mathbf{x}_2\| \quad \text{and} \quad \mathbf{u}_2 = \frac{\mathbf{x}_2}{\|\mathbf{x}_2\|} \right)$$
$$= \left((r^2 - \|\mathbf{x}_2\|^2)^{\frac{1}{2}} | r(1-r_1^2)^{\frac{1}{2}} = \|\mathbf{x}_2\| \right),$$

where we used the fact that r, r_1, and $\mathbf{u}_2$ are independent. Since $1-r_1^2 \sim B\left(\frac{p-q}{2}, \frac{q}{2}\right)$ its p.d.f. is

$$b(t) = \frac{\Gamma\left(\frac{p}{2}\right)}{\Gamma\left(\frac{q}{2}\right)\Gamma\left(\frac{p-q}{2}\right)} t^{\frac{p-q}{2}-1}(1-t)^{\frac{q}{2}-1}, \quad 0 < t < 1.$$

Hence, the p.d.f. of $(1-r_1^2)^{\frac{1}{2}}$ is

$$g(y) = \frac{\Gamma\left(\frac{p}{2}\right)}{\Gamma\left(\frac{q}{2}\right)\Gamma\left(\frac{p-q}{2}\right)} (y^2)^{\frac{p-q}{2}-1}(1-y^2)^{\frac{q}{2}-1} 2y$$
$$= \frac{2\Gamma\left(\frac{p}{2}\right)}{\Gamma\left(\frac{q}{2}\right)\Gamma\left(\frac{p-q}{2}\right)} y^{p-q-1}(1-y^2)^{\frac{q}{2}-1}, \quad 0 < y < 1.$$

Using Lemma 2.2 we obtain a conditional distribution of r given $r(1-r_1^2)^{\frac{1}{2}} = a$.

$$P(r \le u | r(1-r_1^2)^{\frac{1}{2}} = a) \tag{2.17}$$
$$= \begin{cases} \frac{1}{h(a)} \int_{(0,u]} \frac{1}{w} \frac{2\Gamma\left(\frac{p}{2}\right)}{\Gamma\left(\frac{q}{2}\right)\Gamma\left(\frac{p-q}{2}\right)} \left(\frac{a}{w}\right)^{p-q-1} \left(1-\frac{a^2}{w^2}\right)^{\frac{q}{2}-1} dF(w) & \text{if } u \ge 0,\ a > 0, \text{ and } h(a) \ne 0 \\ 1 & \text{if } u \ge 0, \text{ and } a = 0 \\ & \text{or } u \ge 0,\ a > 0 \text{ and } h(a) = 0 \\ 0 & \text{if } u < 0. \end{cases}$$

where

$$h(a) = \int_{(a,\infty)} \frac{1}{w} \frac{2\Gamma\left(\frac{p}{2}\right)}{\Gamma\left(\frac{q}{2}\right)\Gamma\left(\frac{p-q}{2}\right)} \left(\frac{a}{w}\right)^{p-q-1} \left(1-\frac{a^2}{w^2}\right)^{\frac{q}{2}-1} dF(w).$$

Now,

$$\frac{\int_{(a,u]} \frac{1}{w} \frac{2\Gamma\left(\frac{p}{2}\right)}{\Gamma\left(\frac{q}{2}\right)\Gamma\left(\frac{p-q}{2}\right)} \left(\frac{a}{w}\right)^{p-q-1} \left(1-\frac{a^2}{w^2}\right)^{\frac{q}{2}-1} dF(w)}{h(a)}$$
$$= \frac{\int_{(a,u]} (w^2-a^2)^{\frac{q}{2}-1} w^{-(1+p-q-1+q-2)} dF(w)}{\int_{(a,\infty)} (w^2-a^2)^{\frac{q}{2}-1} w^{-(1+p-q-1+q-2)} dF(w)}$$

$$= \frac{\int_{(a,u]} \left(w^2 - a^2\right)^{\frac{q}{2}-1} w^{-(p-2)} dF(w)}{\int_{(a,\infty)} \left(w^2 - a^2\right)^{\frac{q}{2}-1} w^{-(p-2)} dF(w)} . \tag{2.18}$$

We note that $h(a) = 0$ if and only if

$$\int_{(a,\infty)} \left(w^2 - a^2\right)^{\frac{q}{2}-1} w^{-(p-2)} dF(w) , \tag{2.19}$$

and since $\left(w^2 - a^2\right)^{\frac{q}{2}-1} w^{-(p-2)} > 0$ for $w > a$, we see that (2.19) is equivalent to $F(a) = 1$. Therefore, $h(a) = 0$ if and only if $F(a) = 1$.

(a) If $a > 0$ and $F(a) < 1$, then for $r \geq 0$ we have

$$\begin{aligned} P(r_{a^2} \leq y) &= P((r^2 - a^2)^{\frac{1}{2}} \leq y | r(1 - r_1^2)^{\frac{1}{2}} = a) \\ &= P(r \leq (y^2 + a^2)^{\frac{1}{2}} | r(1 - r_1^2)^{\frac{1}{2}} = a) . \end{aligned} \tag{2.20}$$

From (2.17), (2.18), and (2.20) we have

$$\begin{aligned} P(r_{a^2} \leq y) &= P(r \leq (y^2 + a^2)^{\frac{1}{2}} | r(1 - r_1^2)^{\frac{1}{2}} = a) \\ &= \frac{\int_{(a,\sqrt{a^2+y^2}]} \left(w^2 - a^2\right)^{\frac{q}{2}-1} w^{-(p-2)} dF(w)}{\int_{(a,\infty)} \left(w^2 - a^2\right)^{\frac{q}{2}-1} w^{-(p-2)} dF(w)} . \end{aligned}$$

(b) If $r \geq 0$ and $a = 0$ or $r \geq 0$, $a > 0$ and $F(a) = 1$, then from (2.17) we get

$$P(r \leq (y^2 + a^2)^{\frac{1}{2}} | r(1 - r_1^2)^{\frac{1}{2}} = a) = 1 .$$

Take $y = 0$, then we get

$$P(r_{a^2} \leq 0) = P(r \leq a | r(1 - r_1^2)^{\frac{1}{2}} = a) = 1 . \tag{2.21}$$

Now, since $r_{a^2} \geq 0$, (2.21) implies $P(r_{a^2} = 0) = 1$. ■

In order to derive the conditional distribution for the multivariate elliptical distribution we need an additional lemma.

Lemma 2.3. *Let* $\mathbf{x} \sim E_p(\mathbf{m}, \Sigma, \psi)$ *and partition* $\mathbf{x}$, $\mathbf{m}$, Σ *as*

$$\mathbf{x} = \begin{pmatrix} \mathbf{x}_1 \\ \mathbf{x}_2 \end{pmatrix} , \quad \mathbf{m} = \begin{pmatrix} \mathbf{m}_1 \\ \mathbf{m}_2 \end{pmatrix} , \quad \Sigma = \begin{pmatrix} \Sigma_{11} & \Sigma_{12} \\ \Sigma_{21} & \Sigma_{22} \end{pmatrix} ,$$

where $\mathbf{x}_1$, $\mathbf{m}_1$ *are* q*-dimensional vectors and* Σ_{11} *is* $q \times q$, $1 \leq q < p$. *Let* $\mathbf{y} \sim E_p(\mathbf{0}, \mathbf{I}_p, \psi)$ *and partition* $\mathbf{y}$ *as* $\mathbf{y} = \begin{pmatrix} \mathbf{y}_1 \\ \mathbf{y}_2 \end{pmatrix}$, *where* $\mathbf{y}_1$ *is* q*-dimensional. Define* $\Sigma_{11\cdot 2} = \Sigma_{11} - \Sigma_{12}\Sigma_{22}^{-}\Sigma_{21}$ *and let* $\Sigma_{11\cdot 2} = \mathbf{A}\mathbf{A}'$ *and* $\Sigma_{22} = \mathbf{A}_2\mathbf{A}_2'$ *be rank factorizations of* $\Sigma_{11\cdot 2}$ *and* Σ_{22}. *Then*

$$\begin{pmatrix} \mathbf{x}_1 \\ \mathbf{x}_2 \end{pmatrix} \approx \begin{pmatrix} \mathbf{m}_1 + \mathbf{A}\mathbf{y}_1 + \Sigma_{12}\Sigma_{22}^{-}\mathbf{A}_2\mathbf{y}_2 \\ \mathbf{m}_2 + \mathbf{A}_2\mathbf{y}_2 \end{pmatrix}.$$

PROOF: Since $\begin{pmatrix} \mathbf{y}_1 \\ \mathbf{y}_2 \end{pmatrix} \sim E_p(\mathbf{0}, \mathbf{I}_p, \psi)$, we have

$$\begin{pmatrix} \mathbf{m}_1 + \mathbf{A}\mathbf{y}_1 + \Sigma_{12}\Sigma_{22}^{-}\mathbf{A}_2\mathbf{y}_2 \\ \mathbf{m}_2 + \mathbf{A}_2\mathbf{y}_2 \end{pmatrix} = \begin{pmatrix} \mathbf{m}_1 \\ \mathbf{m}_2 \end{pmatrix} + \begin{pmatrix} \mathbf{A} & \Sigma_{12}\Sigma_{22}^{-}\mathbf{A}_2 \\ \mathbf{0} & \mathbf{A}_2 \end{pmatrix} \begin{pmatrix} \mathbf{y}_1 \\ \mathbf{y}_2 \end{pmatrix}$$

$$\sim E_p\left(\begin{pmatrix} \mathbf{m}_1 \\ \mathbf{m}_2 \end{pmatrix}, \begin{pmatrix} \mathbf{A} & \Sigma_{12}\Sigma_{22}^{-}\mathbf{A}_2 \\ \mathbf{0} & \mathbf{A}_2 \end{pmatrix} \begin{pmatrix} \mathbf{A}' & \mathbf{0} \\ \mathbf{A}_2'\Sigma_{22}^{-}\Sigma_{21} & \mathbf{A}_2' \end{pmatrix}, \psi \right)$$

$$= E_p\left(\mathbf{m}, \begin{pmatrix} \mathbf{A}\mathbf{A}' + \Sigma_{12}\Sigma_{22}^{-}\mathbf{A}_2\mathbf{A}_2'\Sigma_{22}^{-}\Sigma_{21} & \Sigma_{12}\Sigma_{22}^{-}\mathbf{A}_2\mathbf{A}_2' \\ \mathbf{A}_2\mathbf{A}_2'\Sigma_{22}^{-}\Sigma_{21} & \mathbf{A}_2\mathbf{A}_2' \end{pmatrix}, \psi \right)$$

$$= E_p\left(\mathbf{m}, \begin{pmatrix} \Sigma_{11} - \Sigma_{12}\Sigma_{22}^{-}\Sigma_{21} + \Sigma_{12}\Sigma_{22}^{-}\Sigma_{22}\Sigma_{22}^{-}\Sigma_{21} & \Sigma_{12}\Sigma_{22}^{-}\Sigma_{22} \\ \Sigma_{22}\Sigma_{22}^{-}\Sigma_{21} & \mathbf{A}_2\mathbf{A}_2' \end{pmatrix}, \psi \right). \quad (2.22)$$

Now we prove that $\Sigma_{12}\Sigma_{22}^{-}\Sigma_{22} = \Sigma_{12}$. If Σ_{22} is of the form $\Sigma_{22} = \begin{pmatrix} \mathbf{L} & \mathbf{0} \\ \mathbf{0} & \mathbf{0} \end{pmatrix}$, where $\mathbf{L}$ is a nonsingular, diagonal $s \times s$ matrix, then Σ_{12} must be of the form $\Sigma_{12} = (\mathbf{K}, \mathbf{0})$ where $\mathbf{K}$ is $q \times s$. Indeed, otherwise there would be numbers i and j such that $1 \leq i \leq q$ and $q + s \leq j \leq p$ and $\sigma_{ij} \neq 0$, $\sigma_{jj} = 0$. Since $\Sigma_{22} \geq \mathbf{0}$, we must have $\begin{vmatrix} \sigma_{ii} & \sigma_{ji} \\ \sigma_{ij} & \sigma_{jj} \end{vmatrix} \geq 0$. However with $\sigma_{ij} \neq 0$, $\sigma_{jj} = 0$, we have $\begin{vmatrix} \sigma_{ii} & \sigma_{ji} \\ \sigma_{ij} & \sigma_{jj} \end{vmatrix} = -\sigma_{ij}^2 < 0$ which is a contradiction. Therefore, $\Sigma_{12} = (\mathbf{K}, \mathbf{0})$.

Let Σ_{22}^{-} be partitioned as

$$\Sigma_{22}^{-} = \begin{pmatrix} \mathbf{A} & \mathbf{B} \\ \mathbf{C} & \mathbf{D} \end{pmatrix}$$

By the definition of a generalized inverse matrix, we must have

$$\begin{pmatrix} \mathbf{L} & \mathbf{0} \\ \mathbf{0} & \mathbf{0} \end{pmatrix} \begin{pmatrix} \mathbf{A} & \mathbf{B} \\ \mathbf{C} & \mathbf{D} \end{pmatrix} \begin{pmatrix} \mathbf{L} & \mathbf{0} \\ \mathbf{0} & \mathbf{0} \end{pmatrix} = \begin{pmatrix} \mathbf{L} & \mathbf{0} \\ \mathbf{0} & \mathbf{0} \end{pmatrix},$$

which gives

$$\begin{pmatrix} \mathbf{LAL} & \mathbf{0} \\ \mathbf{0} & \mathbf{0} \end{pmatrix} = \begin{pmatrix} \mathbf{L} & \mathbf{0} \\ \mathbf{0} & \mathbf{0} \end{pmatrix},$$

So $\mathbf{LAL} = \mathbf{L}$, and since $\mathbf{L}$ is nonsingular, we get $\mathbf{A} = \mathbf{L}^{-1}$. Thus $\Sigma_{22}^{-} = \begin{pmatrix} \mathbf{L}^{-1} & \mathbf{B} \\ \mathbf{C} & \mathbf{D} \end{pmatrix}$. Then,

$$\begin{aligned}
\Sigma_{12}\Sigma_{22}^{-}\Sigma_{22} &= (\mathbf{K}, \mathbf{0}) \begin{pmatrix} \mathbf{L}^{-1} & \mathbf{B} \\ \mathbf{C} & \mathbf{D} \end{pmatrix} \begin{pmatrix} \mathbf{L} & \mathbf{0} \\ \mathbf{0} & \mathbf{0} \end{pmatrix} \\
&= (\mathbf{K}, \mathbf{0}) \begin{pmatrix} \mathbf{I}_s & \mathbf{0} \\ \mathbf{CL} & \mathbf{0} \end{pmatrix} \\
&= (\mathbf{K}, \mathbf{0}) \\
&= \Sigma_{12}.
\end{aligned}$$

If Σ_{22} is not of the form $\Sigma_{22} = \begin{pmatrix} \mathbf{L} & \mathbf{0} \\ \mathbf{0} & \mathbf{0} \end{pmatrix}$, then there exists a $\mathbf{G} \in O(p-q)$ such that $\mathbf{G}\Sigma_{22}\mathbf{G}' = \begin{pmatrix} \mathbf{L} & \mathbf{0} \\ \mathbf{0} & \mathbf{0} \end{pmatrix}$. Now, define

$$\begin{aligned}
\Sigma^* &= \begin{pmatrix} \mathbf{I}_q & \mathbf{0} \\ \mathbf{0} & \mathbf{G} \end{pmatrix} \begin{pmatrix} \Sigma_{11} & \Sigma_{12} \\ \Sigma_{21} & \Sigma_{22} \end{pmatrix} \begin{pmatrix} \mathbf{I}_q & \mathbf{0} \\ \mathbf{0} & \mathbf{G}' \end{pmatrix} \\
&= \begin{pmatrix} \Sigma_{11} & \Sigma_{12}\mathbf{G}' \\ \mathbf{G}\Sigma_{21} & \mathbf{G}\Sigma_{22}\mathbf{G}' \end{pmatrix}.
\end{aligned}$$

Then, we must have

$$\Sigma_{12}\mathbf{G}'(\mathbf{G}\Sigma_{22}\mathbf{G}')^{-}(\mathbf{G}\Sigma_{22}\mathbf{G}') = \Sigma_{12}\mathbf{G}'$$

That is, $\Sigma_{12}\mathbf{G}'\mathbf{G}\Sigma_{22}^{-}\mathbf{G}'\mathbf{G}\Sigma_{22}\mathbf{G}' = \Sigma_{12}\mathbf{G}'$, which is equivalent to $\Sigma_{12}\Sigma_{22}^{-}\Sigma_{22} = \Sigma_{12}$. Using $\Sigma_{12}\Sigma_{22}^{-}\Sigma_{22} = \Sigma_{12}$ in (2.21), we have

$$\begin{aligned}
\begin{pmatrix} \mathbf{m}_1 + \mathbf{A}\mathbf{y}_1 + \Sigma_{12}\Sigma_{22}^{-}\mathbf{A}_2\mathbf{y}_2 \\ \mathbf{m}_2 + \mathbf{A}_2\mathbf{y}_2 \end{pmatrix} &= E_p\left(\mathbf{m}, \begin{pmatrix} \Sigma_{11} - \Sigma_{12}\Sigma_{22}^{-}\Sigma_{21} + \Sigma_{12}\Sigma_{22}^{-}\Sigma_{21} & \Sigma_{12} \\ \Sigma_{21} & \Sigma_{22} \end{pmatrix}, \psi\right) \\
&= E_p\left(\mathbf{m}, \begin{pmatrix} \Sigma_{11} & \Sigma_{12} \\ \Sigma_{21} & \Sigma_{22} \end{pmatrix}, \psi\right) \\
&= E_p(\mathbf{m}, \Sigma, \psi)
\end{aligned}$$

which is the distribution of $\mathbf{x}$. ■

Next, we give the conditional distribution of the multivariate elliptical distribution in two different forms.

Theorem 2.19. *Let* $\mathbf{x} \sim E_p(\mathbf{m}, \Sigma, \psi)$ *with stochastic representation* $\mathbf{m} + r\mathbf{A}\mathbf{u}$. *Let* F *be the distribution function of* r. *Partition* $\mathbf{x}$, $\mathbf{m}$, Σ *as*

$$\mathbf{x} = \begin{pmatrix} \mathbf{x}_1 \\ \mathbf{x}_2 \end{pmatrix}, \quad \mathbf{m} = \begin{pmatrix} \mathbf{m}_1 \\ \mathbf{m}_2 \end{pmatrix}, \quad \Sigma = \begin{pmatrix} \Sigma_{11} & \Sigma_{12} \\ \Sigma_{21} & \Sigma_{22} \end{pmatrix},$$

where $\mathbf{x}_1$, $\mathbf{m}_1$ *are* q*-dimensional vectors and* Σ_{11} *is* $q \times q$, $1 \leq q < p$. *Assume* $rk(\Sigma_{22}) \geq 1$. *Then, a conditional distribution of* $\mathbf{x}_1$ *given* $\mathbf{x}_2$ *is*

$$(\mathbf{x}_1 | \mathbf{x}_2) \sim E_q(\mathbf{m}_1 + \Sigma_{12}\Sigma_{22}^{-}(\mathbf{x}_2 - \mathbf{m}_2), \Sigma_{11 \cdot 2}, \psi_{q(\mathbf{x}_2)}),$$

where

$$\Sigma_{11 \cdot 2} = \Sigma_{11} - \Sigma_{12}\Sigma_{22}^{-}\Sigma_{21}, \quad q(\mathbf{x}_2) = (\mathbf{x}_2 - \mathbf{m}_2)'\Sigma_{22}^{-}(\mathbf{x}_2 - \mathbf{m}_2),$$

and

$$\psi_{q(\mathbf{x}_2)}(u) = \int_0^\infty \Omega_q(r^2 u) dF_{q(\mathbf{x}_2)}(r), \tag{2.23}$$

where

(a)

$$F_{q(\mathbf{x}_2)}(r) = \frac{\int_{(\sqrt{q(\mathbf{x}_2)}, \sqrt{q(\mathbf{x}_2) + r^2}]} (w^2 - q(\mathbf{x}_2))^{\frac{q}{2}-1} w^{-(p-2)} dF(w)}{\int_{(\sqrt{q(\mathbf{x}_2)}, \infty)} (w^2 - q(\mathbf{x}_2))^{\frac{q}{2}-1} w^{-(p-2)} dF(w)} \tag{2.24}$$

for $r \geq 0$ *if* $q(\mathbf{x}_2) > 0$ *and* $F(\sqrt{q(\mathbf{x}_2)}) < 1$, *and*

(b)

$$F_{q(\mathbf{x}_2)}(r) = 1 \quad \textit{for } r \geq 0 \textit{ if } q(\mathbf{x}_2) = 0 \textit{ and } F(\sqrt{q(\mathbf{x}_2)}) = 1. \tag{2.25}$$

PROOF: From Lemma 2.3, we get

$$\begin{pmatrix} \mathbf{x}_1 \\ \mathbf{x}_2 \end{pmatrix} \approx \begin{pmatrix} \mathbf{m}_1 + \mathbf{A}\mathbf{y}_1 + \Sigma_{12}\Sigma_{22}^{-}\mathbf{A}_2\mathbf{y}_2 \\ \mathbf{m}_2 + \mathbf{A}_2\mathbf{y}_2 \end{pmatrix},$$

where $\mathbf{A}\mathbf{A}' = \Sigma_{11 \cdot 2}$ and $\mathbf{A}_2\mathbf{A}_2' = \Sigma_{22}$ are rank factorizations of $\Sigma_{11 \cdot 2}$, Σ_{22}, and $\begin{pmatrix} \mathbf{y}_1 \\ \mathbf{y}_2 \end{pmatrix} \sim E_p(\mathbf{0}, \mathbf{I}_p, \psi)$. Thus,

$$\begin{aligned}(\mathbf{x}_1 | \mathbf{x}_2) &\approx (\mathbf{m}_1 + \mathbf{A}\mathbf{y}_1 + \Sigma_{12}\Sigma_{22}^{-}\mathbf{A}_2\mathbf{y}_2 | \mathbf{m}_2 + \mathbf{A}_2\mathbf{y}_2 = \mathbf{x}_2) \\ &= \mathbf{m}_1 + \Sigma_{12}\Sigma_{22}^{-}(\mathbf{A}_2\mathbf{y}_2 | \mathbf{m}_2 + \mathbf{A}_2\mathbf{y}_2 = \mathbf{x}_2) + \mathbf{A}(\mathbf{y}_1 | \mathbf{m}_2 + \mathbf{A}_2\mathbf{y}_2 = \mathbf{x}_2)\end{aligned}$$

$$= \mathbf{m}_1 + \Sigma_{12}\Sigma_{22}^{-}(\mathbf{A}_2\mathbf{y}_2|\mathbf{A}_2\mathbf{y}_2 = \mathbf{x}_2 - \mathbf{m}_2) + \mathbf{A}(\mathbf{y}_1|\mathbf{A}_2\mathbf{y}_2 = \mathbf{x}_2 - \mathbf{m}_2)$$
$$= \mathbf{m}_1 + \Sigma_{12}\Sigma_{22}^{-}(\mathbf{x}_2 - \mathbf{m}_2) + \mathbf{A}(\mathbf{y}_1|\mathbf{A}_2^{-}\mathbf{A}_2\mathbf{y}_2 = \mathbf{A}_2^{-}(\mathbf{x}_2 - \mathbf{m}_2)). \quad (2.26)$$

Now $\mathbf{A}_2^{-}\mathbf{A}_2 = \mathbf{I}_{rk(\Sigma_{22})}$, and hence we get

$$(\mathbf{y}_1|\mathbf{A}_2^{-}\mathbf{A}_2\mathbf{y}_2 = \mathbf{A}_2^{-}(\mathbf{x}_2 - \mathbf{m}_2)) = (\mathbf{y}_1|\mathbf{y}_2 = \mathbf{A}_2^{-}(\mathbf{x}_2 - \mathbf{m}_2)). \quad (2.27)$$

From Theorem 2.18, we get

$$(\mathbf{y}_1|\mathbf{y}_2 = \mathbf{A}_2^{-}(\mathbf{x}_2 - \mathbf{m}_2)) \sim E_q(\mathbf{0}, \mathbf{I}_q, \psi_{q(\mathbf{x}_2)}),$$

where

$$\begin{aligned} q(\mathbf{x}_2) &= (\mathbf{A}_2^{-}(\mathbf{x}_2 - \mathbf{m}_2))'\mathbf{A}_2^{-}(\mathbf{x}_2 - \mathbf{m}_2) \\ &= (\mathbf{x}_2 - \mathbf{m}_2)'\mathbf{A}_2^{-\prime}\mathbf{A}_2^{-}(\mathbf{x}_2 - \mathbf{m}_2) \\ &= (\mathbf{x}_2 - \mathbf{m}_2)'\Sigma_{22}^{-}(\mathbf{x}_2 - \mathbf{m}_2) \end{aligned}$$

and $\psi_{q(\mathbf{x}_2)}$ is defined by (2.23)–(2.25). Thus,

$$\begin{aligned} \mathbf{A}(\mathbf{y}_1|\mathbf{y}_2 = \mathbf{A}_2^{-}(\mathbf{x}_2 - \mathbf{m}_2)) &\sim E_q(\mathbf{0}, \mathbf{A}\mathbf{A}', \psi_{q(\mathbf{x}_2)}) \\ &= E_q(\mathbf{0}, \Sigma_{11\cdot 2}, \psi_{q(\mathbf{x}_2)}). \end{aligned} \quad (2.28)$$

Finally, from (2.26), (2.27), and (2.28) we get

$$(\mathbf{x}_1|\mathbf{x}_2) \sim E_q(\mathbf{m}_1 + \Sigma_{12}\Sigma_{22}^{-}(\mathbf{x}_2 - \mathbf{m}_2), \Sigma_{11\cdot 2}, \psi_{q(\mathbf{x}_2)}). \quad \blacksquare$$

Another version of the conditional distribution is given in the following theorem.

Theorem 2.20. *Let* $\mathbf{x} \sim E_p(\mathbf{m}, \Sigma, \psi)$ *with stochastic representation* $\mathbf{m} + r\mathbf{A}\mathbf{u}$. *Let* F *be the distribution function of* r. *Partition* $\mathbf{x}$, $\mathbf{m}$, Σ *as*

$$\mathbf{x} = \begin{pmatrix} \mathbf{x}_1 \\ \mathbf{x}_2 \end{pmatrix}, \quad \mathbf{m} = \begin{pmatrix} \mathbf{m}_1 \\ \mathbf{m}_2 \end{pmatrix}, \quad \Sigma = \begin{pmatrix} \Sigma_{11} & \Sigma_{12} \\ \Sigma_{21} & \Sigma_{22} \end{pmatrix},$$

where $\mathbf{x}_1$, $\mathbf{m}_1$ *are* q*-dimensional vectors and* Σ_{11} *is* $q \times q$, $1 \le q < p$. *Assume* $rk(\Sigma_{22}) \ge 1$.

Let S *denote the subspace of* $\mathbb{R}^{p-q}$ *defined by the columns of* Σ_{22}*; that is,* $\mathbf{y} \in S$, *if there exists* $\mathbf{a} \in \mathbb{R}^{p-q}$ *such that* $\mathbf{y} = \Sigma_{22}\mathbf{a}$.

Then, a conditional distribution of $\mathbf{x}_1$ *given* $\mathbf{x}_2$ *is*

(a) $(\mathbf{x}_1|\mathbf{x}_2) \sim E_q(\mathbf{m}_1 + \Sigma_{12}\Sigma_{22}^{-}(\mathbf{x}_2 - \mathbf{m}_2), \Sigma_{11\cdot 2}, \psi_{q(\mathbf{x}_2)})$ *for* $\mathbf{x}_2 \in \mathbf{m}_2 + S$, *where* $q(\mathbf{x}_2) = (\mathbf{x}_2 - \mathbf{m}_2)'\Sigma_{22}^{-}(\mathbf{x}_2 - \mathbf{m}_2)$ *and* $\psi_{q(\mathbf{x}_2)}$ *is defined by (2.23)–(2.25).*
(b) $(\mathbf{x}_1|\mathbf{x}_2) = \mathbf{m}_1$ *for* $\mathbf{x}_2 \notin \mathbf{m}_2 + S$.

PROOF: It suffices to prove that $P(\mathbf{x}_2 \notin \mathbf{m}_2 + S) = 0$ since $(\mathbf{x}_1|\mathbf{x}_2)$ can be arbitrarily defined for $\mathbf{x}_2 \in B$ where B is of measure zero. However, $P(\mathbf{x}_2 \notin \mathbf{m}_2 + S) = 0$ is equivalent to $P(\mathbf{x}_2 \in \mathbf{m}_2 + S) = 1$; that is, $P(\mathbf{x}_2 - \mathbf{m}_2 \in S) = 1$. Now, $\mathbf{x}_2 \sim E_{p-q}(\mathbf{m}_2, \Sigma_{22}, \psi)$ and so $\mathbf{x}_2 - \mathbf{m}_2 \sim E_{p-q}(\mathbf{0}, \Sigma_{22}, \psi)$.

Let $k = rk(\Sigma_{22})$. Let $\mathbf{G} \in O(p-q)$ such that $\mathbf{G}\Sigma_{22}\mathbf{G}' = \begin{pmatrix} \mathbf{L} & \mathbf{0} \\ \mathbf{0} & \mathbf{0} \end{pmatrix}$, where $\mathbf{L}$ is a diagonal and nonsingular $k \times k$ matrix and define $\mathbf{y} = \mathbf{G}(\mathbf{x}_2 - \mathbf{m}_2)$. Then,

$$\mathbf{y} \sim E_{p-q}\left(\mathbf{0}, \begin{pmatrix} \mathbf{L} & \mathbf{0} \\ \mathbf{0} & \mathbf{0} \end{pmatrix}, \psi\right) . \tag{2.29}$$

Partition $\mathbf{y}$ as $\mathbf{y} = \begin{pmatrix} \mathbf{y}_1 \\ \mathbf{y}_2 \end{pmatrix}$, where $\mathbf{y}_1$ is $k \times 1$. We have

$$\begin{aligned}
P(\mathbf{x}_2 - \mathbf{m}_2 \in S) &= P(\mathbf{x}_2 - \mathbf{m}_2 = \Sigma_{22}\mathbf{a} \quad \text{with} \quad \mathbf{a} \in \mathbb{R}^{p-q}) \\
&= P(\mathbf{G}(\mathbf{x}_2 - \mathbf{m}_2) = \mathbf{G}\Sigma_{22}\mathbf{G}'\mathbf{G}\mathbf{a} \quad \text{with} \quad \mathbf{a} \in \mathbb{R}^{p-q}) \\
&= P\left(\mathbf{y} = \begin{pmatrix} \mathbf{L} & \mathbf{0} \\ \mathbf{0} & \mathbf{0} \end{pmatrix}\mathbf{b} \quad \text{with} \quad \mathbf{b} \in \mathbb{R}^{p-q}\right) \\
&= P\left(\mathbf{y} = \begin{pmatrix} \mathbf{L} & \mathbf{0} \\ \mathbf{0} & \mathbf{0} \end{pmatrix}\begin{pmatrix} \mathbf{b}_1 \\ \mathbf{b}_2 \end{pmatrix} \quad \text{with} \quad \mathbf{b}_1 \in \mathbb{R}^k, \mathbf{b}_2 \in \mathbb{R}^{p-q-k}\right) \\
&= P\left(\mathbf{y} = \begin{pmatrix} \mathbf{L}\mathbf{b}_1 \\ \mathbf{0} \end{pmatrix} \quad \text{with} \quad \mathbf{b}_1 \in \mathbb{R}^k\right) \\
&= P\left(\begin{pmatrix} \mathbf{y}_1 \\ \mathbf{y}_2 \end{pmatrix} = \begin{pmatrix} \mathbf{c} \\ \mathbf{0} \end{pmatrix} \quad \text{with} \quad \mathbf{c} \in \mathbb{R}^k\right) \\
&= P(\mathbf{y}_2 = \mathbf{0}) .
\end{aligned}$$

Now, it follows from (2.29) that $\mathbf{y}_2 \sim E_{p-q-k}(\mathbf{0}, \mathbf{0}, \psi)$ and so $P(\mathbf{y}_2 = \mathbf{0}) = 1$. Therefore, $P(\mathbf{x}_2 - \mathbf{m}_2 \in S) = 1$. ■

Now we can derive the conditional distribution for m.e.c. distributions.

Theorem 2.21. *Let* $\mathbf{X} \sim E_{p,n}(\mathbf{M}, \Sigma \otimes \Phi, \psi)$ *with stochastic representation* $\mathbf{M} + r\mathbf{AUB}'$. *Let F be the distribution function of r. Partition* $\mathbf{X}$, $\mathbf{M}$, Σ *as*

$$\mathbf{X} = \begin{pmatrix} \mathbf{X}_1 \\ \mathbf{X}_2 \end{pmatrix}, \quad \mathbf{M} = \begin{pmatrix} \mathbf{M}_1 \\ \mathbf{M}_2 \end{pmatrix}, \quad \text{and} \quad \Sigma = \begin{pmatrix} \Sigma_{11} & \Sigma_{12} \\ \Sigma_{21} & \Sigma_{22} \end{pmatrix},$$

where $\mathbf{X}_1$ *is* $q \times n$, $\mathbf{M}_1$ *is* $q \times n$, *and* Σ_{11} *is* $q \times q$, $1 \le q < p$. *Assume* $rk(\Sigma_{22}) \ge 1$.

Let S denote the subspace of $\mathbb{R}^{(p-q)n}$ *defined by the columns of* $\boldsymbol{\Sigma}_{22} \otimes \boldsymbol{\Phi}$; *that is,* $\mathbf{y} \in S$, *if there exists* $\mathbf{b} \in \mathbb{R}^{(p-q)n}$ *such that* $\mathbf{y} = (\boldsymbol{\Sigma}_{22} \otimes \boldsymbol{\Phi})\mathbf{b}$. *Then, a conditional distribution of* $\mathbf{X}_1$ *given* $\mathbf{X}_2$ *is*

1. $(\mathbf{X}_1|\mathbf{X}_2) \sim E_{q,n}(\mathbf{M}_1 + \Sigma_{12}\Sigma_{22}^{-}(\mathbf{X}_2 - \mathbf{M}_2), \Sigma_{11\cdot 2} \otimes \boldsymbol{\Phi}, \psi_{q(\mathbf{X}_2)})$ *for* $vec(\mathbf{X}_2') \in vec(\mathbf{M}_2') + S$, *where* $q(\mathbf{X}_2) = tr((\mathbf{X}_2 - \mathbf{M}_2)'\Sigma_{22}^{-}(\mathbf{X}_2 - \mathbf{M}_2)\Phi^{-})$,

$$\psi_{q(\mathbf{X}_2)}(u) = \int_0^{\infty} \Omega_{qn}(r^2 u) dF_{q(\mathbf{X}_2)}(r), \tag{2.30}$$

where

(a)

$$F_{a^2}(r) = \frac{\int_{(a,\sqrt{a^2+r^2}]} (w^2 - a^2)^{\frac{qn}{2}-1} w^{-(pn-2)} dF(w)}{\int_{(a,\infty)} (w^2 - a^2)^{\frac{qn}{2}-1} w^{-(pn-2)} dF(w)} \tag{2.31}$$

for $r \geq 0$ *if* $a > 0$ *and* $F(a) < 1$, *and*

(b)

$$F_{a^2}(r) = 1 \quad \text{for } r \geq 0 \text{ if } a = 0 \text{ and } F(a) = 1. \tag{2.32}$$

2. $(\mathbf{X}_1|\mathbf{X}_2) = \mathbf{M}_1$ *for* $vec(\mathbf{X}_2') \notin vec(\mathbf{M}_2') + S$.

PROOF: Define $\mathbf{x} = vec(\mathbf{X}')$, $\mathbf{x}_1 = vec(\mathbf{X}_1')$, $\mathbf{x}_2 = vec(\mathbf{X}_2')$, $\mathbf{m} = vec(\mathbf{M}')$, $\mathbf{m}_1 = vec(\mathbf{M}_1')$, and $\mathbf{m}_2 = vec(\mathbf{M}_2')$. Then $\mathbf{x} = \begin{pmatrix} \mathbf{x}_1 \\ \mathbf{x}_2 \end{pmatrix}$ and $\mathbf{x} \sim E_{pn}(\mathbf{m}, \Sigma \otimes \Phi, \psi)$. Now apply Theorem 2.20.

(1) If $\mathbf{x}_2 \in vec(\mathbf{M}_2') + S$, we have

$$(\mathbf{x}_1|\mathbf{x}_2) \sim E_{qn}\Big(\mathbf{m}_1 + (\Sigma_{12} \otimes \Phi)(\Sigma_{22} \otimes \Phi)^{-}(\mathbf{x}_2 - \mathbf{m}_2),$$
$$(\Sigma_{11} \otimes \Phi) - (\Sigma_{12} \otimes \Phi)(\Sigma_{22} \otimes \Phi)^{-}(\Sigma_{21} \otimes \Phi), \psi_{q(\mathbf{x}_2)}\Big) \tag{2.33}$$

where

$$\begin{aligned} q(\mathbf{x}_2) &= (\mathbf{x}_2 - \mathbf{m}_2)'(\Sigma_{22} \otimes \Phi)^{-}(\mathbf{x}_2 - \mathbf{m}_2) \\ &= (vec((\mathbf{X}_2 - \mathbf{M}_2)'))'(\Sigma_{22}^{-} \otimes \Phi^{-})(vec(\mathbf{X}_2 - \mathbf{M}_2)') \\ &= tr((\mathbf{X}_2 - \mathbf{M}_2)'\Sigma_{22}^{-}(\mathbf{X}_2 - \mathbf{M}_2)\Phi^{-}). \end{aligned}$$

From (2.23) and (2.24) we get (2.30) and (2.31). Since $\mathbf{x}_2 \in \mathbf{m}_2 + S$, there exists $\mathbf{b} \in I\!R^{(p-q)n}$ such that $\mathbf{x}_2 - \mathbf{m}_2 = (\Sigma_{22} \otimes \boldsymbol{\Phi})\mathbf{b}$. Then, we have

$$\begin{aligned}(\Sigma_{12} \otimes \boldsymbol{\Phi})(\Sigma_{22} \otimes \boldsymbol{\Phi})^{-}(\mathbf{x}_2 - \mathbf{m}_2) &= (\Sigma_{12} \otimes \boldsymbol{\Phi})(\Sigma_{22} \otimes \boldsymbol{\Phi})^{-}(\Sigma_{22} \otimes \boldsymbol{\Phi})\mathbf{b} \\ &= (\Sigma_{12}\Sigma_{22}^{-}\Sigma_{22} \otimes \boldsymbol{\Phi}\boldsymbol{\Phi}^{-}\boldsymbol{\Phi})\mathbf{b} \\ &= (\Sigma_{12}\Sigma_{22}^{-}\Sigma_{22} \otimes \boldsymbol{\Phi})\mathbf{b} \\ &= (\Sigma_{12}\Sigma_{22}^{-} \otimes \mathbf{I}_n)(\Sigma_{22} \otimes \boldsymbol{\Phi})\mathbf{b} \\ &= (\Sigma_{12}\Sigma_{22}^{-} \otimes \mathbf{I}_n)(\mathbf{x}_2 - \mathbf{m}_2)\,.\end{aligned}$$

We also have

$$\begin{aligned}&(\Sigma_{11} \otimes \boldsymbol{\Phi}) - (\Sigma_{12} \otimes \boldsymbol{\Phi})(\Sigma_{22} \otimes \boldsymbol{\Phi})^{-}(\Sigma_{21} \otimes \boldsymbol{\Phi}) \\ &= (\Sigma_{11} \otimes \boldsymbol{\Phi}) - (\Sigma_{12}\Sigma_{22}^{-}\Sigma_{21}) \otimes \boldsymbol{\Phi} \\ &= (\Sigma_{11} - \Sigma_{12}\Sigma_{22}^{-}\Sigma_{21}) \otimes \boldsymbol{\Phi}\,.\end{aligned}$$

Therefore, (2.33) can be written as

$$(\mathbf{x}_1|\mathbf{x}_2) \sim E_{qn}(\mathbf{m}_1 + (\Sigma_{12}\Sigma_{22}^{-} \otimes \mathbf{I}_n)(\mathbf{x}_2 - \mathbf{m}_2), \Sigma_{11\cdot 2} \otimes \boldsymbol{\Phi}, \psi_{q(\mathbf{x}_2)}).$$

Hence,

$$(\mathbf{X}_1|\mathbf{X}_2) \sim E_{q,n}(\mathbf{M}_1 + \Sigma_{12}\Sigma_{22}^{-}(\mathbf{X}_2 - \mathbf{M}_2), \Sigma_{11\cdot 2} \otimes \boldsymbol{\Phi}, \psi_{q(\mathbf{X}_2)}).$$

(2) If $\mathbf{x}_2 \notin vec(\mathbf{M}_2') + S$, we get $(\mathbf{x}_1|\mathbf{x}_2) = \mathbf{m}_1$, so $(\mathbf{X}_1|\mathbf{X}_2) = \mathbf{M}_1$. ∎

Corollary 2.5. *With the notation of Theorem 2.21, we have*

$$1 - F(w) = K_{a^2} \int_{(\sqrt{w^2-a^2},\infty)} (r^2 + a^2)^{\frac{pn}{2}-1} r^{-(qn-2)} dF_{a^2}(r)\,, w \geq a\,,$$

where $K_{a^2} = \int_{(a,\infty)} (w^2 - a^2)^{\frac{qn}{2}-1} w^{-(pn-2)} dF(w)$.

PROOF: From (2.31) we get,

$$dF_{a^2}(r) = \frac{1}{K_{a^2}} (r^2)^{\frac{qn}{2}-1} (r^2 + a^2)^{-\frac{pn-2}{2}} \frac{dw}{dr} dF(w)\,,$$

where $r^2 + a^2 = w^2$. Hence,

$$dF(w) = K_{a^2} r^{-(qn-2)} (r^2 + a^2)^{\frac{pn}{2}-1} \frac{dr}{dw} dF_{a^2}(r)\,,$$

where $a < w \le \sqrt{a^2+r^2}$. Therefore,

$$1-F(w) = K_{a^2} \int_{(\sqrt{w^2-a^2},\infty)} (r^2+a^2)^{\frac{pn}{2}-1} r^{-(qn-2)} dF_{a^2}(r), w \ge a. \qquad \blacksquare$$

Theorem 2.22. *Let* $\mathbf{X} \sim E_{p,n}(\mathbf{M}, \Sigma \otimes \boldsymbol{\Phi}, \psi)$ *with stochastic representation* $\mathbf{M} + r\mathbf{AUB}'$. *Let F be the distribution function of r and* $\mathbf{X}$, $\mathbf{M}$, Σ *be partitioned as in Theorem 2.21.*

(1) If $vec(\mathbf{X}_2') \in vec(\mathbf{M}_2') + S$, *where S is defined in Theorem 2.21, and*

(a) If $\mathbf{X}$ *has finite first moment, then*

$$E(\mathbf{X}_1|\mathbf{X}_2) = \mathbf{M}_1 + \Sigma_{12}\Sigma_{22}^{-}(\mathbf{X}_2 - \mathbf{M}_2).$$

(b) If $\mathbf{X}$ *has finite second moment, then*

$$Cov(\mathbf{X}_1|\mathbf{X}_2) = c_1 \Sigma_{11\cdot 2} \otimes \boldsymbol{\Phi},$$

where $c_1 = -2\psi'_{q(\mathbf{X}_2)}(0)$, *and* $\psi_{q(\mathbf{X}_2)}$ *is defined by (2.30), (2.31) and (2.32).*

(2) If $vec(\mathbf{X}_2') \notin vec(\mathbf{M}_2') + S$, *then* $E(\mathbf{X}_1|\mathbf{X}_2) = \mathbf{M}_1$, $Cov(\mathbf{X}_1|\mathbf{X}_2) = \mathbf{0}$.

PROOF: It follows from Theorems 2.11 and 2.21. ■

The next theorem shows that if the distribution of $\mathbf{X}$ is absolutely continuous, then the constant c_1 in Theorem 2.22 can be obtained in a simple way. This was shown by Chu (1973), but his proof applies only to a subclass of absolutely continuous distributions. The following proof, however, works for all absolutely continuous distributions.

Theorem 2.23. *Let* $\mathbf{X} \sim E_{p,n}(\mathbf{M}, \Sigma \otimes \boldsymbol{\Phi}, \psi)$ *and* $\mathbf{X}$, $\mathbf{M}$, Σ *be partitioned as in Theorem 2.21. Assume the distribution of* $\mathbf{X}$ *is absolutely continuous and it has finite second moment.*

Let

$$f_2(\mathbf{X}_2) = \frac{1}{|\Sigma_{22}|^{\frac{n}{2}} |\boldsymbol{\Phi}|^{\frac{p-q}{2}}} h_2\left(tr\left((\mathbf{X}_2-\mathbf{M}_2)'\Sigma_{22}^{-1}(\mathbf{X}_2-\mathbf{M}_2)\boldsymbol{\Phi}^{-1}\right)\right)$$

be the p.d.f. of the submatrix $\mathbf{X}_2$. *Then,*

$$Cov(\mathbf{X}_1|\mathbf{X}_2) = \frac{\int_r^\infty h_2(z)dz}{2h_2(r)} \Sigma_{11\cdot 2} \otimes \boldsymbol{\Phi},$$

where $r = tr\left((\mathbf{X}_2-\mathbf{M}_2)'\Sigma_{22}^{-1}(\mathbf{X}_2-\mathbf{M}_2)\boldsymbol{\Phi}^{-1}\right)$.

PROOF: Step 1. First we prove the theorem for the case $n = 1$, $\mathbf{m} = \mathbf{0}$. From Theorems 2.21 and 2.22, we conclude that $Cov(\mathbf{x}_1|\mathbf{x}_2) = c_1\Sigma_{11\cdot 2}$, where c_1 is determined by p, q, and $\mathbf{x}_2'\Sigma_{22}^{-1}\mathbf{x}_2$. Hence, c_1 does not depend on Σ_{11} and Σ_{12}. Thus without loss of generality, we can assume that $\Sigma_{11} = \mathbf{I}_q$ and $\Sigma_{12} = \mathbf{0}$. Then $\Sigma_{11\cdot 2} = \mathbf{I}_q$. Let $\mathbf{x}_1 = (x_1, x_2 \ldots x_q)'$, and $f_1(x_1, \mathbf{x}_2) = |\Sigma_{22}|^{-\frac{1}{2}} h_1(x_1^2 + \mathbf{x}_2'\Sigma_{22}^{-1}\mathbf{x}_2)$ be the joint p.d.f. of x_1 and $\mathbf{x}_2$. Then,

$$\begin{aligned} c_1 = Var(x_1|\mathbf{x}_2) &= \frac{\int_{-\infty}^{\infty} x_1^2 f_1(x_1, \mathbf{x}_2) dx_1}{f_2(\mathbf{x}_2)} \\ &= \frac{\int_{-\infty}^{\infty} x_1^2 h_1(x_1^2 + \mathbf{x}_2'\Sigma_{22}^{-1}\mathbf{x}_2) dx_1}{h_2(\mathbf{x}_2'\Sigma_{22}^{-1}\mathbf{x}_2)} \\ &= 2\frac{\int_0^{\infty} x_1^2 h_1(x_1^2 + \mathbf{x}_2'\Sigma_{22}^{-1}\mathbf{x}_2) dx_1}{h_2(\mathbf{x}_2'\Sigma_{22}^{-1}\mathbf{x}_2)}. \end{aligned} \tag{2.34}$$

Now, $f_2(\mathbf{x}_2) = \int_{-\infty}^{\infty} f_1(x_1, \mathbf{x}_2) dx_1$, hence

$$\begin{aligned} h_2(\mathbf{x}_2'\Sigma_{22}^{-1}\mathbf{x}_2) &= \int_{-\infty}^{\infty} h_1(x_1^2 + \mathbf{x}_2'\Sigma_{22}^{-1}\mathbf{x}_2) dx_1 \\ &= 2\int_0^{\infty} h_1(x_1^2 + \mathbf{x}_2'\Sigma_{22}^{-1}\mathbf{x}_2) dx_1 . \end{aligned}$$

So, $h_2(z) = 2\int_0^{\infty} h_1(x_1^2 + z) dx_1$, for $z \geq 0$. Hence, for $u \geq 0$, we get

$$\begin{aligned} \int_u^{\infty} h_2(z) dz &= 2\int_u^{\infty}\int_0^{\infty} h_1(x_1^2 + z) dx_1 dz \\ &= 2\int_0^{\infty}\int_0^{\infty} \chi(z \geq u) h_1(x_1^2 + z) dx_1 dz . \end{aligned}$$

Let $w = \sqrt{x_1^2 + z - u}$. Then, $w^2 = x_1^2 + z - u$ and so $J(z \to w) = 2w$. Hence,

$$\begin{aligned} \int_u^{\infty} h_2(z) dz &= 2\int_0^{\infty}\int_0^{\infty} \chi(w^2 - x_1^2 + u \geq u) h_1(w^2 + u) 2w dx_1 dw \\ &= 4\int_0^{\infty}\int_0^{\infty} \chi(w^2 \geq x_1^2) w h_1(w^2 + u) dx_1 dw \\ &= 4\int_0^{\infty}\int_0^{w} w h_1(w^2 + u) dx_1 dw \\ &= 4\int_0^{\infty} \left(w h_1(w^2 + u) \int_0^{w} dx_1 \right) dw \\ &= 4\int_0^{\infty} w h_1(w^2 + u) w dw \\ &= 4\int_0^{\infty} w^2 h_1(w^2 + u) dw . \end{aligned} \tag{2.35}$$

Now from (2.34) and (2.35), we get

$$c_1 = \frac{\int_u^\infty h_2(z)dz}{2h_2(u)}, \quad \text{where} \quad u = \mathbf{x}_2'\Sigma_{22}^{-1}\mathbf{x}_2 .$$

Step 2. Next, let $n = 1$, $\mathbf{m} \neq \mathbf{0}$, and $\mathbf{y} = \mathbf{x} - \mathbf{m}$. Then,

$$Cov(\mathbf{y}_1|\mathbf{y}_2) = \frac{\int_u^\infty h_2(z)dz}{2h_2(u)}\Sigma_{11\cdot 2},$$

where $u = \mathbf{y}_2'\Sigma_{22}^{-1}\mathbf{y}_2$. Therefore,

$$\begin{aligned} Cov(\mathbf{x}_1|\mathbf{x}_2) &= Cov(\mathbf{y}_1 + \mathbf{m}_1|\mathbf{y}_2 = \mathbf{x}_2 - \mathbf{m}_2) \\ &= \frac{\int_u^\infty h_2(z)dz}{2h_2(u)}\Sigma_{11\cdot 2}, \end{aligned}$$

where $u = (\mathbf{x}_2 - \mathbf{m}_2)'\Sigma_{22}^{-1}(\mathbf{x}_2 - \mathbf{m}_2)$.

Step 3. Finally, let $\mathbf{X} \sim E_{p,n}(\mathbf{M}, \Sigma \otimes \Phi, \psi)$. Define $\mathbf{x} = vec(\mathbf{X}')$. Now, for $\mathbf{x}$ we can use Step 2. Therefore,

$$Cov(\mathbf{X}_1|\mathbf{X}_2) = \frac{\int_r^\infty h_2(z)dz}{2h_2(r)}\Sigma_{11\cdot 2} \otimes \Phi,$$

where $r = tr\left((\mathbf{X}_2 - \mathbf{M}_2)'\Sigma_{22}^{-1}(\mathbf{X}_2 - \mathbf{M}_2)\Phi^{-1}\right)$. ■

2.7 Examples

In this section we give some examples of the elliptically contoured distributions. We also give a method to generate elliptically contoured distributions.

2.7.1 One-Dimensional Case

Let $p = n = 1$. Then, the class $E_1(m, \sigma, \psi)$, coincides with the class of one-dimensional distributions which are symmetric about a point. More precisely, $x \sim E_1(m, \sigma, \psi)$ if and only if $P(x \leq r) = P(x \geq m - r)$ for every $r \in \mathbb{R}$. Some examples are: uniform, normal, Cauchy, double exponential, Student's t-distribution, and the distribution with the p.d.f.

$$f(x) = \frac{\sqrt{2}}{\pi\sigma\left(1+\left(\frac{x}{\sigma}\right)^4\right)}, \qquad \sigma > 0.$$

2.7.2 *Vector Variate Case*

The definitions and results here are taken from Fang, Kotz and Ng (1990). Let $p > 1$ and $n = 1$.

2.7.2.1 Multivariate Uniform Distribution

The p-dimensional random vector $\mathbf{u}$ is said to have a multivariate uniform distribution if it is uniformly distributed on the unit sphere in $I\!R^p$.

Theorem 2.24. *Let $x = (x_1, x_2, \ldots, x_p)'$ have a p-variate uniform distribution. Then the p.d.f. of $(x_1, x_2, \ldots, x_{p-1})$ is*

$$\frac{\Gamma\left(\frac{p}{2}\right)}{\pi^{\frac{p}{2}}}\left(1-\sum_{i=1}^{p-1} x_i^2\right)^{-\frac{1}{2}}, \qquad \sum_{i=1}^{p-1} x_i^2 < 1.$$

PROOF: It follows from Theorem 2.17 that $\left(x_1^2, x_2^2, \ldots, x_{p-1}^2\right) \sim D\left(\frac{1}{2}, \ldots, \frac{1}{2}; \frac{1}{2}\right)$. Hence, the p.d.f. of $\left(x_1^2, x_2^2, \ldots, x_{p-1}^2\right)$ is

$$\frac{\Gamma\left(\frac{p}{2}\right)}{\left(\Gamma\left(\frac{1}{2}\right)\right)^p} \prod_{i=1}^{p-1} (x_i^2)^{-\frac{1}{2}} \left(1-\sum_{i=1}^{p-1} x_i^2\right)^{-\frac{1}{2}}$$

Since the Jacobian of the transformation $(x_1, x_2, \ldots, x_{p-1}) \to (|x_1|, |x_2|, \ldots, |x_{p-1}|)$ is $2^{p-1}\prod_{i=1}^{p-1} |x_i|$, the p.d.f. of $(|x_1|, |x_2|, \ldots, |x_{p-1}|)$ is

$$\frac{2^{p-1}\Gamma\left(\frac{p}{2}\right)}{\pi^{\frac{p}{2}}}\left(1-\sum_{i=1}^{p-1} x_i^2\right)^{-\frac{1}{2}}$$

Now because of the spherical symmetry of $\mathbf{x}$, the p.d.f. of $(x_1, x_2, \ldots, x_{p-1})$ is the p.d.f. of $(|x_1|, |x_2|, \ldots, |x_{p-1}|)$ divided by 2^{p-1}; that is

$$\frac{\Gamma\left(\frac{p}{2}\right)}{\pi^{\frac{p}{2}}}\left(1-\sum_{i=1}^{p-1} x_i^2\right)^{-\frac{1}{2}}, \qquad \sum_{i=1}^{p-1} x_i^2 < 1.$$

■

2.7.2.2 Symmetric Kotz Type Distribution

The p-dimensional random vector $\mathbf{x}$ is said to have a symmetric Kotz type distribution with parameters q, r, $s \in I\!R$, μ: p-dimensional vector, Σ: $p \times p$ matrix, $r > 0$, $s > 0$, $2q + p > 2$, and $\Sigma > \mathbf{0}$ if its p.d.f. is

$$f(\mathbf{x}) = \frac{sr^{\frac{2q+p-2}{2s}}\Gamma\left(\frac{p}{2}\right)}{\pi^{\frac{p}{2}}\Gamma\left(\frac{2q+p-2}{2s}\right)|\Sigma|^{\frac{1}{2}}}[(\mathbf{x}-\mu)'\Sigma^{-1}(\mathbf{x}-\mu)]^{q-1}\exp\{-r[(\mathbf{x}-\mu)'\Sigma^{-1}(\mathbf{x}-\mu)]^{s}\}.$$

As a special case, take $q = s = 1$ and $r = \frac{1}{2}$. Then, we get the multivariate normal distribution with p.d.f.

$$f(\mathbf{x}) = \frac{1}{(2\pi)^{\frac{p}{2}}|\Sigma|^{\frac{1}{2}}}\exp\left\{-\frac{(\mathbf{x}-\mu)'\Sigma^{-1}(\mathbf{x}-\mu)}{2}\right\},$$

and its characteristic function is

$$\phi_{\mathbf{x}}(\mathbf{t}) = \exp(i\mathbf{t}'\mu)\exp\left(-\frac{1}{2}\mathbf{t}'\Sigma\mathbf{t}\right). \tag{2.36}$$

The multivariate normal distribution is denoted by $N_p(\mu,\Sigma)$.

Remark 2.6. The distribution, $N_p(\mu,\Sigma)$, can be defined by its characteristic function (2.36). Then, Σ does not have to be positive definite; it suffices to assume that $\Sigma \geq \mathbf{0}$.

2.7.2.3 Symmetric Multivariate Pearson Type II Distribution

The p-dimensional random vector $\mathbf{x}$ is said to have a symmetric multivariate Pearson type II distribution with parameters $q \in I\!R$, μ: p-dimensional vector, Σ: $p \times p$ matrix with $q > -1$, and $\Sigma > \mathbf{0}$ if its p.d.f. is

$$f(\mathbf{x}) = \frac{\Gamma\left(\frac{p}{2}+q+1\right)}{\pi^{\frac{p}{2}}\Gamma(q+1)|\Sigma|^{\frac{1}{2}}}(1-(\mathbf{x}-\mu)'\Sigma^{-1}(\mathbf{x}-\mu))^{q},$$

where $(\mathbf{x}-\mu)'\Sigma^{-1}(\mathbf{x}-\mu) \leq 1$.

2.7.2.4 Symmetric Multivariate Pearson Type VII Distribution

The p-dimensional random vector $\mathbf{x}$ is said to have a symmetric multivariate Pearson type VII distribution with parameters q, $r \in I\!R$, μ: p-dimensional vector, Σ: $p \times p$ matrix with $r > 0$, $q > \frac{p}{2}$, and $\Sigma > \mathbf{0}$ if its p.d.f. is

$$f(\mathbf{x}) = \frac{\Gamma(q)}{(\pi r)^{\frac{p}{2}} \Gamma\left(q - \frac{p}{2}\right) |\Sigma|^{\frac{1}{2}}} \left(1 + \frac{(\mathbf{x}-\mu)'\Sigma^{-1}(\mathbf{x}-\mu)}{r}\right)^{-q}. \tag{2.37}$$

As a special case, when $q = \frac{p+r}{2}$, $\mathbf{x}$ is said to have a multivariate t-distribution with r degrees of freedom and it is denoted by $Mt_p(r, \mu, \Sigma)$.

Theorem 2.25. *The class of symmetric multivariate Pearson type VII distributions coincides with the class of multivariate t-distributions.*

PROOF: Clearly, the multivariate t-distribution is Pearson type VII distribution. We only have to prove that all Pearson type VII distributions are multivariate t-distributions.

Assume $\mathbf{x}$ has p.d.f. (2.37). Define

$$r_0 = 2\left(q - \frac{p}{2}\right) \quad \text{and} \quad \Sigma_0 = \Sigma \frac{r}{r_0}.$$

Then, $Mt_p(r_0, \mu, \Sigma_0)$ is the same distribution as the one with p.d.f. (2.37). ■

The special case of multivariate t-distribution when $r = 1$; that is, $Mt_p(1, \mu, \Sigma)$ is called multivariate Cauchy distribution.

2.7.2.5 Symmetric Multivariate Bessel Distribution

The p-dimensional random vector $\mathbf{x}$ is said to have a symmetric multivariate Bessel distribution with parameters $q, r \in \mathbb{R}$, μ: p-dimensional vector, Σ: $p \times p$ matrix with $r > 0$, $q > -\frac{p}{2}$, and $\Sigma > \mathbf{0}$ if its p.d.f. is

$$f(\mathbf{x}) = \frac{[(\mathbf{x}-\mu)'\Sigma^{-1}(\mathbf{x}-\mu)]^{\frac{q}{2}}}{2^{q+p-1}\pi^{\frac{p}{2}} r^{p+q} \Gamma\left(q + \frac{p}{2}\right) |\Sigma|^{\frac{1}{2}}} K_q\left(\frac{[(\mathbf{x}-\mu)'\Sigma^{-1}(\mathbf{x}-\mu)]^{\frac{1}{2}}}{r}\right),$$

where $K_q(z)$ is the modified Bessel function of the third kind; that is $K_q(z) = \frac{\pi}{2} \frac{I_{-q}(z) - I_q(z)}{sin(q\pi)}$, $|arg(z)| < \pi$, q is integer and

$$I_q(z) = \sum_{k=0}^{\infty} \frac{1}{k!\Gamma(k+q+1)} \left(\frac{z}{2}\right)^{q+2k}, \quad |z| < \infty, \quad |arg(z)| < \pi.$$

If $q = 0$ and $r = \frac{\sigma}{\sqrt{2}}$, $\sigma > 0$, then $\mathbf{x}$ is said to have a multivariate Laplace distribution.

2.7.2.6 Symmetric Multivariate Logistic Distribution

The p-dimensional random vector $\mathbf{x}$ is said to have an elliptically symmetric logistic distribution with parameters μ: p-dimensional vector, Σ: $p \times p$ matrix with $\Sigma > \mathbf{0}$ if its p.d.f. is

$$f(\mathbf{x}) = \frac{c}{|\Sigma|^{\frac{1}{2}}} \frac{\exp\{-(\mathbf{x}-\mu)'\Sigma^{-1}(\mathbf{x}-\mu)\}}{(1+\exp\{-(\mathbf{x}-\mu)'\Sigma^{-1}(\mathbf{x}-\mu)\})^2}$$

with

$$c = \frac{\pi^{\frac{p}{2}}}{\Gamma\left(\frac{p}{2}\right)} \int_0^\infty z^{\frac{p}{2}-1} \frac{e^{-z}}{(1+e^{-z})^2} dz.$$

2.7.2.7 Symmetric Multivariate Stable Law

The p-dimensional random vector $\mathbf{x}$ is said to follow a symmetric multivariate stable law with parameters q, $r \in I\!R$, μ: p-dimensional vector, Σ: $p \times p$ matrix with $0 < q \leq 1$, $r > 0$, and $\Sigma \geq \mathbf{0}$ if its characteristic function is

$$\phi_{\mathbf{x}}(\mathbf{t}) = \exp(i\mathbf{t}'\mu - r(\mathbf{t}'\Sigma\mathbf{t})^q).$$

2.7.3 *General Matrix Variate Case*

The matrix variate elliptically contoured distributions listed here are the matrix variate versions of the multivariate distributions given in Sect. 2.7.2. Let $p \geq 1$ and $n \geq 1$.

2.7.3.1 Matrix Variate Symmetric Kotz Type Distribution

The $p \times n$ random matrix $\mathbf{X}$ is said to have a matrix variate symmetric Kotz type distribution with parameters q, r, $s \in I\!R$, $\mathbf{M} : p \times n$, $\Sigma : p \times p$, $\Phi : n \times n$ with $r > 0$, $s > 0$, $2q + pn > 2$, $\Sigma > \mathbf{0}$, and $\Phi > \mathbf{0}$ if its p.d.f. is

$$f(\mathbf{X}) = \frac{sr^{\frac{2q+pn-2}{2s}}\Gamma\left(\frac{pn}{2}\right)}{\pi^{\frac{pn}{2}}\Gamma\left(\frac{2q+pn-2}{2s}\right)|\Sigma|^{\frac{n}{2}}|\Phi|^{\frac{p}{2}}}[tr((\mathbf{X}-\mathbf{M})'\Sigma^{-1}(\mathbf{X}-\mathbf{M})\Phi^{-1})]^{q-1}$$
$$\times \exp\{-r[tr((\mathbf{X}-\mathbf{M})'\Sigma^{-1}(\mathbf{X}-\mathbf{M})\Phi^{-1})]^s\}.$$

If we take $q = s = 1$ and $r = \frac{1}{2}$, we obtain the p.d.f. of the absolutely continuous matrix variate normal distribution:

$$f(\mathbf{X}) = \frac{1}{(2\pi)^{\frac{pn}{2}} |\Sigma|^{\frac{n}{2}} |\Phi|^{\frac{p}{2}}} etr\left\{ -\frac{(\mathbf{X}-\mathbf{M})'\Sigma^{-1}(\mathbf{X}-\mathbf{M})\Phi^{-1}}{2} \right\}.$$

The characteristic function of this distribution is

$$\phi_{\mathbf{X}}(\mathbf{T}) = etr(i\mathbf{T}'\mathbf{M})etr\left(-\frac{1}{2}\mathbf{T}'\Sigma\mathbf{M}\Phi\right). \tag{2.38}$$

Remark 2.7. If we define $N_{p,n}(\mathbf{M}, \Sigma \otimes \Phi)$ through its characteristic function (2.38), then $\Sigma > \mathbf{0}$ is not required, instead it suffices to assume that $\Sigma \geq \mathbf{0}$.

2.7.3.2 Matrix Variate Pearson Type II Distribution

The $p \times n$ random matrix $\mathbf{X}$ is said to have a matrix variate symmetric Pearson type II distribution with parameters $q \in \mathbb{R}$, $\mathbf{M} : p \times n$, $\Sigma : p \times p$, $\Phi : n \times n$ with $q > -1$, $\Sigma > \mathbf{0}$, and $\Phi > \mathbf{0}$ if its p.d.f. is

$$f(\mathbf{X}) = \frac{\Gamma\left(\frac{pn}{2}+q+1\right)}{\pi^{\frac{pn}{2}}\Gamma(q+1)|\Sigma|^{\frac{n}{2}}|\Phi|^{\frac{p}{2}}}(1-tr((\mathbf{X}-\mathbf{M})'\Sigma^{-1}(\mathbf{X}-\mathbf{M})\Phi^{-1}))^q,$$

where $tr((\mathbf{X}-\mathbf{M})'\Sigma^{-1}(\mathbf{X}-\mathbf{M})\Phi^{-1}) \leq 1$.

2.7.3.3 Matrix Variate Pearson Type VII Distribution

The $p \times n$ random matrix $\mathbf{X}$ is said to have a matrix variate symmetric Pearson type VII distribution with parameters $q, r \in \mathbb{R}$, $\mathbf{M} : p \times n$, $\Sigma : p \times p$, $\Phi : n \times n$ with $r > 0$, $q > \frac{pn}{2}$, $\Sigma > \mathbf{0}$, and $\Phi > \mathbf{0}$ if its p.d.f. is

$$f(\mathbf{X}) = \frac{\Gamma(q)}{(\pi r)^{\frac{pn}{2}}\Gamma\left(q-\frac{pn}{2}\right)|\Sigma|^{\frac{n}{2}}|\Phi|^{\frac{p}{2}}}\left(1+\frac{tr((\mathbf{X}-\mathbf{M})'\Sigma^{-1}(\mathbf{X}-\mathbf{M})\Phi^{-1})}{r}\right)^{-q}. \tag{2.39}$$

Particularly, when $q = \frac{pn+r}{2}$, $\mathbf{X}$ is said to have a matrix variate t-distribution with r degrees of freedom and it is denoted by $Mt_{p,n}(r, \mathbf{M}, \Sigma \otimes \Phi)$. It follows, from Theorem 2.25, that the class of matrix variate symmetric Pearson type VII distributions coincides with the class of matrix variate t-distributions.

When $r = 1$, in the definition of matrix variate t-distribution, i.e., $Mt_{p,n}(1, \mathbf{M}, \Sigma \otimes \Phi)$, then $\mathbf{X}$ is said to have a matrix variate Cauchy distribution.

2.7.3.4 Matrix Variate Symmetric Bessel Distribution

The $p \times n$ random matrix $\mathbf{X}$ is said to have a matrix variate symmetric Bessel distribution with parameters $q, r \in I\!R$, $\mathbf{M} : p \times n$, $\Sigma : p \times p$, $\Phi : n \times n$ with $r > 0$, $q > -\frac{pn}{2}$, $\Sigma > \mathbf{0}$, and $\Phi > \mathbf{0}$ if its p.d.f. is

$$f(\mathbf{X}) = \frac{[tr((\mathbf{X}-\mathbf{M})'\Sigma^{-1}(\mathbf{X}-\mathbf{M})\Phi^{-1})]^{\frac{1}{2}}}{2^{q+pn-1}\pi^{\frac{pn}{2}} r^{pn+q}\Gamma\left(q+\frac{pn}{2}\right)|\Sigma|^{\frac{n}{2}}|\Phi|^{\frac{p}{2}}} \times K_q\left(\frac{[tr((\mathbf{X}-\mathbf{M})'\Sigma^{-1}(\mathbf{X}-\mathbf{M})\Phi^{-1})]^{\frac{1}{2}}}{r}\right),$$

where $K_q(z)$ is the modified Bessel function of the third kind as defined in Sect. 2.7.2.5. For $q = 0$ and $r = \frac{\sigma}{\sqrt{2}}$, $\sigma > 0$, this distribution is known as the matrix variate Laplace distribution.

2.7.3.5 Matrix Variate Symmetric Logistic Distribution

The $p \times n$ random matrix $\mathbf{X}$ is said to have a matrix variate symmetric logistic distribution with parameters $\mathbf{M} : p \times n$, $\Sigma : p \times p$, $\Phi : n \times n$ with $\Sigma > \mathbf{0}$, and $\Phi > \mathbf{0}$ if its p.d.f. is

$$f(\mathbf{X}) = \frac{c}{|\Sigma|^{\frac{n}{2}}|\Phi|^{\frac{p}{2}}} \frac{etr(-(\mathbf{X}-\mathbf{M})'\Sigma^{-1}(\mathbf{X}-\mathbf{M})\Phi^{-1})}{(1+etr(-(\mathbf{X}-\mathbf{M})'\Sigma^{-1}(\mathbf{X}-\mathbf{M})\Phi^{-1}))^2}$$

with

$$c = \frac{\pi^{\frac{pn}{2}}}{\Gamma\left(\frac{pn}{2}\right)} \int_0^\infty z^{\frac{pn}{2}-1} \frac{e^{-z}}{(1+e^{-z})^2} dz.$$

2.7.3.6 Matrix Variate Symmetric Stable Law

The $p \times n$ random matrix $\mathbf{X}$ is said to follow a matrix variate symmetric stable law with parameters $q, r \in I\!R$, $\mathbf{M} : p \times n$, $\Sigma : p \times p$, $\Phi : n \times n$ with $0 < q \leq 1$, $r > 0$, $\Sigma \geq \mathbf{0}$, and $\Phi \geq \mathbf{0}$ if its characteristic function is

$$\phi_{\mathbf{X}}(\mathbf{T}) = etr\left(i\mathbf{T}'\mathbf{M}\right)\exp\left(-r(tr(\mathbf{T}'\Sigma\mathbf{M}\Phi))^q\right).$$

2.7.4 Generating Elliptically Contoured Distributions

If we have a m.e.c. distribution, based on it we can easily generate other m.e.c. distributions. For vector variate elliptical distributions, this is given in Muirhead (1982).

Theorem 2.26. *Let* $\mathbf{X} \sim E_{p,n}(\mathbf{M}, \Sigma \otimes \Phi, \psi)$ *have the p.d.f.*

$$f(\mathbf{X}) = \frac{1}{|\Sigma|^{\frac{n}{2}}|\Phi|^{\frac{p}{2}}} h(tr((\mathbf{X}-\mathbf{M})'\Sigma^{-1}(\mathbf{X}-\mathbf{M})\Phi^{-1})).$$

Suppose $G(z)$ *is a distribution function on* $(0,\infty)$. *Let*

$$g(\mathbf{X}) = \frac{1}{|\Sigma|^{\frac{n}{2}}|\Phi|^{\frac{p}{2}}} \int_0^{\infty} z^{-\frac{pn}{2}} h\left(\frac{1}{z} tr((\mathbf{X}-\mathbf{M})'\Sigma^{-1}(\mathbf{X}-\mathbf{M})\Phi^{-1})\right) dG(z).$$

Then, $g(\mathbf{X})$ *is also the p.d.f. of a m.e.c. distribution.*

PROOF: Clearly, $g(\mathbf{X}) \geq 0$. Moreover,

$$\begin{aligned}
&\int_{\mathbb{R}^{p\times n}} g(\mathbf{X}) d\mathbf{X} \\
&= \int_{\mathbb{R}^{p\times n}} \frac{1}{|\Sigma|^{\frac{n}{2}}|\Phi|^{\frac{p}{2}}} \int_0^{\infty} z^{-\frac{pn}{2}} h\left(\frac{1}{z} tr((\mathbf{X}-\mathbf{M})'\Sigma^{-1}(\mathbf{X}-\mathbf{M})\Phi^{-1})\right) dG(z) d\mathbf{X} \\
&= \int_0^{\infty} \int_{\mathbb{R}^{p\times n}} \frac{1}{|z\Sigma|^{\frac{n}{2}}|\Phi|^{\frac{p}{2}}} h\left(tr((\mathbf{X}-\mathbf{M})'(z\Sigma)^{-1}(\mathbf{X}-\mathbf{M})\Phi^{-1})\right) d\mathbf{X} dG(z) \\
&= \int_0^{\infty} 1 dG(z) = 1.
\end{aligned}$$

Hence, $g(\mathbf{X})$ is a p.d.f. Let $r(w) = \int_0^{\infty} z^{-\frac{pn}{2}} h\left(\frac{w}{z}\right) dG(z)$. Then,

$$g(\mathbf{X}) = \frac{1}{|\Sigma|^{\frac{n}{2}}|\Phi|^{\frac{p}{2}}} r(tr((\mathbf{X}-\mathbf{M})'\Sigma^{-1}(\mathbf{X}-\mathbf{M})\Phi^{-1})).$$

Therefore, $g(\mathbf{X})$ is the p.d.f. of an elliptically contoured distribution. ∎

Corollary 2.6. *Let* $h(u) = (2\pi)^{-\frac{pn}{2}} \exp\left(-\frac{u}{2}\right)$ *in Theorem 2.26. Then, for any distribution function* $G(z)$ *on* $(0,\infty)$,

$$g(\mathbf{X}) = \frac{1}{(2\pi)^{\frac{pn}{2}}|\Sigma|^{\frac{n}{2}}|\Phi|^{\frac{p}{2}}} \int_0^{\infty} z^{-\frac{pn}{2}} \exp\left(-\frac{1}{2z} tr((\mathbf{X}-\mathbf{M})'\Sigma^{-1}(\mathbf{X}-\mathbf{M})\Phi^{-1})\right) dG(z).$$

defines the p.d.f. of a m.e.c. distribution. In this case, the distribution of $\mathbf{X}$ *is said to be a mixture of normal distributions.*

In particular, if $G(1) = 1 - \varepsilon$ and $G(\sigma^2) = \varepsilon$ with $0 < \varepsilon < 1$, $\sigma^2 > 0$, we obtain the ε-contaminated matrix variate normal distribution. It has the p.d.f.

$$f(\mathbf{X}) = \frac{1}{(2\pi)^{\frac{pn}{2}}|\Sigma|^{\frac{n}{2}}|\Phi|^{\frac{p}{2}}}\left[(1-\varepsilon)etr\left(-\frac{1}{2}(\mathbf{X}-\mathbf{M})'\Sigma^{-1}(\mathbf{X}-\mathbf{M})\Phi^{-1}\right)\right.$$
$$\left.+\frac{\varepsilon}{\sigma^{pn}}etr\left(-\frac{1}{2\sigma^2}(\mathbf{X}-\mathbf{M})'\Sigma^{-1}(\mathbf{X}-\mathbf{M})\Phi^{-1}\right)\right].$$

Chapter 3
Probability Density Function and Expected Values

3.1 Probability Density Function

The m.e.c. probability density function has some interesting properties which will be given in this chapter. These results are taken from Kelker (1970), Cambanis, Huang, and Simons (1981), Fang, Kotz, and Ng (1990), and Gupta and Varga (1994c).

The first remarkable property is that the marginal distributions of a m.e.c. distribution are absolutely continuous unless the original distribution has an atom of positive weight at zero. Even if the original distribution has an atom of positive weight at zero, the marginal density is absolutely continuous outside zero. This is shown for multivariate elliptical distributions in the following theorem due to Cambanis, Huang, and Simons (1981).

Theorem 3.1. *Let* $\mathbf{x} \sim E_p(\mathbf{0}, \mathbf{I}_p, \psi)$ *have the stochastic representation* $\mathbf{x} \approx r\mathbf{u}$. *Let* $F(r)$ *be the distribution function of* r. *Assume* $\mathbf{x}$ *is partitioned as* $\mathbf{x} = \begin{pmatrix} \mathbf{x}_1 \\ \mathbf{x}_2 \end{pmatrix}$, *where* $\mathbf{x}_1$ *is* q*-dimensional,* $1 \leq q < p$. *Let* $\mathbf{x}_1$ *have the stochastic representation* $\mathbf{x}_1 \approx r_1 \mathbf{u}_1$. *Then, the distribution of* r_1 *has an atom of weight* $F(0)$ *at zero and it is absolutely continuous on* $(0, \infty)$ *with p.d.f.*

$$g_q(s) = \frac{2\Gamma\left(\frac{p}{2}\right) s^{q-1}}{\Gamma\left(\frac{q}{2}\right)\Gamma\left(\frac{p-q}{2}\right)} \int_s^{\infty} r^{-(p-2)} (r^2 - s^2)^{\frac{p-q}{2}-1} dF(r), \quad 0 < s < \infty. \tag{3.1}$$

PROOF: From Corollary 2.4, it follows that $r_1 \approx r r_0$, where $r_0^2 \sim B\left(\frac{q}{2}, \frac{p-q}{2}\right)$. Therefore, $P(r_1 = 0) = P(r = 0) = F(0)$ and

$$P(0 < r_1 \leq t) = P(0 < r r_0 \leq t)$$
$$= \int_{(0,\infty)} P(0 < r r_0 \leq t) dF(r)$$

A.K. Gupta et al., *Elliptically Contoured Models in Statistics and Portfolio Theory*, DOI 10.1007/978-1-4614-8154-6_3, © Springer Science+Business Media New York 2013

$$= \int_{(0,\infty)} P\left(0 < r_0 \leq \frac{t}{r}\right) dF(r)$$

$$= \int_{(0,\infty)} P\left(0 < r_0^2 \leq \frac{t^2}{r^2}\right) dF(r)$$

$$= \int_{(0,\infty)} \frac{\Gamma\left(\frac{p}{2}\right)}{\Gamma\left(\frac{q}{2}\right)\Gamma\left(\frac{p-q}{2}\right)} \int_{\left(0,\min\left(1,\frac{t^2}{r^2}\right)\right]} x^{\frac{q}{2}-1}(1-x)^{\frac{p-q}{2}-1} dx dF(r).$$

Let $x = \frac{s^2}{r^2}$. Then, $J(x \to s) = \frac{r^2}{2s}$ and we have

$$\int_{(0,\infty)} \frac{\Gamma\left(\frac{p}{2}\right)}{\Gamma\left(\frac{q}{2}\right)\Gamma\left(\frac{p-q}{2}\right)} \int_{\left(0,\min\left(1,\frac{t^2}{r^2}\right)\right]} x^{\frac{q}{2}-1}(1-x)^{\frac{p-q}{2}-1} dx dF(r)$$

$$= \int_{(0,\infty)} \frac{\Gamma\left(\frac{p}{2}\right)}{\Gamma\left(\frac{q}{2}\right)\Gamma\left(\frac{p-q}{2}\right)} \int_{(0,\min(r,t)]} \frac{2s}{r^2}\left(\frac{s}{r}\right)^{q-2}\left(1-\frac{s^2}{r^2}\right)^{\frac{p-q}{2}-1} ds dF(r)$$

$$= \frac{2\Gamma\left(\frac{p}{2}\right)}{\Gamma\left(\frac{q}{2}\right)\Gamma\left(\frac{p-q}{2}\right)} \int_{(0,\infty)} \int_{(0,\min(r,t)]} s^{q-1} r^{-2-q+2-(p-q)+2}(r^2-s^2)^{\frac{p-q}{2}-1} ds dF(r)$$

$$= \int_{(0,t]} \frac{2\Gamma\left(\frac{p}{2}\right) s^{q-1}}{\Gamma\left(\frac{q}{2}\right)\Gamma\left(\frac{p-q}{2}\right)} \int_{(0,\min(r,t)]} r^{-(p-2)}(r^2-s^2)^{\frac{p-q}{2}-1} dF(r) ds,$$

from which (3.1) follows. ∎

Corollary 3.1. *Let* $\mathbf{x} \sim E_p(\mathbf{0}, \mathbf{I}_p, \psi)$ *and assume that* $P(\mathbf{x} = \mathbf{0}) = 0$. *Let* $\mathbf{x} \approx r\mathbf{u}$ *be the stochastic representation of* $\mathbf{x}$ *and* $F(r)$ *the distribution function of* r. *Partition* $\mathbf{x}$ *into* $\mathbf{x} = \begin{pmatrix} \mathbf{x}_1 \\ \mathbf{x}_2 \end{pmatrix}$, *where* $\mathbf{x}_1$ *is* q*-dimensional,* $1 \leq q < p$. *Then,* $\mathbf{x}_1$ *is absolutely continuous with p.d.f.*

$$f_q(\mathbf{y}) = \frac{\Gamma\left(\frac{p}{2}\right)}{\pi^{\frac{q}{2}}\Gamma\left(\frac{p-q}{2}\right)} \int_{(\mathbf{y}'\mathbf{y})^{\frac{1}{2}}}^{\infty} r^{-(p-2)}(r^2 - \mathbf{y}'\mathbf{y})^{\frac{p-q}{2}-1} dF(r). \tag{3.2}$$

PROOF: From Theorem 3.1, it follows that $\mathbf{x}_1$ is absolutely continuous and if $r_1\mathbf{u}_1$ is the stochastic representation of $\mathbf{x}_1$ then r_1 has the p.d.f.

$$g_q(s) = \frac{2\Gamma\left(\frac{p}{2}\right) s^{q-1}}{\Gamma\left(\frac{q}{2}\right)\Gamma\left(\frac{p-q}{2}\right)} \int_s^{\infty} r^{-(p-2)}(r^2 - s^2)^{\frac{p-q}{2}-1} dF(r), \quad 0 < s < \infty. \tag{3.3}$$

Since, $\mathbf{x}_1$ also has a m.e.c. distribution, its p.d.f. is of the form

$$f_q(\mathbf{y}) = h_q(\mathbf{y}'\mathbf{y}). \tag{3.4}$$

From Theorem 2.16, it follows that

$$g_q(s) = \frac{2\pi^{\frac{q}{2}}}{\Gamma\left(\frac{q}{2}\right)} s^{q-1} h_q(s^2), s \geq 0, \tag{3.5}$$

and from (3.3) and (3.5), it follows that

$$h_q(s^2) = \frac{\Gamma\left(\frac{p}{2}\right)}{\pi^{\frac{q}{2}}\Gamma\left(\frac{p-q}{2}\right)} \int_s^\infty r^{-(p-2)}(r^2 - s^2)^{\frac{p-q}{2}-1} dF(r). \tag{3.6}$$

Now, (3.2) follows from (3.4) and (3.6). ■

The following result was given by Fang, Kotz, and Ng (1990).

Theorem 3.2. *Let* $\mathbf{x} \sim E_p(\mathbf{0}, \mathbf{I}_p, \psi)$ *with p.d.f.* $f(\mathbf{x}) = h(\mathbf{x}'\mathbf{x})$. *Let* $\mathbf{x}$ *be partitioned as* $\mathbf{x} = \begin{pmatrix} \mathbf{x}_1 \\ \mathbf{x}_2 \end{pmatrix}$, *where* $\mathbf{x}_1$ *is q-dimensional,* $1 \leq q < p$. *Then,* $\mathbf{x}_1$ *is absolutely continuous and its p.d.f. is*

$$f_q(\mathbf{y}) = \frac{\pi^{\frac{p-q}{2}}}{\Gamma\left(\frac{p-q}{2}\right)} \int_{\mathbf{y}'\mathbf{y}}^\infty (u - \mathbf{y}'\mathbf{y})^{\frac{p-q}{2}-1} h(u) du. \tag{3.7}$$

PROOF: Let $r\mathbf{u}$ be the stochastic representation of $\mathbf{x}$ and F be the distribution function of r. Then, from Theorem 2.16, we get the p.d.f. of r as

$$g(r) = \frac{2\pi^{\frac{p}{2}}}{\Gamma\left(\frac{p}{2}\right)} r^{p-1} h(r^2). \tag{3.8}$$

From (3.2) and (3.8), we have

$$\begin{aligned} f_q(y) &= \frac{\Gamma\left(\frac{p}{2}\right)}{\pi^{\frac{q}{2}}\Gamma\left(\frac{p-q}{2}\right)} \int_{(\mathbf{y}'\mathbf{y})^{\frac{1}{2}}}^\infty r^{-(p-2)}(r^2 - \mathbf{y}'\mathbf{y})^{\frac{p-q}{2}-1} \frac{2\pi^{\frac{p}{2}}}{\Gamma\left(\frac{p}{2}\right)} r^{p-1} h(r^2) dr \\ &= \frac{2\pi^{\frac{p-q}{2}}}{\Gamma\left(\frac{p-q}{2}\right)} \int_{(\mathbf{y}'\mathbf{y})^{\frac{1}{2}}}^\infty r(r^2 - \mathbf{y}'\mathbf{y})^{\frac{p-q}{2}-1} h(r^2) dr. \end{aligned}$$

Let $u = r^2$. Then, $J(r \to u) = \frac{1}{2r}$ and we have

$$f_q(\mathbf{y}) = \frac{\pi^{\frac{p-q}{2}}}{\Gamma\left(\frac{p-q}{2}\right)} \int_{\mathbf{y}'\mathbf{y}}^\infty (u - \mathbf{y}'\mathbf{y})^{\frac{p-q}{2}-1} h(u) du.$$

■

Marginal densities have certain continuity and differentiability properties. These are given in the following theorems. The first theorem is due to Kelker (1970).

Theorem 3.3. *Let* $\mathbf{x} \sim E_p(\mathbf{0}, \mathbf{I}_p, \psi)$, $1 < p$, *with p.d.f.* $f(\mathbf{x}) = h(\mathbf{x}'\mathbf{x})$. *Let* $\mathbf{x}$ *be partitioned as* $\mathbf{x} = \begin{pmatrix} \mathbf{x}_1 \\ x_2 \end{pmatrix}$, *where* $\mathbf{x}_1$ *is* $p-1$ *dimensional. Let* $\mathbf{x}_1$ *have the p.d.f.* $f_{p-1}(\mathbf{y}) = h_{p-1}(\mathbf{y}'\mathbf{y})$. *If* $h_{p-1}(z)$ *is bounded in a neighborhood of* z_0 *then* $h_{p-1}(z)$ *is continuous at* $z = z_0$.

PROOF: From (3.7), we get

$$h_{p-1}(z) = \frac{\pi^{\frac{1}{2}}}{\Gamma\left(\frac{1}{2}\right)} \int_z^\infty (u-z)^{-\frac{1}{2}} h(u) du .$$

Thus,

$$h_{p-1}(z) = \int_z^\infty (u-z)^{-\frac{1}{2}} h(u) du . \tag{3.9}$$

Choose any $\eta > 0$. There exist $k > 0$ and $K > 0$ such that if $|z - z_0| < k$ then $h(z) < K$. Let $\varepsilon = \min\left(k, \frac{\eta^2}{64}\right)$. Further let δ be such that $0 < \delta < \varepsilon$, and $z_0 < z < z_0 + \delta$. Then, we have

$$\begin{aligned}
|h_{p-1}(z) - h_{p-1}(z_0)| &= \left| \int_z^\infty (u-z)^{-\frac{1}{2}} h(u) du - \int_{z_0}^\infty (u-z_0)^{-\frac{1}{2}} h(u) du \right| \\
&= \left| \int_{z_0+\varepsilon}^\infty \left((u-z)^{-\frac{1}{2}} - (u-z_0)^{-\frac{1}{2}} \right) h(u) du \right. \\
&\quad + \int_z^{z_0+\varepsilon} \left((u-z)^{-\frac{1}{2}} - (u-z_0)^{-\frac{1}{2}} \right) h(u) du \\
&\quad \left. - \int_{z_0}^z (u-z_0)^{-\frac{1}{2}} h(u) du \right| \\
&\le \int_{z_0+\varepsilon}^\infty \left((u-z)^{-\frac{1}{2}} - (u-z_0)^{-\frac{1}{2}} \right) h(u) du \\
&\quad + \int_z^{z_0+\varepsilon} \left((u-z)^{-\frac{1}{2}} - (u-z_0)^{-\frac{1}{2}} \right) h(u) du \\
&\quad + \int_{z_0}^z (u-z_0)^{-\frac{1}{2}} h(u) du .
\end{aligned} \tag{3.10}$$

Now,

$$\int_z^{z_0+\varepsilon} \left((u-z)^{-\frac{1}{2}} - (u-z_0)^{-\frac{1}{2}} \right) h(u) du$$

$$\leq \int_{z}^{z_0+\varepsilon} (u-z)^{-\frac{1}{2}} h(u)du$$
$$\leq K \int_{z}^{z_0+\varepsilon} (u-z)^{-\frac{1}{2}} du$$
$$\leq 2K(\varepsilon + z_0 - z)^{\frac{1}{2}} \leq 2\sqrt{\varepsilon} \tag{3.11}$$

and

$$\int_{z_0}^{z} (u-z_0)^{-\frac{1}{2}} h(u)du \leq 2K(z-z_0)^{-\frac{1}{2}}$$
$$\leq 2\sqrt{\varepsilon}. \tag{3.12}$$

Furthermore,

$$\int_{z_0+\varepsilon}^{\infty} \left((u-z)^{-\frac{1}{2}} - (u-z_0)^{-\frac{1}{2}}\right) h(u)du$$
$$\leq \int_{z_0+\varepsilon}^{\infty} \left((u-(z_0+\delta))^{-\frac{1}{2}} - (u-z_0)^{-\frac{1}{2}}\right) h(u)du. \tag{3.13}$$

We have

$$\lim_{\delta \to 0} \left[(u-(z_0+\delta))^{-\frac{1}{2}} - (u-z_0)^{-\frac{1}{2}}\right] = 0 \quad \text{for} \quad u \geq z_0 + \varepsilon,$$

and

$$0 \leq (u-(z_0+\delta))^{-\frac{1}{2}} - (u-z_0)^{-\frac{1}{2}}$$
$$\leq (u-(z_0+\varepsilon))^{-\frac{1}{2}} - (u-z_0)^{-\frac{1}{2}}$$
$$\leq (u-(z_0+\varepsilon))^{-\frac{1}{2}} \quad \text{for} \quad u \geq z_0 + \varepsilon.$$

Since,

$$\int_{z_0+\varepsilon}^{\infty} (u-(z_0+\varepsilon))^{-\frac{1}{2}} h(u)du = h_{p-1}(z_0+\varepsilon) < \infty,$$

we can use the dominated Lebesgue convergence theorem to get

$$\lim_{\delta \to 0} \int_{z_0+\varepsilon}^{\infty} \left((u-z)^{-\frac{1}{2}} - (u-z_0)^{-\frac{1}{2}}\right) h(u)du = 0.$$

Therefore, there exists $\upsilon > 0$ such that if $0 < \delta < \upsilon$ then

$$\int_{z_0+\varepsilon}^{\infty} \left((u-z)^{-\frac{1}{2}} - (u-z_0)^{-\frac{1}{2}}\right) h(u)du < \frac{\eta}{2}. \tag{3.14}$$

Hence, if $\delta = \min(\varepsilon, \upsilon)$ then for $z_0 < z < z_0 + \delta$, (3.10)–(3.14) give

$$|h_{p-1}(z) - h_{p-1}(z_0)| \leq 4\sqrt{\varepsilon} + \frac{\eta}{2} \leq \eta .$$

Therefore, $h_{p-1}(z)$ is continuous from the right at z_0. In a similar way, it can be proved that $h_{p-1}(z)$ is continuous from the left at z_0. ■

The next theorem shows that the $p-2$ dimensional marginal density of a p-dimensional absolutely continuous multivariate elliptical distribution is differentiable. This theorem is due to Kelker (1970).

Theorem 3.4. *Let* $\mathbf{x} \sim E_p(\mathbf{0}, \mathbf{I}_p, \psi)$, $2 < p$, *with p.d.f.* $f(\mathbf{x}) = h(\mathbf{x}'\mathbf{x})$. *Let* $\mathbf{x}$ *be partitioned as* $\mathbf{x} = \begin{pmatrix} \mathbf{x}_1 \\ \mathbf{x}_2 \end{pmatrix}$, *where* $\mathbf{x}_1$ *is* $p-2$ *dimensional. Let* $\mathbf{x}_1$ *have the p.d.f.* $f_{p-2}(\mathbf{y}) = h_{p-2}(\mathbf{y}'\mathbf{y})$. *Then,* $h_{p-2}(z)$ *is differentiable and*

$$h'_{p-2}(z) = -\pi h(z). \tag{3.15}$$

PROOF: From Theorem 3.2, we get

$$h_{p-2}(z) = \pi \int_z^\infty h(u)du, \quad z \geq 0. \tag{3.16}$$

Hence, $h'_{p-2}(z) = -\pi h(z)$. ■

Remark 3.1. Theorem 3.4 shows that if $p \geq 3$ and the one-dimensional marginal density of a p-dimensional absolutely continuous spherical distribution is known, then all the marginal densities and also the density of the parent distribution can be obtained easily. In fact, if $f_j(\mathbf{y}) = h_j(\mathbf{y}'\mathbf{y})$, $\mathbf{y} \in I\!R^j$, denotes the p.d.f. of the j-dimensional marginal, then from (3.15) we get

$$h_{2j+1}(z) = \left(-\frac{1}{\pi}\right)^j h_1^{(j)}(z), \quad z \geq 0.$$

From (3.9), we have

$$h_2(z) = \int_z^\infty (u-z)^{-\frac{1}{2}} h_3(u)du .$$

From $h_2(z)$, we can obtain the other marginals using

$$h_{2j}(z) = \left(-\frac{1}{\pi}\right)^{j-1} h_2^{(j-1)}(z), \quad z \geq 0.$$

Remark 3.2. Assume $\mathbf{x} \sim E_p(\mathbf{0}, \mathbf{I}_p, \psi)$ is absolutely continuous, $p \geq 3$. If $f_j(\mathbf{y}) = h_j(\mathbf{y}'\mathbf{y})$, $\mathbf{y} \in I\!R^j$, denotes the p.d.f. of the j-dimensional marginal then from (3.16) it follows that $h_j(z)$, $z \geq 0$ is nondecreasing for $j = 1,2,\ldots,p-2$. Since $f_1(z) = h_1(z^2)$, the p.d.f. of the one-dimensional marginal, is nondecreasing on $(-\infty,0]$ and nonincreasing on $[0,\infty)$.

The results given in this chapter so far referred to the vector variate elliptically contoured distributions. Their extension to the matrix variate case is straightforward since $\mathbf{X} \sim E_{p,n}(\mathbf{0},\mathbf{I}_p \otimes \mathbf{I}_n,\psi)$ is equivalent to $\mathbf{x} = vec(\mathbf{X}') \sim E_{pn}(\mathbf{0},\mathbf{I}_{pn},\psi)$. The following theorems can be easily derived.

Theorem 3.5. *Let $\mathbf{X} \sim E_{p,n}(\mathbf{0},\mathbf{I}_p \otimes \mathbf{I}_n,\psi)$ have the stochastic representation $\mathbf{X} \approx r\mathbf{U}$. Let $F(r)$ be the distribution function of r. Assume $\mathbf{X}$ is partitioned as $\mathbf{X} = \begin{pmatrix}\mathbf{X}_1\\ \mathbf{X}_2\end{pmatrix}$, where $\mathbf{X}_1$ is $q \times n$. Let $\mathbf{X}_1$ have the stochastic representation $\mathbf{X}_1 \approx r_1\mathbf{U}_1$. Then, the distribution of r_1 has an atom of weight $F(0)$ at zero and it is absolutely continuous on $(0,\infty)$ with p.d.f.*

$$g_{q,n}(s) = \frac{2\Gamma\left(\frac{pn}{2}\right)s^{qn-1}}{\Gamma\left(\frac{qn}{2}\right)\Gamma\left(\frac{(p-q)n}{2}\right)}\int_s^\infty r^{-(pn-2)}(r^2-s^2)^{\frac{(p-q)n}{2}-1}dF(r),$$

$$0<s<\infty.$$

Corollary 3.2. *With the notation of Theorem 3.5 we get $P(\mathbf{X}_1 = \mathbf{0}) = P(\mathbf{X} = \mathbf{0})$.*

Theorem 3.6. *Let $\mathbf{X} \sim E_{p,n}(\mathbf{0},\mathbf{I}_p \otimes \mathbf{I}_n,\psi)$ and assume that $P(\mathbf{X}=\mathbf{0})$. Let $\mathbf{X} \approx r\mathbf{U}$ be the stochastic representation of $\mathbf{X}$ and $F(r)$ the distribution function of r. Partition $\mathbf{X}$ into $\mathbf{X} = \begin{pmatrix}\mathbf{X}_1\\ \mathbf{X}_2\end{pmatrix}$, where $\mathbf{X}_1$ is $q \times n$, $1 \leq q < p$. Then, $\mathbf{X}_1$ is absolutely continuous with p.d.f.*

$$f_{q,n}(\mathbf{Y}) = \frac{\Gamma\left(\frac{pn}{2}\right)}{\pi^{\frac{qn}{2}}\Gamma\left(\frac{(p-q)n}{2}\right)}\int_{(tr(\mathbf{Y}'\mathbf{Y}))^{\frac{1}{2}}}^\infty r^{-(pn-2)}(r^2-tr(\mathbf{Y}'\mathbf{Y}))^{\frac{(p-q)n}{2}-1}dF(r).$$

Theorem 3.7. *Let $\mathbf{X} \sim E_{p,n}(\mathbf{0},\mathbf{I}_p \otimes \mathbf{I}_n,\psi)$ with p.d.f. $f(\mathbf{X}) = h(tr(\mathbf{X}'\mathbf{X}))$. Let $\mathbf{X}$ be partitioned as $\mathbf{X} = \begin{pmatrix}\mathbf{X}_1\\ \mathbf{X}_2\end{pmatrix}$, where $\mathbf{X}_1$ is $q \times n$, $1 \leq q < p$. Then, $\mathbf{X}_1$ is absolutely continuous with p.d.f.*

$$f_{q,n}(\mathbf{Y}) = \frac{\pi^{\frac{(p-q)n}{2}}}{\Gamma\left(\frac{(p-q)n}{2}\right)}\int_{tr(\mathbf{Y}'\mathbf{Y})}^\infty (u-tr(\mathbf{Y}'\mathbf{Y}))^{\frac{(p-q)n}{2}-1}h(u)du.$$

Theorem 3.8. *Let $\mathbf{X} \sim E_{p,n}(\mathbf{0},\mathbf{I}_p \otimes \mathbf{I}_n,\psi)$, $1 < p$, $1 < n$, with p.d.f. $f(\mathbf{X}) = h(tr(\mathbf{X}'\mathbf{X}))$. Let $\mathbf{X}$ be partitioned as $\mathbf{X} = \begin{pmatrix}\mathbf{X}_1\\ \mathbf{X}_2\end{pmatrix}$, where $\mathbf{X}_1$ is $q \times n$, $1 \leq q < p$. Let $\mathbf{X}_1$ have the p.d.f. $f_{q,n}(\mathbf{Y}) = h_{q,n}(tr(\mathbf{Y}'\mathbf{Y}))$. Then, $h_{q,n}(z)$ is differentiable and*

(i) If $(p-q)n$ is even, $(p-q)n = 2j$, then

$$h_{q,n}^{(j)}(z) = (-\pi)^j h(z),$$

(ii) If $(p-q)n$ is odd, $(p-q)n = 2j+1$, then

$$\frac{\partial^j}{\partial z^j}\left(\int_z^\infty (u-z)^{-\frac{1}{2}} h'_{q,n}(u)du\right) = (-\pi)^{j+1} h(z).$$

Theorem 3.9. *Let $\mathbf{X} \sim E_{p,n}(\mathbf{0}, \mathbf{I}_p \otimes \mathbf{I}_n, \psi)$, $pn \geq 3$ with p.d.f. $f(\mathbf{X}) = h(tr(\mathbf{X}'\mathbf{X}))$. Let $f_1(y) = h_1(y^2)$ denote the p.d.f. of a one-dimensional marginal of $\mathbf{X}$. Then, if we know $h_1(z)$, we can obtain $h(z)$ in the following way.*

(i) If pn is odd, $pn = 2j+1$, then

$$h(z) = \frac{h_1^{(j)}(z)}{(-\pi)^j}.$$

(ii) If pn is even, $pn = 2j$, then

$$h(z) = \left(-\frac{1}{\pi}\right)^j \frac{\partial^{j-1}}{\partial z^{j-1}} \int_z^\infty (u-z)^{-\frac{1}{2}} h'_1(u)du.$$

Theorem 3.10. *Assume $\mathbf{X} \sim E_{p,n}(\mathbf{0}, \mathbf{I}_p \otimes \mathbf{I}_n, \psi)$ is absolutely continuous, $p > 1$, $n > 1$. If $f_{j,n}(\mathbf{Y}) = h_{j,n}(tr(\mathbf{Y}'\mathbf{Y}))$, $\mathbf{Y} \in \mathbb{R}^{j \times n}$, denotes the p.d.f. of the $j \times n$ dimensional marginal, then $h_j(z)$, $z \geq 0$ is nondecreasing for $j = 1, 2, \ldots, p-1$. Moreover, the p.d.f. of the one-dimensional marginal is nondecreasing on $(-\infty, 0]$ and nonincreasing on $[0, \infty)$.*

3.2 More on Expected Values

The stochastic representation can be used to compute the expected values of matrix variate functions when the underlying distribution is elliptically contoured. In order to simplify the expressions for the expected values, we need the following theorem which is a special case of Berkane and Bentler (1986a).

Theorem 3.11. *Let $\phi(t) = \psi(t^2)$ be a characteristic function of a one-dimensional elliptical distribution. Assume the distribution has finite mth moment. Then,*

$$\phi^{(m)}(0) = \begin{cases} \frac{m!}{\left(\frac{m}{2}\right)!} \psi^{\left(\frac{m}{2}\right)}(0) & \text{if } m \text{ is even} \\ 0 & \text{if } m \text{ is odd.} \end{cases} \quad (3.17)$$

PROOF: If the mth moment exists, say μ_m then, $\phi^{(m)}(t)$ also exists and $\phi^{(m)}(0) = i^m \mu_m$. First we prove that

$$\phi^{(k)}(t) = \sum_{n=\left[\frac{k+1}{2}\right]}^{k} K_n^k t^{2n-k} \psi^{(n)}(t^2), \tag{3.18}$$

where $0 \leq k \leq m$, $[z]$ denotes the integer part of z, and K_n^k is a constant not depending on ψ. We prove this by induction.

If $k=0$, we have $\phi(t) = \psi(t^2)$, so $K_0^0 = 1$. If $k=1$ we get $\phi'(t) = 2t\psi'(t^2)$, and the statement is true in this case, with $K_1^1 = 2$. Assume the theorem is true up to $l < m$. Then, for $k = l+1$ we get

$$\begin{aligned}
\phi^{(k)}(t) &= \sum_{n=\left[\frac{l+1}{2}\right]}^{l} (2n-l) K_n^l t^{2n-l-1} \psi^{(n)}(t^2) \\
&+ \sum_{n=\left[\frac{l+1}{2}\right]}^{l} 2K_n^l t^{2n-l-1} \psi^{(n+1)}(t^2) \\
&= \sum_{n=\left[\frac{k}{2}\right]}^{k-1} (2n-k+1) K_n^{k-1} t^{2n-k} \psi^{(n)}(t^2) \\
&+ \sum_{n=\left[\frac{k}{2}\right]+1}^{k} 2K_{n-1}^{k-1} t^{2n-k} \psi^{(n)}(t^2) \\
&= \left(2\left[\frac{k}{2}\right] - k + 1\right) K_{\left[\frac{k}{2}\right]}^{k-1} t^{2\left[\frac{k}{2}\right]-k} \psi^{\left(\left[\frac{k}{2}\right]\right)}(t^2) \\
&+ \sum_{n=\left[\frac{k}{2}\right]+1}^{k-1} \left(2K_{n-1}^{k-1} + (2n-k+1)K_n^{k-1}\right) t^{2n-k} \psi^{(n)}(t^2) \\
&+ 2K_{k-1}^{k-1} t^{2k-k} \psi^{(k)}(t^2).
\end{aligned} \tag{3.19}$$

We have to distinguish between two cases.

(a) k even.
Then, $\left[\frac{k+1}{2}\right] = \left[\frac{k}{2}\right]$ and $2\left[\frac{k}{2}\right] - k + 1 = 1$. Hence, (3.19) gives

$$\phi^{(k)}(t) = \sum_{n=\left[\frac{k+1}{2}\right]}^{k} K_n^k t^{2n-k} \psi^{(n)}(t^2),$$

with

$$K_n^k = \begin{cases} K_{\left[\frac{k+1}{2}\right]}^{k-1} & \text{if } n = \left[\frac{k+1}{2}\right] \\ 2K_{n-1}^{k-1} + (2n-k+1)K_n^{k-1} & \text{if } \left[\frac{k+1}{2}\right] < n < k \\ 2K_{k-1}^{k-1} & \text{if } n = k\,. \end{cases}$$

(b) k odd.
Then, $\left[\frac{k+1}{2}\right] = \left[\frac{k}{2}\right] + 1$ and $2\left[\frac{k}{2}\right] - k + 1 = 0$. From (3.19) we get

$$\phi^{(k)}(t) = \sum_{n=\left[\frac{k+1}{2}\right]}^{k} K_n^k t^{2n-k} \psi^{(n)}(t^2)$$

with

$$K_n^k = \begin{cases} 2K_{n-1}^{k-1} + (2n-k+1)K_n^{k-1} & \text{if } \left[\frac{k+1}{2}\right] \le n < k \\ 2K_{k-1}^{k-1} & \text{if } n = k\,. \end{cases}$$

Hence (3.18) is established. Taking $k = m$ and $t = 0$ in (3.18), we get

$$\phi^{(m)}(0) = \begin{cases} K_{\frac{m}{2}}^m \psi^{\left(\frac{m}{2}\right)}(0) & \text{if } m \text{ is even} \\ 0 & \text{if } m \text{ is odd.} \end{cases} \tag{3.20}$$

If m is even, $m = 2s$ say, then we have $\phi^{(2s)}(0) = i^{2s}\mu_{2s}$, that is $\phi^{(2s)}(0) = (-1)^s \mu_{2s}$. Let $x \sim N_1(0,1)$, then $\mu_{2s} = \frac{(2s)!}{2^s s!}$, $\psi(z) = \exp\left(-\frac{z}{2}\right)$, and $\psi^{(s)}(z) = \left(-\frac{1}{2}\right)^s \exp\left(-\frac{z}{2}\right)$, from which $\psi^{(s)}(0) = \left(-\frac{1}{2}\right)^s$ follows. Therefore,

$$\phi^{(2s)}(0) = K_s^{2s}\psi^{(s)}(0) = K_s^{2s}\left(-\frac{1}{2}\right)^s.$$

Thus, we get

$$K_s^{2s}\left(-\frac{1}{2}\right)^s = (-1)^s \frac{(2s)!}{2^s s!},$$

from which it follows that $K_s^{2s} = \frac{(2s)!}{s!}$. Comparing this with (3.20), we obtain (3.17). ∎

Corollary 3.3. *Let $\phi(t) = \psi(t^2)$ be as in Theorem 3.11, then*

(i) $\phi(0) = 1$,
(ii) If $\psi(t)$ is differentiable, $\phi'(0) = 0$,
(iii) If $\psi(t)$ has second derivative, $\phi''(0) = \psi'(0)$,
(iv) If $\psi(t)$ has third derivative, $\phi'''(0) = 0$,
(v) If $\psi(t)$ has fourth derivative, $\phi^{iv}(0) = 12\psi''(0)$.

Corollary 3.4. *Let $\phi(t) = \exp\left(-\frac{t^2}{2}\right)$. Then,*

$$\phi^{(m)}(0) = \begin{cases} \left(-\frac{1}{2}\right)^{\frac{m}{2}} \frac{m!}{\left(\frac{m}{2}\right)!} & \text{if } m \text{ is even} \\ 0 & \text{if } m \text{ is odd.} \end{cases}$$

For example,

(i) $\phi(0) = 1$,
(ii) $\phi'(0) = 0$,
(iii) $\phi''(0) = -1$,
(iv) $\phi'''(0) = 0$,
(v) $\phi^{iv}(0) = 3$.

Now, we can derive the following results.

Theorem 3.12. *Let $\mathbf{X} \sim E_{p,n}(\mathbf{0}, \Sigma \otimes \Phi, \psi)$ with $P(\mathbf{X} = \mathbf{0}) = 0$. Let $q = rk(\Sigma)$, $m = rk(\Phi)$ and $r\mathbf{AUB}'$ be the stochastic representation of $\mathbf{X}$. Assume $\mathbf{Y} \sim N_{p,n}(\mathbf{0}, \Sigma \otimes \Phi)$. Let F be a subset of the $p \times n$ real matrices such that if $\mathbf{Z} \in \mathbb{R}^{p\times n}$, $\mathbf{Z} \in \mathscr{F}$, and $a > 0$, then $a\mathbf{Z} \in \mathscr{F}$ and $P(X \notin F) = P(Y \notin F) = 0$. Let $K(\mathbf{Z})$ be a function defined on $\mathscr{F}$ such that if $\mathbf{Z} \in \mathscr{F}$ and $a > 0$, then $K(a\mathbf{Z}) = a^k K(\mathbf{Z})$ where $k > -qm$. Assuming $E(K(\mathbf{X}))$ and $E(K(\mathbf{Y}))$ exist, we get*

(a) $E(K(\mathbf{X})) = E(K(\mathbf{Y}))\frac{E(r^k)\Gamma\left(\frac{qm}{2}\right)}{2^{\frac{k}{2}}\Gamma\left(\frac{qm+k}{2}\right)}$,
(b) *If $\mathbf{X}$ has the p.d.f. $f(\mathbf{X}) = \frac{1}{|\Sigma|^{\frac{n}{2}}|\Phi|^{\frac{p}{2}}} h(tr(\mathbf{X}'\Sigma^{-1}\mathbf{X}\Phi^{-1}))$ then*

$$E(K(\mathbf{X})) = E(K(\mathbf{Y}))\frac{\pi^{\frac{pn}{2}} \int_0^\infty z^{pn+k-1} h(z^2) dz}{2^{\frac{k}{2}-1}\Gamma\left(\frac{pn+k}{2}\right)},$$

(c) *If $k = 0$, then $E(K(\mathbf{X})) = E(K(\mathbf{Y}))$,*
(d) *If k is a positive even integer, then*

$$E(K(\mathbf{X})) = E(K(\mathbf{Y}))(-2)^{\frac{k}{2}} \psi^{\frac{k}{2}}(0),$$

(e) *If k is a positive odd integer and $K(aZ) = a^k K(Z)$ holds for all $a \geq 0$, then $E(K(\mathbf{X})) = \mathbf{0}$.*

PROOF:

(a) $K(\mathbf{X})$ and $K(\mathbf{Y})$ are defined if $\mathbf{X} \in \mathscr{F}$ and $\mathbf{Y} \in \mathscr{F}$. Since

$$P(X \notin F) = P(Y \notin F) = 0,$$

we see that $K(\mathbf{X})$ and $K(\mathbf{Y})$ are defined with probability one.

Let $r_2\mathbf{AU}_2\mathbf{B}'$ be the stochastic representation of $\mathbf{Y}$. From the conditions of the theorem, it follows that if $a\mathbf{Z} \in \mathscr{F}$ and $a > 0$, then $\mathbf{Z} \in \mathscr{F}$. Since, $P(\mathbf{X} = \mathbf{0}) = 0$, we have $P(r = 0) = 0$, and from $P(r\mathbf{AU}_1\mathbf{B}' \in \mathscr{F}) = 1$, we get $P(\mathbf{AUB}' \in \mathscr{F}) = 1$. Thus $K(\mathbf{AUB}')$ is defined with probability one. Moreover, $P(K(r\mathbf{AUB}') = r^k K(\mathbf{AUB}')) = 1$. Therefore $E(K(r\mathbf{AUB}')) = E(r^k K(\mathbf{AUB}'))$. Since, r and $\mathbf{U}$ are independent, we get

$$E(K(r\mathbf{AUB}')) = E(r^k)E(K(\mathbf{AUB}')). \tag{3.21}$$

Similarly,

$$E(K(r_2\mathbf{AU}_2\mathbf{B}')) = E(r_2^k)E(K(\mathbf{AU}_2\mathbf{B}')). \tag{3.22}$$

However, $\mathbf{AUB}' \approx \mathbf{AU}_2\mathbf{B}'$, hence

$$E(K(\mathbf{AUB}')) = E(K(\mathbf{AU}_2\mathbf{B}')). \tag{3.23}$$

Now Theorem 2.16 shows that r_2 has the p.d.f.

$$q_2(r_2) = \frac{1}{2^{\frac{qm}{2}-1}\Gamma\left(\frac{qm}{2}\right)} r_2^{qm-1} \exp\left(-\frac{r_2^2}{2}\right), \quad r_2 \geq 0.$$

Therefore,

$$E(r_2^k) = \int_0^\infty \frac{1}{2^{\frac{qm}{2}-1}\Gamma\left(\frac{qm}{2}\right)} r_2^{qm+k-1} \exp\left(-\frac{r_2^2}{2}\right) dr_2.$$

Let $z = r_2^2$. Then $\frac{dr_2}{dz} = \frac{1}{2\sqrt{z}}$ and we have

$$\begin{aligned}
E(r_2^k) &= \frac{1}{2^{\frac{qm}{2}-1}\Gamma\left(\frac{qm}{2}\right)} \int_0^\infty z^{\frac{qm+k-1}{2}} \exp\left(-\frac{z}{2}\right) \frac{1}{2\sqrt{z}} dz \\
&= \frac{1}{2^{\frac{qm}{2}}\Gamma\left(\frac{qm}{2}\right)} \int_0^\infty z^{\frac{qm+k}{2}-1} \exp\left(-\frac{z}{2}\right) dz \\
&= \frac{2^{\frac{k}{2}}\Gamma\left(\frac{qm+k}{2}\right)}{\Gamma\left(\frac{qm}{2}\right)}.
\end{aligned} \tag{3.24}$$

Now, from (3.22) and (3.24) we get

$$E\left(K\left(\boldsymbol{\Sigma}^{\frac{1}{2}}\mathbf{U}_2\boldsymbol{\Phi}^{\frac{1}{2}}\right)\right) = \frac{E(K(\mathbf{Y}))}{E(r_2^k)}$$

$$= \frac{\Gamma\left(\frac{qm}{2}\right)}{2^{\frac{k}{2}}\Gamma\left(\frac{qm+k}{2}\right)}E(K(\mathbf{Y})). \tag{3.25}$$

Using (3.21), (3.23) and (3.25) we get

$$E(K(\mathbf{X})) = E(K(\mathbf{Y}))\frac{E(r^k)\Gamma\left(\frac{qm}{2}\right)}{2^{\frac{k}{2}}\Gamma\left(\frac{qm+k}{2}\right)}. \tag{3.26}$$

(b) Using Theorem 2.16, the p.d.f. of r is

$$q_1(r) = \frac{2\pi^{\frac{pn}{2}}}{\Gamma\left(\frac{pn}{2}\right)}r^{pn-1}h(r^2), \quad r \geq 0.$$

Therefore,

$$E(r^k) = \int_0^\infty \frac{2\pi^{\frac{pn}{2}}}{\Gamma\left(\frac{pn}{2}\right)}z^{pn+k-1}h(z^2)dz. \tag{3.27}$$

From (3.26) and (3.27), we get

$$E(K(\mathbf{X})) = E(K(\mathbf{Y}))\frac{\pi^{\frac{pn}{2}}\int_0^\infty z^{pn+k-1}h(z^2)dz}{2^{\frac{k}{2}-1}\Gamma\left(\frac{pn+k}{2}\right)},$$

(c) This is a special case of (a).

(d) It follows from part (a), that $E(K(\mathbf{X})) = c_{q,m}(\psi,k)E(K(\mathbf{Y}))$ where $c_{q,m}(\psi,k)$ is a constant depending on q, m, k and ψ only. So, in order to determine $c_{q,m}(\psi,k)$, we can choose $\mathbf{X} \sim E_{q,m}(\mathbf{0}, \mathbf{I}_q \otimes \mathbf{I}_m, \psi)$, $\mathbf{Y} \sim N_{q,m}(\mathbf{0}, \mathbf{I}_q \otimes \mathbf{I}_m)$ and $K(\mathbf{Z}) = z_{11}^k$ where z_{11} is the $(1,1)$th element of the $q \times m$ matrix $\mathbf{Z}$. Then $K(a\mathbf{Z}) = a^k K(\mathbf{Z})$, $a > 0$, is obviously satisfied. Now, $x_{11}^2 \leq tr(\mathbf{X}'\mathbf{X})$ and hence $|x_{11}|^k \leq (tr(\mathbf{X}'\mathbf{X}))^{\frac{k}{2}}$. Here $r \approx (tr(\mathbf{X}'\mathbf{X}))^{\frac{1}{2}}$ and since r^k is integrable, $(tr(\mathbf{X}'\mathbf{X}))^{\frac{1}{2}}$ is also integrable over $[0,\infty)$. Therefore, $E(x_{11}^k)$ exists. Similarly, $E(y_{11}^k)$ also exists. Hence, we can write

$$E(x_{11}^k) = c_{q,m}(\psi,k)E(y_{11}^k). \tag{3.28}$$

However, $x_{11} \sim E_1(0,1,\psi)$ and $y_{11} \sim N_1(0,1)$. Then, from Theorem 3.11, it follows that $E(x_{11}^k) = \frac{k!}{i^k\left(\frac{k}{2}\right)!}\psi^{\left(\frac{k}{2}\right)}(0)$ and from Corollary 3.4 we get that $E(y_{11}^k) = \frac{k!}{i^k\left(\frac{k}{2}\right)!}\left(-\frac{1}{2}\right)^{\left(\frac{k}{2}\right)}$. Hence $c_{q,m}(\psi,k) = (-2)^{\left(\frac{k}{2}\right)}\psi^{\left(\frac{k}{2}\right)}(0)$.

(e) Take $a = -1$. Then, we have $K(-\mathbf{Z}) = (-1)^k K(\mathbf{Z})$ and since k is odd, we get $K(-\mathbf{Z}) = -K(\mathbf{Z})$. However, $\mathbf{X} \approx -\mathbf{X}$ and so $K(\mathbf{X}) \approx K(-\mathbf{X})$. Therefore, $E(K(\mathbf{X})) = E(K(-\mathbf{X})) = -E(K(\mathbf{X}))$ and hence $E(K(\mathbf{X})) = O$. ■

In the next theorem we examine expected values of functions which are defined on the whole $p \times n$ dimensional real space. In contrast to Theorem 3.12, here we do not require that the underlying distribution assign probability zero to the zero matrix.

Theorem 3.13. *Let* $\mathbf{X} \sim E_{p,n}(\mathbf{0}, \Sigma \otimes \Phi, \psi)$*. Let* $q = rk(\Sigma)$, $m = rk(\Phi)$ *and* $r\mathbf{AUB}'$ *be the stochastic representation of* $\mathbf{X}$*. Assume* $\mathbf{Y} \sim N_{p,n}(\mathbf{0}, \Sigma \otimes \Phi)$*. Let* $K(\mathbf{Z})$ *be a function defined on* $\mathbb{R}^{p \times n}$ *such that if* $\mathbf{Z} \in \mathbb{R}^{p \times n}$ *and* $a \geq 0$ *then* $K(a\mathbf{Z}) = a^k K(\mathbf{Z})$ *where* $k > -qm$*. Assume* $E(K(\mathbf{X}))$ *and* $E(K(\mathbf{Y}))$ *exist. Then,*

(i) $E(K(\mathbf{X})) = E(K(\mathbf{Y})) \frac{E(r^k)\Gamma\left(\frac{qm}{2}\right)}{2^{\frac{k}{2}}\Gamma\left(\frac{qm+k}{2}\right)}$,
(ii) If $k = 0$ *then* $E(K(\mathbf{X})) = E(K(\mathbf{Y}))$,
(iii) If k *is a positive even integer, then*

$$E(K(\mathbf{X})) = E(K(\mathbf{Y}))(-2)^{\left(\frac{k}{2}\right)} \psi^{\left(\frac{k}{2}\right)}(0),$$

(iv) If k *is a positive odd integer and* $K(a\mathbf{Z}) = a^k K(\mathbf{Z})$ *holds for all* $a \neq 0$*, then*

$$E(K(\mathbf{X})) = \mathbf{0}.$$

PROOF: Let $r_2\mathbf{AU}_2\mathbf{B}'$ be the stochastic representation of $\mathbf{Y}$. Then, we have

$$E(K(r\mathbf{AUB}')) = E(r^k)E(K(\mathbf{AUB}')) \tag{3.29}$$

and

$$E(K(r_2\mathbf{AU}_2\mathbf{B}')) = E(r_2^k)E(K(\mathbf{AU}_2\mathbf{B}')). \tag{3.30}$$

However (3.29) and (3.30) are the same as (3.21) and (3.22). Therefore, the proof can be completed in exactly the same way as the proof of Theorem 3.12. ∎

Since moments of normal random variables are well known, using Theorem 3.14, we can obtain the moments of m.e.c. distributions.

Theorem 3.14. *Let* $\mathbf{X} \sim E_{p,n}(\mathbf{0}, \mathbf{I}_p \otimes \mathbf{I}_n, \psi)$*. Let* $r\mathbf{U}$ *be the stochastic representation of* $\mathbf{X}$*. Then, provided the moments exist, we get*

$$E\left(\prod_{i=1}^{p}\prod_{j=1}^{n} x_{ij}^{2s_{ij}}\right) = E(r^{2s}) \frac{\Gamma\left(\frac{pn}{2}\right)}{\pi^{\frac{pn}{2}}\Gamma\left(\frac{pn}{2}+s\right)} \prod_{i=1}^{p}\prod_{j=1}^{n} \Gamma\left(\frac{1}{2}+s_{ij}\right), \tag{3.31}$$

where s_{ij} *are nonnegative integers and* $s = \sum_{i=1}^{p}\sum_{j=1}^{n} s_{ij}$*. We also have*

$$E\left(\prod_{i=1}^{p}\prod_{j=1}^{n} x_{ij}^{2s_{ij}}\right) = \psi^{(s)}(0) \frac{(-1)^s 2^{2s}}{\pi^{\frac{pn}{2}}} \prod_{i=1}^{p}\prod_{j=1}^{n} \Gamma\left(\frac{1}{2}+s_{ij}\right). \tag{3.32}$$

PROOF: We use Theorem 3.13. Let $\mathbf{Z} \in I\!R^{p\times n}$ and $K(\mathbf{Z}) = \prod_{i=1}^{p}\prod_{j=1}^{n} z_{ij}^{2s_{ij}}$. If $a \geq 0$, then

$$K(a\mathbf{Z}) = a^{2s}K(\mathbf{Z}). \tag{3.33}$$

If $\mathbf{Y} \sim N_{p,n}(\mathbf{0}, \mathbf{I}_p \otimes \mathbf{I}_n)$, then the elements of $\mathbf{Y}$ are independently and identically distributed standard normal variables, hence,

$$E\left(\prod_{i=1}^{p}\prod_{j=1}^{n} y_{ij}^{2s_{ij}}\right) = \prod_{i=1}^{p}\prod_{j=1}^{n} E\left(y_{ij}^{2s_{ij}}\right), \tag{3.34}$$

and

$$\begin{aligned} E\left(y_{ij}^{2s_{ij}}\right) &= \frac{(2s_{ij})!}{s_{ij}!2^{s_{ij}}} \\ &= 2^{s_{ij}}\frac{\Gamma\left(\frac{1}{2}\right)}{\pi^{\frac{1}{2}}}\frac{1}{2}\frac{3}{2}\cdots\frac{2s_{ij}-1}{2} \\ &= \frac{2^{s_{ij}}\Gamma\left(\frac{1}{2}+s_{ij}\right)}{\pi^{\frac{1}{2}}}. \end{aligned} \tag{3.35}$$

Now, from (3.33), (3.34), (3.35) and part (a) of Theorem 3.13 we obtain (3.31). On the other hand, (3.33)–(3.35), and part (c) of Theorem 3.13 yield (3.31). ■

The formula (3.31) is given in Fang, Kotz, and Ng (1990).

Theorem 3.15. *Let* $\mathbf{X} \sim E_{p,n}(\mathbf{0}, \Sigma \otimes \Phi, \psi)$. *Then, provided the left-hand sides exist,*

(i) $E(x_{i_1j_1}) = 0.$
(ii) $E(x_{i_1j_1}x_{i_2j_2}) = -2\psi'(0)\sigma_{i_1i_2}\phi_{j_1j_2}.$
(iii) $E(x_{i_1j_1}x_{i_2j_2}x_{i_3j_3}) = 0.$
(iv)

$$\begin{aligned} E(x_{i_1j_1}x_{i_2j_2}x_{i_3j_3}x_{i_4j_4}) &= 4\psi''(0)(\sigma_{i_1i_2}\phi_{j_1j_2}\sigma_{i_3i_4}\phi_{j_3j_4} + \sigma_{i_1i_3}\phi_{j_1j_3}\sigma_{i_2i_4}\phi_{j_2j_4} \\ &\quad + \sigma_{i_1i_4}\phi_{j_1j_4}\sigma_{i_2i_3}\phi_{j_2j_3}). \end{aligned}$$

PROOF: Step 1. Let $\mathbf{Y} \sim N_{p,n}(\mathbf{0}, \mathbf{I}_p \otimes \mathbf{I}_n)$. Then, the elements of $\mathbf{Y}$ are independent, standard normal variables, so

$$\begin{aligned} E(y_{i_1j_1}) &= 0, \\ E(y_{i_1j_1}y_{i_2j_2}) &= \delta_{i_1i_2}\delta_{j_1j_2}, \\ E(y_{i_1j_1}y_{i_2j_2}y_{i_3j_3}) &= 0, \\ E(y_{i_1j_1}y_{i_2j_2}y_{i_3j_3}y_{i_4j_4}) &= \delta_{i_1i_2}\delta_{j_1j_2}\delta_{i_3i_4}\delta_{j_3j_4} + \delta_{i_1i_3}\delta_{j_1j_3}\delta_{i_2i_4}\delta_{j_2j_4} \\ &\quad + \delta_{i_1i_4}\delta_{j_1j_4}\delta_{i_2i_3}\delta_{j_2j_3}. \end{aligned}$$

Step 2. Let $\mathbf{X} \sim N_{p,n}(\mathbf{0}, \Sigma \otimes \Phi)$. Let $q = rk(\Sigma)$, $m = rk(\Phi)$ and $r\mathbf{AUB}'$ be the stochastic representation of $\mathbf{X}$. Then, we can write $\mathbf{X} = \mathbf{AYB}'$, where $\mathbf{Y} \sim N_{p,n}(\mathbf{0}, \mathbf{I}_q \otimes \mathbf{I}_n)$, $x_{ij} = \sum_{l=1}^{n} \sum_{k=1}^{p} a_{ik} y_{kl} b_{lj}$, $\Sigma = \mathbf{AA}'$, and $\Phi = \mathbf{BB}'$.

Using the result of Step 1, we get

$$\begin{aligned} E(x_{i_1 j_1}) &= E\left(\sum_{l=1}^{n} \sum_{k=1}^{p} a_{i_1 k} y_{kl} b_{l j_1} \right) \\ &= \sum_{l,k} a_{i_1 k} E(y_{kl}) b_{l j_1} \\ &= 0 . \end{aligned}$$

$$\begin{aligned} E(x_{i_1 j_1} x_{i_2 j_2}) &= E\left(\left(\sum_{l=1}^{n} \sum_{k=1}^{p} a_{i_1 k} y_{kl} b_{l j_1} \right) \left(\sum_{t=1}^{n} \sum_{s=1}^{p} a_{i_2 s} y_{st} b_{t j_2} \right) \right) \\ &= \sum_{l,k,t,s} a_{i_1 k} b_{l j_1} a_{i_2 s} b_{t j_2} E(y_{kl} y_{st}) \\ &= \sum_{l,k,t,s} a_{i_1 k} b_{l j_1} a_{i_2 s} b_{t j_2} \delta_{ks} \delta_{lt} \\ &= \sum_{l,k} a_{i_1 k} a_{i_2 k} b_{l j_1} b_{l j_2} \\ &= \sigma_{i_1 i_2} \phi_{j_1 j_2} . \end{aligned}$$

$$\begin{aligned} E(x_{i_1 j_1} x_{i_2 j_2} x_{i_3 j_3}) &= E\left(\left(\sum_{l=1}^{n} \sum_{k=1}^{p} a_{i_1 k} y_{kl} b_{l j_1} \right) \left(\sum_{t=1}^{n} \sum_{s=1}^{p} a_{i_2 s} y_{st} b_{t j_2} \right) \right. \\ &\quad \times \left. \left(\sum_{q=1}^{n} \sum_{r=1}^{p} a_{i_3 r} y_{rq} b_{q j_3} \right) \right) \\ &= \sum_{l,k,t,s,q,r} a_{i_1 k} b_{l j_1} a_{i_2 s} b_{t j_2} a_{i_3 r} b_{q j_3} E(y_{kl} y_{st} y_{rq}) \\ &= 0 . \end{aligned}$$

$$\begin{aligned} E(x_{i_1 j_1} x_{i_2 j_2} x_{i_3 j_3} x_{i_4 j_4}) &= E\left(\left(\sum_{l=1}^{n} \sum_{k=1}^{p} a_{i_1 k} y_{kl} b_{l j_1} \right) \left(\sum_{t=1}^{n} \sum_{s=1}^{p} a_{i_2 s} y_{st} b_{t j_2} \right) \right. \\ &\quad \times \left. \left(\sum_{q=1}^{n} \sum_{r=1}^{p} a_{i_3 r} y_{rq} b_{q j_3} \right) \left(\sum_{w=1}^{n} \sum_{u=1}^{p} a_{i_4 u} y_{uw} b_{w j_4} \right) \right) \\ &= \sum_{l,k,t,s,q,r,w,u} a_{i_1 k} b_{l j_1} a_{i_2 s} b_{t j_2} \end{aligned}$$

$$
\begin{aligned}
&\times a_{i_3 r} b_{q j_3} a_{i_4 u} b_{w j_4} E(y_{kl} y_{st} y_{rq} y_{uw}) \\
&= \sum_{l,k,t,s,q,r,w,u} a_{i_1 k} b_{l j_1} a_{i_2 s} b_{t j_2} \\
&\times a_{i_3 r} b_{q j_3} a_{i_4 u} b_{w j_4} (\delta_{ks} \delta_{lt} \delta_{ru} \delta_{qw} \\
&+ \delta_{kr} \delta_{lq} \delta_{su} \delta_{tw} + \delta_{ku} \delta_{lw} \delta_{sr} \delta_{tq}) \\
&= \sum_{k,l,r,q} a_{i_1 k} a_{i_2 k} b_{l j_1} b_{l j_2} a_{i_3 r} a_{i_4 r} b_{q j_3} b_{q j_4} \\
&+ \sum_{k,l,s,t} a_{i_1 k} a_{i_3 k} b_{l j_1} b_{l j_3} a_{i_2 s} a_{i_4 s} b_{t j_2} b_{t j_4} \\
&+ \sum_{k,l,s,t} a_{i_1 k} a_{i_4 k} b_{l j_1} b_{l j_4} a_{i_2 s} a_{i_3 s} b_{t j_2} b_{t j_3} \\
&= \sigma_{i_1 i_2} \phi_{j_1 j_2} \sigma_{i_3 i_4} \phi_{j_3 j_4} + \sigma_{i_1 i_3} \phi_{j_1 j_3} \sigma_{i_2 i_4} \phi_{j_2 j_4} \\
&+ \sigma_{i_1 i_4} \phi_{j_1 j_4} \sigma_{i_2 i_3} \phi_{j_2 j_3} .
\end{aligned}
$$

Step 3. Let $\mathbf{X} \sim E_{p,n}(\mathbf{0}, \Sigma \otimes \Phi, \psi)$. Define, on the set of $p \times n$ dimensional matrices, the functions

$$
\begin{aligned}
K_{i_1,j_1}(\mathbf{Z}) &= z_{i_1 j_1} \\
K_{i_1,j_1,i_2,j_2}(\mathbf{Z}) &= z_{i_1 j_1} z_{i_2 j_2} \\
K_{i_1,j_1,i_2,j_2,i_3,j_3}(\mathbf{Z}) &= z_{i_1 j_1} z_{i_2 j_2} z_{i_3 j_3} \\
K_{i_1,j_1,i_2,j_2,i_3,j_3,i_4,j_4}(\mathbf{Z}) &= z_{i_1 j_1} z_{i_2 j_2} z_{i_3 j_3} z_{i_4 j_4} .
\end{aligned}
$$

Now, the results follow from Theorem 3.13. ∎

Theorem 3.16. *Let $\mathbf{X} \sim E_{p,n}(\mathbf{M}, \Sigma \otimes \Phi, \psi)$. Then, provided the left-hand sides exist,*

(i) $E(x_{i_1 j_1}) = m_{i_1 j_1}$.
(ii) $E(x_{i_1 j_1} x_{i_2 j_2}) = -2\psi'(0) \sigma_{i_1 i_2} \phi_{j_1 j_2} + m_{i_1 j_1} m_{i_2 j_2}$.
(iii)

$$
\begin{aligned}
E(x_{i_1 j_1} x_{i_2 j_2} x_{i_3 j_3}) &= -2\psi'(0)(\sigma_{i_1 i_2} \phi_{j_1 j_2} m_{i_3 j_3} + \sigma_{i_1 i_3} \phi_{j_1 j_3} m_{i_2 j_2} \\
&+ \sigma_{i_2 i_3} \phi_{j_2 j_3} m_{i_1 j_1}) + m_{i_1 j_1} m_{i_2 j_2} m_{i_3 j_3} .
\end{aligned}
$$

(iv)

$$
\begin{aligned}
E(x_{i_1 j_1} x_{i_2 j_2} x_{i_3 j_3} x_{i_4 j_4}) &= 4\psi''(0)(\sigma_{i_1 i_2} \phi_{j_1 j_2} \sigma_{i_3 i_4} \phi_{j_3 j_4} + \sigma_{i_1 i_3} \phi_{j_1 j_3} \sigma_{i_2 i_4} \phi_{j_2 j_4} \\
&+ \sigma_{i_1 i_4} \phi_{j_1 j_4} \sigma_{i_2 i_3} \phi_{j_2 j_3}) - 2\psi'(0)(m_{i_1 j_1} m_{i_2 j_2} \sigma_{i_3 i_4} \phi_{j_3 j_4} \\
&+ m_{i_1 j_1} m_{i_3 j_3} \sigma_{i_2 i_4} \phi_{j_2 j_4} + m_{i_1 j_1} m_{i_4 j_4} \sigma_{i_2 i_3} \phi_{j_2 j_3}
\end{aligned}
$$

$$+ m_{i_2j_2}m_{i_3j_3}\sigma_{i_1i_4}\phi_{j_1j_4} + m_{i_2j_2}m_{i_4j_4}\sigma_{i_1i_3}\phi_{j_1j_3}$$
$$+ m_{i_3j_3}m_{i_4j_4}\sigma_{i_1i_2}\phi_{j_1j_2}) + m_{i_1j_1}m_{i_2j_2}m_{i_3j_3}m_{i_4j_4}\,.$$

PROOF: Let $\mathbf{Y} = \mathbf{X} - \mathbf{M}$. Then, $\mathbf{Y} \sim E_{p,n}(\mathbf{0}, \Sigma \otimes \Phi, \psi)$ and using Theorem 3.15, we obtain the first four moments of $\mathbf{Y}$. Therefore,

(i) $E(x_{i_1j_1}) = E(y_{i_1j_1} + m_{i_1j_1}) = m_{i_1j_1}$.
(ii)

$$\begin{aligned} E(x_{i_1j_1}x_{i_2j_2}) &= E((y_{i_1j_1} + m_{i_1j_1})(y_{i_2j_2} + m_{i_2j_2})) \\ &= E(y_{i_1j_1}y_{i_2j_2}) + E(y_{i_1j_1})m_{i_2j_2} + E(y_{i_2j_2})m_{i_1j_1} \\ &+ m_{i_1j_1}m_{i_2j_2} \\ &= -2\psi'(0)\sigma_{i_1i_2}\phi_{j_1j_2} + m_{i_1j_1}m_{i_2j_2}\,. \end{aligned}$$

(iii)

$$\begin{aligned} E(x_{i_1j_1}x_{i_2j_2}x_{i_3j_3}) &= E((y_{i_1j_1} + m_{i_1j_1})(y_{i_2j_2} + m_{i_2j_2})(y_{i_3j_3} + m_{i_3j_3})) \\ &= E(y_{i_1j_1}y_{i_2j_2}y_{i_3j_3}) + E(y_{i_1j_1}y_{i_2j_2})m_{i_3j_3} \\ &+ E(y_{i_1j_1}y_{i_3j_3})m_{i_2j_2} + E(y_{i_2j_2}y_{i_3j_3})m_{i_1j_1} \\ &+ m_{i_1j_1}m_{i_2j_2}m_{i_3j_3} \\ &= -2\psi'(0)(\sigma_{i_1i_2}\phi_{j_1j_2}m_{i_3j_3} + \sigma_{i_1i_3}\phi_{j_1j_3}m_{i_2j_2} \\ &+ \sigma_{i_2i_3}\phi_{j_2j_3}m_{i_1j_1}) + m_{i_1j_1}m_{i_2j_2}m_{i_3j_3}\,. \end{aligned}$$

(iv)

$$\begin{aligned} E(x_{i_1j_1}x_{i_2j_2}x_{i_3j_3}x_{i_4j_4}) &= E((y_{i_1j_1} + m_{i_1j_1})(y_{i_2j_2} + m_{i_2j_2}) \\ &\times (y_{i_3j_3} + m_{i_3j_3})(y_{i_4j_4} + m_{i_4j_4})) \\ &= E(y_{i_1j_1}y_{i_2j_2}y_{i_3j_3}y_{i_4j_4}) + E(y_{i_1j_1}y_{i_2j_2}y_{i_3j_3})m_{i_4j_4} \\ &+ E(y_{i_1j_1}y_{i_2j_2}y_{i_4j_4})m_{i_3j_3} + E(y_{i_1j_1}y_{i_3j_3}y_{i_4j_4})m_{i_2j_2} \\ &+ E(y_{i_2j_2}y_{i_3j_3}y_{i_4j_4})m_{i_1j_1} \\ &+ E(y_{i_1j_1}y_{i_2j_2})m_{i_3j_3}m_{i_4j_4} \\ &+ E(y_{i_1j_1}y_{i_3j_3})m_{i_2j_2}m_{i_4j_4} \\ &+ E(y_{i_1j_1}y_{i_4j_4})m_{i_2j_2}m_{i_3j_3} \\ &+ E(y_{i_2j_2}y_{i_3j_3})m_{i_1j_1}m_{i_4j_4} \\ &+ E(y_{i_2j_2}y_{i_4j_4})m_{i_1j_1}m_{i_3j_3} \end{aligned}$$

$$+ E(y_{i_3j_3}y_{i_4j_4})m_{i_1j_1}m_{i_2j_2} + m_{i_1j_1}m_{i_2j_2}m_{i_3j_3}m_{i_4j_4}$$
$$= 4\psi''(0)(\sigma_{i_1i_2}\phi_{j_1j_2}\sigma_{i_3i_4}\phi_{j_3j_4} + \sigma_{i_1i_3}\phi_{j_1j_3}\sigma_{i_2i_4}\phi_{j_2j_4}$$
$$+ \sigma_{i_1i_4}\phi_{j_1j_4}\sigma_{i_2i_3}\phi_{j_2j_3}) - 2\psi'(0)(m_{i_1j_1}m_{i_2j_2}\sigma_{i_3i_4}\phi_{j_3j_4}$$
$$+ m_{i_1j_1}m_{i_3j_3}\sigma_{i_2i_4}\phi_{j_2j_4} + m_{i_1j_1}m_{i_4j_4}\sigma_{i_2i_3}\phi_{j_2j_3}$$
$$+ m_{i_2j_2}m_{i_3j_3}\sigma_{i_1i_4}\phi_{j_1j_4} + m_{i_2j_2}m_{i_4j_4}\sigma_{i_1i_3}\phi_{j_1j_3}$$
$$+ m_{i_3j_3}m_{i_4j_4}\sigma_{i_1i_2}\phi_{j_1j_2}) + m_{i_1j_1}m_{i_2j_2}m_{i_3j_3}m_{i_4j_4}\,. \quad \blacksquare$$

Remark 3.3. The derivation of (i) and (ii) of Theorem 3.16 provides another proof of Theorem 2.11.

Theorem 3.17. *Let* $\mathbf{X} \sim E_{p,n}(\mathbf{M}, \Sigma \otimes \Phi, \psi)$ *with finite second order moments. Let* $c_0 = -2\psi'(0)$. *Then, for any constant matrix* $\mathbf{A}$, *if the expressions on the left-hand sides are defined, we have*

(i) $E(\mathbf{XAX}) = c_0\Sigma\mathbf{A}'\Phi + \mathbf{MAM}$,
(ii) $E(\mathbf{XAX}') = c_0\Sigma tr(\mathbf{A}'\Phi) + \mathbf{MAM}'$,
(iii) $E(\mathbf{X}'\mathbf{AX}) = c_0\Phi tr(\Sigma\mathbf{A}') + \mathbf{M}'\mathbf{AM}$,
(iv) $E(\mathbf{X}'\mathbf{AX}') = c_0\Phi\mathbf{A}'\Sigma + \mathbf{M}'\mathbf{AM}'$.

PROOF:

(i)

$$E(\mathbf{XAX})_{ij} = E\left(\sum_{l=1}^{p}\sum_{k=1}^{n} x_{ik}a_{kl}x_{lj}\right)$$
$$= \sum_{l,k}(c_0\sigma_{il}\phi_{kj} + m_{ik}m_{jl})a_{kl}$$
$$= c_0\sum_{l,k}\sigma_{il}a_{kl}\phi_{kj} + \sum_{l,k}m_{ik}a_{kl}m_{lj}$$
$$= (c_0\Sigma\mathbf{A}'\Phi + \mathbf{MAM})_{ij}\,.$$

(ii)

$$E(\mathbf{XAX}')_{ij} = E\left(\sum_{l=1}^{p}\sum_{k=1}^{n} x_{ik}a_{kl}x_{jl}\right)$$
$$= \sum_{l,k}(c_0\sigma_{ij}\phi_{kl} + m_{ik}m_{jl})a_{kl}$$
$$= c_0\sigma_{ij}\sum_{l,k}\phi_{kl}a_{kl} + \sum_{l,k}m_{ik}a_{kl}m_{jl}$$
$$= (c_0\Sigma tr(\mathbf{A}'\Phi) + \mathbf{MAM}')_{ij}\,.$$

(iii) From Theorem 2.3, it follows that $\mathbf{X}' \sim E_{n,p}(\mathbf{M}', \Phi \otimes \Sigma, \psi)$. Using (ii), we get $E(\mathbf{X}'\mathbf{A}\mathbf{X}) = c_0 \Phi tr(\mathbf{A}'\Sigma) + \mathbf{M}'\mathbf{A}\mathbf{M}$.

(iv) Since, $\mathbf{X}' \sim E_{n,p}(\mathbf{M}', \Phi \otimes \Sigma, \psi)$, from (i) we have $E(\mathbf{X}'\mathbf{A}\mathbf{X}') = c_0 \Phi \mathbf{A}'\Sigma + \mathbf{M}'\mathbf{A}\mathbf{M}'$. ∎

Theorem 3.18. *Let* $\mathbf{X} \sim E_{p,n}(\mathbf{M}, \Sigma \otimes \Phi, \psi)$ *with finite second order moment. Let* $c0 = -2\psi'(0)$. *Then for any constant matrices* $\mathbf{A}$, $\mathbf{B}$, *if the express ions on the left-hand sides are defined, we have*

(i) $E(\mathbf{X}tr(\mathbf{A}\mathbf{X})) = c_0 \Sigma \mathbf{A}'\Phi + \mathbf{M}tr(\mathbf{A}\mathbf{M})$,
(ii) $E(\mathbf{X}tr(\mathbf{A}\mathbf{X}')) = c_0 \Sigma \mathbf{A}\Phi + \mathbf{M}tr(\mathbf{A}\mathbf{M}')$,
(iii) $E(\mathbf{X}'tr(\mathbf{A}\mathbf{X})) = c_0 \Phi \mathbf{A}\Sigma + \mathbf{M}'tr(\mathbf{A}\mathbf{M})$,
(iv) $E(\mathbf{X}'tr(\mathbf{A}\mathbf{X}')) = c_0 \Phi \mathbf{A}'\Sigma + \mathbf{M}'tr(\mathbf{A}\mathbf{M}')$,
(v) $E(tr(\mathbf{X}\mathbf{A}\mathbf{X}\mathbf{B})) = c_0 tr(\Sigma \mathbf{A}'\Phi \mathbf{B}) + tr(\mathbf{M}\mathbf{A}\mathbf{M}\mathbf{B})$,
(vi) $E(tr(\mathbf{X}\mathbf{A}\mathbf{X}'\mathbf{B})) = c_0 tr(\Sigma \mathbf{B}) tr(\Phi \mathbf{A}') + tr(\mathbf{M}\mathbf{A}\mathbf{M}'\mathbf{B})$,
(vii) $E(tr(\mathbf{X}\mathbf{A})tr(\mathbf{X}\mathbf{B})) = c_0 tr(\Sigma \mathbf{B}'\Phi \mathbf{A}) + tr(\mathbf{M}\mathbf{A})tr(\mathbf{M}\mathbf{B})$,
(viii) $E(tr(\mathbf{X}\mathbf{A})tr(\mathbf{X}'\mathbf{B})) = c_0 tr(\Sigma \mathbf{B}\Phi \mathbf{A}) + tr(\mathbf{M}\mathbf{A})tr(\mathbf{M}'\mathbf{B})$.

PROOF:

(i)

$$\begin{aligned} E(\mathbf{X}tr(\mathbf{A}\mathbf{X}))_{ij} &= E\left(x_{ij} \sum_{k=1}^{n} \sum_{l=1}^{p} a_{kl} x_{kl}\right) \\ &= \sum_{l,k} (c_0 \sigma_{il} \phi_{jk} + m_{ij} m_{lk}) a_{kl} \\ &= c_0 \sum_{l,k} \sigma_{il} a_{kl} \phi_{kj} + m_{ij} \sum_{l,k} a_{kl} m_{lk} \\ &= (c_0 \Sigma \mathbf{A}'\Phi + \mathbf{M}tr(\mathbf{A}\mathbf{M}))_{ij}. \end{aligned}$$

(ii)

$$\begin{aligned} E(\mathbf{X}tr(\mathbf{A}\mathbf{X}')) &= E(\mathbf{X}tr(\mathbf{X}\mathbf{A}')) \\ &= E(\mathbf{X}tr(\mathbf{A}'\mathbf{X})) \\ &= c_0 \Sigma \mathbf{A}\Phi + \mathbf{M}tr(\mathbf{A}'\mathbf{M}) \\ &= c_0 \Sigma \mathbf{A}\Phi + \mathbf{M}tr(\mathbf{A}\mathbf{M}'). \end{aligned}$$

(iii) $\mathbf{X}' \sim E_{n,p}(\mathbf{M}', \Phi \otimes \Sigma, \psi)$.

$$\begin{aligned} E(\mathbf{X}'tr(\mathbf{A}\mathbf{X})) &= E(\mathbf{X}'tr(\mathbf{A}'\mathbf{X}')) \\ &= c_0 \Phi \mathbf{A}\Sigma + \mathbf{M}'tr(\mathbf{A}'\mathbf{M}') \\ &= c_0 \Phi \mathbf{A}\Sigma + \mathbf{M}'tr(\mathbf{A}\mathbf{M}). \end{aligned}$$

(iv) $\mathbf{X}' \sim E_{n,p}(\mathbf{M}', \boldsymbol{\Phi} \otimes \boldsymbol{\Sigma}, \psi)$.

$$E(\mathbf{X}'tr(\mathbf{A}\mathbf{X}')) = c_0\boldsymbol{\Phi}\mathbf{A}'\boldsymbol{\Sigma} + \mathbf{M}'tr(\mathbf{A}\mathbf{M}').$$

(v)

$$\begin{aligned}
E(tr(\mathbf{XAXB})) &= E\left(\sum_{i=1}^{p}\sum_{j=1}^{n}\sum_{k=1}^{p}\sum_{l=1}^{n} x_{ij}a_{jk}x_{kl}b_{li}\right) \\
&= \sum_{i,j,k,l} (c_0\sigma_{ik}\phi_{jl} + m_{ij}m_{kl})a_{jk}b_{li} \\
&= c_0\left(\sum_{i,j,k,l} \sigma_{ik}a_{jk}\phi_{jl}b_{li}\right) + \sum_{i,j,k,l} m_{ij}a_{jk}m_{kl}b_{li} \\
&= c_0 tr(\boldsymbol{\Sigma}\mathbf{A}'\boldsymbol{\Phi}\mathbf{B}) + tr(\mathbf{MAMB}).
\end{aligned}$$

(vi)

$$\begin{aligned}
E(tr(\mathbf{XAX}'\mathbf{B})) &= E\left(\sum_{i=1}^{p}\sum_{j=1}^{n}\sum_{k=1}^{n}\sum_{l=1}^{p} x_{ij}a_{jk}x_{lk}b_{li}\right) \\
&= \sum_{i,j,k,l} (c_0\sigma_{il}\phi_{jk} + m_{ij}m_{lk})a_{jk}b_{li} \\
&= c_0\left(\sum_{i,l}\sigma_{il}b_{li}\right)\left(\sum_{i,l}\phi_{jk}a_{jk}\right) \\
&\quad + \sum_{i,j,k,l} m_{ij}a_{jk}m_{lk}b_{li} \\
&= c_0 tr(\boldsymbol{\Sigma}\mathbf{B})tr(\boldsymbol{\Phi}\mathbf{A}) + tr(\mathbf{MAM}'\mathbf{B}) \\
&= c_0 tr(\boldsymbol{\Sigma}\mathbf{B})tr(\boldsymbol{\Phi}\mathbf{A}') + tr(\mathbf{MAM}'\mathbf{B}).
\end{aligned}$$

(vii)

$$\begin{aligned}
E(tr(\mathbf{XA})tr(\mathbf{XB})) &= E\left(\left(\sum_{i=1}^{p}\sum_{l=1}^{n} x_{il}a_{li}\right)\left(\sum_{j=1}^{p}\sum_{k=1}^{n} x_{jk}b_{kj}\right)\right) \\
&= \sum_{i,j,k,l} (c_0\sigma_{ij}\phi_{lk} + m_{il}m_{jk})a_{li}b_{kj} \\
&= c_0\sum_{i,j,k,l} \sigma_{ij}b_{kj}\phi_{lk}a_{li}
\end{aligned}$$

$$+\left(\sum_{i,l} m_{il}a_{li}\right)\left(\sum_{j,k} m_{jk}a_{kj}\right)$$
$$= c_0 tr(\Sigma \mathbf{B}'\Phi \mathbf{A}) + tr(\mathbf{MA})tr(\mathbf{MB}).$$

(viii)

$$\begin{aligned} E(tr(\mathbf{XA})tr(\mathbf{X}'\mathbf{B})) &= E(tr(\mathbf{XA})tr(\mathbf{XB}')) \\ &= c_0 tr(\Sigma \mathbf{B}\Phi \mathbf{A}) + tr(\mathbf{MA})tr(\mathbf{MB}') \\ &= c_0 tr(\Sigma \mathbf{B}\Phi \mathbf{A}) + tr(\mathbf{MA})tr(\mathbf{M}'\mathbf{B}). \end{aligned}$$
■

Theorem 3.19. *Let* $\mathbf{X} \sim E_{p,n}(\mathbf{M}, \Sigma \otimes \Phi, \psi)$ *with finite third order moment. Let* $c_0 = -2\psi'(0)$. *Then, for any constant matrices* $\mathbf{A}$ *and* $\mathbf{B}$, *if the expressions on the left-hand sides are defined, we have*

(i) $E(\mathbf{XAXBX}) = c_0(\mathbf{MA}\Sigma\mathbf{B}'\Phi + \Sigma\mathbf{B}'\mathbf{M}'\mathbf{A}'\Phi + \Sigma\mathbf{A}'\Phi\mathbf{BM}) + \mathbf{MAMBM}$,

(ii)

$$E(\mathbf{X}'\mathbf{AXBX}) = c_0(\mathbf{M}'\mathbf{A}\Sigma\mathbf{B}'\Phi + \Phi tr(\Sigma\mathbf{B}'\mathbf{M}'\mathbf{A}') + \Phi\mathbf{BM}tr(\mathbf{A}\Sigma)) + \mathbf{M}'\mathbf{AMBM},$$

(iii)

$$E(\mathbf{X}'\mathbf{AX}'\mathbf{BX}) = c_0(\mathbf{M}'\mathbf{A}\Phi tr(\Sigma\mathbf{B}') + \Phi tr(\mathbf{AM}'\mathbf{B}\Sigma) + \Phi\mathbf{A}'\Sigma\mathbf{BM}) + \mathbf{M}'\mathbf{AM}'\mathbf{BM},$$

(iv)

$$E(\mathbf{X}'\mathbf{AXBX}') = c_0(\mathbf{M}'\mathbf{A}\Sigma tr(\mathbf{B}\Phi) + \Phi\mathbf{B}'\mathbf{M}'\mathbf{A}'\Sigma + \Phi\mathbf{BM}'tr(\mathbf{A}\Sigma)) + \mathbf{M}'\mathbf{AMBM}',$$

(v)

$$E(\mathbf{XAX}'\mathbf{BX}') = c_0(\mathbf{MA}\Phi\mathbf{B}'\Sigma + \Sigma tr(\mathbf{AM}'\mathbf{B}\Phi) + \Sigma\mathbf{BM}'tr(\mathbf{A}\Phi)) + \mathbf{MAM}'\mathbf{BM}',$$

(vi)

$$E(\mathbf{X}'\mathbf{AX}'\mathbf{BX}') = c_0(\mathbf{M}'\mathbf{A}\Phi\mathbf{B}'\Sigma + \Phi\mathbf{B}'\mathbf{MA}'\Sigma + \Phi\mathbf{A}'\Sigma\mathbf{BM}') + \mathbf{M}'\mathbf{AM}'\mathbf{BM}',$$

(vii)

$$E(\mathbf{XAX'BX}) = c_0(\mathbf{MA}\boldsymbol{\Phi} tr(\mathbf{B}\boldsymbol{\Sigma}) + \boldsymbol{\Sigma}\mathbf{B'MA'}\boldsymbol{\Phi} + \boldsymbol{\Sigma}\mathbf{BM}tr(\mathbf{A}\boldsymbol{\Phi})) + \mathbf{MAM'BM},$$

(viii)

$$E(\mathbf{XAXBX'}) = c_0(\mathbf{MA}\boldsymbol{\Sigma} tr(\boldsymbol{\Phi}\mathbf{B'}) + \boldsymbol{\Sigma} tr(\mathbf{AMB}\boldsymbol{\Phi}) + \boldsymbol{\Sigma}\mathbf{A'}\boldsymbol{\Phi}\mathbf{BM'}) + \mathbf{MAMBM'}.$$

PROOF:

(i)

$$\begin{aligned}
E(\mathbf{XAXBX})_{ij} &= E\left(\sum_{k=1}^{n}\sum_{l=1}^{p}\sum_{r=1}^{n}\sum_{q=1}^{p} x_{ik}a_{kl}x_{lr}b_{rq}x_{qj}\right)\\
&= \sum_{k,l,r,q}[c_0(\sigma_{il}\phi_{kr}m_{qj} + \sigma_{iq}\phi_{kj}m_{lr} + \sigma_{lq}\phi_{rj}m_{ik})\\
&+ m_{ik}m_{lr}m_{qj}]a_{kl}b_{rq}\\
&= c_0\left[\sum_{k,l,r,q}\sigma_{il}a_{kl}\phi_{kr}b_{rq}m_{qj} + \sum_{k,l,r,q}\sigma_{iq}b_{rq}m_{lr}a_{kl}\phi_{kj}\right.\\
&\left. + \sum_{k,l,r,q} m_{ik}a_{kl}\sigma_{lq}b_{rq}\phi_{rj}\right]\\
&+ \sum_{k,l,r,q} m_{ik}a_{kl}m_{lr}b_{rq}m_{qj}\\
&= (c_0(\boldsymbol{\Sigma}\mathbf{A'}\boldsymbol{\Phi}\mathbf{BM} + \boldsymbol{\Sigma}\mathbf{B'M'A'}\boldsymbol{\Phi} + \mathbf{MA}\boldsymbol{\Sigma}\mathbf{B'}\boldsymbol{\Phi})\\
&+ \mathbf{MAMBM})_{ij}.
\end{aligned}$$

(ii)

$$\begin{aligned}
E(\mathbf{X'AXBX})_{ij} &= E\left(\sum_{k=1}^{p}\sum_{l=1}^{p}\sum_{r=1}^{n}\sum_{q=1}^{p} x_{ki}a_{kl}x_{lr}b_{rq}x_{qj}\right)\\
&= \sum_{k,l,r,q}[c_0(\sigma_{kl}\phi_{ir}m_{qj} + \sigma_{kq}\phi_{ij}m_{lr} + \sigma_{lq}\phi_{rj}m_{ki})\\
&+ m_{ki}m_{lr}m_{qj}]a_{kl}b_{rq}
\end{aligned}$$

$$
\begin{aligned}
&= c_0\left[\left(\sum_{k,l} a_{kl}\sigma_{lk}\right)\left(\sum_{r,q}\phi_{ir}b_{rq}m_{qj}\right)\right.\\
&+ \phi_{ij}\sum_{k,l,r,q}\sigma_{kq}b_{rq}m_{lr}a_{kl}\\
&\left.+ \sum_{k,l,r,q} m_{ki}a_{kl}\sigma_{lq}b_{rq}\phi_{rj}\right] + \sum_{k,l,r,q} m_{ki}a_{kl}m_{lr}b_{rq}m_{qj}\\
&= (c_0(tr(\mathbf{A}\Sigma)\Phi\mathbf{B}\mathbf{M} + \Phi tr(\Sigma\mathbf{B}'\mathbf{M}'\mathbf{A}') + \mathbf{M}'\mathbf{A}\Sigma\mathbf{B}'\Phi)\\
&+ \mathbf{M}'\mathbf{A}\mathbf{M}\mathbf{B}\mathbf{M})_{ij}.
\end{aligned}
$$

(iii)

$$
\begin{aligned}
E(\mathbf{X}'\mathbf{A}\mathbf{X}'\mathbf{B}\mathbf{X}) &= E(\mathbf{X}'\mathbf{B}'\mathbf{X}\mathbf{A}'\mathbf{X})'\\
&= (c_0(tr(\mathbf{B}'\Sigma)\Phi\mathbf{A}'\mathbf{M} + \Phi tr(\Sigma\mathbf{A}\mathbf{M}'\mathbf{B}) + \mathbf{M}'\mathbf{B}'\Sigma\mathbf{A}\Phi)\\
&+ \mathbf{M}'\mathbf{B}'\mathbf{M}\mathbf{A}'\mathbf{M})'\\
&= c_0(\mathbf{M}'\mathbf{A}\Phi tr(\Sigma\mathbf{B}') + \Phi tr(\mathbf{A}\mathbf{M}'\mathbf{B}\Sigma) + \Phi\mathbf{A}'\Sigma\mathbf{B}\mathbf{M})\\
&+ \mathbf{M}'\mathbf{A}\mathbf{M}'\mathbf{B}\mathbf{M}.
\end{aligned}
$$

(iv)

$$
\begin{aligned}
E(\mathbf{X}'\mathbf{A}\mathbf{X}\mathbf{B}\mathbf{X}')_{ij} &= E\left(\sum_{k=1}^{p}\sum_{l=1}^{p}\sum_{r=1}^{n}\sum_{q=1}^{n} x_{ki}a_{kl}x_{lr}b_{rq}x_{jq}\right)\\
&= \sum_{k,l,r,q}[c_0(\sigma_{kl}\phi_{ir}m_{jq} + \sigma_{kj}\phi_{iq}m_{lr} + \sigma_{lj}\phi_{rq}m_{ki})\\
&+ m_{ki}m_{lr}m_{jq}]a_{kl}b_{rq}\\
&= c_0\left[\left(\sum_{k,l} a_{kl}\sigma_{lk}\right)\left(\sum_{r,q}\phi_{ir}b_{rq}m_{jq}\right)\right.\\
&+ \sum_{k,l,r,q}\phi_{iq}b_{rq}m_{lr}a_{kl}\sigma_{kj}\\
&\left.+ \left(\sum_{k,l} m_{ki}a_{kl}\sigma_{lj}\right)\left(\sum_{r,q} b_{rq}\phi_{qr}\right)\right]\\
&+ \sum_{k,l,r,q} m_{ki}a_{kl}m_{lr}b_{rq}m_{jq}\\
&= (c_0(\Phi\mathbf{B}\mathbf{M}'tr(\mathbf{A}\Sigma) + \Phi\mathbf{B}'\mathbf{M}'\mathbf{A}'\Sigma + \mathbf{M}'\mathbf{A}\Sigma tr(\mathbf{B}\Phi))\\
&+ \mathbf{M}'\mathbf{A}\mathbf{M}\mathbf{B}\mathbf{M}')_{ij}.
\end{aligned}
$$

(v) $\mathbf{X}' \sim E_{n,p}(\mathbf{M}', \Phi \otimes \Sigma, \psi)$.

$$\begin{aligned} E(\mathbf{XAX'BX'}) &= c_0(\mathbf{MA}\Phi\mathbf{B}'\Sigma + \Sigma tr(\Phi\mathbf{B}'\mathbf{MA}') + \Sigma\mathbf{BM}'tr(\mathbf{A}\Phi)) \\ &+ \mathbf{MAM'BM'} \\ &= c_0(\mathbf{MA}\Phi\mathbf{B}'\Sigma + \Sigma tr(\mathbf{AM'B}\Phi) + \Sigma\mathbf{BM}'tr(\mathbf{A}\Phi)) \\ &+ \mathbf{MAM'BM'}. \end{aligned}$$

(vi) $\mathbf{X}' \sim E_{n,p}(\mathbf{M}', \Phi \otimes \Sigma, \psi)$.

$$E(\mathbf{X'AX'BX'}) = c_0(\mathbf{M'A}\Phi\mathbf{B}'\Sigma + \Phi\mathbf{B'MA}'\Sigma + \Phi\mathbf{A}'\Sigma\mathbf{BM}') + \mathbf{M'AM'BM'}.$$

(vii)

$$\begin{aligned} E(\mathbf{XAX'BX}) &= (E(\mathbf{X'B'XA'X'}))' \\ &= (c_0(\mathbf{M'B}'\Sigma tr(\mathbf{A}'\Phi) + \Phi\mathbf{AM'B}\Sigma + \Phi\mathbf{A'M}'tr(\mathbf{B}'\Sigma)) \\ &+ \mathbf{M'B'MA'M'})' \\ &= c_0(\Sigma\mathbf{BM}tr(\mathbf{A}\Phi) + \Sigma\mathbf{B'MA}'\Phi + \mathbf{MA}\Phi tr(\mathbf{B}\Sigma)) \\ &+ \mathbf{MAM'BM}. \end{aligned}$$

(viii)

$$\begin{aligned} E(\mathbf{XAXBX'}) &= (E(\mathbf{XB'X'A'X'}))' \\ &= (c_0(\mathbf{MB}'\Phi\mathbf{A}\Sigma + \Sigma tr(\mathbf{B'M'A}'\Phi) + \Sigma\mathbf{A'M}'tr(\mathbf{B}'\Phi)) \\ &+ \mathbf{MB'M'A'M'})' \\ &= c_0(\Sigma\mathbf{A}'\Phi\mathbf{BM}' + \Sigma tr(\mathbf{AMB}\Phi) + \mathbf{MA}\Sigma tr(\Phi\mathbf{B}')) \\ &+ \mathbf{MAMBM'}. \end{aligned}$$

■

Theorem 3.20. *Let* $\mathbf{X} \sim E_{p,n}(\mathbf{M}, \Sigma \otimes \Phi, \psi)$ *with finite third order moment. Let* $c_0 = -2\psi'(0)$. *Then, for any constant matrices* $\mathbf{A}$ *and* $\mathbf{B}$, *if the expressions on the left-hand sides are defined, we have*

(i)

$$\begin{aligned} E(\mathbf{X}tr(\mathbf{X'AXB})) &= c_0(\mathbf{M}tr(\mathbf{A}'\Sigma)tr(\mathbf{B}'\Phi) + \Sigma\mathbf{A'MB}'\Phi + \Sigma\mathbf{AMB}\Phi) \\ &+ \mathbf{M}tr(\mathbf{M'AMB}), \end{aligned}$$

(ii)

$$E(\mathbf{XBX}tr(\mathbf{AX})) = c_0(\mathbf{MB\Sigma A'}\boldsymbol{\Phi} + \boldsymbol{\Sigma}\mathbf{A'}\boldsymbol{\Phi}\mathbf{BM} + \boldsymbol{\Sigma}\mathbf{B'}\boldsymbol{\Phi}tr(\mathbf{AM})) + \mathbf{MBM}tr(\mathbf{AM}),$$

(iii)

$$E(\mathbf{X'BX}tr(\mathbf{AX})) = c_0(\mathbf{M'B\Sigma A'}\boldsymbol{\Phi} + \boldsymbol{\Phi}\mathbf{A\Sigma BM} + \boldsymbol{\Phi}tr(\mathbf{AM})tr(\mathbf{B\Sigma})) + \mathbf{M'BM}tr(\mathbf{AM}).$$

PROOF:

(i)

$$\begin{aligned}
E(\mathbf{X}tr(\mathbf{X'AXB}))_{ij} &= E\left(x_{ij}\sum_{k=1}^{n}\sum_{l=1}^{p}\sum_{q=1}^{n}\sum_{r=1}^{p}x_{lk}a_{lr}x_{rq}b_{qk}\right)\\
&= \sum_{k,l,r,q}[c_0(\sigma_{il}\phi_{jk}m_{rq} + \sigma_{ir}\phi_{jq}m_{lk} + \sigma_{lr}\phi_{kq}m_{ij})\\
&+ m_{ij}m_{lk}m_{rq}]a_{lr}b_{qr}\\
&= c_0\left[\sum_{k,l,r,q}\sigma_{il}a_{lr}m_{rq}b_{qk}\phi_{jk}\right.\\
&+ \sum_{k,l,r,q}\sigma_{ir}a_{lr}m_{lk}b_{qk}\phi_{jq}\\
&+ \left. m_{ij}\left(\sum_{k,q}b_{qk}\phi_{qk}\right)\left(\sum_{r,l}a_{lr}\sigma_{lr}\right)\right]\\
&+ m_{ij}\sum_{k,l,r,q}m_{lk}a_{lr}m_{rq}b_{qk}\\
&= (c_0(\boldsymbol{\Sigma}\mathbf{AMB}\boldsymbol{\Phi} + \boldsymbol{\Sigma}\mathbf{A'MB'}\boldsymbol{\Phi} + \mathbf{M}tr(\mathbf{A'}\boldsymbol{\Sigma})tr(\mathbf{B'}\boldsymbol{\Phi}))\\
&+ \mathbf{M}tr(\mathbf{M'AMB}))_{ij}.
\end{aligned}$$

(ii)

$$\begin{aligned}
E(\mathbf{XBX}tr(\mathbf{AX}))_{ij} &= E\left(\left(\sum_{k=1}^{n}\sum_{l=1}^{p}x_{ik}b_{kl}x_{lj}\right)\left(\sum_{r=1}^{n}\sum_{q=1}^{p}a_{rq}x_{qr}\right)\right)\\
&= \sum_{k,l,r,q}[c_0(\sigma_{il}\phi_{kj}m_{qr} + \sigma_{iq}\phi_{kr}m_{lj} + \sigma_{lq}\phi_{jr}m_{ik})\\
&+ m_{ik}m_{lj}m_{qr}]b_{kl}a_{rq}
\end{aligned}$$

$$= c_0\left[\left(\sum_{k,l}\sigma_{il}b_{kl}\phi_{kj}\right)\left(\sum_{r,q}a_{rq}m_{qr}\right)\right.$$

$$+ \sum_{k,l,r,q}\sigma_{iq}a_{rq}\phi_{kr}b_{kl}m_{lj}$$

$$\left. + \sum_{k,l,r,q}m_{ik}b_{kl}\sigma_{lq}a_{rq}\phi_{jr}\right]$$

$$+ \left(\sum_{k,l}m_{ik}b_{kl}m_{lj}\right)\left(\sum_{r,q}a_{rq}m_{qr}\right)$$

$$= (c_0(\boldsymbol{\Sigma}\mathbf{B}'\boldsymbol{\Phi}tr(\mathbf{AM}) + \boldsymbol{\Sigma}\mathbf{A}'\boldsymbol{\Phi}\mathbf{BM} + \mathbf{MB}\boldsymbol{\Sigma}\mathbf{A}'\boldsymbol{\Phi})$$

$$+ \mathbf{MBM}tr(\mathbf{AM}))_{ij}.$$

(iii)

$$E(\mathbf{X}'\mathbf{BX}tr(\mathbf{AX}))_{ij} = E\left(\left(\sum_{k=1}^{p}\sum_{l=1}^{p}x_{ki}b_{kl}x_{lj}\right)\left(\sum_{r=1}^{n}\sum_{q=1}^{p}a_{rq}x_{qr}\right)\right)$$

$$= \sum_{k,l,r,q}[c_0(\sigma_{kl}\phi_{ij}m_{qr} + \sigma_{kq}\phi_{ir}m_{lj} + \sigma_{lq}\phi_{jr}m_{ki})$$

$$+ m_{ki}m_{lj}m_{qr}]b_{kl}a_{rq}$$

$$= c_0\left[\phi_{ij}\left(\sum_{k,l}b_{kl}\sigma_{lk}\right)\left(\sum_{r,q}a_{rq}m_{qr}\right)\right.$$

$$+ \sum_{k,l,r,q}\phi_{ir}a_{rq}\sigma_{kq}b_{kl}m_{lj}$$

$$\left. + \sum_{k,l,r,q}m_{ki}b_{kl}\sigma_{lq}a_{rq}\phi_{jr}\right]$$

$$+ \left(\sum_{k,l}m_{ki}b_{kl}m_{lj}\right)\left(\sum_{r,q}a_{rq}m_{qr}\right)$$

$$= (c_0(\boldsymbol{\Phi}tr(\mathbf{B}\boldsymbol{\Sigma})tr(\mathbf{AM}) + \boldsymbol{\Phi}\mathbf{A}\boldsymbol{\Sigma}\mathbf{BM} + \mathbf{M}'\mathbf{B}\boldsymbol{\Sigma}\mathbf{A}'\boldsymbol{\Phi})$$

$$+ \mathbf{M}'\mathbf{BM}tr(\mathbf{AM}))_{ij}.$$ ■

Theorem 3.21. *Let $\mathbf{X} \sim E_{p,n}(\mathbf{M}, \boldsymbol{\Sigma} \otimes \boldsymbol{\Phi}, \psi)$ with finite fourth order moments. Let $c_0 = -2\psi'(0)$ and $k_0 = 4\psi''(0)$. Then, for any constant matrices $\mathbf{A}$, $\mathbf{B}$, and $\mathbf{C}$, if the expressions on the left-hand sides are defined, we have*

(i)

$$E(\mathbf{XAXBXCX}) = k_0(\boldsymbol{\Sigma}\mathbf{C}'\boldsymbol{\Phi}\mathbf{BA}'\boldsymbol{\Phi} + \boldsymbol{\Sigma}\mathbf{A}'\boldsymbol{\Phi}\mathbf{B}\boldsymbol{\Sigma}\mathbf{C}'\boldsymbol{\Phi} + \boldsymbol{\Sigma}\mathbf{B}'\boldsymbol{\Phi}tr(\mathbf{A}\boldsymbol{\Sigma}\mathbf{C}'\boldsymbol{\Phi}))$$

$$+ c_0(\mathbf{MAMB}\boldsymbol{\Sigma}\mathbf{C}'\boldsymbol{\Phi} + \mathbf{MA}\boldsymbol{\Sigma}\mathbf{C}'\mathbf{M}'\mathbf{B}'\boldsymbol{\Phi}$$

$$+ \boldsymbol{\Sigma}\mathbf{C}'\mathbf{M}'\mathbf{B}'\mathbf{M}'\mathbf{A}'\boldsymbol{\Phi} + \mathbf{M}\mathbf{A}\boldsymbol{\Sigma}\mathbf{B}'\boldsymbol{\Phi}\mathbf{C}\mathbf{M}$$
$$+ \boldsymbol{\Sigma}\mathbf{B}'\mathbf{M}'\mathbf{A}'\boldsymbol{\Phi}\mathbf{C}\mathbf{M} + \boldsymbol{\Sigma}\mathbf{A}'\boldsymbol{\Phi}\mathbf{B}\mathbf{M}\mathbf{C}\mathbf{M})$$
$$+ \mathbf{MAMBMCM},$$

(ii)

$$\begin{aligned} E(\mathbf{X}'\mathbf{AXBXCX}) &= k_0(\boldsymbol{\Phi} tr(\boldsymbol{\Sigma}\mathbf{C}'\boldsymbol{\Phi}\mathbf{B}\boldsymbol{\Sigma}\mathbf{A}') + \boldsymbol{\Phi}\mathbf{B}\boldsymbol{\Sigma}\mathbf{C}'\boldsymbol{\Phi} tr(\mathbf{A}\boldsymbol{\Sigma}) \\ &+ \boldsymbol{\Phi}\mathbf{C}\boldsymbol{\Sigma}\mathbf{A}'\boldsymbol{\Sigma}\mathbf{B}'\boldsymbol{\Phi}) + c_0(\mathbf{M}'\mathbf{AMB}\boldsymbol{\Sigma}\mathbf{C}'\boldsymbol{\Phi} \\ &+ \mathbf{M}'\mathbf{A}\boldsymbol{\Sigma}\mathbf{C}'\mathbf{M}'\mathbf{B}'\boldsymbol{\Phi} + \boldsymbol{\Phi} tr(\mathbf{AMBMC}\boldsymbol{\Sigma}) \\ &+ \mathbf{M}'\mathbf{A}\boldsymbol{\Sigma}\mathbf{B}'\boldsymbol{\Phi}\mathbf{CM} + \boldsymbol{\Phi}\mathbf{CM} tr(\mathbf{AMB}\boldsymbol{\Sigma}) \\ &+ \boldsymbol{\Phi}\mathbf{BMCM} tr(\mathbf{A}\boldsymbol{\Sigma})) + \mathbf{M}'\mathbf{AMBMCM}, \end{aligned}$$

(iii)

$$\begin{aligned} E(\mathbf{X}'\mathbf{AX}'\mathbf{BXCX}) &= k_0(\boldsymbol{\Phi} tr(\boldsymbol{\Sigma}\mathbf{C}'\boldsymbol{\Phi}\mathbf{A}') tr(\mathbf{B}\boldsymbol{\Sigma}) + \boldsymbol{\Phi}\mathbf{A}'\boldsymbol{\Sigma}\mathbf{B}\boldsymbol{\Sigma}\mathbf{C}'\boldsymbol{\Phi} \\ &+ \boldsymbol{\Phi}\mathbf{C}\boldsymbol{\Sigma}\mathbf{B}\boldsymbol{\Sigma}\mathbf{A}\boldsymbol{\Phi}) + c_0(\mathbf{M}'\mathbf{AM}'\mathbf{B}\boldsymbol{\Sigma}\mathbf{C}'\boldsymbol{\Phi} \\ &+ \mathbf{M}'\mathbf{A}\boldsymbol{\Phi} tr(\mathbf{MC}\boldsymbol{\Sigma}\mathbf{B}) + \boldsymbol{\Phi} tr(\mathbf{AM}'\mathbf{BMC}\boldsymbol{\Sigma}) \\ &+ \mathbf{M}'\mathbf{A}\boldsymbol{\Phi}\mathbf{CM} tr(\mathbf{B}\boldsymbol{\Sigma}) + \boldsymbol{\Phi}\mathbf{CM} tr(\mathbf{AM}'\mathbf{B}\boldsymbol{\Sigma}) \\ &+ \boldsymbol{\Phi}\mathbf{A}'\boldsymbol{\Sigma}\mathbf{BMCM}) + \mathbf{M}'\mathbf{AM}'\mathbf{BMCM}, \end{aligned}$$

(iv)

$$\begin{aligned} E(\mathbf{X}'\mathbf{AXBX}'\mathbf{CX}) &= k_0(\boldsymbol{\Phi} tr(\boldsymbol{\Sigma}\mathbf{C}'\boldsymbol{\Sigma}\mathbf{A}') tr(\mathbf{B}\boldsymbol{\Phi}) + \boldsymbol{\Phi}\mathbf{B}\boldsymbol{\Phi} tr(\mathbf{A}\boldsymbol{\Sigma}) tr(\mathbf{C}\boldsymbol{\Sigma}) \\ &+ \boldsymbol{\Phi}\mathbf{B}'\boldsymbol{\Phi} tr(\mathbf{A}\boldsymbol{\Sigma}\mathbf{C}'\boldsymbol{\Sigma})) + c_0(\mathbf{M}'\mathbf{AMB}\boldsymbol{\Phi} tr(\boldsymbol{\Sigma}\mathbf{C}) \\ &+ \mathbf{M}'\mathbf{A}\boldsymbol{\Sigma}\mathbf{C}'\mathbf{MB}'\boldsymbol{\Phi} + \boldsymbol{\Phi} tr(\mathbf{AMBM}'\mathbf{C}\boldsymbol{\Sigma}) \\ &+ \mathbf{M}'\mathbf{A}\boldsymbol{\Sigma}\mathbf{CM} tr(\mathbf{B}\boldsymbol{\Phi}) + \boldsymbol{\Phi}\mathbf{B}'\mathbf{M}'\mathbf{A}'\boldsymbol{\Sigma}\mathbf{CM} \\ &+ \boldsymbol{\Phi}\mathbf{BM}'\mathbf{CM} tr(\mathbf{A}\boldsymbol{\Sigma})) + \mathbf{M}'\mathbf{AMBM}'\mathbf{CM}, \end{aligned}$$

(v)

$$\begin{aligned} E(\mathbf{XAX}'\mathbf{BXCX}) &= k_0(\boldsymbol{\Sigma}\mathbf{B}\boldsymbol{\Sigma}\mathbf{C}'\boldsymbol{\Phi} tr(\boldsymbol{\Phi}\mathbf{A}) + \boldsymbol{\Sigma}\mathbf{B}'\boldsymbol{\Sigma}\mathbf{C}'\boldsymbol{\Phi}\mathbf{A}'\boldsymbol{\Phi} \\ &+ \boldsymbol{\Sigma}\mathbf{C}'\boldsymbol{\Phi}\mathbf{A}' tr(\boldsymbol{\Sigma}\mathbf{B})) + c_0(\mathbf{MAM}'\mathbf{B}\boldsymbol{\Sigma}\mathbf{C}'\boldsymbol{\Phi} \\ &+ \mathbf{MA}\boldsymbol{\Phi} tr(\mathbf{BMC}\boldsymbol{\Sigma}) + \mathbf{MA}\boldsymbol{\Phi}\mathbf{CM} tr(\boldsymbol{\Sigma}\mathbf{B}) \\ &+ \boldsymbol{\Sigma}\mathbf{C}'\mathbf{M}'\mathbf{B}'\mathbf{MA}'\boldsymbol{\Phi} + \boldsymbol{\Sigma}\mathbf{B}'\mathbf{MA}'\boldsymbol{\Phi}\mathbf{CM} \\ &+ \boldsymbol{\Sigma}\mathbf{BMCM} tr(\boldsymbol{\Phi}\mathbf{A})) + \mathbf{MAM}'\mathbf{BMCM}, \end{aligned}$$

(vi)

$$E(\mathbf{X}'\mathbf{AXBXCX}') = k_0(\boldsymbol{\Phi}\mathbf{B}\Sigma tr(\Sigma\mathbf{A})tr(\boldsymbol{\Phi}\mathbf{C}) + \boldsymbol{\Phi}\mathbf{C}\boldsymbol{\Phi}\mathbf{B}\Sigma\mathbf{A}\Sigma$$
$$+ \boldsymbol{\Phi}\mathbf{C}'\boldsymbol{\Phi}\mathbf{B}\Sigma\mathbf{A}'\Sigma) + c_0(\mathbf{M}'\mathbf{AMB}\Sigma tr(\boldsymbol{\Phi}\mathbf{C})$$
$$+ \mathbf{M}'\mathbf{A}\Sigma tr(\mathbf{BMC}\boldsymbol{\Phi}) + \mathbf{M}'\mathbf{A}\Sigma\mathbf{B}'\boldsymbol{\Phi}\mathbf{CM}'$$
$$+ \boldsymbol{\Phi}\mathbf{C}'\mathbf{M}'\mathbf{B}'\mathbf{M}'\mathbf{A}'\Sigma + \boldsymbol{\Phi}\mathbf{CM}'tr(\mathbf{AMB}\Sigma)$$
$$+ \boldsymbol{\Phi}\mathbf{BMCM}'tr(\Sigma\mathbf{A})) + \mathbf{M}'\mathbf{AMBMCM}'.$$

PROOF:

(i)

$$E(\mathbf{XAXBXCX})_{ij} = E\left(\sum_{q=1}^{p}\sum_{r=1}^{n}\sum_{l=1}^{p}\sum_{k=1}^{n}\sum_{t=1}^{p}\sum_{s=1}^{n} x_{is}a_{st}x_{tk}b_{kl}x_{lr}c_{rq}x_{qj}\right)$$
$$= \sum_{\substack{q,r,l\\k,t,s}} \{k_0[\sigma_{it}\phi_{sk}\sigma_{lq}\phi_{rj} + \sigma_{il}\phi_{sr}\sigma_{tq}\phi_{kj} + \sigma_{iq}\phi_{sj}\sigma_{tl}\phi_{kr}]$$
$$+ c_0[m_{is}m_{tk}\sigma_{lq}\phi_{rj} + m_{is}m_{lr}\sigma_{tq}\phi_{kj}$$
$$+ m_{is}m_{qj}\sigma_{tl}\phi_{kr} + m_{tk}m_{lr}\sigma_{iq}\phi_{sj} + m_{tk}m_{qj}\sigma_{il}\phi_{sr}$$
$$+ m_{lr}m_{qj}\sigma_{it}\phi_{sk}] + m_{is}m_{tk}m_{lr}m_{qj}\}a_{st}b_{kl}c_{rq}$$
$$= \sum_{\substack{q,r,l\\k,t,s}} \{k_0[\sigma_{it}a_{st}\phi_{sk}b_{kl}\sigma_{lq}c_{rq}\phi_{rj}$$
$$+ (\sigma_{il}b_{kl}\phi_{kj})(\phi_{sr}c_{rq}\sigma_{tq}a_{st})$$
$$+ \sigma_{iq}c_{rq}\phi_{kr}b_{kl}\sigma_{tl}a_{st}\phi_{sj}]$$
$$+ c_0[m_{is}a_{st}m_{tk}b_{kl}\sigma_{lq}c_{rq}\phi_{rj}$$
$$+ m_{is}a_{st}\sigma_{tq}c_{rq}m_{lr}b_{kl}\phi_{kj}$$
$$+ m_{is}a_{st}\sigma_{tl}b_{kl}\phi_{kr}c_{rq}m_{qj}$$
$$+ \sigma_{iq}c_{rq}m_{lr}b_{kl}m_{tk}a_{st}\phi_{sj}$$
$$+ \sigma_{il}b_{kl}m_{tk}a_{st}\phi_{sr}c_{rq}m_{qj}$$
$$+ \sigma_{it}a_{st}\phi_{sk}b_{kl}m_{lr}c_{rq}m_{qj}]$$
$$+ m_{is}a_{st}m_{tk}b_{kl}m_{lr}c_{rq}m_{qj}\}$$
$$= (k_0[\Sigma\mathbf{A}'\boldsymbol{\Phi}\mathbf{B}\Sigma\mathbf{C}'\boldsymbol{\Phi} + (\Sigma\mathbf{B}'\boldsymbol{\Phi})tr(\boldsymbol{\Phi}\mathbf{C}\Sigma\mathbf{A}')$$
$$+ \Sigma\mathbf{C}'\boldsymbol{\Phi}\mathbf{BA}'\boldsymbol{\Phi}] + c_0[\mathbf{MAMB}\Sigma\mathbf{C}'\boldsymbol{\Phi}$$
$$+ \mathbf{MA}\Sigma\mathbf{C}'\mathbf{M}'\mathbf{B}'\boldsymbol{\Phi} + \mathbf{MA}\Sigma\mathbf{B}'\boldsymbol{\Phi}\mathbf{CM}$$

$$+ \boldsymbol{\Sigma}\mathbf{C}'\mathbf{M}'\mathbf{B}'\mathbf{M}'\mathbf{A}'\boldsymbol{\Phi} + \boldsymbol{\Sigma}\mathbf{B}'\mathbf{M}'\mathbf{A}'\boldsymbol{\Phi}\mathbf{C}\mathbf{M}$$
$$+ \boldsymbol{\Sigma}\mathbf{A}'\boldsymbol{\Phi}\mathbf{B}\mathbf{M}\mathbf{C}\mathbf{M}] + \mathbf{M}\mathbf{A}\mathbf{M}\mathbf{B}\mathbf{M}\mathbf{C}\mathbf{M})_{ij}.$$

(ii)

$$\begin{aligned}
E(\mathbf{X}'\mathbf{A}\mathbf{X}\mathbf{B}\mathbf{X}\mathbf{C}\mathbf{X})_{ij} &= E\left(\sum_{q=1}^{p}\sum_{r=1}^{n}\sum_{l=1}^{p}\sum_{k=1}^{n}\sum_{t=1}^{p}\sum_{s=1}^{p} x_{si}a_{st}x_{tk}b_{kl}x_{lr}c_{rq}x_{qj}\right)\\
&= \sum_{\substack{q,r,l\\k,t,s}} \{k_0[\sigma_{st}\phi_{ik}\sigma_{lq}\phi_{rj} + \sigma_{sl}\phi_{ir}\sigma_{tq}\phi_{kj}\\
&+ \sigma_{sq}\phi_{ij}\sigma_{tl}\phi_{kr}] + c_0[m_{si}m_{tk}\sigma_{lq}\phi_{rj}\\
&+ m_{si}m_{lr}\sigma_{tq}\phi_{kj} + m_{si}m_{qj}\sigma_{tl}\phi_{kr}\\
&+ m_{tk}m_{lr}\sigma_{sq}\phi_{ij} + m_{tk}m_{qj}\sigma_{sl}\phi_{ir}\\
&+ m_{lr}m_{qj}\sigma_{st}\phi_{ik}] + m_{si}m_{tk}m_{lr}m_{qj}\}a_{st}b_{kl}c_{rq}\\
&= \sum_{\substack{q,r,l\\k,t,s}} k_0[(\phi_{ik}b_{kl}\sigma_{lq}c_{rq}\phi_{rj})(\sigma_{st}a_{st})\\
&+ \phi_{ir}c_{rq}\sigma_{tq}a_{st}\sigma_{sl}b_{kl}\phi_{kj}\\
&+ \phi_{ij}(\sigma_{sq}a_{st}\sigma_{tl}b_{kl}\phi_{kr}c_{rq})]\\
&+ c_0[m_{si}a_{st}m_{tk}b_{kl}\sigma_{lq}c_{rq}\phi_{rj}\\
&+ m_{si}a_{st}\sigma_{tq}c_{rq}m_{lr}b_{kl}\phi_{kj}\\
&+ m_{si}a_{st}\sigma_{tl}b_{kl}\phi_{kr}c_{rq}m_{qj}\\
&+ \phi_{ij}(\sigma_{sq}c_{rq}m_{lr}b_{kl}m_{tk}a_{st})\\
&+ (\phi_{ir}c_{rq}m_{qj})(\sigma_{sl}b_{kl}m_{tk}a_{st})\\
&+ (\phi_{ik}b_{kl}m_{lr}c_{rq}m_{qj})(\sigma_{st}a_{st})]\\
&+ m_{si}a_{st}m_{tk}b_{kl}m_{lr}c_{rq}m_{qj}\\
&= (k_0[\boldsymbol{\Phi}\mathbf{B}\boldsymbol{\Sigma}\mathbf{C}'\boldsymbol{\Phi}tr(\boldsymbol{\Sigma}\mathbf{A}) + \boldsymbol{\Phi}\mathbf{C}\boldsymbol{\Sigma}\mathbf{A}'\boldsymbol{\Sigma}\mathbf{B}'\boldsymbol{\Phi}\\
&+ \boldsymbol{\Phi}tr(\boldsymbol{\Sigma}\mathbf{A}\boldsymbol{\Sigma}\mathbf{B}'\boldsymbol{\Phi}\mathbf{C})] + c_0[\mathbf{M}'\mathbf{A}\mathbf{M}\mathbf{B}\boldsymbol{\Sigma}\mathbf{C}'\boldsymbol{\Phi}\\
&+ \mathbf{M}'\mathbf{A}\boldsymbol{\Sigma}\mathbf{C}'\mathbf{M}'\mathbf{B}'\boldsymbol{\Phi} + \mathbf{M}'\mathbf{A}\boldsymbol{\Sigma}\mathbf{B}'\boldsymbol{\Phi}\mathbf{C}\mathbf{M}\\
&+ \boldsymbol{\Phi}tr(\boldsymbol{\Sigma}\mathbf{C}'\mathbf{M}'\mathbf{B}'\mathbf{M}'\mathbf{A}') + \boldsymbol{\Phi}\mathbf{C}\mathbf{M}tr(\boldsymbol{\Sigma}\mathbf{B}'\mathbf{M}'\mathbf{A}')\\
&+ \boldsymbol{\Phi}\mathbf{B}\mathbf{M}\mathbf{C}\mathbf{M}tr(\boldsymbol{\Sigma}\mathbf{A})] + \mathbf{M}'\mathbf{A}\mathbf{M}\mathbf{B}\mathbf{M}\mathbf{C}\mathbf{M})_{ij}.
\end{aligned}$$

(iii)

$$
\begin{aligned}
E(\mathbf{X}'\mathbf{A}\mathbf{X}'\mathbf{B}\mathbf{X}\mathbf{C}\mathbf{X})_{ij} &= E\left(\sum_{q=1}^{p}\sum_{r=1}^{n}\sum_{l=1}^{p}\sum_{k=1}^{p}\sum_{t=1}^{n}\sum_{s=1}^{p} x_{si}a_{st}x_{kt}b_{kl}x_{lr}c_{rq}x_{qj}\right)\\
&= \sum_{\substack{q,r,l\\k,t,s}} \{k_0[\sigma_{sk}\phi_{it}\sigma_{lq}\phi_{rj} + \sigma_{sl}\phi_{ir}\sigma_{kq}\phi_{tj}\\
&+ \sigma_{sq}\phi_{ij}\sigma_{kl}\phi_{tr}] + c_0[m_{si}m_{kt}\sigma_{lq}\phi_{rj}\\
&+ m_{si}m_{lr}\sigma_{kq}\phi_{tj} + m_{si}m_{qj}\sigma_{kl}\phi_{tr}\\
&+ m_{kt}m_{lr}\sigma_{sq}\phi_{ij} + m_{kt}m_{qj}\sigma_{sl}\phi_{ir}\\
&+ m_{lr}m_{qj}\sigma_{sk}\phi_{it}] + m_{si}m_{kt}m_{lr}m_{qj}\}a_{st}b_{kl}c_{rq}\\
&= \sum_{\substack{q,r,l\\k,t,s}} k_0[\phi_{it}a_{st}\sigma_{sk}b_{kl}\sigma_{lq}c_{rq}\phi_{rj}\\
&+ \phi_{ir}c_{rq}\sigma_{kq}b_{kl}\sigma_{sl}a_{st}\phi_{tj}\\
&+ \phi_{ij}(\sigma_{sq}c_{rq}\phi_{tr}a_{st})(\sigma_{kl}b_{kl})]\\
&+ c_0[m_{si}a_{st}m_{kt}b_{kl}\sigma_{lq}c_{rq}\phi_{rj}\\
&+ (m_{si}a_{st}\phi_{tj})(\sigma_{kq}c_{rq}m_{lr}b_{kl})\\
&+ (m_{si}a_{st}\phi_{tr}c_{rq}m_{qj})(\sigma_{kl}b_{kl})\\
&+ \phi_{ij}(\sigma_{sq}c_{rq}m_{lr}b_{kl}m_{kt}a_{st})\\
&+ (\phi_{ir}c_{rq}m_{qj})(\sigma_{sl}b_{kl}m_{kt}a_{st})\\
&+ \phi_{it}a_{st}\sigma_{sk}b_{kl}m_{lr}c_{rq}m_{qj}]\\
&+ m_{si}a_{st}m_{kt}b_{kl}m_{lr}c_{rq}m_{qj}\\
&= (k_0[\boldsymbol{\Phi}\mathbf{A}'\boldsymbol{\Sigma}\mathbf{B}\boldsymbol{\Sigma}\mathbf{C}'\boldsymbol{\Phi} + \boldsymbol{\Phi}\mathbf{C}\boldsymbol{\Sigma}\mathbf{B}\boldsymbol{\Sigma}\mathbf{A}\boldsymbol{\Phi}\\
&+ \boldsymbol{\Phi} tr(\boldsymbol{\Sigma}\mathbf{C}'\boldsymbol{\Phi}\mathbf{A}')tr(\boldsymbol{\Sigma}\mathbf{B})] + c_0[\mathbf{M}'\mathbf{A}\mathbf{M}'\mathbf{B}\boldsymbol{\Sigma}\mathbf{C}'\boldsymbol{\Phi}\\
&+ \mathbf{M}'\mathbf{A}\boldsymbol{\Phi} tr(\boldsymbol{\Sigma}\mathbf{C}'\mathbf{M}'\mathbf{B}') + \mathbf{M}'\mathbf{A}\boldsymbol{\Phi}\mathbf{C}\mathbf{M} tr(\boldsymbol{\Sigma}\mathbf{B})\\
&+ \boldsymbol{\Phi} tr(\boldsymbol{\Sigma}\mathbf{C}'\mathbf{M}'\mathbf{B}'\mathbf{M}\mathbf{A}') + \boldsymbol{\Phi}\mathbf{C}\mathbf{M} tr(\boldsymbol{\Sigma}\mathbf{B}'\mathbf{M}\mathbf{A}')\\
&+ \boldsymbol{\Phi}\mathbf{A}'\boldsymbol{\Sigma}\mathbf{B}\mathbf{M}\mathbf{C}\mathbf{M}] + \mathbf{M}'\mathbf{A}\mathbf{M}'\mathbf{B}\mathbf{M}\mathbf{C}\mathbf{M})_{ij}.
\end{aligned}
$$

(iv)

$$
E(\mathbf{X}'\mathbf{A}\mathbf{X}\mathbf{B}\mathbf{X}'\mathbf{C}\mathbf{X})_{ij} = E\left(\sum_{q=1}^{p}\sum_{r=1}^{p}\sum_{l=1}^{n}\sum_{k=1}^{n}\sum_{t=1}^{p}\sum_{s=1}^{p} x_{si}a_{st}x_{tk}b_{kl}x_{rl}c_{rq}x_{qj}\right)
$$

$$
\begin{aligned}
&= \sum_{\substack{q,r,l\\k,t,s}} \{k_0[\sigma_{st}\phi_{ik}\sigma_{rq}\phi_{lj} + \sigma_{sr}\phi_{il}\sigma_{tq}\phi_{kj} \\
&+ \sigma_{sq}\phi_{ij}\sigma_{tr}\phi_{kl}] + c_0[m_{si}m_{tk}\sigma_{rq}\phi_{lj} \\
&+ m_{si}m_{rl}\sigma_{tq}\phi_{kj} + m_{si}m_{qj}\sigma_{tr}\phi_{kl} \\
&+ m_{tk}m_{rl}\sigma_{sq}\phi_{ij} + m_{tk}m_{qj}\sigma_{sr}\phi_{il} \\
&+ m_{rl}m_{qj}\sigma_{st}\phi_{ik}] + m_{si}m_{tk}m_{rl}m_{qj}\}a_{st}b_{kl}c_{rq} \\
&= \sum_{\substack{q,r,l\\k,t,s}} k_0[(\phi_{ik}b_{kl}\phi_{lj})(\sigma_{st}a_{st})(\sigma_{rq}c_{rq}) \\
&+ (\phi_{il}b_{kl}\phi_{kj})(\sigma_{sr}c_{rq}\sigma_{tq}a_{st}) \\
&+ \phi_{ij}(\sigma_{sq}c_{rq}\sigma_{tr}a_{st})(\phi_{kl}b_{kl})] \\
&+ c_0[(m_{si}a_{st}m_{tk}b_{kl}\phi_{lj})(\sigma_{rq}c_{rq}) \\
&+ m_{si}a_{st}\sigma_{tq}c_{rq}m_{rl}b_{kl}\phi_{kj} \\
&+ (m_{si}a_{st}\sigma_{tr}c_{rq}m_{qj})(\phi_{kl}b_{kl}) \\
&+ \phi_{ij}(\sigma_{sq}c_{rq}m_{lr}b_{kl}m_{tk}a_{st}) \\
&+ \phi_{il}b_{kl}m_{tk}a_{st}\sigma_{sr}c_{rq}m_{qj} \\
&+ (\phi_{ik}b_{kl}m_{rl}c_{rq}m_{qj})(\sigma_{st}a_{st})] \\
&+ m_{si}a_{st}m_{tk}b_{kl}m_{rl}c_{rq}m_{qj} \\
&= (k_0[\boldsymbol{\Phi}\mathbf{B}\boldsymbol{\Phi}tr(\boldsymbol{\Sigma}\mathbf{A})tr(\boldsymbol{\Sigma}\mathbf{C}) + (\boldsymbol{\Phi}\mathbf{B}'\boldsymbol{\Phi})tr(\boldsymbol{\Sigma}\mathbf{C}\boldsymbol{\Sigma}\mathbf{A}') \\
&+ \boldsymbol{\Phi}tr(\boldsymbol{\Sigma}\mathbf{C}'\boldsymbol{\Sigma}\mathbf{A}')tr(\boldsymbol{\Phi}\mathbf{B})] + c_0[\mathbf{M}'\mathbf{A}\mathbf{M}\mathbf{B}\boldsymbol{\Phi}tr(\boldsymbol{\Sigma}\mathbf{C}) \\
&+ \mathbf{M}'\mathbf{A}\boldsymbol{\Sigma}\mathbf{C}'\mathbf{M}\mathbf{B}'\boldsymbol{\Phi} + \mathbf{M}'\mathbf{A}\boldsymbol{\Sigma}\mathbf{C}\mathbf{M}tr(\boldsymbol{\Phi}\mathbf{B}) \\
&+ \boldsymbol{\Phi}tr(\boldsymbol{\Sigma}\mathbf{C}'\mathbf{M}\mathbf{B}'\mathbf{M}'\mathbf{A}') + \boldsymbol{\Phi}\mathbf{B}'\mathbf{M}'\mathbf{A}'\boldsymbol{\Sigma}\mathbf{C}\mathbf{M} \\
&+ \boldsymbol{\Phi}\mathbf{B}\mathbf{M}'\mathbf{C}\mathbf{M}tr(\boldsymbol{\Sigma}\mathbf{A})] + \mathbf{M}'\mathbf{A}\mathbf{M}\mathbf{B}\mathbf{M}'\mathbf{C}\mathbf{M})_{ij}.
\end{aligned}
$$

(v)

$$
\begin{aligned}
E(\mathbf{X}\mathbf{A}\mathbf{X}'\mathbf{B}\mathbf{X}\mathbf{C}\mathbf{X})_{ij} &= E\left(\sum_{q=1}^{p}\sum_{r=1}^{n}\sum_{l=1}^{p}\sum_{k=1}^{p}\sum_{t=1}^{n}\sum_{s=1}^{n} x_{is}a_{st}x_{kt}b_{kl}x_{lr}c_{rq}x_{qj}\right) \\
&= \sum_{\substack{q,r,l\\k,t,s}} \{k_0[\sigma_{ik}\phi_{st}\sigma_{lq}\phi_{rj} + \sigma_{il}\phi_{sr}\sigma_{kq}\phi_{tj} \\
&+ \sigma_{iq}\phi_{sj}\sigma_{kl}\phi_{tr}] + c_0[m_{is}m_{kt}\sigma_{lq}\phi_{rj} \\
&+ m_{is}m_{lr}\sigma_{kq}\phi_{tj} + m_{is}m_{qj}\sigma_{kl}\phi_{tr}
\end{aligned}
$$

$$
\begin{aligned}
&+ m_{kt}m_{lr}\sigma_{iq}\phi_{sj} + m_{kt}m_{qj}\sigma_{il}\phi_{sr} \\
&+ m_{lr}m_{qj}\sigma_{ik}\phi_{st}] + m_{is}m_{kt}m_{lr}m_{qj}\}a_{st}b_{kl}c_{rq} \\
&= \sum_{\substack{q,r,l\\k,t,s}} k_0[(\sigma_{ik}b_{kl}\sigma_{lq}c_{rq}\phi_{rj})(\phi_{st}a_{st}) \\
&+ \sigma_{il}b_{kl}\sigma_{kq}c_{rq}\phi_{sr}a_{st}\phi_{tj} \\
&+ (\sigma_{iq}c_{rq}\phi_{tr}a_{st}\phi_{sj})(\sigma_{kl}b_{kl})] \\
&+ c_0[m_{is}a_{st}m_{kt}b_{kl}\sigma_{lq}c_{rq}\phi_{rj} \\
&+ (m_{is}a_{st}\phi_{tj})(\sigma_{kq}c_{rq}m_{lr}b_{kl}) \\
&+ (m_{is}a_{st}\phi_{tr}c_{rq}m_{qj})(\sigma_{kl}b_{kl}) \\
&+ \sigma_{iq}c_{rq}m_{lr}b_{kl}m_{kt}a_{st}\phi_{sj} \\
&+ \sigma_{il}b_{kl}m_{kt}a_{st}\phi_{sr}c_{rq}m_{qj} \\
&+ (\sigma_{ik}b_{kl}m_{lr}c_{rq}m_{qj})(\phi_{st}a_{st})] \\
&+ m_{is}a_{st}m_{kt}b_{kl}m_{lr}c_{rq}m_{qj} \\
&= (k_0[\boldsymbol{\Sigma}\mathbf{B}\boldsymbol{\Sigma}\mathbf{C}'\boldsymbol{\Phi}tr(\boldsymbol{\Phi}\mathbf{A}) + \boldsymbol{\Sigma}\mathbf{B}'\boldsymbol{\Sigma}\mathbf{C}'\boldsymbol{\Phi}\mathbf{A}\boldsymbol{\Phi} \\
&+ \boldsymbol{\Sigma}\mathbf{C}'\boldsymbol{\Phi}\mathbf{A}'\boldsymbol{\Phi}tr(\boldsymbol{\Sigma}\mathbf{B})] + c_0[\mathbf{M}\mathbf{A}\mathbf{M}'\mathbf{B}\boldsymbol{\Sigma}\mathbf{C}'\boldsymbol{\Phi} \\
&+ \mathbf{M}\mathbf{A}\boldsymbol{\Phi}tr(\boldsymbol{\Sigma}\mathbf{C}'\mathbf{M}'\mathbf{B}') + \mathbf{M}\mathbf{A}\boldsymbol{\Phi}\mathbf{C}\mathbf{M}tr(\boldsymbol{\Sigma}\mathbf{B}) \\
&+ \boldsymbol{\Sigma}\mathbf{C}'\mathbf{M}'\mathbf{B}'\mathbf{M}\mathbf{A}'\boldsymbol{\Phi} + \boldsymbol{\Sigma}\mathbf{B}'\mathbf{M}\mathbf{A}'\boldsymbol{\Phi}\mathbf{C}\mathbf{M} \\
&+ \boldsymbol{\Sigma}\mathbf{B}\mathbf{M}\mathbf{C}\mathbf{M}tr(\boldsymbol{\Phi}\mathbf{A})] + \mathbf{M}\mathbf{A}\mathbf{M}'\mathbf{B}\mathbf{M}\mathbf{C}\mathbf{M})_{ij}.
\end{aligned}
$$

(vi)

$$
\begin{aligned}
E(\mathbf{X}'\mathbf{A}\mathbf{X}\mathbf{B}\mathbf{X}\mathbf{C}\mathbf{X}')_{ij} &= E\left(\sum_{q=1}^{n}\sum_{r=1}^{n}\sum_{l=1}^{p}\sum_{k=1}^{n}\sum_{t=1}^{p}\sum_{s=1}^{p} x_{si}a_{st}x_{tk}b_{kl}x_{lr}c_{rq}x_{jq}\right) \\
&= \sum_{\substack{q,r,l\\k,t,s}} \{k_0[\sigma_{st}\phi_{ik}\sigma_{lj}\phi_{rq} + \sigma_{sl}\phi_{ir}\sigma_{tj}\phi_{kq} \\
&+ \sigma_{sj}\phi_{iq}\sigma_{tl}\phi_{kr}] + c_0[m_{si}m_{tk}\sigma_{lj}\phi_{rq} \\
&+ m_{si}m_{lr}\sigma_{tj}\phi_{kq} + m_{si}m_{jq}\sigma_{tl}\phi_{kr} \\
&+ m_{tk}m_{lr}\sigma_{sj}\phi_{iq} + m_{tk}m_{jq}\sigma_{sl}\phi_{ir} \\
&+ m_{lr}m_{jq}\sigma_{st}\phi_{ik}] + m_{si}m_{tk}m_{lr}m_{jq}\}a_{st}b_{kl}c_{rq} \\
&= \sum_{\substack{q,r,l\\k,t,s}} k_0[(\phi_{ik}b_{kl}\sigma_{lj})(\sigma_{st}a_{st})(c_{rq}\phi_{rq})
\end{aligned}
$$

$$
\begin{aligned}
&+ \phi_{ir}c_{rq}\phi_{kq}b_{kl}\sigma_{sl}a_{st}\sigma_{tj} \\
&+ \phi_{iq}c_{rq}\phi_{kr}b_{kl}\sigma_{tl}a_{st}\sigma_{sj}] \\
&+ c_0[(m_{si}a_{st}m_{tk}b_{kl}\sigma_{lj})(\phi_{rq}c_{rq}) \\
&+ (m_{si}a_{st}\sigma_{tj})(\phi_{kq}c_{rq}m_{lr}b_{kl}) \\
&+ m_{si}a_{st}\sigma_{tl}b_{kl}\phi_{kr}c_{rq}m_{jq} \\
&+ \phi_{iq}c_{rq}m_{lr}b_{kl}m_{tk}a_{st}\sigma_{sj} \\
&+ (\phi_{ir}c_{rq}m_{jq})(\sigma_{sl}b_{kl}m_{tk}a_{st}) \\
&+ (\phi_{ik}b_{kl}m_{lr}c_{rq}m_{jq})(\sigma_{st}a_{st})] \\
&+ m_{si}a_{st}m_{tk}b_{kl}m_{lr}c_{rq}m_{jq} \\
&= (k_0[\boldsymbol{\Phi}\mathbf{B}\boldsymbol{\Sigma}tr(\boldsymbol{\Sigma}\mathbf{A})tr(\boldsymbol{\Phi}\mathbf{C}) + \boldsymbol{\Phi}\mathbf{C}\boldsymbol{\Phi}\mathbf{B}\boldsymbol{\Sigma}\mathbf{A}\boldsymbol{\Sigma} \\
&+ \boldsymbol{\Phi}\mathbf{C}'\boldsymbol{\Phi}\mathbf{B}\boldsymbol{\Sigma}\mathbf{A}'\boldsymbol{\Sigma}] + c_0[\mathbf{M}'\mathbf{A}\mathbf{M}\mathbf{B}\boldsymbol{\Sigma}tr(\boldsymbol{\Phi}\mathbf{C}) \\
&+ \mathbf{M}'\mathbf{A}\boldsymbol{\Sigma}tr(\boldsymbol{\Phi}\mathbf{C}'\mathbf{M}'\mathbf{B}') + \mathbf{M}'\mathbf{A}\boldsymbol{\Sigma}\mathbf{B}'\boldsymbol{\Phi}\mathbf{C}\mathbf{M}' \\
&+ \boldsymbol{\Phi}\mathbf{C}'\mathbf{M}'\mathbf{B}'\mathbf{M}'\mathbf{A}'\boldsymbol{\Sigma} + \boldsymbol{\Phi}\mathbf{C}\mathbf{M}'tr(\boldsymbol{\Sigma}\mathbf{B}'\mathbf{M}'\mathbf{A}') \\
&+ \boldsymbol{\Phi}\mathbf{B}\mathbf{M}\mathbf{C}\mathbf{M}'tr(\boldsymbol{\Sigma}\mathbf{A})] + \mathbf{M}'\mathbf{A}\mathbf{M}\mathbf{B}\mathbf{M}\mathbf{C}\mathbf{M}')_{ij}. \qquad \blacksquare
\end{aligned}
$$

Theorem 3.22. *Let* $\mathbf{X} \sim E_{p,n}(\mathbf{M}, \boldsymbol{\Sigma} \otimes \boldsymbol{\Phi}, \psi)$ *with finite fourth order moments. Let* $c_0 = -2\psi'(0)$ *and* $k_0 = 4\psi''(0)$*. Then, for any constant matrices* $\mathbf{A}$, $\mathbf{B}$, *and* $\mathbf{C}$*, if the expressions on the left-hand sides are defined, we have*

(i)

$$
\begin{aligned}
E(\mathbf{X}\mathbf{A}tr(\mathbf{X}\mathbf{B}\mathbf{X}\mathbf{C}\mathbf{X}')) &= k_0(\boldsymbol{\Sigma}^2\mathbf{B}'\boldsymbol{\Phi}\mathbf{C}\boldsymbol{\Phi}\mathbf{A} + \boldsymbol{\Sigma}^2\mathbf{B}'\boldsymbol{\Phi}\mathbf{A}tr(\mathbf{C}\boldsymbol{\Phi}) \\
&+ \boldsymbol{\Sigma}\mathbf{B}'\boldsymbol{\Phi}\mathbf{C}'\boldsymbol{\Phi}\mathbf{A}tr(\boldsymbol{\Sigma})) + c_0(\mathbf{M}\mathbf{A}tr(\mathbf{M}\mathbf{B}\boldsymbol{\Sigma})tr(\mathbf{C}\boldsymbol{\Phi}) \\
&+ \mathbf{M}\mathbf{A}tr(\mathbf{M}\mathbf{C}\boldsymbol{\Phi}\mathbf{B})tr(\boldsymbol{\Sigma}) + \boldsymbol{\Sigma}\mathbf{M}\mathbf{B}\mathbf{M}\mathbf{C}\boldsymbol{\Phi}\mathbf{A} \\
&+ \mathbf{M}\mathbf{A}tr(\mathbf{M}\mathbf{C}'\boldsymbol{\Phi}\mathbf{B}\boldsymbol{\Sigma}) + \boldsymbol{\Sigma}\mathbf{B}'\mathbf{M}'\mathbf{M}\mathbf{C}'\boldsymbol{\Phi}\mathbf{A} \\
&+ \boldsymbol{\Sigma}\mathbf{M}\mathbf{C}'\mathbf{M}'\mathbf{B}\boldsymbol{\Phi}\mathbf{A}) + \mathbf{M}\mathbf{A}tr(\mathbf{M}\mathbf{B}\mathbf{M}\mathbf{C}\mathbf{M}'),
\end{aligned}
$$

(ii)

$$
\begin{aligned}
E(\mathbf{X}\mathbf{A}\mathbf{X}tr(\mathbf{B}\mathbf{X}\mathbf{C}\mathbf{X}')) &= k_0(\boldsymbol{\Sigma}\mathbf{B}\boldsymbol{\Sigma}\mathbf{A}'\boldsymbol{\Phi}\mathbf{C}'\boldsymbol{\Phi} + \boldsymbol{\Sigma}\mathbf{A}'\boldsymbol{\Phi}tr(\mathbf{B}\boldsymbol{\Sigma})tr(\mathbf{C}\boldsymbol{\Phi}) \\
&+ \boldsymbol{\Sigma}\mathbf{B}'\boldsymbol{\Sigma}\mathbf{A}'\boldsymbol{\Phi}\mathbf{C}\boldsymbol{\Phi}) + c_0(\mathbf{M}\mathbf{A}\mathbf{M}tr(\mathbf{B}\boldsymbol{\Sigma})tr(\mathbf{C}\boldsymbol{\Phi}) \\
&+ \mathbf{M}\mathbf{A}\boldsymbol{\Sigma}\mathbf{B}\mathbf{M}\mathbf{C}\boldsymbol{\Phi} + \boldsymbol{\Sigma}\mathbf{M}\mathbf{B}\mathbf{M}\mathbf{C}\boldsymbol{\Phi}\mathbf{A}\mathbf{M} \\
&+ \mathbf{M}\mathbf{A}\boldsymbol{\Sigma}\mathbf{B}'\mathbf{M}\mathbf{C}'\boldsymbol{\Phi} + \boldsymbol{\Sigma}\mathbf{B}'\mathbf{M}\mathbf{C}'\boldsymbol{\Phi}\mathbf{A}\mathbf{M} \\
&+ \boldsymbol{\Sigma}\mathbf{A}'\boldsymbol{\Phi}tr(\mathbf{M}\mathbf{C}'\mathbf{M}'\mathbf{B}')) + \mathbf{M}\mathbf{A}\mathbf{M}tr(\mathbf{B}\mathbf{M}\mathbf{C}\mathbf{M}'),
\end{aligned}
$$

(iii)

$$\begin{aligned}E(\mathbf{X}'\mathbf{A}\mathbf{X}tr(\mathbf{B}\mathbf{X}\mathbf{C}\mathbf{X}')) &= k_0(\Phi\mathbf{C}'\Phi tr(\Sigma\mathbf{B}\Sigma\mathbf{A}') + \Phi tr(\mathbf{A}\Sigma)tr(\mathbf{B}\Sigma)tr(\mathbf{C}\Phi)\\ &+ \Phi\mathbf{C}\Phi tr(\Sigma\mathbf{B}'\Sigma\mathbf{A}')) + c_0(\mathbf{M}'\mathbf{A}\mathbf{M}tr(\mathbf{B}\Sigma)tr(\mathbf{C}\Phi)\\ &+ \mathbf{M}'\mathbf{A}\Sigma\mathbf{B}\mathbf{M}\mathbf{C}\Phi + \Phi\mathbf{C}'\mathbf{M}'\mathbf{B}'\Sigma\mathbf{A}\mathbf{M}\\ &+ \mathbf{M}'\mathbf{A}\Sigma\mathbf{B}'\mathbf{M}\mathbf{C}'\Phi + \Phi\mathbf{C}\mathbf{M}'\mathbf{B}\Sigma\mathbf{A}\mathbf{M}\\ &+ \Phi tr(\mathbf{A}\Sigma)tr(\mathbf{M}\mathbf{C}\mathbf{M}'\mathbf{B})) + \mathbf{M}'\mathbf{A}\mathbf{M}tr(\mathbf{B}\mathbf{M}\mathbf{C}\mathbf{M}'),\end{aligned}$$

(iv)

$$\begin{aligned}E(\mathbf{X}\mathbf{B}\mathbf{X}\mathbf{C}\mathbf{X}tr(\mathbf{A}\mathbf{X})) &= k_0(\Sigma\mathbf{B}'\Phi\mathbf{C}\Sigma\mathbf{A}'\Phi + \Sigma\mathbf{A}'\Phi\mathbf{B}\Sigma\mathbf{C}'\Phi\\ &+ \Sigma\mathbf{C}'\Phi\mathbf{A}\Sigma\mathbf{B}'\Phi) + c_0(\mathbf{M}\mathbf{B}\Sigma\mathbf{C}'\Phi tr(\mathbf{A}\mathbf{M})\\ &+ \Sigma\mathbf{C}'\mathbf{M}'\mathbf{B}'\Phi tr(\mathbf{A}\mathbf{M}) + \mathbf{M}\mathbf{B}\mathbf{M}\mathbf{C}\Sigma\mathbf{A}'\Phi\\ &+ \Sigma\mathbf{B}'\Phi\mathbf{C}\mathbf{M}tr(\mathbf{M}\mathbf{A}) + \mathbf{M}\mathbf{B}\Sigma\mathbf{A}'\Phi\mathbf{C}\mathbf{M}\\ &+ \Sigma\mathbf{A}'\Phi\mathbf{B}\mathbf{M}\mathbf{C}\mathbf{M}) + \mathbf{M}\mathbf{B}\mathbf{M}\mathbf{C}\mathbf{M}tr(\mathbf{A}\mathbf{M}),\end{aligned}$$

(v)

$$\begin{aligned}E(\mathbf{X}'\mathbf{B}\mathbf{X}\mathbf{C}\mathbf{X}tr(\mathbf{A}\mathbf{X})) &= k_0(\Phi\mathbf{C}\Sigma\mathbf{A}'\Phi tr(\mathbf{B}\Sigma) + \Phi\mathbf{A}\Sigma\mathbf{B}\Sigma\mathbf{C}'\Phi\\ &+ \Phi tr(\Sigma\mathbf{B}'\Sigma\mathbf{C}'\Phi\mathbf{A})) + c_0(\mathbf{M}'\mathbf{B}\Sigma\mathbf{C}'\Phi tr(\mathbf{M}\mathbf{A})\\ &+ \Phi tr(\mathbf{M}\mathbf{A})tr(\mathbf{M}\mathbf{C}\Sigma\mathbf{B}) + \mathbf{M}'\mathbf{B}\mathbf{M}\mathbf{C}\Sigma\mathbf{A}'\Phi\\ &+ \Phi\mathbf{C}\mathbf{M}tr(\mathbf{B}\Sigma)tr(\mathbf{M}\mathbf{A}) + \mathbf{M}'\mathbf{B}\Sigma\mathbf{A}'\Phi\mathbf{C}\mathbf{M}\\ &+ \Phi\mathbf{A}\Sigma\mathbf{B}\mathbf{M}\mathbf{C}\mathbf{M}) + \mathbf{M}'\mathbf{B}\mathbf{M}\mathbf{C}\mathbf{M}tr(\mathbf{M}\mathbf{A}),\end{aligned}$$

(vi)

$$\begin{aligned}E(\mathbf{X}'\mathbf{B}\mathbf{X}'\mathbf{C}\mathbf{X}tr(\mathbf{A}\mathbf{X})) &= k_0(\Phi\mathbf{B}'\Sigma\mathbf{C}\Sigma\mathbf{A}'\Phi + \Phi\mathbf{A}\Sigma\mathbf{B}\Phi tr(\Sigma\mathbf{C})\\ &+ \Phi tr(\Sigma\mathbf{C}\Sigma\mathbf{B}\Phi\mathbf{A})) + c_0(\mathbf{M}'\mathbf{B}\Phi tr(\mathbf{M}\mathbf{A})tr(\Sigma\mathbf{C})\\ &+ \Phi tr(\mathbf{M}\mathbf{A})tr(\mathbf{M}\mathbf{B}'\Sigma\mathbf{C}') + \mathbf{M}'\mathbf{B}\mathbf{M}'\mathbf{C}\Sigma\mathbf{A}'\Phi\\ &+ \Phi\mathbf{B}'\Sigma\mathbf{C}\mathbf{M}tr(\mathbf{M}\mathbf{A}) + \mathbf{M}'\mathbf{B}\Phi\mathbf{A}\Sigma\mathbf{C}\mathbf{M}\\ &+ \Phi\mathbf{A}\Sigma\mathbf{B}\mathbf{M}'\mathbf{C}\mathbf{M}) + \mathbf{M}'\mathbf{B}\mathbf{M}'\mathbf{C}\mathbf{M}tr(\mathbf{A}\mathbf{M}).\end{aligned}$$

PROOF:

(i)

$$E(\mathbf{X}\mathbf{A}tr(\mathbf{X}\mathbf{B}\mathbf{X}\mathbf{C}\mathbf{X}'))_{ij} = E\left(\left(\sum_{s=1}^{n} x_{is}a_{sj}\right)\left(\sum_{q=1}^{n}\sum_{r=1}^{n}\sum_{l=1}^{p}\sum_{k=1}^{n}\sum_{t=1}^{p} x_{tk}b_{kl}x_{lr}c_{rq}x_{tq}\right)\right)$$

$$
\begin{aligned}
&= E\left(\sum_{\substack{q,r,l\\k,t,s}} x_{is}a_{sj}x_{tk}b_{kl}x_{lr}c_{rq}x_{tq}\right)\\
&= \sum_{\substack{q,r,l\\k,t,s}} \{k_0[\sigma_{it}\phi_{sk}\sigma_{lt}\phi_{rq} + \sigma_{il}\phi_{sr}\sigma_{tt}\phi_{kq} + \sigma_{it}\phi_{sq}\sigma_{tl}\phi_{kr}]\\
&+ c_0[m_{is}m_{tk}\sigma_{lt}\phi_{rq} + m_{is}m_{lr}\sigma_{tt}\phi_{kq}\\
&+ m_{is}m_{tq}\sigma_{tl}\phi_{kr} + m_{tk}m_{lr}\sigma_{it}\phi_{sq} + m_{tk}m_{tq}\sigma_{il}\phi_{sr}]\\
&+ m_{is}m_{tk}m_{lr}m_{tq}\}a_{sj}b_{kl}c_{rq}\\
&= \sum_{\substack{q,r,l\\k,t,s}} k_0[(\sigma_{it}\sigma_{tl}b_{kl}\phi_{sk}a_{sj})(\phi_{rq}c_{rq})\\
&+ (\sigma_{il}b_{kl}\phi_{kq}c_{rq}\phi_{sr}a_{sj})\sigma_{tt}\\
&+ \sigma_{it}\sigma_{tl}b_{kl}\phi_{kr}c_{rq}\phi_{sq}a_{sj}]\\
&+ c_0[(m_{is}a_{sj})(m_{tk}b_{kl}\sigma_{lt})(\phi_{rq}c_{rq})\\
&+ (m_{is}a_{is})\sigma_{tt}(\phi_{kq}c_{rq}m_{lr}b_{kl})\\
&+ (m_{is}a_{sj})(\sigma_{tl}b_{kl}\phi_{kr}c_{rq}m_{tq})\\
&+ \sigma_{it}m_{tk}b_{kl}m_{lr}c_{rq}\phi_{sq}a_{sj}\\
&+ \sigma_{il}b_{kl}m_{tk}m_{tq}c_{rq}\phi_{sr}a_{sj}\\
&+ \sigma_{it}m_{tq}c_{rq}m_{lr}b_{kl}\phi_{sk}a_{sj}]\\
&+ (m_{is}a_{sj})(m_{tk}b_{kl}m_{lr}c_{rq}m_{qj})\\
&= (k_0[\boldsymbol{\Sigma}^2\mathbf{B}'\boldsymbol{\Phi}\mathbf{A}tr(\boldsymbol{\Phi}\mathbf{C}) + \boldsymbol{\Sigma}\mathbf{B}'\boldsymbol{\Phi}\mathbf{C}'\boldsymbol{\Phi}\mathbf{A}tr(\boldsymbol{\Sigma})\\
&+ \boldsymbol{\Sigma}^2\mathbf{B}'\boldsymbol{\Phi}\mathbf{C}\boldsymbol{\Phi}\mathbf{A}] + c_0[\mathbf{M}\mathbf{A}tr(\mathbf{M}\mathbf{B}\boldsymbol{\Sigma})tr(\boldsymbol{\Phi}\mathbf{C})\\
&+ \mathbf{M}\mathbf{A}tr(\boldsymbol{\Sigma})tr(\boldsymbol{\Phi}\mathbf{C}'\mathbf{M}'\mathbf{B}') + \mathbf{M}\mathbf{A}tr(\boldsymbol{\Sigma}\mathbf{B}'\boldsymbol{\Phi}\mathbf{C}\mathbf{M}')\\
&+ \boldsymbol{\Sigma}\mathbf{M}\mathbf{B}\mathbf{M}\mathbf{C}\boldsymbol{\Phi}\mathbf{A} + \boldsymbol{\Sigma}\mathbf{B}'\mathbf{M}'\mathbf{M}\mathbf{C}'\boldsymbol{\Phi}\mathbf{A}\\
&+ \boldsymbol{\Sigma}\mathbf{M}\mathbf{C}'\mathbf{M}'\mathbf{B}'\boldsymbol{\Phi}\mathbf{A}] + \mathbf{M}\mathbf{A}tr(\mathbf{M}\mathbf{B}\mathbf{M}\mathbf{C}\mathbf{M}'))_{ij}.
\end{aligned}
$$

(ii)

$$
\begin{aligned}
E(\mathbf{X}\mathbf{A}\mathbf{X}tr(\mathbf{B}\mathbf{X}\mathbf{C}\mathbf{X}'))_{ij} &= E\left(\left(\sum_{t=1}^{p}\sum_{s=1}^{n} x_{is}a_{st}x_{tj}\right)\left(\sum_{q=1}^{n}\sum_{r=1}^{n}\sum_{l=1}^{p}\sum_{k=1}^{p} b_{kl}x_{lr}c_{rq}x_{kq}\right)\right)\\
&= E\left(\sum_{\substack{q,r,l\\k,t,s}} x_{is}a_{st}x_{tj}b_{kl}x_{lr}c_{rq}x_{kq}\right)
\end{aligned}
$$

$$
\begin{aligned}
&= \sum_{\substack{q,r,l\\k,t,s}} \{k_0[\sigma_{it}\phi_{sj}\sigma_{lk}\phi_{rq} + \sigma_{il}\phi_{sr}\sigma_{tk}\phi_{jq} + \sigma_{ik}\phi_{sq}\sigma_{tl}\phi_{jr}] \\
&+ c_0[m_{is}m_{tj}\sigma_{lk}\phi_{rq} + m_{is}m_{lr}\sigma_{tk}\phi_{jq} \\
&+ m_{is}m_{kq}\sigma_{tl}\phi_{jr} + m_{tj}m_{lr}\sigma_{is}\phi_{sq} + m_{tj}m_{kq}\sigma_{il}\phi_{sr} \\
&+ m_{lr}m_{kq}\sigma_{it}\phi_{sj}] + m_{is}m_{tj}m_{lr}m_{kq}\}a_{st}b_{kl}c_{rq} \\
&= \sum_{\substack{q,r,l\\k,t,s}} k_0[(\sigma_{it}a_{st}\phi_{sj})(\sigma_{lk}b_{kl})(\phi_{rq}c_{rq}) \\
&+ \sigma_{il}b_{kl}\sigma_{tk}a_{st}\phi_{sr}c_{rq}\phi_{jq} \\
&+ \sigma_{ik}b_{kl}\sigma_{tl}a_{st}\phi_{sq}c_{rq}\phi_{jr}] \\
&+ c_0[(m_{is}a_{st}m_{tj})(\sigma_{lk}b_{kl})(\phi_{rq}c_{rq}) \\
&+ m_{is}a_{st}\sigma_{tk}b_{kl}m_{lr}c_{rq}\phi_{jq} \\
&+ m_{is}a_{st}\sigma_{tl}b_{kl}m_{kq}c_{rq}\phi_{jr} \\
&+ \sigma_{ik}b_{kl}m_{lr}c_{rq}\phi_{sq}a_{st}m_{tj} \\
&+ \sigma_{il}b_{kl}m_{kq}c_{rq}\phi_{sr}a_{st}m_{tj} \\
&+ (\sigma_{it}a_{st}\phi_{sj})(m_{kq}c_{rq}m_{lr}b_{kl})] \\
&+ (m_{is}a_{st}m_{tj})(m_{kq}c_{rq}m_{lr}b_{kl}) \\
&= (k_0[\mathbf{\Sigma A}'\boldsymbol{\Phi} tr(\mathbf{\Sigma B})tr(\boldsymbol{\Phi}\mathbf{C}) + \mathbf{\Sigma B}'\mathbf{\Sigma A}'\boldsymbol{\Phi}\mathbf{C}\boldsymbol{\Phi} \\
&+ \mathbf{\Sigma B \Sigma A}'\boldsymbol{\Phi}\mathbf{C}'\boldsymbol{\Phi}] + c_0[\mathbf{MAM}tr(\mathbf{\Sigma B})tr(\boldsymbol{\Phi}\mathbf{C}) \\
&+ \mathbf{MA\Sigma BMC}\boldsymbol{\Phi} + \mathbf{MA\Sigma B}'\mathbf{MC}'\boldsymbol{\Phi} \\
&+ \mathbf{\Sigma BMC}\boldsymbol{\Phi}\mathbf{AM} + \mathbf{\Sigma B}'\mathbf{MC}'\boldsymbol{\Phi}\mathbf{AM} \\
&+ \mathbf{\Sigma A}'\boldsymbol{\Phi} tr(\mathbf{MC}'\mathbf{M}'\mathbf{B}')] + \mathbf{MAM}tr(\mathbf{MC}'\mathbf{M}'\mathbf{B}'))_{ij}.
\end{aligned}
$$

(iii)

$$
\begin{aligned}
E(\mathbf{X}'\mathbf{AX}tr(\mathbf{BXCX}'))_{ij} &= E\left(\left(\sum_{t=1}^{p}\sum_{s=1}^{p} x_{si}a_{st}x_{tj}\right)\left(\sum_{q=1}^{n}\sum_{r=1}^{n}\sum_{l=1}^{p}\sum_{k=1}^{p} b_{kl}x_{lr}c_{rq}x_{kq}\right)\right) \\
&= E\left(\sum_{\substack{q,r,l\\k,t,s}} x_{si}a_{st}x_{tj}b_{kl}x_{lr}c_{rq}x_{kq}\right) \\
&= \sum_{\substack{q,r,l\\k,t,s}} \{k_0[\sigma_{st}\phi_{ij}\sigma_{lk}\phi_{rq} + \sigma_{sl}\phi_{ir}\sigma_{tk}\phi_{jq} + \sigma_{sk}\phi_{iq}\sigma_{tl}\phi_{jr}] \\
&+ c_0[m_{si}m_{tj}\sigma_{lk}\phi_{rq} + m_{si}m_{lr}\sigma_{tk}\phi_{jq}
\end{aligned}
$$

$$
\begin{aligned}
&+ m_{si}m_{kq}\sigma_{tl}\phi_{jr} + m_{tj}m_{lr}\sigma_{sk}\phi_{iq} + m_{tj}m_{kq}\sigma_{sl}\phi_{ir} \\
&+ m_{lr}m_{kq}\sigma_{st}\phi_{ij}] + m_{si}m_{tj}m_{lr}m_{kq}\}a_{st}b_{kl}c_{rq} \\
&= \sum_{\substack{q,r,l\\k,t,s}} k_0[\phi_{ij}(\sigma_{st}a_{st})(\sigma_{lk}b_{kl})(\phi_{rq}c_{rq}) \\
&+ (\phi_{ir}c_{rq}\phi_{jq})(\sigma_{sl}b_{kl}\sigma_{tk}a_{st}) \\
&+ (\phi_{iq}c_{rq}\phi_{jr})(\sigma_{sk}b_{kl}\sigma_{tl}a_{st})] \\
&+ c_0[(m_{si}a_{st}m_{tj})(\sigma_{kl}b_{kl})(\phi_{rq}c_{rq}) \\
&+ m_{si}a_{st}\sigma_{tk}b_{kl}m_{lr}c_{rq}\phi_{jq} \\
&+ m_{si}a_{st}\sigma_{tl}b_{kl}m_{kq}c_{rq}\phi_{jr} \\
&+ \phi_{iq}c_{rq}m_{lr}b_{kl}\sigma_{sk}a_{st}m_{tj} \\
&+ \phi_{ir}c_{rq}m_{kq}b_{kl}\sigma_{sl}a_{st}m_{tj} \\
&+ \phi_{ij}(\sigma_{st}a_{st})(m_{kq}c_{rq}m_{lr}b_{kl})] \\
&+ (m_{si}a_{st}m_{tj})(m_{kq}c_{rq}m_{lr}b_{kl}) \\
&= (k_0[\boldsymbol{\Phi} tr(\boldsymbol{\Sigma}\mathbf{A})tr(\boldsymbol{\Sigma}\mathbf{B})tr(\boldsymbol{\Phi}\mathbf{C}) + \boldsymbol{\Phi}\mathbf{C}\boldsymbol{\Phi} tr(\boldsymbol{\Sigma}\mathbf{B}'\boldsymbol{\Sigma}\mathbf{A}') \\
&+ \boldsymbol{\Phi}\mathbf{C}'\boldsymbol{\Phi} tr(\boldsymbol{\Sigma}\mathbf{B}\boldsymbol{\Sigma}\mathbf{A}')] + c_0[\mathbf{M}'\mathbf{A}\mathbf{M} tr(\boldsymbol{\Sigma}\mathbf{B})tr(\boldsymbol{\Phi}\mathbf{C}) \\
&+ \mathbf{M}'\mathbf{A}\boldsymbol{\Sigma}\mathbf{B}\mathbf{M}\mathbf{C}\boldsymbol{\Phi} + \mathbf{M}'\mathbf{A}\boldsymbol{\Sigma}\mathbf{B}'\mathbf{M}\mathbf{C}'\boldsymbol{\Phi} \\
&+ \boldsymbol{\Phi}\mathbf{C}'\mathbf{M}'\mathbf{B}'\boldsymbol{\Sigma}\mathbf{A}\mathbf{M} + \boldsymbol{\Phi}\mathbf{C}\mathbf{M}'\mathbf{B}\boldsymbol{\Sigma}\mathbf{A}\mathbf{M} \\
&+ \boldsymbol{\Phi} tr(\boldsymbol{\Sigma}\mathbf{A})tr(\mathbf{M}\mathbf{C}'\mathbf{M}'\mathbf{B}')] + \mathbf{M}'\mathbf{A}\mathbf{M} tr(\mathbf{M}\mathbf{C}'\mathbf{M}'\mathbf{B}'))_{ij}.
\end{aligned}
$$

(iv)

$$
\begin{aligned}
E(\mathbf{XBXCX}tr(\mathbf{AX}))_{ij} &= E\left(\left(\sum_{t=1}^{p}\sum_{s=1}^{n} x_{ts}a_{st}\right)\left(\sum_{q=1}^{p}\sum_{r=1}^{n}\sum_{l=1}^{p}\sum_{k=1}^{n} x_{ik}b_{kl}x_{lr}c_{rq}x_{qj}\right)\right) \\
&= E\left(\sum_{\substack{q,r,l\\k,t,s}} x_{ts}a_{st}x_{ik}b_{kl}x_{lr}c_{rq}x_{qj}\right) \\
&= \sum_{\substack{q,r,l\\k,t,s}} \{k_0[\sigma_{ti}\phi_{sk}\sigma_{lq}\phi_{rj} + \sigma_{tl}\phi_{sr}\sigma_{iq}\phi_{kj} + \sigma_{tq}\phi_{sj}\sigma_{il}\phi_{kr}] \\
&+ c_0[m_{ts}m_{ik}\sigma_{lq}\phi_{rj} + m_{ts}m_{lr}\sigma_{iq}\phi_{kj} \\
&+ m_{ts}m_{qj}\sigma_{il}\phi_{kr} + m_{ik}m_{lr}\sigma_{tq}\phi_{sj} + m_{ik}m_{qj}\sigma_{tl}\phi_{sr} \\
&+ m_{lr}m_{qj}\sigma_{ti}\phi_{sk}] + m_{ts}m_{ik}m_{lr}m_{qj}\}a_{st}b_{kl}c_{rq}
\end{aligned}
$$

$$
\begin{aligned}
&= \sum_{\substack{q,r,l\\k,t,s}} k_0[\sigma_{ti}a_{st}\phi_{sk}b_{kl}\sigma_{lq}c_{rq}\phi_{rj}\\
&+ \sigma_{iq}c_{rq}\phi_{sr}a_{st}\sigma_{tl}b_{kl}\phi_{kj}\\
&+ \sigma_{il}b_{kl}\phi_{kr}c_{rq}\sigma_{tq}a_{st}\phi_{sj}]\\
&+ c_0[(m_{ik}b_{kl}\sigma_{lq}c_{rq}\phi_{rj})(m_{ts}a_{st})\\
&+ (\sigma_{iq}c_{rq}m_{lr}b_{kl}\phi_{kj})(m_{ts}a_{st})\\
&+ (\sigma_{il}b_{kl}\phi_{kr}c_{rq}m_{qj})(m_{ts}a_{st})\\
&+ m_{ik}b_{kl}m_{lr}c_{rq}\sigma_{tq}a_{st}\phi_{sj}\\
&+ m_{ik}b_{kl}\sigma_{tl}a_{st}\phi_{sr}c_{rq}m_{qj}\\
&+ \sigma_{ti}a_{st}\phi_{sk}b_{kl}m_{lr}c_{rq}m_{qj}]\\
&+ (m_{ik}b_{kl}m_{lr}c_{rq}m_{qj})(m_{ts}a_{st})\\
&= (k_0[\mathbf{\Sigma A'}\boldsymbol{\Phi}\mathbf{B\Sigma C'}\boldsymbol{\Phi} + \mathbf{\Sigma C'}\boldsymbol{\Phi}\mathbf{A\Sigma B'}\boldsymbol{\Phi}\\
&+ \mathbf{\Sigma B'}\boldsymbol{\Phi}\mathbf{C\Sigma A'}\boldsymbol{\Phi}] + c_0[\mathbf{MB\Sigma C'}\boldsymbol{\Phi} tr(\mathbf{MA})\\
&+ \mathbf{\Sigma C'M'B'}\boldsymbol{\Phi} tr(\mathbf{MA}) + \mathbf{\Sigma B'}\boldsymbol{\Phi}\mathbf{CM} tr(\mathbf{MA})\\
&+ \mathbf{MBMC\Sigma A'}\boldsymbol{\Phi} + \mathbf{MB\Sigma A'}\boldsymbol{\Phi}\mathbf{CM}\\
&+ \mathbf{\Sigma A'}\boldsymbol{\Phi}\mathbf{BMCM}] + \mathbf{MBMCM} tr(\mathbf{AM}))_{ij}.
\end{aligned}
$$

(v)

$$
\begin{aligned}
E(\mathbf{X'BXCX} tr(\mathbf{AX}))_{ij} &= E\left(\left(\sum_{t=1}^{p}\sum_{s=1}^{n} x_{ts}a_{st}\right)\left(\sum_{q=1}^{p}\sum_{r=1}^{n}\sum_{l=1}^{p}\sum_{k=1}^{p} x_{ki}b_{kl}x_{lr}c_{rq}x_{qj}\right)\right)\\
&= E\left(\sum_{\substack{q,r,l\\k,t,s}} x_{ts}a_{st}x_{ki}b_{kl}x_{lr}c_{rq}x_{qj}\right)\\
&= \sum_{\substack{q,r,l\\k,t,s}} \{k_0[\sigma_{tk}\phi_{si}\sigma_{lq}\phi_{rj} + \sigma_{tl}\phi_{sr}\sigma_{kq}\phi_{ij} + \sigma_{tq}\phi_{sj}\sigma_{kl}\phi_{ir}]\\
&+ c_0[m_{ts}m_{ki}\sigma_{lq}\phi_{rj} + m_{ts}m_{lr}\sigma_{kq}\phi_{ij}\\
&+ m_{ts}m_{qj}\sigma_{kl}\phi_{ir} + m_{ki}m_{lr}\sigma_{tq}\phi_{sj} + m_{ki}m_{qj}\sigma_{tl}\phi_{sr}\\
&+ m_{lr}m_{qj}\sigma_{tk}\phi_{si}] + m_{ts}m_{ki}m_{lr}m_{qj}\}a_{st}b_{kl}c_{rq}\\
&= \sum_{\substack{q,r,l\\k,t,s}} k_0[\phi_{si}a_{st}\sigma_{tk}b_{kl}\sigma_{lq}c_{rq}\phi_{rj}\\
&+ \phi_{ij}(\phi_{sr}c_{rq}\sigma_{kq}b_{kl}\sigma_{tl}a_{st})
\end{aligned}
$$

$$
\begin{aligned}
&+ (\phi_{ir}c_{rq}\sigma_{tq}a_{st}\phi_{sj})(\sigma_{kl}b_{kl})] \\
&+ c_0[(m_{ki}b_{kl}\sigma_{lq}c_{rq}\phi_{rj})(m_{ts}a_{st}) \\
&+ \phi_{ij}(m_{ts}a_{st})(\sigma_{kq}c_{rq}m_{lr}b_{kl}) \\
&+ (\phi_{ir}c_{rq}m_{qj})(m_{ts}a_{st})(\sigma_{kl}b_{kl}) \\
&+ m_{ki}b_{kl}m_{lr}c_{rq}\sigma_{tq}a_{st}\phi_{sj} \\
&+ m_{ki}b_{kl}\sigma_{tl}a_{st}\phi_{sr}c_{rq}m_{qj} \\
&+ \phi_{si}a_{st}\sigma_{tk}b_{kl}m_{lr}c_{rq}m_{qj}] \\
&+ (m_{ki}b_{kl}m_{lr}c_{rq}m_{qj})(m_{ts}a_{st}) \\
&= (k_0[\boldsymbol{\Phi}\mathbf{A}\boldsymbol{\Sigma}\mathbf{B}\boldsymbol{\Sigma}\mathbf{C}'\boldsymbol{\Phi} + \boldsymbol{\Phi}tr(\boldsymbol{\Phi}\mathbf{C}\boldsymbol{\Sigma}\mathbf{B}\boldsymbol{\Sigma}\mathbf{A}') \\
&+ \boldsymbol{\Phi}\mathbf{C}\boldsymbol{\Sigma}\mathbf{A}'\boldsymbol{\Phi}tr(\boldsymbol{\Sigma}\mathbf{B})] + c_0[\mathbf{M}'\mathbf{B}\boldsymbol{\Sigma}\mathbf{C}'\boldsymbol{\Phi}tr(\mathbf{M}\mathbf{A}) \\
&+ \boldsymbol{\Phi}tr(\mathbf{M}\mathbf{A})tr(\boldsymbol{\Sigma}\mathbf{C}'\mathbf{M}'\mathbf{B}') + \boldsymbol{\Phi}\mathbf{C}\mathbf{M}tr(\mathbf{M}\mathbf{A})tr(\boldsymbol{\Sigma}\mathbf{B}') \\
&+ \mathbf{M}'\mathbf{B}\mathbf{M}\mathbf{C}\boldsymbol{\Sigma}\mathbf{A}'\boldsymbol{\Phi} + \mathbf{M}'\mathbf{B}\boldsymbol{\Sigma}\mathbf{A}'\boldsymbol{\Phi}\mathbf{C}\mathbf{M} \\
&+ \boldsymbol{\Phi}\mathbf{A}\boldsymbol{\Sigma}\mathbf{B}\mathbf{M}\mathbf{C}\mathbf{M}] + \mathbf{M}'\mathbf{B}\mathbf{M}\mathbf{C}\mathbf{M}tr(\mathbf{M}\mathbf{A}))_{ij}.
\end{aligned}
$$

(vi)

$$
\begin{aligned}
E(\mathbf{X}'\mathbf{B}\mathbf{X}'\mathbf{C}\mathbf{X}tr(\mathbf{A}\mathbf{X}))_{ij} &= E\left(\left(\sum_{t=1}^{p}\sum_{s=1}^{n} x_{ts}a_{st}\right)\left(\sum_{q=1}^{p}\sum_{r=1}^{p}\sum_{l=1}^{n}\sum_{k=1}^{p} x_{ki}b_{kl}x_{rl}c_{rq}x_{qj}\right)\right) \\
&= E\left(\sum_{\substack{q,r,l\\k,t,s}} x_{ts}a_{st}x_{ki}b_{kl}x_{rl}c_{rq}x_{qj}\right) \\
&= \sum_{\substack{q,r,l\\k,t,s}} \{k_0[\sigma_{tk}\phi_{si}\sigma_{rq}\phi_{lj} + \sigma_{tr}\phi_{sl}\sigma_{kq}\phi_{ij} + \sigma_{tq}\phi_{sj}\sigma_{kr}\phi_{il}] \\
&+ c_0[m_{ts}m_{ki}\sigma_{rq}\phi_{lj} + m_{ts}m_{rl}\sigma_{kq}\phi_{ij} \\
&+ m_{ts}m_{qj}\sigma_{kr}\phi_{il} + m_{ki}m_{rl}\sigma_{tq}\phi_{sj} + m_{ki}m_{qj}\sigma_{tr}\phi_{sl} \\
&+ m_{rl}m_{qj}\sigma_{tk}\phi_{si}] + m_{ts}m_{ki}m_{rl}m_{qj}\}a_{st}b_{kl}c_{rq} \\
&= \sum_{\substack{q,r,l\\k,t,s}} k_0[(\phi_{si}a_{st}\sigma_{tk}b_{kl}\phi_{lj})(\sigma_{rq}c_{rq}) \\
&+ \phi_{ij}(\phi_{sl}b_{kl}\sigma_{kq}c_{rq}\sigma_{tr}a_{st}) \\
&+ \phi_{il}b_{kl}\sigma_{kr}c_{rq}\sigma_{tq}a_{st}\phi_{sj}] \\
&+ c_0[(m_{ki}b_{kl}\phi_{lj})(m_{ts}a_{st})(\sigma_{rq}c_{rq}) \\
&+ \phi_{ij}(m_{ts}a_{st})(\sigma_{kq}c_{rq}m_{rl}b_{kl})
\end{aligned}
$$

$$+ (\phi_{il}b_{kl}\sigma_{kr}c_{rq}m_{qj})(m_{ts}a_{st})$$
$$+ m_{ki}b_{kl}m_{rl}c_{rq}\sigma_{tq}a_{st}\phi_{sj}$$
$$+ m_{ki}b_{kl}\phi_{sl}a_{st}\sigma_{tr}c_{rq}m_{qj}$$
$$+ \phi_{si}a_{st}\sigma_{tk}b_{kl}m_{rl}c_{rq}m_{qj}]$$
$$+ (m_{ki}b_{kl}m_{rl}c_{rq}m_{qj})(m_{ts}a_{st})$$
$$= (k_0[\boldsymbol{\Phi}\mathbf{A}\boldsymbol{\Sigma}\mathbf{B}\boldsymbol{\Phi}tr(\boldsymbol{\Sigma}\mathbf{C}) + \boldsymbol{\Phi}tr(\boldsymbol{\Phi}\mathbf{B}'\boldsymbol{\Sigma}\mathbf{C}'\boldsymbol{\Sigma}\mathbf{A}')$$
$$+ \boldsymbol{\Phi}\mathbf{B}'\boldsymbol{\Sigma}\mathbf{C}\boldsymbol{\Sigma}\mathbf{A}'\boldsymbol{\Phi}] + c_0[\mathbf{M}'\mathbf{B}\boldsymbol{\Phi}tr(\mathbf{M}\mathbf{A})tr(\boldsymbol{\Sigma}\mathbf{C})$$
$$+ \boldsymbol{\Phi}tr(\mathbf{M}\mathbf{A})tr(\boldsymbol{\Sigma}\mathbf{C}'\mathbf{M}\mathbf{B}') + \boldsymbol{\Phi}\mathbf{B}'\boldsymbol{\Sigma}\mathbf{C}\mathbf{M}tr(\mathbf{M}\mathbf{A})$$
$$+ \mathbf{M}'\mathbf{B}\mathbf{M}'\mathbf{C}\boldsymbol{\Sigma}\mathbf{A}'\boldsymbol{\Phi} + \mathbf{M}'\mathbf{B}\boldsymbol{\Phi}\mathbf{A}\boldsymbol{\Sigma}\mathbf{C}\mathbf{M}$$
$$+ \boldsymbol{\Phi}\mathbf{A}\boldsymbol{\Sigma}\mathbf{B}\mathbf{M}'\mathbf{C}\mathbf{M}] + \mathbf{M}'\mathbf{B}\mathbf{M}'\mathbf{C}\mathbf{M}tr(\mathbf{A}\mathbf{M}))_{ij}.$$

Remark 3.4. The expected values of many of the expressions in Theorems 3.18–3.22 were computed by Nel (1977) for the case where $\mathbf{X}$ has matrix variate normal distribution. If $\mathbf{X} \sim N_{p,n}(\mathbf{M}, \boldsymbol{\Sigma} \otimes \boldsymbol{\Phi})$, then $-2\psi'(0) = 1$ and $4\psi''(0) = 1$. Therefore taking $c_0 = k_0 = 1$, our results give the expected values for the normal case, and so Nel's results can be obtained as special cases of the formulae presented here.

Next, we give some applications of Theorems 3.18–3.22.

Theorem 3.23. *Let* $\mathbf{X} \sim E_{p,n}(\mathbf{0}, \boldsymbol{\Sigma} \otimes \mathbf{I}_n, \psi)$ *with finite fourth order moments. Let* $\mathbf{Y} = \mathbf{X}\left(\mathbf{I}_n - \frac{\mathbf{e}_n\mathbf{e}_n'}{n}\right)\mathbf{X}'$. *Let* $c_0 = -2\psi'(0)$ *and* $k_0 = 4\psi''(0)$. *Then,*

(i)

$$E(\mathbf{Y}) = c_0(n-1)\boldsymbol{\Sigma}, \tag{3.36}$$

(ii)

$$Cov(y_{ij}, y_{kl}) = k_0(n-1)(\sigma_{il}\sigma_{jk} + \sigma_{ik}\sigma_{jl}) + (n-1)^2(k_0 - c_0^2)\sigma_{ij}\sigma_{kl}, \tag{3.37}$$

(iii)

$$Var(y_{ij}) = (nk_0 - (n-1)c_0^2)(n-1)\sigma_{ij}^2 + k_0(n-1)\sigma_{ii}\sigma_{jj}. \tag{3.38}$$

PROOF: Let $\mathbf{A} = \mathbf{I}_n - \frac{\mathbf{e}_n\mathbf{e}_n'}{n}$.

(i) Using Theorem 3.18, we get

$$E(\mathbf{X}\mathbf{A}\mathbf{X}') = c_0\boldsymbol{\Sigma}tr\left(\left(\mathbf{I}_n - \frac{\mathbf{e}_n\mathbf{e}_n'}{n}\right)\mathbf{I}_n\right) + \mathbf{0} = c_0(n-1)\boldsymbol{\Sigma}.$$

(ii) Let $\mathbf{q}^m$ be a p-dimensional column vector such that

$$\mathbf{q}_i^m = \begin{cases} 1 \text{ if } i = m \\ 0 \text{ if } i \neq m \end{cases}, \quad m = 1,2,\ldots,p; \quad i = 1,2,\ldots,p.$$

Then, $y_{ij} = \mathbf{q}^{i\prime}\mathbf{Y}\mathbf{q}^j = \mathbf{q}^{i\prime}\mathbf{X}\mathbf{A}\mathbf{X}'\mathbf{q}^j$ and $y_{kl} = \mathbf{q}^{k\prime}\mathbf{Y}\mathbf{q}^l = \mathbf{q}^{k\prime}\mathbf{X}\mathbf{A}\mathbf{X}'\mathbf{q}^l$. Now from (3.36), it follows that

$$E(y_{ij}) = c_0(n-1)\sigma_{ij} \quad \text{and} \quad E(y_{kl}) = c_0(n-1)\sigma_{kl}.$$

Since, $\mathbf{X} \sim E_{p,n}(\mathbf{0}, \Sigma \otimes \mathbf{I}_n, \psi)$, we have $\mathbf{X}' \sim E_{n,p}(\mathbf{0}, \mathbf{I}_n \otimes \Sigma, \psi)$. Using Theorem 3.22, we get

$$\begin{aligned} E(y_{ij}y_{kl}) &= E(\mathbf{q}^{i\prime}\mathbf{X}\mathbf{A}\mathbf{X}'\mathbf{q}^j\,\mathbf{q}^{k\prime}\mathbf{X}\mathbf{A}\mathbf{X}'\mathbf{q}^l) \\ &= \mathbf{q}^{i\prime}E(\mathbf{X}\mathbf{A}\mathbf{X}'\mathbf{q}^j\,\mathbf{q}^{k\prime}\mathbf{X}\mathbf{A}\mathbf{X}')\mathbf{q}^l \\ &= \mathbf{q}^{i\prime}\Big(k_0(\Sigma tr(\mathbf{I}_n\mathbf{A}\mathbf{I}_n\mathbf{A})tr(\mathbf{q}^j\mathbf{q}^{k\prime}\Sigma) \\ &+ \Sigma\mathbf{q}^j\mathbf{q}^{k\prime}\Sigma tr(\mathbf{A}\mathbf{I}_n)tr(\mathbf{A}\mathbf{I}_n) + \Sigma(\mathbf{q}^j\mathbf{q}^{k\prime})'\Sigma tr(\mathbf{A}\mathbf{I}_n\mathbf{A}\mathbf{I}_n))\Big)\mathbf{q}^l \\ &= k_0\Big((n-1)\mathbf{q}^{i\prime}\Sigma\mathbf{q}^l tr(\mathbf{q}^{k\prime}\Sigma\mathbf{q}^j) + \mathbf{q}^{i\prime}\Sigma\mathbf{q}^j\mathbf{q}^{k\prime}\Sigma\mathbf{q}^l(n-1)^2 \\ &+ \mathbf{q}^{i\prime}\Sigma\mathbf{q}^k\,\mathbf{q}^{j\prime}\Sigma\mathbf{q}^l(n-1)\Big) \\ &= k_0\left((n-1)\sigma_{il}\sigma_{kj} + (n-1)^2\sigma_{ij}\sigma_{kl} + (n-1)\sigma_{ik}\sigma_{jl}\right). \end{aligned}$$

Then,

$$\begin{aligned} Cov(y_{ij}, y_{kl}) &= E(y_{ij}y_{kl}) - E(y_{ij})E(y_{kl}) \\ &= k_0\left((n-1)\sigma_{il}\sigma_{kj} + (n-1)^2\sigma_{ij}\sigma_{kl} + (n-1)\sigma_{ik}\sigma_{jl}\right) \\ &- c_0^2(n-1)^2\sigma_{ij}\sigma_{kl} \\ &= k_0(n-1)(\sigma_{il}\sigma_{jk} + \sigma_{ik}\sigma_{jl}) + (n-1)^2(k_0 - c_0^2)\sigma_{ij}\sigma_{kl} \end{aligned}$$

which proves (3.37).

(iii) Take $k = i$ and $l = j$ in (3.37). Then,

$$\begin{aligned} Var(y_{ij}) &= k_0(n-1)(\sigma_{ij}^2 + \sigma_{ii}\sigma_{jj}) + (n-1)^2(k_0 - c_0^2)\sigma_{ij}^2 \\ &= (n-1)(nk_0 - (n-1)c_0^2)\sigma_{ij}^2 + k_0(n-1)\sigma_{ii}\sigma_{jj}. \end{aligned}$$

■

Example 3.1. Let $\mathbf{X} \sim Mt_{p,n}(m, \mathbf{0}, \Sigma \otimes \Phi)$, $m > 4$, and $\mathbf{Y} = \mathbf{X}\left(\mathbf{I}_n - \frac{\mathbf{e}_n\mathbf{e}_n'}{n}\right)\mathbf{X}'$. We want to find $E(\mathbf{Y})$, $Cov(y_{ij}, y_{kl})$ and $Var(y_{ij})$. In order to compute them, we need to know $c_0 = -2\psi'(0)$ and $k_0 = 4\psi''(0)$. Let $u = \frac{x_{11}}{\sqrt{\sigma_{11}}}$. Then, $u \sim Mt_{1,1}(m, 0, 1)$;

that is u has a one-dimensional Student's t-distribution with m degrees of freedom. (This will be shown in Chap. 4.) Hence, using Theorem 3.16 we get $E(u^2) = c_0$ and $E(u^4) = 3k_0$. It is known that

$$E(u^r) = \frac{m^{\frac{r}{2}} \beta\left(\frac{r+1}{2}, \frac{m-r}{2}\right)}{\beta\left(\frac{1}{2}, \frac{m}{2}\right)} \quad \text{for } m > r \text{ and } r \text{ even},$$

(see Mood, Graybill, and Boes, 1974, p. 543). In particular, we have

$$\begin{aligned} E(u^2) &= \frac{m\beta\left(\frac{2+1}{2}, \frac{m-2}{2}\right)}{\beta\left(\frac{1}{2}, \frac{m}{2}\right)} \\ &= m\frac{\frac{1}{2}\Gamma\left(\frac{1}{2}\right)\Gamma\left(\frac{m-2}{2}\right)}{\Gamma\left(\frac{1}{2}\right)\left(\frac{m-2}{2}\right)\Gamma\left(\frac{m-2}{2}\right)} \\ &= \frac{m}{m-2}, \end{aligned}$$

and

$$\begin{aligned} E(u^4) &= \frac{m^2\beta\left(\frac{4+1}{2}, \frac{m-4}{2}\right)}{\beta\left(\frac{1}{2}, \frac{m}{2}\right)} \\ &= m^2\frac{\frac{3}{2}\frac{1}{2}\Gamma\left(\frac{1}{2}\right)\Gamma\left(\frac{m-4}{2}\right)}{\Gamma\left(\frac{1}{2}\right)\left(\frac{m-2}{2}\right)\left(\frac{m-4}{2}\right)\Gamma\left(\frac{m-2}{4}\right)} \\ &= \frac{3m^2}{(m-2)(m-4)}. \end{aligned}$$

Hence, $c_0 = \frac{m}{m-2}$ and $k_0 = \frac{m^2}{(m-2)(m-4)}$. Now, using Theorem 3.23, we get

$$E(\mathbf{Y}) = (n-1)\frac{m}{m-2}\boldsymbol{\Sigma},$$

$$\begin{aligned} Cov(y_{ij}, y_{kl}) &= (n-1)\left[\frac{m^2}{(m-2)(m-4)}(\sigma_{il}\sigma_{jk} + \sigma_{ik}\sigma_{jl})\right. \\ &\quad \left. + (n-1)\left(\frac{m^2}{(m-2)(m-4)} - \frac{m^2}{(m-2)^2}\right)\sigma_{ij}\sigma_{kl}\right] \\ &= (n-1)\frac{m^2}{(m-2)(m-4)} \\ &\quad \times \left[\sigma_{il}\sigma_{jk} + \sigma_{ik}\sigma_{jl} + (n-1)\left(1 - \frac{m-4}{m-2}\right)\sigma_{ij}\sigma_{kl}\right] \end{aligned}$$

$$= (n-1)\frac{m^2}{(m-2)(m-4)}$$
$$\times \left[\sigma_{il}\sigma_{jk} + \sigma_{ik}\sigma_{jl} + \frac{2(n-1)}{m-2}\sigma_{ij}\sigma_{kl}\right],$$

and

$$Var(y_{ij}) = (n-1)\frac{m^2}{(m-2)(m-4)}\left[\sigma_{ij}^2 + \sigma_{ii}\sigma_{jj} + \frac{2(n-1)}{m-2}\sigma_{ij}^2\right]$$
$$= (n-1)\frac{m^2}{(m-2)(m-4)}\left[\sigma_{ii}\sigma_{jj} + \frac{2n+m-4}{m-2}\sigma_{ij}^2\right].$$

Chapter 4
Mixtures of Normal Distributions

4.1 Mixture by Distribution Function

Muirhead (1982) gave a definition of scale mixture of vector variate normal distributions. Using Corollary 2.6, the scale mixture of matrix variate normal distributions can be defined as follows (Gupta and Varga, 1995a).

Definition 4.1. Let $\mathbf{M}: p \times n$, $\Sigma: p \times p$, and $\Phi: n \times n$ be constant matrices such that $\Sigma > \mathbf{0}$ and $\Phi > \mathbf{0}$. Assume $G(z)$ is a distribution function on $(0, \infty)$. Let $\mathbf{X} \in \mathbb{R}^{p \times n}$ and define

$$g(\mathbf{X}) = \frac{1}{(2\pi)^{\frac{pn}{2}} |\Sigma|^{\frac{n}{2}} |\Phi|^{\frac{p}{2}}} \times \int_0^\infty z^{-\frac{pn}{2}} etr\left(-\frac{1}{2z}(\mathbf{X}-\mathbf{M})'\Sigma^{-1}(\mathbf{X}-\mathbf{M})\Phi^{-1}\right) dG(z). \tag{4.1}$$

Then the m.e.c. distribution whose p.d.f. is $g(\mathbf{X})$ is called a scale mixture of matrix variate normal distributions.

Remark 4.1. In this chapter we will denote the p.d.f. of $\mathbf{X} \sim N_{p,n}(\mathbf{M}, \Sigma \otimes \Phi)$ by $f_{N_{p,n}(\mathbf{M},\Sigma\otimes\Phi)}(\mathbf{X})$. With this notation, (4.1) can be written as

$$g(\mathbf{X}) = \int_0^\infty f_{N_{p,n}(\mathbf{M},z\Sigma\otimes\Phi)}(\mathbf{X}) dG(z).$$

Remark 4.2. Let $\mathbf{X}$ be a $p \times n$ random matrix. Then, $\mathbf{X} \sim E_{p,n}(\mathbf{M}, \Sigma \otimes \Phi, \psi)$, $\Sigma > \mathbf{0}$, $\Phi > \mathbf{0}$ has the p.d.f. defined by (4.1) if and only if its characteristic function is

$$\phi_{\mathbf{X}}(\mathbf{T}) = etr(i\mathbf{T}'\mathbf{M}) \int_0^\infty etr\left(-\frac{z\mathbf{T}'\Sigma\mathbf{T}\Phi}{2}\right) dG(z), \tag{4.2}$$

A.K. Gupta et al., *Elliptically Contoured Models in Statistics and Portfolio Theory*, DOI 10.1007/978-1-4614-8154-6_4, © Springer Science+Business Media New York 2013

that is

$$\psi(v) = \int_0^\infty \exp\left(-\frac{zv}{2}\right) dG(z).$$

This statement follows from a more general result proved later in Theorem 4.5.

Remark 4.3. From (4.2), it follows that $\mathbf{X} \sim E_{p,n}(\mathbf{M}, \Sigma \otimes \Phi, \psi)$, $\Sigma > \mathbf{0}$, $\Phi > \mathbf{0}$ has the p.d.f. defined by (4.1) if and only if $\mathbf{X} \approx z^{\frac{1}{2}}\mathbf{Y}$ where $\mathbf{Y} \sim N_{p,n}(\mathbf{M}, \Sigma \otimes \Phi)$, z has the distribution $G(z)$, and z and $\mathbf{Y}$ are independent.

The relationship between the characteristic function of a scale mixture of normal distributions and its stochastic representation is pointed out in the next theorem, due to Cambanis, Huang, and Simons (1981).

Theorem 4.1. *Let* $\mathbf{X} \sim E_{p,n}(\mathbf{0}, \mathbf{I}_p \otimes \mathbf{I}_n, \psi)$ *have the stochastic representation* $\mathbf{X} \approx r\mathbf{U}$. *Let* $G(z)$ *be a distribution function on* $(0, \infty)$. *Then,*

$$\psi(v) = \int_0^\infty \exp\left(-\frac{zv}{2}\right) dG(z) \tag{4.3}$$

if and only if r is absolutely continuous with p.d.f.

$$l(r) = \frac{1}{2^{\frac{pn}{2}-1}\Gamma\left(\frac{pn}{2}\right)} r^{pn-1} \int_0^\infty z^{-\frac{pn}{2}} \exp\left(-\frac{r^2}{2z}\right) dG(z). \tag{4.4}$$

PROOF: First, assume (4.3) holds. Then, with the notation of Remark 4.3, we can write $\mathbf{X} \approx z^{\frac{1}{2}}\mathbf{Y}$. Hence, $r\mathbf{U} \approx z^{\frac{1}{2}}\mathbf{Y}$, and $tr(r^2\mathbf{U}'\mathbf{U}) \approx tr(z\mathbf{Y}'\mathbf{Y})$. Consequently,

$$r^2 \approx z tr(\mathbf{Y}'\mathbf{Y}). \tag{4.5}$$

Since $\mathbf{Y} \sim N_{p,n}(\mathbf{0}, \mathbf{I}_p \otimes \mathbf{I}_n)$, and hence $tr(\mathbf{Y}'\mathbf{Y}) \sim \chi^2_{pn}$. Now, χ^2_{pn} has the p.d.f.

$$f_1(w) = \frac{1}{2^{\frac{pn}{2}}\Gamma\left(\frac{pn}{2}\right)} w^{\frac{pn}{2}-1} e^{-\frac{w}{2}}, \quad w > 0.$$

Denoting r^2 by s, from (4.5) we obtain the p.d.f. of s as

$$\begin{aligned} f_2(s) &= \int_0^\infty \frac{1}{z} f_1\left(\frac{s}{z}\right) dG(z) \\ &= \int_0^\infty \frac{1}{z} \frac{1}{2^{\frac{pn}{2}}\Gamma\left(\frac{pn}{2}\right)} \left(\frac{s}{z}\right)^{\frac{pn}{2}-1} e^{-\frac{s}{2z}} dG(z) \\ &= \frac{1}{2^{\frac{pn}{2}}\Gamma\left(\frac{pn}{2}\right)} s^{\frac{pn}{2}-1} \int_0^\infty z^{-\frac{pn}{2}} e^{-\frac{s}{2z}} dG(z). \end{aligned}$$

Since $r^2 = s$, we have $J(s \to r) = 2r$ and so the p.d.f. of r is

$$l(r) = \frac{1}{2^{\frac{pn}{2}-1}\Gamma\left(\frac{pn}{2}\right)} r^{pn-1} \int_0^\infty z^{-\frac{pn}{2}} \exp\left(-\frac{r^2}{2z}\right) dG(z)$$

which is (4.4).

The other direction of the theorem follows from the fact that r and ψ determine each other. ■

The question arises when the p.d.f. of a m.e.c. distribution can be expressed as a scale mixture of matrix variate normal distributions. With the help of the next theorem, we can answer this question. This theorem was first derived by Schoenberg (1938) and the proof given here is due to Fang, Kotz, and Ng (1990) who made use of a derivation of Kingman (1972).

Theorem 4.2. *Let $\psi : [0,\infty) \to I\!R$ be a real function. Then, $\psi(\mathbf{t}'\mathbf{t})$, $\mathbf{t} \in I\!R^k$ is a characteristic function for every $k \geq 1$ if and only if*

$$\psi(u) = \int_0^\infty \exp\left(-\frac{uz}{2}\right) dG(z), \tag{4.6}$$

where $G(z)$ is a distribution function on $[0,\infty)$.

PROOF: First, assume (4.6) holds. Let $k \geq 1$ be an integer and $\mathbf{x}$ be a k-dimensional random vector with p.d.f.

$$g(\mathbf{x}) = \frac{1}{(2\pi)^{\frac{k}{2}}} \int_0^\infty z^{-\frac{k}{2}} \exp\left(-\frac{\mathbf{x}'\mathbf{x}}{2z}\right) dG(z).$$

Then, the characteristic function of $\mathbf{x}$ is

$$\begin{aligned}
\phi_{\mathbf{x}}(\mathbf{t}) &= \int_{I\!R^k} \exp(i\mathbf{t}'\mathbf{x}) g(\mathbf{x}) d\mathbf{x} \\
&= \int_{I\!R^k} \exp(i\mathbf{t}'\mathbf{x}) \frac{1}{(2\pi)^{\frac{k}{2}}} \int_0^\infty z^{-\frac{k}{2}} \exp\left(-\frac{\mathbf{x}'\mathbf{x}}{2z}\right) dG(z) d\mathbf{x} \\
&= \int_0^\infty \int_{I\!R^k} \exp(i\mathbf{t}'\mathbf{x}) \frac{1}{(2\pi z)^{\frac{k}{2}}} \exp\left(-\frac{\mathbf{x}'\mathbf{x}}{2z}\right) d\mathbf{x} dG(z) \\
&= \int_0^\infty \exp\left(-\frac{\mathbf{t}'\mathbf{t}z}{2}\right) dG(z),
\end{aligned}$$

where we used the fact that $\int_{I\!R^k} \exp(i\mathbf{t}'\mathbf{x}) \frac{1}{(2\pi z)^{\frac{k}{2}}} \exp\left(-\frac{\mathbf{x}'\mathbf{x}}{2z}\right) d\mathbf{x}$ is the characteristic function of $N_k(\mathbf{0}, z\mathbf{I}_k)$ and hence is equal to $\exp\left(-\frac{\mathbf{t}'\mathbf{t}z}{2}\right)$. Hence, $\psi(\mathbf{t}'\mathbf{t})$ is the characteristic function of $\mathbf{x}$, where $\mathbf{t} \in I\!R^k$.

Conversely, assume $\psi(\mathbf{t}'\mathbf{t})$ is a characteristic function for every $k \geq 1$. Then, we can choose an infinite sequence of random variables $(x_1, x_2, \ldots)$, such that for every $k \geq 1$ integer, the characteristic function of $(x_1, x_2, \ldots, x_k)$ is $\psi(\mathbf{t}'\mathbf{t})$, where $\mathbf{t} \in \mathbb{R}^k$. Let $\{\pi(1), \pi(2), \ldots, \pi(k)\}$ be a permutation of the numbers $\{1, 2, \ldots, k\}$. Then the characteristic function of $(x_{\pi(1)}, x_{\pi(2)}, \ldots, x_{\pi(k)})$ is also $\psi(\mathbf{t}'\mathbf{t})$. Hence, $(x_1, x_2, \ldots, x_k)$ and $(x_{\pi(1)}, x_{\pi(2)}, \ldots, x_{\pi(k)})$ are identically distributed. Thus the infinite sequence $(x_1, x_2, \ldots)$ is exchangeable.

Let $(\Omega, \mathscr{A}, P)$ be the probability space on which $x_1, x_2, \ldots, x_n, \ldots$ are defined. From De Finetti's theorem (see Billingsley, 1979, p. 425), we know that there exists a sub σ-field of $\mathscr{A}$ say $\mathscr{F}$ such that conditional upon $\mathscr{F}$, x_i's are identically distributed and conditionally independent. The conditional independence means that for every integer n

$$P(x_i \in M_i,\ i = 1, 2, \ldots, n|\mathscr{F}) = \prod_{i=1}^{n} P(x_i \in M_i|\mathscr{F}) \tag{4.7}$$

where $M_i \in \mathscr{B}(\mathbb{R})$, $i = 1, 2, \ldots, n$. Moreover, $\mathscr{F}$ has the property that for every permutation of $1, 2, \ldots, n$, say $\{\pi(1), \pi(2), \ldots, \pi(k)\}$,

$$P(x_i \in M_i,\ i = 1, 2, \ldots, n|\mathscr{F}) = P(x_{\pi(i)} \in M_i,\ i = 1, 2, \ldots, n|\mathscr{F}). \tag{4.8}$$

Let $(x_i|\mathscr{F})_\omega$ be a regular conditional distribution of x_i given $\mathscr{F}$ and $E(x_i|\mathscr{F})_\omega$ be a conditional expectation of x_i given $\mathscr{F}$, $i = 1, 2, \ldots$. Here, $\omega \in \Omega$ (see Billingsley, 1979, p. 390). Then, for $g : \mathbb{R} \to \mathbb{R}$ integrable, we have

$$E(g(x_i)|\mathscr{F})_\omega = \int_0^\infty g(x_i) d(x_i|\mathscr{F})_\omega \tag{4.9}$$

(see Billingsley, 1979, p. 399). Now, it follows from (4.7) and (4.8) that for fixed $\omega \in \Omega$,

$$\begin{aligned} P(x_1 \in M|\mathscr{F})_\omega &= P(x_1 \in M, x_i \in \mathbb{R}, i = 2, \ldots, k|\mathscr{F})_\omega \\ &= P(x_k \in M, x_1 \in \mathbb{R}, x_i \in \mathbb{R}, i = 2, \ldots, k-1|\mathscr{F})_\omega \\ &= P(x_k \in M|\mathscr{F})_\omega. \end{aligned}$$

Hence, for fixed $\omega \in \Omega$, and any positive integer k,

$$(x_1|\mathscr{F})_\omega = (x_k|\mathscr{F})_\omega \quad \text{almost everywhere.} \tag{4.10}$$

Define

$$\phi(t)_\omega = \int_{\mathbb{R}} e^{itx_1} d(x_1|\mathscr{F})_\omega, \tag{4.11}$$

where $t \in \mathbb{R}$, and $\omega \in \Omega$. Then, from (4.10) we get

$$\phi(t)_\omega = \int_{\mathbb{R}} e^{itx_k} d(x_k|\mathscr{F})_\omega. \tag{4.12}$$

For fixed $t \in \mathbb{R}$, $\phi(t)_\omega$ is a $\mathscr{F}$-measurable random variable. On the other hand, for fixed ω, $\phi(t)_\omega$ is a continuous function of t since it is the characteristic function of the distribution defined by $(x_1|\mathscr{F})_\omega$. Since $\phi(t)_\omega$ is a characteristic function, we have $|\phi(t)_\omega| \leq 1$ and $\phi(0)_\omega = 1$. We also have

$$\phi(-t)_\omega = E(e^{i(-t)x_1}|\mathscr{F})_\omega = \overline{E(e^{itx_1}|\mathscr{F})_\omega} = \overline{\phi(t)_\omega}.$$

From (4.9) and (4.12), we see that for any positive integer k,

$$\phi(t)_\omega = E(e^{itx_k}|\mathscr{F})_\omega. \tag{4.13}$$

Using (4.7) and (4.13), we get

$$\begin{aligned} E\left(\exp\left(i\sum_{j=1}^n t_j x_j\right)|\mathscr{F}\right)_\omega &= \prod_{j=1}^n E\left(e^{it_jx_j}|\mathscr{F}\right)_\omega \\ &= \prod_{j=1}^n \phi(t_j)_\omega. \end{aligned} \tag{4.14}$$

Therefore,

$$E\left(\exp\left(i\sum_{j=1}^n t_j x_j\right)\right) = E\left(\prod_{j=1}^n \phi(t_j)\right).$$

The left-hand side of the last expression is the characteristic function of $(x_1,x_2,\ldots,x_n)$. Hence, we get

$$E\left(\prod_{j=1}^n \phi(t_j)\right) = \psi\left(\sum_{j=1}^n t_j^2\right). \tag{4.15}$$

Let u and v be real numbers and define $w = (u^2+v^2)^{\frac{1}{2}}$. Then, we can write

$$\begin{aligned} E(|\phi(w)-\phi(u)\phi(v)|^2) &= E((\phi(w)-\phi(u)\phi(v))\overline{(\phi(w)-\phi(u)\phi(v))}) \\ &= E((\phi(w)-\phi(u)\phi(v))(\phi(-w)-\phi(-u)\phi(-v))) \\ &= E(\phi(w)\phi(-w)) + E(\phi(u)\phi(-u)\phi(v)\phi(-v)) \\ &\quad - E(\phi(w)\phi(-u)\phi(-v)) - E(\phi(-w)\phi(u)\phi(v)). \end{aligned}$$

Using (4.15), we see that all four terms in the last expression equal $\psi(2w^2)$, hence,

$$E(|\phi(w) - \phi(u)\phi(v)|^2) = 0.$$

That means, $\phi(u)\phi(v) = \phi(w)$ with probability one, or equivalently $\phi(u)_\omega\phi(v)_\omega = \phi(w)_\omega$ for $\omega \in C(u,v)$, where $C(u,v) \subset \Omega$ and $P(C(u,v)) = 1$.

Now, we have

$$\phi(u)_\omega\phi(0)_\omega = \phi(|u|)_\omega \quad \text{for} \quad \omega \in C(u,0)$$

and

$$\phi(-u)_\omega\phi(0)_\omega = \phi(|u|)_\omega \quad \text{for} \quad \omega \in C(-u,0).$$

But $\phi(0)_\omega = 1$ and so $\phi(u)_\omega = \phi(\underline{-u)_\omega}$ for $\omega \in C(u,0) \cap C(\underline{-u,0})$. However we have already shown that $\phi(-u)_\omega = \overline{\phi(u)_\omega}$. Hence, $\phi(u)_\omega = \overline{\phi(u)_\omega}$ and therefore $\phi(u)_\omega$ is real for $\omega \in C(u,0) \cap C(-u,0)$. Similarly, $\phi(v)_\omega$ is real for $\omega \in C(0,v) \cap C(0,-v)$.

Define

$$C = \bigcap_{u,v \text{ rational numbers}} \{C(u,v) \cap C(u,0) \cap C(-u,0) \cap C(0,v) \cap C(0,-v)\}.$$

Then, $P(C) = 1$, $\phi(u)_\omega\phi(v)_\omega = \phi(\sqrt{u^2+v^2})_\omega$ and $\phi(u)_\omega$ is real for u, v rational and $\omega \in C$. However we have already shown that $\phi(t)_\omega$ is continuous in t for fixed $\omega \in \Omega$. Hence $\phi(t)_\omega$ is real for all $t \in \mathbb{R}$ and $\omega \in C$. Moreover, with the notation $\xi(t)_\omega = \phi\left(t^{\frac{1}{2}}\right)_\omega$, we have

$$\xi(t_1)_\omega\xi(t_2)_\omega = \xi(t_1+t_2)_\omega, \quad t_1 \geq 0,\ t_2 \geq 0 \tag{4.16}$$

for $t_1^{\frac{1}{2}}$, $t_2^{\frac{1}{2}}$ rational. Since $\xi(t)_\omega$ is continuous in t, we conclude that (4.16) holds for all nonnegative numbers t_1 and t_2. Now using Corollary 1.1 we find that the solution of (4.16) is

$$\xi(t)_\omega = e^{-k(\omega)t}, \quad t \geq 0$$

where $k(\omega)$ is a positive number depending on ω (see Feller, 1957, p.413). So, we can write

$$\phi(t)_\omega = e^{-\frac{z(\omega)}{2}t^2}, \quad t \geq 0$$

where $z(\omega)$ depends on ω. This is also true if $t < 0$ since $\phi(-t)_\omega = \phi(t)_\omega$. Now, $z(\omega)$ defines a random variable z with probability one. Therefore, we can write

$$\phi(t) = e^{-\frac{z}{2}t^2} \tag{4.17}$$

and hence $z = -2log\phi(1)$. Since, $\phi(1)$ is $\mathscr{F}$-measurable, so is z, and we have

$$E(y|z) = E(E(y|\mathscr{F})|z) \tag{4.18}$$

For any random variable y. Now take $y = e^{itx_1}$, then using (4.13), (4.17) and (4.18) we get

$$\begin{aligned} E(e^{itx_1}|z) &= E(E(e^{itx_1}|\mathscr{F})|z) \\ &= E(\phi(t)|z) \\ &= E\left(e^{-\frac{z}{2}t^2}|z\right) \\ &= e^{-\frac{z}{2}t^2}. \end{aligned}$$

Hence, the characteristic function of x_1 is

$$\begin{aligned} \psi(t^2) &= E(e^{itx_1}) \\ &= E(E(e^{itx_1}|z)) \\ &= E\left(e^{-\frac{z}{2}t^2}\right) \\ &= \int_0^\infty \exp\left(-\frac{zt^2}{2}\right) dG(z), \end{aligned}$$

where $G(z)$ denotes the distribution function of z. Thus,

$$\psi(u) = \int_0^\infty \exp\left(-\frac{zu}{2}\right) dG(z)$$

which proves (4.6). ■

We also need the following lemma.

Lemma 4.1. *Let* $\mathbf{X} \sim E_{p,n}(\mathbf{0}, \Sigma \otimes \Phi, \psi)$. *Let* $\mathbf{Y}$ *be a* $q \times m$ *submatrix of* $\mathbf{X}$ *such that* $qm < pn$ *and* $P(\mathbf{Y} = \mathbf{0}) = 0$. *Then* $\mathbf{Y}$ *is absolutely continuous.*

PROOF: From Theorem 3.6, it follows that $\mathbf{Y}$ is absolutely continuous. ■

Now, we can prove the following theorem.

Theorem 4.3. *Let* $\mathbf{X} \sim E_{p,n}(\mathbf{M}, \Sigma \otimes \Phi, \psi)$ *such that* $P(\mathbf{X} = \mathbf{M}) = 0$. *Then, the distribution of* $\mathbf{X}$ *is absolutely continuous and the p.d.f. of* $\mathbf{X}$ *can be written as a scale mixture of matrix variate normal distributions if and only if for every integer* $k > pn$ *there exists* $\mathbf{Y} \sim E_{q,m}(\mathbf{M}_1, \Sigma_1 \otimes \Phi_1, \psi_1)$, *such that* $qm \geq k$, $\Sigma_1 > \mathbf{0}$, $\Phi_1 > \mathbf{0}$ *and* $\mathbf{Y}$ *has a submatrix* Y_0 *with* $Y_0 \approx X$.

PROOF: First, assume that for every integer $k > pn$ there exists $\mathbf{Y} \sim E_{q,m}(\mathbf{M}_1, \Sigma_1 \otimes \Phi_1, \psi_1)$ such that $qm \geq k$ and $\mathbf{Y}$ has a submatrix $\mathbf{Y}_0$ with $\mathbf{Y}_0 \approx \mathbf{X}$. Then, for fixed k, from Remark 2.3, it follows that $\psi_1 = \psi$. Moreover, from Lemma 4.1, it follows that the distribution of $\mathbf{Y}_0$ and consequently that of $\mathbf{X}$ is absolutely continuous. Let $\mathbf{w} = vec\left(\Sigma_1^{-\frac{1}{2}}(\mathbf{Y}-\mathbf{M}_1)\Phi_1^{-\frac{1}{2}}\right)'$, then $\mathbf{w} \sim E_{qm}(\mathbf{0}, \mathbf{I}_{qm}, \psi)$. Let $\mathbf{v}$ be a k-dimensional subvector of $\mathbf{w}$. Then, $\mathbf{v} \sim E_k(\mathbf{0}, \mathbf{I}_k, \psi)$ and the characteristic function of $\mathbf{v}$ is $\phi_{\mathbf{v}}(\mathbf{t}) = \psi(\mathbf{t}'\mathbf{t})$, where $\mathbf{t} \in I\!R^k$. Using Theorem 4.2, we get $\psi(u) = \int_0^\infty \exp\left(-\frac{zu}{2}\right) dG(z)$. Therefore, the p.d.f. of $\mathbf{X}$ is

$$g(\mathbf{X}) = \int_0^\infty f_{N_{p,n}(\mathbf{M}, z\Sigma\otimes\Phi)}(\mathbf{X}) dG(z).$$

Next, assume that $\mathbf{X}$ can be written as a scale mixture of matrix variate normal distributions; that is, the p.d.f. of $\mathbf{X}$ is $g(\mathbf{X}) = \int_0^\infty f_{N_{p,n}(\mathbf{M}, z\Sigma\otimes\Phi)}(\mathbf{X}) dG(z)$. Then, we have

$$\psi(u) = \int_0^\infty \exp\left(-\frac{zu}{2}\right) dG(z).$$

It follows from Theorem 4.2, that $\psi(\mathbf{t}'\mathbf{t})$, $\mathbf{t} \in I\!R^k$ is a characteristic function for every $k \geq 1$. Choose $k > pn$, and let $q \geq p$, $m \geq n$, such that $qm \geq k$. Define a qm-dimensional random vector $\mathbf{w}$ such that $\mathbf{w} \sim E_{qm}(\mathbf{0}, \mathbf{I}_{qm}, \psi)$. Let $\mathbf{w} = vec(\mathbf{S}')$ where $\mathbf{S}$ is $q \times m$ matrix, then $\mathbf{S} \sim E_{q,m}(\mathbf{0}, \mathbf{I}_q \otimes \mathbf{I}_m, \psi)$. Further define

$$\Sigma_1 = \begin{pmatrix} \Sigma & \mathbf{0} \\ \mathbf{0} & \mathbf{I}_{q-p} \end{pmatrix}, \quad \Phi_1 = \begin{pmatrix} \Phi & \mathbf{0} \\ \mathbf{0} & \mathbf{I}_{m-n} \end{pmatrix}, \quad \text{and } \mathbf{M}_1 = \begin{pmatrix} \mathbf{M} & \mathbf{0} \\ \mathbf{0} & \mathbf{0} \end{pmatrix},$$

where $\mathbf{M}_1$ is $q \times m$. Let $\mathbf{Y} = \Sigma_1^{\frac{1}{2}}\mathbf{S}\Phi_1^{\frac{1}{2}} + \mathbf{M}_1$. Then,

$$\mathbf{Y} \sim E_{q,m}\left(\begin{pmatrix} \mathbf{M} & \mathbf{0} \\ \mathbf{0} & \mathbf{0} \end{pmatrix}, \begin{pmatrix} \Sigma & \mathbf{0} \\ \mathbf{0} & \mathbf{I}_{q-p} \end{pmatrix} \otimes \begin{pmatrix} \Phi & \mathbf{0} \\ \mathbf{0} & \mathbf{I}_{m-n} \end{pmatrix}, \psi\right).$$

Partition $\mathbf{Y}$ into $\mathbf{Y} = \begin{pmatrix} \mathbf{Y}_{11} & \mathbf{Y}_{12} \\ \mathbf{Y}_{21} & \mathbf{Y}_{22} \end{pmatrix}$ where $\mathbf{Y}_{11}$ is $p \times n$. Then, $\mathbf{Y}_{11} \sim E_{p,n}(\mathbf{M}, \Sigma \otimes \Phi, \psi)$ and hence $\mathbf{X} \approx \mathbf{Y}_{11}$. ■

Example 4.1. Here, we list some m.e.c. distributions together with the distribution function $G(z)$ which generates the p.d.f. through (4.1).

(i) Matrix variate normal distribution:

$$G(1) = 1.$$

(ii) ε-contaminated matrix variate normal distribution:

$$G(1) = 1 - \varepsilon, \quad G(\sigma^2) = \varepsilon.$$

(iii) Matrix variate Cauchy distribution:

$$G(z) = \frac{1}{4\pi} \int_0^z t^{-\frac{3}{2}} e^{-\frac{1}{2t}} dt.$$

(iv) Matrix variate t-distribution with m degrees of freedom:

$$G(z) = \frac{\left(\frac{m}{2}\right)^{\frac{m}{2}}}{\Gamma\left(\frac{m}{2}\right)} \int_0^z t^{-\left(1+\frac{m}{2}\right)} e^{-\frac{m}{2t}} dt.$$

Here, (a) and (b) are obvious, and (c) and (d) will be shown in Sect. 4.2.

Next, we give an example which shows that the p.d.f. of an absolutely continuous elliptically contoured distribution is not always expressible as the scale mixture of normal distributions.

Example 4.2. Let x be a one-dimensional random variable with p.d.f. $f(x) = \frac{\sqrt{2}}{\pi} \frac{1}{1+x^4}$. Assume $f(x)$ has a scale mixture representation. Then, from Theorem 4.3, there exists a $q \times m$ dimensional elliptical distribution $\mathbf{Y} \sim E_{q,m}(\mathbf{M}_1, \Sigma_1 \otimes \Phi_1, \psi_1)$, such that $qm > 5$ and one element of $\mathbf{Y}$ is identically distributed as x. Therefore, there exists a 5-dimensional random vector $\mathbf{w}$ such that $\mathbf{w} \sim E_5(\mathbf{m}_2, \Sigma_2, \psi)$ and $w_1 \approx x$.

Now, $f(x) = h(x^2)$ where $h(z) = \frac{\sqrt{2}}{\pi} \frac{1}{1+z^2}$. Let the p.d.f. of $\mathbf{w}$ be $f_1(\mathbf{w}) = h_1((\mathbf{w} - \mathbf{m}_2)' \Sigma_2^{-1} (\mathbf{w} - \mathbf{m}_2))$. It follows, from Theorem 3.4, that $h_1(z) = \frac{1}{\pi^2} \frac{\partial^2 h(z)}{\partial z^2}$. We have

$$\frac{\partial h(z)}{\partial z} = \frac{\sqrt{2}}{\pi} \frac{\partial}{\partial z} \frac{1}{1+z^2}$$
$$= -\frac{2\sqrt{2}}{\pi} \frac{z}{(1+z^2)^2},$$

and

$$\frac{\partial^2 h(z)}{\partial z^2} = -\frac{2\sqrt{2}}{\pi} \frac{1-3z^2}{(1+z^2)^3}.$$

Consequently we get $h_1(z) = \frac{2\sqrt{2}}{\pi^3} \frac{3z^2-1}{(1+z^2)^3}$. However, $h_1(z) < 0$ for $0 < z < \frac{1}{\sqrt{3}}$ and hence $h(x^2)$ cannot be a p.d.f. This is a contradiction. Therefore, $f(x)$ cannot be written as a scale mixture of normal distributions.

Next, we prove some important theorems about scale mixture representations.

Theorem 4.4. *Let $\lambda : \mathbb{R}^{p\times n} \to \mathbb{R}^{q\times m}$ be a Borel-measurable matrix variate function. Assume that if $\mathbf{X} \sim N_{p,n}(\mathbf{M}, \Sigma \otimes \Phi)$ then the p.d.f. of $\mathbf{W} = \lambda(\mathbf{X})$ is $l^{\lambda}_{N_{p,n}(\mathbf{M},\Sigma\otimes\Phi)}(\mathbf{W})$. Then, if $\mathbf{X} \sim E_{p,n}(\mathbf{M}, \Sigma \otimes \Phi, \psi)$ with p.d.f. $g(\mathbf{X}) = \int_0^\infty f_{N_{p,n}(\mathbf{M},z\Sigma\otimes\Phi)}(\mathbf{X})dG(z)$, the p.d.f. of $\mathbf{W} = \lambda(\mathbf{X})$ is*

$$l(\mathbf{W}) = \int_0^\infty l^{\lambda}_{N_{p,n}(\mathbf{M},z\Sigma\otimes\Phi)}(\mathbf{W})dG(z)\,.$$

PROOF: Let $A \subset \mathbb{R}^{q\times m}$. Then,

$$\begin{aligned}
\int_A l(\mathbf{W})d\mathbf{W} &= \int_A \int_0^\infty l^{\lambda}_{N_{p,n}(\mathbf{M},z\Sigma\otimes\Phi)}(\mathbf{W})dG(z)d\mathbf{W}\\
&= \int_0^\infty \int_A l^{\lambda}_{N_{p,n}(\mathbf{M},z\Sigma\otimes\Phi)}(\mathbf{W})d\mathbf{W}dG(z)\\
&= \int_0^\infty P(\lambda(\mathbf{X}) \in A|\mathbf{X} \sim N_{p,n}(\mathbf{M}, z\Sigma \otimes \Phi))dG(z)\\
&= \int_0^\infty \int_{\mathbb{R}^{p\times n}} \chi_A(\lambda(\mathbf{X}))f_{N_{p,n}(\mathbf{M},z\Sigma\otimes\Phi)}(\mathbf{X})\mathbf{X}dG(z)\\
&= \int_{\mathbb{R}^{p\times n}} \chi_A(\lambda(\mathbf{X})) \int_0^\infty f_{N_{p,n}(\mathbf{M},z\Sigma\otimes\Phi)}(\mathbf{X})dG(z)\mathbf{X}\\
&= \int_{\mathbb{R}^{p\times n}} \chi_A(\lambda(\mathbf{X}))g(\mathbf{X})\mathbf{X}\\
&= P(\lambda(\mathbf{X}) \in A|\mathbf{X} \sim E_{p,n}(\mathbf{M}, \Sigma \otimes \Phi, \psi))\,.
\end{aligned}$$

■

Corollary 4.1. *Let $\mathbf{X} \sim E_{p,n}(\mathbf{M}, \Sigma \otimes \Phi, \psi)$ with p.d.f.*

$$g(\mathbf{X}) = \int_0^\infty f_{N_{p,n}(\mathbf{M},z\Sigma\otimes\Phi)}(\mathbf{X})dG(z).$$

Let $\mathbf{C} : q \times m$, $\mathbf{A} : q \times p$, and $\mathbf{B} : n \times m$ be constant matrices, such that $rk(\mathbf{A}) = q$ and $rk(\mathbf{B}) = m$. Then, from Theorems 4.4 and 2.2, it follows that the p.d.f. of $\mathbf{AXB} + \mathbf{C}$ is

$$g^*(\mathbf{X}) = \int_0^\infty f_{N_{p,n}(\mathbf{AMB}+\mathbf{C},z(\mathbf{A}\Sigma\mathbf{A}')\otimes\mathbf{B}\Phi\mathbf{B}')}(\mathbf{X})dG(z)\,.$$

Furthermore, if $\mathbf{X}$, $\mathbf{M}$, and Σ are partitioned into

$$\mathbf{X} = \begin{pmatrix}\mathbf{X}_1\\ \mathbf{X}_2\end{pmatrix}, \quad \mathbf{M} = \begin{pmatrix}\mathbf{M}_1\\ \mathbf{M}_2\end{pmatrix}, \quad \text{and} \quad \Sigma = \begin{pmatrix}\Sigma_{11} & \Sigma_{12}\\ \Sigma_{21} & \Sigma_{22}\end{pmatrix},$$

where $\mathbf{X}_1 : q \times n$, $\mathbf{M}_1 : q \times n$, and $\Sigma_{11} : q \times q$, $1 \le q < p$, then the p.d.f. of $\mathbf{X}_1$ is

$$g_1(\mathbf{X}) = \int_0^\infty f_{N_{p,n}(\mathbf{M}_1, z\Sigma_{11}\otimes\Phi)}(\mathbf{X})dG(z).$$

Corollary 4.2. *Let* $\mathbf{X} \sim E_{p,n}(\mu\mathbf{e}_n', \Sigma\otimes\mathbf{I}_n, \psi)$ *with p.d.f.* $g(\mathbf{X})$*, where*

$$g(\mathbf{X}) = \int_0^\infty f_{N_{p,n}(\mathbf{M}, z\Sigma\otimes\Phi)}(\mathbf{X})dG(z).$$

Then,

(a) The p.d.f. of $\mathbf{y}_1 = \frac{\mathbf{X}\mathbf{e}_n}{n}$ *is*

$$g_1(\mathbf{y}_1) = \int_0^\infty f_{N_p(\mu, z\Sigma/n)}(\mathbf{y}_1)dG(z)$$

(b) The p.d.f. of $\mathbf{Y}_2 = \mathbf{X}\left(\mathbf{I}_n - \frac{\mathbf{e}_n\mathbf{e}_n'}{n}\right)\mathbf{X}'$ *is*

$$g_2(\mathbf{Y}_2) = \int_0^\infty f_{W_p(z\Sigma, n-1)}(\mathbf{Y}_2)dG(z),$$

(c) The p.d.f. of $\mathbf{Y}_3 = \mathbf{X}\mathbf{X}'$*, for* $\mu = \mathbf{0}$*, is*

$$g_3(\mathbf{Y}_3) = \int_0^\infty f_{W_p(z\Sigma, n)}(\mathbf{Y}_3)dG(z),$$

Theorem 4.5. *Let* $\lambda : \mathbb{R}^{pn} \to \mathbb{R}^{q\times m}$ *be a Borel-measurable matrix variate function. Assume that if* $\mathbf{X} \sim N_{p,n}(\mathbf{M}, \Sigma\otimes\Phi)$*, then* $E(\lambda(\mathbf{X}))$ *exists and it is denoted by* $E_{N_{p,n}(\mathbf{M},\Sigma\otimes\Phi)}(\lambda(\mathbf{X}))$*. Then, if* $\mathbf{X} \sim E_{p,n}(\mathbf{M}, \Sigma\otimes\Phi, \psi)$ *with p.d.f.*

$$g(\mathbf{X}) = \int_0^\infty f_{N_{p,n}(\mathbf{M}, z\Sigma\otimes\Phi)}(\mathbf{X})dG(z),$$

such that $E(\lambda(\mathbf{X}))$ *exists and it is denoted by* $E_{E_{p,n}(\mathbf{M},\Sigma\otimes\Phi,\psi)}(\lambda(\mathbf{X}))$*, we have*

$$E_{E_{p,n}(\mathbf{M},\Sigma\otimes\Phi,\psi)}(\lambda(\mathbf{X})) = \int_0^\infty E_{N_{p,n}(\mathbf{M},z\Sigma\otimes\Phi)}((\lambda(\mathbf{X})))dG(z).$$

PROOF:

$$\begin{aligned}
E_{E_{p,n}(\mathbf{M},\Sigma\otimes\Phi,\psi)}(\lambda(\mathbf{X})) &= \int_{\mathbb{R}^{p\times n}} \lambda(\mathbf{X})g(\mathbf{X})d\mathbf{X} \\
&= \int_{\mathbb{R}^{p\times n}} \lambda(\mathbf{X}) \int_0^\infty f_{N_{p,n}(\mathbf{M},z\Sigma\otimes\Phi)}(\mathbf{X})dG(z)d\mathbf{X} \\
&= \int_0^\infty \int_{\mathbb{R}^{p\times n}} \lambda(\mathbf{X}) f_{N_{p,n}(\mathbf{M},z\Sigma\otimes\Phi)}(\mathbf{X})d\mathbf{X}dG(z) \\
&= \int_0^\infty E_{N_{p,n}(\mathbf{M},z\Sigma\otimes\Phi)}(\lambda(\mathbf{X}))dG(z).
\end{aligned}$$

■

Corollary 4.3. *With the notations of Theorem 4.5, if $Cov(\mathbf{X})$ exists, then*

$$Cov(\mathbf{X}) = \left(\int_0^\infty z dG(z)\right) \Sigma \otimes \Phi.$$

Next, we give a theorem which shows the relationship between the characteristic function of a scale mixture of normal distributions and the characteristic function of a conditional distribution. This theorem is due to Cambanis, Huang, and Simons (1981).

Theorem 4.6. *Let $\mathbf{X} \sim E_{p,n}(\mathbf{M}, \Sigma \otimes \Phi, \psi)$ with p.d.f.*

$$g(\mathbf{X}) = \int_0^\infty f_{N_{p,n}(\mathbf{M}, z\Sigma\otimes\Phi)}(\mathbf{X}) dG(z).$$

Let $\mathbf{X}$, $\mathbf{M}$, and Σ be partitioned as

$$\mathbf{X} = \begin{pmatrix} \mathbf{X}_1 \\ \mathbf{X}_2 \end{pmatrix}, \quad \mathbf{M} = \begin{pmatrix} \mathbf{M}_1 \\ \mathbf{M}_2 \end{pmatrix}, \quad \text{and} \quad \Sigma = \begin{pmatrix} \Sigma_{11} & \Sigma_{12} \\ \Sigma_{21} & \Sigma_{22} \end{pmatrix},$$

where $\mathbf{X}_1$ and $\mathbf{M}_1$ are $q \times n$ and $\Sigma_{11} : q \times q$. Then, the conditional p.d.f. of $\mathbf{X}_1|\mathbf{X}_2$ can be written as

$$g_1(\mathbf{X}_1|\mathbf{X}_2) = \int_0^\infty f_{N_{q,n}(\mathbf{M}_1 + \Sigma_{12}\Sigma_{22}^{-1}(\mathbf{X}_2 - \mathbf{M}_2), z\Sigma_{11\cdot2}\otimes\Phi)}(\mathbf{X}) dG_{q(\mathbf{X}_2)}(z), \tag{4.19}$$

where $q(\mathbf{X}_2) = tr((\mathbf{X}_2 - \mathbf{M}_2)'\Sigma_{22}^{-1}(\mathbf{X}_2 - \mathbf{M}_2)\Phi^{-1})$ and

$$G_{a^2}(z) = \frac{\int_0^z v^{-\frac{(p-q)n}{2}} \exp\left(-\frac{a^2}{2v}\right) dG(v)}{\int_0^\infty v^{-\frac{(p-q)n}{2}} \exp\left(-\frac{a^2}{2v}\right) dG(v)} \quad \text{if} \quad a > 0, \quad z \geq 0, \tag{4.20}$$

and $G_0(z) = 1$ if $z \geq 0$.

PROOF: Let $\mathbf{X} \approx r\Sigma^{\frac{1}{2}}\mathbf{U}\Phi^{\frac{1}{2}} + \mathbf{M}$ be the stochastic representation of $\mathbf{X}$. Then, (4.4) gives the p.d.f. of r. It follows, from Theorem 2.21, that the stochastic representation of $\mathbf{X}_1|\mathbf{X}_2$ has the form

$$\mathbf{X}_1|\mathbf{X}_2 \approx r_{q(\mathbf{X}_2)}\Sigma_{11}^{\frac{1}{2}}\mathbf{U}_1\Phi^{\frac{1}{2}} + (\mathbf{M}_1 + (\mathbf{X}_2 - \mathbf{M}_2)\Sigma_{22}^{-1}\Sigma_{21}).$$

Here, $vec(\mathbf{U}_1')$ is uniformly distributed on S_{qn}. It follows, from (2.31) and (4.4), that

$$P(r_{a^2} \leq c) = \frac{\int_a^{\sqrt{c^2+a^2}} (r^2 - a^2)^{\frac{qn}{2}-1} r^{-(pn-2)} r^{pn-1} \int_0^\infty s^{-\frac{pn}{2}} \exp\left(-\frac{r^2}{2s}\right) dG(s) dr}{\int_a^\infty (r^2 - a^2)^{\frac{qn}{2}-1} r^{-(pn-2)} r^{pn-1} \int_0^\infty s^{-\frac{pn}{2}} \exp\left(-\frac{r^2}{2s}\right) dG(s) dr}.$$

Let $y^2 = r^2 - a^2$, then $J(r \to y) = \frac{r}{y}$, and we have

$$P(r_{a^2} \le c) = \frac{\int_0^c y^{qn-1} \int_0^\infty s^{-\frac{pn}{2}} \exp\left(-\frac{y^2+a^2}{2s}\right) dG(s)dy}{\int_0^\infty y^{qn-1} \int_0^\infty s^{-\frac{pn}{2}} \exp\left(-\frac{y^2+a^2}{2s}\right) dG(s)dy}$$
$$= \frac{\int_0^c y^{qn-1} \int_0^\infty s^{-\frac{pn}{2}} \exp\left(-\frac{y^2}{2s}\right) \exp\left(-\frac{a^2}{2s}\right) dG(s)dy}{\int_0^\infty s^{-\frac{pn}{2}} \exp\left(-\frac{a^2}{2s}\right) \int_0^\infty y^{qn-1} \exp\left(-\frac{y^2}{2s}\right) dydG(s)}. \quad (4.21)$$

In order to compute $\int_0^\infty y^{qn-1} \exp\left(-\frac{y^2}{2s}\right) dy$ we substitute $t = \frac{y^2}{2s}$. Then, $J(y \to t) = \frac{s}{t}$, and hence

$$\int_0^\infty y^{qn-1} \exp\left(-\frac{y^2}{2s}\right) dy = (2s)^{\frac{qn}{2}-1} s \int_0^\infty t^{\frac{qn}{2}-1} \exp(-t)dt$$
$$= 2^{\frac{qn}{2}-1} s^{\frac{qn}{2}} \Gamma\left(\frac{qn}{2}\right).$$

Substituting this into (4.21), we get

$$P(r_{a^2} \le c) = \frac{\int_0^c y^{qn-1} \int_0^\infty s^{-\frac{pn}{2}} \exp\left(-\frac{y^2}{2s}\right) \exp\left(-\frac{a^2}{2s}\right) dG(s)dy}{\int_0^\infty 2^{\frac{qn}{2}-1} \Gamma\left(\frac{qn}{2}\right) s^{-\frac{(p-q)n}{2}} \exp\left(-\frac{a^2}{2s}\right) dG(s)} \quad (4.22)$$
$$= \int_0^c \frac{1}{2^{\frac{qn}{2}-1} \Gamma\left(\frac{qn}{2}\right)} y^{qn-1} \int_0^\infty \frac{\exp\left(-\frac{y^2}{2s}\right)}{s^{\frac{qn}{2}}} \frac{s^{-\frac{(p-q)n}{2}} \exp\left(-\frac{a^2}{2s}\right)}{\int_0^\infty s^{-\frac{(p-q)n}{2}} \exp\left(-\frac{a^2}{2s}\right) dG(s)} dG(s)dy.$$

Define the distribution function

$$G_{a^2}(z) = \frac{\int_0^z s^{-\frac{(p-q)n}{2}} \exp\left(-\frac{a^2}{2s}\right) dG(s)}{\int_0^\infty s^{-\frac{(p-q)n}{2}} \exp\left(-\frac{a^2}{2s}\right) dG(s)}.$$

Then, from (4.22), we get

$$P(r_{a^2} \le c) = \int_0^c \frac{1}{2^{\frac{qn}{2}-1} \Gamma\left(\frac{qn}{2}\right)} y^{qn-1} \int_0^\infty \frac{\exp\left(-\frac{y^2}{2z}\right)}{z^{\frac{qn}{2}}} d(G_{a^2}(z))dy.$$

Hence the p.d.f. of r_{a^2} is

$$l^*(r_{a^2}) = \frac{1}{2^{\frac{qn}{2}-1}\Gamma\left(\frac{qn}{2}\right)} r_a^{qn-1} \int_0^\infty \frac{\exp\left(-\frac{r_a^2}{2z}\right)}{z^{\frac{qn}{2}}} dG(z)$$

Now, from Theorem 4.1, this means that the p.d.f. of $\mathbf{X}_1|\mathbf{X}_2$ has the form (4.19) with $G_{a^2}(z)$ defined by (4.20). ■

4.2 Mixture by Weighting Function

Chu (1973) showed another way to obtain the p.d.f. of a m.e.c. distribution from the density functions of matrix variate normal distributions. For this purpose, he used Laplace transform. We recall here that if $f(t)$ is a real function defined on the set of nonnegative real numbers, then its Laplace transform, $\mathscr{L}[f(t)]$ is defined by

$$\begin{aligned} g(s) &= \mathscr{L}[f(t)] \\ &= \int_0^\infty e^{-st} f(t)dt. \end{aligned}$$

Moreover, the inverse Laplace transform of a function $g(s)$ (see Abramowitz and Stegun, 1965, p.1020) is defined by

$$\begin{aligned} f(t) &= \mathscr{L}^{-1}[g(s)] \\ &= \frac{1}{2\pi i} \int_{c-i\infty}^{c+i\infty} e^{st} g(s)ds, \end{aligned}$$

where c is an appropriately chosen real number. It is known that $\mathscr{L}^{-1}[g(s)]$ exists if $g(s)$ is differentiable for sufficiently large s and $g(s) = o(s^{-k})$ as $s \to \infty$, $k > 1$.

The following theorem was proved by Chu (1973) for the vector variate case.

Theorem 4.7. *Let* $\mathbf{X} \sim E_{p,n}(\mathbf{M}, \Sigma \otimes \Phi, \psi)$ *with p.d.f.* $g(\mathbf{X})$ *where*

$$g(\mathbf{X}) = |\Sigma|^{-\frac{n}{2}} |\Phi|^{-\frac{p}{2}} h(tr((\mathbf{X}-\mathbf{M})'\Sigma^{-1}(\mathbf{X}-\mathbf{M})\Phi^{-1})).$$

If $h(t)$, $t \in [0,\infty)$ *has the inverse Laplace transform, then we have*

$$g(\mathbf{X}) = \int_0^\infty f_{N_{p,n}(\mathbf{M}, z^{-1}\Sigma\otimes\Phi)}(\mathbf{X}) w(z)dz, \tag{4.23}$$

where

$$w(z) = (2\pi)^{\frac{pn}{2}} z^{-\frac{pn}{2}} \mathscr{L}^{-1}[h(2t)]. \tag{4.24}$$

PROOF: From (4.24), we get

$$\begin{aligned} h(2t) &= \mathscr{L}[(2\pi)^{-\frac{pn}{2}} z^{\frac{pn}{2}} w(z)] \\ &= \int_0^\infty e^{-tz} (2\pi)^{-\frac{pn}{2}} z^{\frac{pn}{2}} w(z) dz. \end{aligned}$$

Hence,

$$\begin{aligned} g(\mathbf{X}) &= |\Sigma|^{-\frac{n}{2}} |\boldsymbol{\Phi}|^{-\frac{p}{2}} \int_0^\infty etr\left(-\frac{1}{2} tr((\mathbf{X}-\mathbf{M})'\Sigma^{-1}(\mathbf{X}-\mathbf{M})\boldsymbol{\Phi}^{-1})z\right)\left(\frac{z}{2\pi}\right)^{\frac{pn}{2}} w(z)dz \\ &= \int_0^\infty \frac{1}{(2\pi)^{\frac{pn}{2}} |z^{-1}\Sigma|^{\frac{n}{2}} |\boldsymbol{\Phi}|^{\frac{p}{2}}} etr\left(-\frac{1}{2} tr((\mathbf{X}-\mathbf{M})'\Sigma^{-1}(\mathbf{X}-\mathbf{M})\boldsymbol{\Phi}^{-1})z\right) w(z)dz \\ &= \int_0^\infty f_{N_{p,n}(\mathbf{M}, z^{-1}\boldsymbol{\Sigma}\otimes\boldsymbol{\Phi})}(\mathbf{X}) w(z) dz. \end{aligned}$$

■

Remark 4.4. Let $g(\mathbf{x}) = \int_0^\infty f_{N_p(\mathbf{0}, z^{-1}\boldsymbol{\Sigma})}(\mathbf{x}) w(z) dz$. Assume, $w(z)$ is a function (not a functional), and define $u(z) = \frac{1}{z^2} w\left(\frac{1}{z}\right)$. Then,

$$g(\mathbf{X}) = \int_0^\infty f_{N_{p,n}(\mathbf{M}, z\boldsymbol{\Sigma}\otimes\boldsymbol{\Phi})}(\mathbf{X}) u(z) dz. \tag{4.25}$$

Indeed, let $t = \frac{1}{z}$. Then, $J(z \to t) = \frac{1}{t^2}$ and so (4.23) can be rewritten as

$$\begin{aligned} g(\mathbf{X}) &= \int_0^\infty f_{N_{p,n}(\mathbf{M}, t\boldsymbol{\Sigma}\otimes\boldsymbol{\Phi})}(\mathbf{X}) w\left(\frac{1}{t}\right) \frac{1}{t^2} dt \\ &= \int_0^\infty f_{N_{p,n}(\mathbf{M}, t\boldsymbol{\Sigma}\otimes\boldsymbol{\Phi})}(\mathbf{X}) u(t) dt. \end{aligned}$$

Remark 4.5. Even if $w(z)$ is a functional the representation (4.25) may exist as parts (a) and (b) of the next example show.

Example 4.3. Here, we list some m.e.c. distributions together with the functions $w(z)$ and $u(z)$ which generate the p.d.f. through (4.23) and (4.25).

(a) Matrix variate normal distribution;

$$\begin{aligned} w(z) &= \delta(z-1), \quad \text{and} \\ u(z) &= \delta(z-1). \end{aligned}$$

(b) ε-contaminated matrix variate normal distribution;

$$w(z) = (1-\varepsilon)\delta(z-1) + \varepsilon\delta(z-\sigma^2), \quad \text{and}$$
$$u(z) = (1-\varepsilon)\delta(z-1) + \varepsilon\delta(z-\sigma^2).$$

(c) Matrix variate Cauchy distribution:

$$w(z) = \frac{1}{4\pi\sqrt{ze^z}}, \quad \text{and}$$
$$u(z) = \frac{1}{4\pi z\sqrt{ze^{\frac{1}{z}}}}.$$

(d) Matrix variate t-distribution with m degrees of freedom;

$$w(z) = \frac{\left(\frac{mz}{2}\right)^{\frac{m}{2}} e^{-\frac{mz}{2}}}{z\Gamma\left(\frac{m}{2}\right)}, \quad \text{and}$$
$$u(z) = \frac{\left(\frac{m}{2z}\right)^{\frac{m}{2}} e^{-\frac{m}{2z}}}{z\Gamma\left(\frac{m}{2}\right)}, \delta(z-1).$$

(e) The one-dimensional distribution with p.d.f.

$$g(x) = \frac{\sqrt{2}}{\pi\sigma}\frac{1}{1+\left(\frac{x}{\sigma}\right)^4};$$
$$w(z) = \frac{1}{\sqrt{\pi z}} sin\frac{z}{2}, \quad \text{and}$$
$$u(z) = \frac{1}{\sqrt{\pi z}} sin\frac{1}{2z}.$$

The functions $w(z)$ in parts (a) and (c)–(e) are given in Chu (1973) and $u(z)$ can be easily computed from $w(z)$.

Remark 4.6. It may be noted that $u(z)$ is not always nonnegative as part (e) of Example 4.3 shows. However, if it is nonnegative then defining $G(z) = \int_0^z u(s)ds$, (4.25) yields $g(\mathbf{X}) = \int_0^\infty f_{N_{p,n}(\mathbf{M}, z\boldsymbol{\Sigma}\otimes\boldsymbol{\Phi})}(\mathbf{X})dG(z)$ which is the expression given in Remark 4.1. We have to see that $\int_0^\infty u(s)ds = 1$ but this will follow from the next theorem if we take $v = 0$ in (4.26). Therefore, using Example 4.3 we obtain the results in parts (c) and (d) of Example 4.1.

Now, we can state theorems similar to those in Sect. 4.1.

Theorem 4.8. *Let* $\mathbf{X} \sim E_{p,n}(\mathbf{M}, \Sigma \otimes \Phi, \psi)$ *be absolutely continuous. Then,* $\mathbf{X}$ *has the p.d.f. defined by (4.23) if and only if the characteristic function of* $\mathbf{X}$ *is*

$$\phi_{\mathbf{X}}(\mathbf{T}) = etr(i\mathbf{T}'\mathbf{M}) \int_0^\infty etr\left(-\frac{\mathbf{T}'\Sigma\mathbf{T}\Phi}{2z}\right) w(z)dz,$$

that is

$$\psi(v) = \int_0^\infty \exp\left(-\frac{v}{2z}\right) w(z)dz.$$

Also, $\mathbf{X}$ *has the p.d.f. defined by (4.25) if and only if*

$$\phi_{\mathbf{X}}(\mathbf{T}) = etr(i\mathbf{T}'\mathbf{M}) \int_0^\infty etr\left(-\frac{z(\mathbf{T}'\Sigma\mathbf{T}\Phi)}{2}\right) u(z)dz,$$

that is

$$\psi(v) = \int_0^\infty \exp\left(-\frac{zv}{2}\right) u(z)dz. \tag{4.26}$$

Theorem 4.9. *Let* $\lambda : \mathbb{R}^{p\times n} \rightarrow \mathbb{R}^{q\times m}$ *be a Borel-measurable matrix variate function. Assume that if* $\mathbf{X} \sim N_{p,n}(\mathbf{M}, \Sigma \otimes \Phi)$ *then the p.d.f. of* $\mathbf{W} = \lambda(\mathbf{X})$ *is* $l^\lambda_{N_{p,n}(\mathbf{M},\Sigma\otimes\Phi)}(\mathbf{W})$*. Then, if* $\mathbf{X} \sim E_{p,n}(\mathbf{M}, \Sigma \otimes \Phi, \psi)$ *with p.d.f.* $g(\mathbf{X}) = \int_0^\infty f_{N_{p,n}(\mathbf{M},z^{-1}\Sigma\otimes\Phi)}(\mathbf{X})w(z)dz$*, the p.d.f. of* $\mathbf{W} = \lambda(\mathbf{X})$ *is*

$$l(\mathbf{W}) = \int_0^\infty l^\lambda_{N_{p,n}(\mathbf{M},z^{-1}\Sigma\otimes\Phi)}(\mathbf{W})w(z)dz. \tag{4.27}$$

If $g(\mathbf{X}) = \int_0^\infty f_{N_{p,n}(\mathbf{M},z\Sigma\otimes\Phi)}(\mathbf{X})u(z)dz$*, the p.d.f. of* $\mathbf{W} = \lambda(\mathbf{X})$ *is*

$$l(\mathbf{W}) = \int_0^\infty l^\lambda_{N_{p,n}(\mathbf{M},z\Sigma\otimes\Phi)}(\mathbf{W})u(z)dz. \tag{4.28}$$

PROOF: In the proof of Theorem 4.4 if $dG(z)$ is replaced by $u(z)dz$, we obtain (4.28). In the same proof if we replace $dG(z)$ by $w(z)dz$ and $N_{p,n}(\mathbf{M}, z\Sigma \otimes \Phi)$ by $N_{p,n}(\mathbf{M}, z^{-1}\Sigma \otimes \Phi)$ we obtain (4.27). ∎

Corollary 4.4. *Let* $\mathbf{X} \sim E_{p,n}(\mathbf{M}, \Sigma \otimes \Phi, \psi)$ *with p.d.f.*

$$g(\mathbf{X}) = \int_0^\infty f_{N_{p,n}(\mathbf{M},z^{-1}\Sigma\otimes\Phi)}(\mathbf{X})w(z)dz.$$

Let $\mathbf{C} : q \times m$*,* $\mathbf{A} : q \times p$*, and* $\mathbf{B} : n \times m$ *be constant matrices, such that* $rk(\mathbf{A}) = q$ *and* $rk(\mathbf{B}) = m$*. Then, from Theorems 4.9 and 2.2, it follows that the p.d.f. of* $\mathbf{AXB} + \mathbf{C}$ *is*

$$g^*(\mathbf{X}) = \int_0^\infty f_{N_{p,n}(\mathbf{AMB}+\mathbf{C},z^{-1}(\mathbf{A}\Sigma\mathbf{A}')\otimes(\mathbf{B}\Phi\mathbf{B}'))}(\mathbf{X})w(z)dz.$$

If

$$g(\mathbf{X}) = \int_0^\infty f_{N_{p,n}(\mathbf{M},z\Sigma\otimes\Phi)}(\mathbf{X})u(z)dz \tag{4.29}$$

then

$$g^*(\mathbf{X}) = \int_0^\infty f_{N_{p,n}(\mathbf{AMB}+\mathbf{C},z(\mathbf{A}\Sigma\mathbf{A}')\otimes(\mathbf{B}\Phi\mathbf{B}'))}(\mathbf{X})u(z)dz.$$

If $\mathbf{X}$, $\mathbf{M}$, *and* Σ *are partitioned as*

$$\mathbf{X} = \begin{pmatrix} \mathbf{X}_1 \\ \mathbf{X}_2 \end{pmatrix}, \mathbf{M} = \begin{pmatrix} \mathbf{M}_1 \\ \mathbf{M}_2 \end{pmatrix}, \textit{and } \Sigma = \begin{pmatrix} \Sigma_{11} & \Sigma_{12} \\ \Sigma_{21} & \Sigma_{22} \end{pmatrix},$$

where $\mathbf{X}_1 : q \times n$, $\mathbf{M}_1 : q \times n$, *and* $\Sigma_{11} : q \times q$, $1 \le q < p$, *then the p.d.f. of* $\mathbf{X}_1$ *is*

$$g_1(\mathbf{X}_1) = \int_0^\infty f_{N_{p,n}(\mathbf{M}_1,z^{-1}\Sigma_{11}\otimes\Phi)}(\mathbf{X}_1)w(z)dz.$$

and if (4.29) holds, then

$$g_1(\mathbf{X}_1) = \int_0^\infty f_{N_{p,n}(\mathbf{M}_1,z\Sigma_{11}\otimes\Phi)}(\mathbf{X}_1)u(z)dz.$$

Corollary 4.5. *Let* $\mathbf{X} \sim E_{p,n}(\mu\mathbf{e}_n', \Sigma\otimes\mathbf{I}_n, \psi)$ *with p.d.f.*

$$g(\mathbf{X}) = \int_0^\infty f_{N_{p,n}(\mu\mathbf{e}_n',z^{-1}\Sigma\otimes\mathbf{I}_n)}(\mathbf{X})w(z)dz$$

and $\mu \in I\!R^p$. *Then,*

(a) The p.d.f. of $\mathbf{y}_1 = \frac{\mathbf{X}\mathbf{e}_n}{n}$ *is*

$$g_1(\mathbf{y}_1) = \int_0^\infty f_{N_p(\mu,z^{-1}\Sigma/n)}(\mathbf{y}_1)w(z)dz$$

(b) The p.d.f. of $\mathbf{Y}_2 = \mathbf{X}\left(\mathbf{I}_n - \frac{\mathbf{e}_n\mathbf{e}_n'}{n}\right)\mathbf{X}'$, *for* $p \le n-1$, *is*

$$g_2(\mathbf{Y}_2) = \int_0^\infty f_{W_p(z^{-1}\Sigma,n-1)}(\mathbf{Y}_2)w(z)dz,$$

(c) The p.d.f. of $\mathbf{Y}_3 = \mathbf{X}\mathbf{X}'$, *for* $p \le n$ *and* $\mu = \mathbf{0}$, *is*

$$g_3(\mathbf{Y}_3) = \int_0^\infty f_{W_p(z^{-1}\Sigma,n)}(\mathbf{Y}_3)w(z)dz.$$

If $g(\mathbf{X}) = \int_0^\infty f_{N_{p,n}(\mu \mathbf{e}_n', z\Sigma\otimes\mathbf{I}_n)}(\mathbf{X})u(z)dz$ *and* $\mu \in I\!R^p$. *Then,*

(a) The p.d.f. of $\mathbf{y}_1 = \frac{\mathbf{X}\mathbf{e}_n}{n}$ *is*

$$g_1(\mathbf{y}_1) = \int_0^\infty f_{N_p(\mu, z\Sigma/n)}(\mathbf{y}_1)u(z)dz$$

(b) The p.d.f. of $\mathbf{Y}_2 = \mathbf{X}\left(\mathbf{I}_n - \frac{\mathbf{e}_n\mathbf{e}_n'}{n}\right)\mathbf{X}'$ *is*

$$g_2(\mathbf{Y}_2) = \int_0^\infty f_{W_p(z\Sigma,n-1)}(\mathbf{Y}_2)u(z)dz,$$

(c) The p.d.f. of $\mathbf{Y}_3 = \mathbf{X}\mathbf{X}'$, *for* $\mu = \mathbf{0}$, *is*

$$g_3(\mathbf{Y}_3) = \int_0^\infty f_{W_p(z\Sigma,n)}(\mathbf{Y}_3)u(z)dz.$$

Remark 4.7. It follows, from Example 4.3 and Corollary 4.4, that any submatrix of a random matrix with Cauchy distribution also has Cauchy distribution. Also any submatrix of a random matrix having t-distribution with m degrees of freedom has t-distribution with m degrees of freedom.

Example 4.4. Let $\mathbf{X} \sim E_{p,n}(\mu\mathbf{e}_n', \Sigma\otimes\mathbf{I}_n, \psi)$ have matrix variate t-distribution with m degrees of freedom. Then applying Corollary 4.5 with $w(z) = \frac{m\left(\frac{mz}{2}\right)^{\frac{m}{2}-1}e^{-\frac{mz}{2}}}{2\Gamma\left(\frac{m}{2}\right)}$, we see that

(a) The p.d.f. of $\mathbf{y}_1 = \frac{\mathbf{X}\mathbf{e}_n}{n}$ is

$$g_1(\mathbf{y}_1) = \frac{m^{\frac{m}{2}} n^{\frac{p}{2}} \Gamma\left(\frac{m+p}{2}\right)}{\pi^{\frac{p}{2}} \Gamma\left(\frac{m}{2}\right) |\Sigma|^{\frac{1}{2}}} (m + n(\mathbf{y}_1 - \mu)'\Sigma^{-1}(\mathbf{y}_1 - \mu))^{-\frac{m+p}{2}}$$

(b) The p.d.f. of $\mathbf{Y}_2 = \mathbf{X}\left(\mathbf{I}_n - \frac{\mathbf{e}_n\mathbf{e}_n'}{n}\right)\mathbf{X}'$ is

$$g_2(\mathbf{Y}_2) = \frac{m^{\frac{m}{2}} \Gamma\left(\frac{m+p(n-1)}{2}\right)}{\Gamma\left(\frac{m}{2}\right) \Gamma_p\left(\frac{n-1}{2}\right) |\Sigma|^{\frac{n-1}{2}}} (m + tr(\Sigma^{-1}\mathbf{Y}_2)^{-\frac{m+p(n-1)}{2}} |\mathbf{Y}_2|^{\frac{n-p-2}{2}},$$

(c) The p.d.f. of $\mathbf{Y}_3 = \mathbf{X}\mathbf{X}'$, if $\mu = \mathbf{0}$, is

$$g_3(\mathbf{Y}_3) = \frac{m^{\frac{m}{2}} \Gamma\left(\frac{m+pn}{2}\right)}{\Gamma\left(\frac{m}{2}\right) \Gamma_p\left(\frac{n}{2}\right) |\Sigma|^{\frac{n}{2}}} (m + tr(\Sigma^{-1}\mathbf{Y}_2)^{-\frac{m+pn}{2}} |\mathbf{Y}_2|^{\frac{n-p-1}{2}}.$$

Here, $\Gamma_p(t) = \pi^{\frac{p(p-1)}{4}} \prod_{i=1}^p \Gamma\left(t - \frac{i-1}{2}\right)$.

The mixture representation of the p.d.f. of a m.e.c. distribution makes it possible to derive monotone likelihood ratio (MLR) properties. To do this, we also need the following lemma (Karlin, 1956) given in Eaton (1972), p. B.2.

Lemma 4.2. *Assume $p(x,r)$ and $q(r,\theta)$ are functions which have monotone likelihood ratios. Then $g(x,\theta) = \int p(x,r)q(r,\theta)dr$ also has monotone likelihood ratio.*

Theorem 4.10. *Let $\mathbf{X} \sim E_n(\mu\mathbf{e}_n, \sigma^2\mathbf{I}_n, \psi)$ with p.d.f. $g(x,\sigma)$ where $\mu \in I\!R$ and $g(x,\sigma) = \int_0^\infty f_{N_n(\mu\mathbf{e}_n, z\sigma^2\mathbf{I}_n)}(\mathbf{x})u(z)dz$. Assume*

$$u(c_1r_1)u(c_2r_2) \leq u(c_1r_2)u(c_2r_1), \tag{4.30}$$

for $0 < c_1 < c_2$, $0 < r_1 < r_2$.

(a) If $n > p$ and $g_1(y,\sigma)$ denotes the p.d.f. of $\mathbf{y} = \mathbf{x}'\left(\mathbf{I}_n - \frac{\mathbf{e}_n\mathbf{e}_n'}{n}\right)\mathbf{x}$, then

$$g_1(y_1,\sigma_1)g_1(y_2,\sigma_2) \geq g_1(y_1,\sigma_2)g_1(y_2,\sigma_1),$$

for $0 < \sigma_1 < \sigma_2$, $0 < y_1 < y_2$; that is, $g_1(y,\sigma)$ has MLR.

(b) If $n \geq p$, $\mu = 0$, and $g_2(v,\sigma)$ denotes the p.d.f. of $v = \mathbf{x}'\mathbf{x}$, then

$$g_2(v_1,\sigma_1)g_2(v_2,\sigma_2) \geq g_2(v_1,\sigma_2)g_2(v_2,\sigma_1),$$

for $0 < \sigma_1 < \sigma_2$, $0 < v_1 < v_2$; that is, $g_2(v,\sigma)$ has MLR.

PROOF:

(a) We know, from Corollary 4.5, that

$$g_1(y,\sigma) = \int_0^\infty f_{W_1(z\sigma^2,n-1)}(y)u(z)dz.$$

Let $r = z\sigma^2$. Then, $z = \frac{r}{\sigma^2}$ and $J(z \to r) = \frac{1}{\sigma^2}$. Thus,

$$g_1(y,\sigma) = \int_0^\infty f_{W_1(z\sigma^2,n-1)}(y)u\left(\frac{r}{\sigma^2}\right)\frac{1}{\sigma^2}dr.$$

Let

$$p(y,r) = f_{W_1(z\sigma^2,n-1)}(y) = \frac{1}{2^{\frac{n-1}{2}}\Gamma\left(\frac{n-1}{2}\right)} \frac{y^{\frac{n-1}{2}-1}e^{-\frac{y}{2r}}}{r^{\frac{n-1}{2}}}$$

and $q(r,\sigma) = \left(\frac{r}{\sigma^2}\right)\frac{1}{\sigma^2}$.

It is easy to see that $p(y_1,r_1)p(y_2,r_2) \geq p(y_1,r_2)p(y_2,r_1)$, if $0 < r_1 < r_2$, $0 < y_1 < y_2$. Thus $p(y,r)$ has MLR. It follows, from (4.30), that $q(r,\sigma)$ also has MLR. Using Lemma 4.2, we obtain the desired result.

(b) It follows, from Corollary 4.5, that

$$g_2(y,\sigma) = \int_0^\infty f_{W_1(z\sigma^2,n)}(y)u(z)dz.$$

Then, proceeding in a similar way as in the proof of part (a), we find that $g_2(v,\sigma)$ has MLR. ■

The following theorem was given by Chu (1973) for the vector variate case.

Theorem 4.11. *Let $\lambda : I\!R^{pn} \to I\!R^{q\times m}$ be a Borel-measurable matrix variate function. Assume that if $\mathbf{X} \sim N_{p,n}(\mathbf{M},\Sigma\otimes\Phi)$, then $E(\lambda(\mathbf{X}))$ exists and it is denoted by $E_{N_{p,n}(\mathbf{M},\Sigma\otimes\Phi)}(\lambda(\mathbf{X}))$. Then, if $\mathbf{X} \sim E_{p,n}(\mathbf{M},\Sigma\otimes\Phi,\psi)$ with p.d.f.*

$$g(\mathbf{X}) = \int_0^\infty f_{N_{p,n}(\mathbf{M},z^{-1}\Sigma\otimes\Phi)}(\mathbf{X})w(z)dz,$$

such that $E(\lambda(\mathbf{X}))$ exists and it is denoted by $E_{E_{p,n}(\mathbf{M},\Sigma\otimes\Phi,\psi)}(\lambda(\mathbf{X}))$, we have

$$E_{E_{p,n}(\mathbf{M},\Sigma\otimes\Phi,\psi)}(\lambda(\mathbf{X})) = \int_0^\infty E_{N_{p,n}(\mathbf{M},z^{-1}\Sigma\otimes\Phi)}(\lambda(\mathbf{X}))w(z)dz. \tag{4.31}$$

If $g(\mathbf{X}) = \int_0^\infty f_{N_{p,n}(\mathbf{M},z\Sigma\otimes\Phi)}(\mathbf{X})u(z)dz$ and $E(\lambda(\mathbf{X}))$ exists, then

$$E_{E_{p,n}(\mathbf{M},\Sigma\otimes\Phi,\psi)}(\lambda(\mathbf{X})) = \int_0^\infty E_{N_{p,n}(\mathbf{M},z\Sigma\otimes\Phi)}(\lambda(\mathbf{X}))u(z)dz. \tag{4.32}$$

PROOF: In the proof of Theorem 4.5 if $dG(z)$ is replaced by $u(z)dz$, we obtain (4.31). In the same proof if we replace $dG(z)$ by $w(z)dz$ and $N_{p,n}(\mathbf{M},z\Sigma\otimes\Phi)$ by $N_{p,n}(\mathbf{M},z^{-1}\Sigma\otimes\Phi)$ we obtain (4.32). ■

Corollary 4.6. *With the notation of Theorem 4.10, if $Cov(\mathbf{X})$ exists, then*

$$Cov(\mathbf{X}) = \left(\int_0^\infty \frac{w(s)}{s}ds\right)\Sigma\otimes\Phi$$

and also

$$Cov(\mathbf{X}) = \left(\int_0^\infty su(s)ds\right)\Sigma\otimes\Phi.$$

Chapter 5
Quadratic Forms and Other Functions of Elliptically Contoured Matrices

5.1 Extension of Cochran's Theorem to Multivariate Elliptically Contoured Distributions

Anderson and Fang (1987) studied how results, similar to Cochran's theorem can be derived for m.e.c. distributions. This section presents their results. Results from Anderson and Fang (1982b) are also used.

We will need the following lemma.

Lemma 5.1. *Let* $\mathbf{X} : p \times n$ *be a random matrix with p.d.f.* $f(\mathbf{XX}')$. *Let* $\mathbf{A} = \mathbf{XX}'$, *then the p.d.f. of* $\mathbf{A}$ *is*

$$\frac{\pi^{\frac{pn}{2}}}{\Gamma_p\left(\frac{n}{2}\right)} |\mathbf{A}|^{\frac{n-p-1}{2}} f(\mathbf{A}), \quad \mathbf{A} > \mathbf{0},$$

where $\Gamma_p(t) = \pi^{\frac{p(p-1)}{4}} \prod_{i=1}^{p} \Gamma\left(t - \frac{i-1}{2}\right)$.

PROOF: See Anderson (2003), p. 539.

The next lemma generalizes the result of Lemma 5.1.

Lemma 5.2. *Let* $\mathbf{X}$ *be a random* $p \times n$ *matrix, and write* $\mathbf{X} = (\mathbf{X}_1, \mathbf{X}_2, \ldots, \mathbf{X}_m)$ *where* $\mathbf{X}_i$ *is* $p \times n_i$, $i = 1, \ldots, m$. *Assume* $\mathbf{X}$ *has the p.d.f.* $p(\mathbf{X}) = f(\mathbf{X}_1\mathbf{X}_1', \mathbf{X}_2\mathbf{X}_2', \ldots, \mathbf{X}_m\mathbf{X}_m')$. *Further let* $\mathbf{W}_i = \mathbf{X}_i\mathbf{X}_i'$, $i = 1, \ldots, m$. *Then, the p.d.f. of* $(\mathbf{W}_1, \mathbf{W}_2, \ldots, \mathbf{W}_m)$ *is*

$$\frac{\pi^{\frac{pn}{2}}}{\prod_{i=1}^{m} \Gamma_p\left(\frac{n_i}{2}\right)} \prod_{i=1}^{m} |\mathbf{W}_i|^{\frac{n_i-p-1}{2}} f(\mathbf{W}_1, \mathbf{W}_2, \ldots, \mathbf{W}_m), \quad \mathbf{W}_i > \mathbf{0},\ i = 1, \ldots, m. \tag{5.1}$$

PROOF: We prove, by induction, that the p.d.f. of $(\mathbf{W}_1, \ldots, \mathbf{W}_k, \mathbf{X}_{k+1}, \ldots, \mathbf{X}_m)$ is

A.K. Gupta et al., *Elliptically Contoured Models in Statistics and Portfolio Theory*,
DOI 10.1007/978-1-4614-8154-6_5, © Springer Science+Business Media New York 2013

$$\frac{\pi^{\frac{p\sum_{i=1}^{k} n_i}{2}}}{\prod_{i=1}^{k}\Gamma_p\left(\frac{n_i}{2}\right)}\prod_{i=1}^{k}|\mathbf{W}_i|^{\frac{n_i-p-1}{2}}\,p(\mathbf{X}_1,\ldots,\mathbf{X}_k,\mathbf{X}_{k+1},\ldots,\mathbf{X}_m)|_{\mathbf{X}_1\mathbf{X}_1'=\mathbf{W}_1,\ldots,\mathbf{X}_k\mathbf{X}_k'=\mathbf{W}_k}, \tag{5.2}$$

for $k=1,\ldots,m$. In the proof, $p(\mathbf{Y})$ will denote the p.d.f. of any random matrix $\mathbf{Y}$.

If $k=1$, we can write

$$\begin{aligned}
p(\mathbf{W}_1,\mathbf{X}_2,\ldots,\mathbf{X}_m) &= p(\mathbf{W}_1|\mathbf{X}_2,\ldots,\mathbf{X}_m)p(\mathbf{X}_2,\ldots,\mathbf{X}_m)\\
&= \frac{\pi^{\frac{pn_1}{2}}}{\Gamma_p\left(\frac{n_1}{2}\right)}|\mathbf{W}_1|^{\frac{n_1-p-1}{2}}\,p(\mathbf{X}_1|\mathbf{X}_2,\ldots,\mathbf{X}_m)|_{\mathbf{X}_1\mathbf{X}_1'=\mathbf{W}_1}\,p(\mathbf{X}_2,\ldots,\mathbf{X}_m)\\
&= \frac{\pi^{\frac{pn_1}{2}}}{\Gamma_p\left(\frac{n_1}{2}\right)}|\mathbf{W}_1|^{\frac{n_1-p-1}{2}}\,p(\mathbf{X}_1,\mathbf{X}_2,\ldots,\mathbf{X}_m)|_{\mathbf{X}_1\mathbf{X}_1'=\mathbf{W}_1},
\end{aligned}$$

where we used Lemma 5.1.

Now, assume the statement is true for $k=l<m$. Then, for $k=l+1$, we get

$$\begin{aligned}
&p(\mathbf{W}_1,\ldots,\mathbf{W}_{l+1},\mathbf{X}_{l+2},\ldots,\mathbf{X}_m)\\
&= p(\mathbf{W}_{l+1}|\mathbf{W}_1,\ldots,\mathbf{W}_l,\mathbf{X}_{l+2},\ldots,\mathbf{X}_m)p(\mathbf{W}_1,\ldots,\mathbf{W}_l,\mathbf{X}_{l+2},\ldots,\mathbf{X}_m)\\
&= \frac{\pi^{\frac{pn_{l+1}}{2}}}{\Gamma_p\left(\frac{n_{l+1}}{2}\right)}|\mathbf{W}_{l+1}|^{\frac{n_{l+1}-p-1}{2}}\\
&\times p(\mathbf{X}_{l+1}|\mathbf{W}_1,\ldots,\mathbf{W}_l,\mathbf{X}_{l+2},\ldots,\mathbf{X}_m)_{\mathbf{X}_{l+1}\mathbf{X}_{l+1}'=\mathbf{W}_{l+1}}\,p(\mathbf{W}_1,\ldots,\mathbf{W}_l,\mathbf{X}_{l+2},\ldots,\mathbf{X}_m)\\
&= \frac{\pi^{\frac{pn_{l+1}}{2}}}{\Gamma_p\left(\frac{n_{l+1}}{2}\right)}|\mathbf{W}_{l+1}|^{\frac{n_{l+1}-p-1}{2}}\\
&\times p(\mathbf{W}_1,\ldots,\mathbf{W}_l,\mathbf{X}_{l+1},\mathbf{X}_{l+2},\ldots,\mathbf{X}_m)_{\mathbf{X}_{l+1}\mathbf{X}_{l+1}'=\mathbf{W}_{l+1}}\\
&= \frac{\pi^{\frac{pn_{l+1}}{2}}}{\Gamma_p\left(\frac{n_{l+1}}{2}\right)}|\mathbf{W}_{l+1}|^{\frac{n_{l+1}-p-1}{2}}\,\frac{\pi^{\frac{p\sum_{i=1}^{l} n_i}{2}}}{\prod_{i=1}^{l}\Gamma_p\left(\frac{n_i}{2}\right)}\prod_{i=1}^{l}|\mathbf{W}_i|^{\frac{n_i-p-1}{2}}\\
&\times p(\mathbf{X}_1,\ldots,\mathbf{X}_l,\mathbf{X}_{l+1},\mathbf{X}_{l+2},\ldots,\mathbf{X}_m)|_{\mathbf{X}_1\mathbf{X}_1'=\mathbf{W}_1,\ldots,\mathbf{X}_{l+1}\mathbf{X}_{l+1}'=\mathbf{W}_{l+1}}\\
&= \frac{\pi^{\frac{p\sum_{i=1}^{l+1} n_i}{2}}}{\prod_{i=1}^{l+1}\Gamma_p\left(\frac{n_i}{2}\right)}\prod_{i=1}^{l+1}|\mathbf{W}_i|^{\frac{n_i-p-1}{2}}\\
&\times p(\mathbf{X}_1,\ldots,\mathbf{X}_{l+1},\mathbf{X}_{l+2},\ldots,\mathbf{X}_m)|_{\mathbf{X}_1\mathbf{X}_1'=\mathbf{W}_1,\ldots,\mathbf{X}_{l+1}\mathbf{X}_{l+1}'=\mathbf{W}_{l+1}}
\end{aligned}$$

where we used Lemma 5.1 and the induction hypothesis. Taking $k=m$ in (5.2) we obtain (5.1). ■

Definition 5.1. Let $\mathbf{X} \sim E_{p,n}(\mathbf{0}, \Sigma \otimes \mathbf{I}_n, \psi)$, $\Sigma > \mathbf{0}$. Partition $\mathbf{X}$ as $\mathbf{X} = (\mathbf{X}_1, \mathbf{X}_2, \ldots, \mathbf{X}_m)$, where $\mathbf{X}_i$ is $p \times n_i$, $i = 1, \ldots, m$. Then, $G_{p,m}\left(\Sigma, \frac{n_1}{2}, \ldots, \frac{n_m}{2}, \psi\right)$ denotes the distribution of $(\mathbf{X}_1\mathbf{X}_1', \mathbf{X}_2\mathbf{X}_2', \ldots, \mathbf{X}_m\mathbf{X}_m')$.

Remark 5.1. If in Definition 5.1, $\Sigma = \mathbf{I}_p$, we also use the notation $G_{p,m}(\frac{n_1}{2}, \ldots, \frac{n_m}{2}, \psi)$; that is, $\mathbf{I}_p$ can be dropped from the notation.

Remark 5.2. Definition 5.1 generalizes the Wishart distribution. In fact, if $m = 1$ and $\psi(z) = \exp\left(-\frac{z}{2}\right)$, then $G_{p,1}\left(\Sigma, \frac{n}{2}, \psi\right)$ is the same as $W_p(\Sigma, n)$.

Theorem 5.1. *Let $\mathbf{X} \sim E_{p,n}(\mathbf{0}, \Sigma \otimes \mathbf{I}_n, \psi)$, $\Sigma > \mathbf{0}$ and $r\mathbf{AU}$ be the stochastic representation of $\mathbf{X}$. Partition $\mathbf{X}$ as $\mathbf{X} = (\mathbf{X}_1, \mathbf{X}_2, \ldots, \mathbf{X}_m)$, where $\mathbf{X}_i$ is $p \times n_i$, $i = 1, \ldots, m$. Then,*

$$(\mathbf{X}_1\mathbf{X}_1', \mathbf{X}_2\mathbf{X}_2', \ldots, \mathbf{X}_m\mathbf{X}_m') \approx r^2\mathbf{A}(z_1\mathbf{V}_1, z_2\mathbf{V}_2, \ldots, z_m\mathbf{V}_m)\mathbf{A}', \tag{5.3}$$

where $(z_1, z_2, \ldots, z_{m-1}) \sim D\left(\frac{n_1}{2}, \frac{n_2}{2}, \ldots, \frac{n_{m-1}}{2}; \frac{n_m}{2}\right)$, $\sum_{i=1}^m z_i = 1$, $\mathbf{V}_i = \mathbf{U}_i\mathbf{U}_i'$ with $vec(\mathbf{U}_i')$ uniformly distributed on S_{pn_i}, and r, $\mathbf{V}_1, \mathbf{V}_2, \ldots, \mathbf{V}_m$, $(z_1, z_2, \ldots, z_m)$ are independent.

PROOF: From Theorem 2.17, it follows that

$$(\mathbf{X}_1, \mathbf{X}_2, \ldots, \mathbf{X}_m) \approx r\mathbf{A}(\sqrt{z_1}\mathbf{U}_1, \sqrt{z_2}\mathbf{U}_2, \ldots, \sqrt{z_m}\mathbf{U}_m),$$

from which (5.3) follows immediately. ∎

Theorem 5.2. *Let $(\mathbf{W}_1, \mathbf{W}_2, \ldots, \mathbf{W}_m) \sim G_{p,m}\left(\Sigma, \frac{n_1}{2}, \frac{n_2}{2}, \ldots, \frac{n_m}{2}, \psi\right)$, where $\mathbf{W}_i$ is $p \times p$, $i = 1, \ldots, m$. Then, for $1 \leq l \leq m$,*

$$(\mathbf{W}_1, \mathbf{W}_2, \ldots, \mathbf{W}_l) \sim G_{p,l}\left(\Sigma, \frac{n_1}{2}, \frac{n_2}{2}, \ldots, \frac{n_l}{2}, \psi\right).$$

PROOF: Define $\mathbf{X} \sim E_{p,n}(\mathbf{0}, \Sigma \otimes \mathbf{I}_n, \psi)$ and partition $\mathbf{X}$ as $\mathbf{X} = (\mathbf{X}_1, \mathbf{X}_2, \ldots, \mathbf{X}_m)$, where $\mathbf{X}_i$ is $p \times n_i$ dimensional, $i = 1, \ldots, m$. Then, by Definition 5.1 we have

$$(\mathbf{X}_1\mathbf{X}_1', \mathbf{X}_2\mathbf{X}_2', \ldots, \mathbf{X}_m\mathbf{X}_m') \approx (\mathbf{W}_1, \mathbf{W}_2, \ldots, \mathbf{W}_m).$$

Hence, $(\mathbf{X}_1\mathbf{X}_1', \mathbf{X}_2\mathbf{X}_2', \ldots, \mathbf{X}_l\mathbf{X}_l') \approx (\mathbf{W}_1, \mathbf{W}_2, \ldots, \mathbf{W}_l)$. Let $\mathbf{Y} = (\mathbf{X}_1, \mathbf{X}_2, \ldots, \mathbf{X}_l)$. Then, $\mathbf{Y} \sim E_{p,n^*}(\mathbf{0}, \Sigma \otimes \mathbf{I}_{n^*}, \psi)$ with $n^* = \sum_{i=1}^l n_i$. Therefore,

$$(\mathbf{X}_1\mathbf{X}_1', \mathbf{X}_2\mathbf{X}_2', \ldots, \mathbf{X}_l\mathbf{X}_l') \sim G_{p,l}\left(\Sigma, \frac{n_1}{2}, \frac{n_2}{2}, \ldots, \frac{n_l}{2}, \psi\right),$$

which completes the proof. ∎

Theorem 5.3. *Let* $(\mathbf{W}_1, \mathbf{W}_2, \ldots, \mathbf{W}_m) \sim G_{p,m}\left(\Sigma, \frac{n_1}{2}, \frac{n_2}{2}, \ldots, \frac{n_m}{2}, \psi\right)$, *where* $m > 1$, $\mathbf{W}_i$ *is* $p \times p$, $i = 1, \ldots, m$. *Then,*

$$(\mathbf{W}_1 + \mathbf{W}_2, \mathbf{W}_3, \ldots, \mathbf{W}_m) \sim G_{p,m-1}\left(\Sigma, \frac{n_1+n_2}{2}, \frac{n_3}{2}, \ldots, \frac{n_m}{2}, \psi\right).$$

PROOF: Let $\mathbf{X}$ be defined as in the proof of Theorem 5.2. Define $\mathbf{X}_0 = (\mathbf{X}_1, \mathbf{X}_2)$ and $\mathbf{Y} = (\mathbf{X}_0, \mathbf{X}_3, \ldots, \mathbf{X}_m)$. Then,

$$\begin{aligned}(\mathbf{X}_0\mathbf{X}_0', \mathbf{X}_3\mathbf{X}_3', \ldots, \mathbf{X}_m\mathbf{X}_m') &= (\mathbf{X}_1\mathbf{X}_1' + \mathbf{X}_2\mathbf{X}_2', \mathbf{X}_3\mathbf{X}_3', \ldots, \mathbf{X}_m\mathbf{X}_m') \\ &\approx (\mathbf{W}_1 + \mathbf{W}_2, \mathbf{W}_3, \ldots, \mathbf{W}_m).\end{aligned}$$

We also have

$$\mathbf{Y} = (\mathbf{X}_0, \mathbf{X}_3, \ldots, \mathbf{X}_m) \sim E_{p,n}(\mathbf{0}, \Sigma \otimes \mathbf{I}_n, \psi).$$

Hence, $(\mathbf{X}_0\mathbf{X}_0', \mathbf{X}_3\mathbf{X}_3', \ldots, \mathbf{X}_m\mathbf{X}_m') \sim G_{p,m-1}\left(\Sigma, \frac{n_1+n_2}{2}, \frac{n_3}{2}, \ldots, \frac{n_m}{2}, \psi\right)$ which completes the proof. ■

Theorem 5.4. *Let* $\mathbf{X} \sim E_{p,n}(\mathbf{0}, \Sigma \otimes \mathbf{I}_n, \psi)$, $\Sigma > \mathbf{0}$ *and* $P(\mathbf{X} = \mathbf{0}) = 0$ *and stochastic representation* $\mathbf{X} \approx r\Sigma^{\frac{1}{2}}\mathbf{U}$. *Partition* $\mathbf{X}$ *as* $\mathbf{X} = (\mathbf{X}_1, \mathbf{X}_2, \ldots, \mathbf{X}_m)$, *where* $\mathbf{X}_i$ *is* $p \times n_i$, $i = 1, \ldots, m$, $1 < m \leq n$, $p \leq n_i$, $i = 1, 2, \ldots, m-1$, $1 \leq n_m$. *Let* $\mathbf{W}_i = \mathbf{X}_i\mathbf{X}_i'$, $i = 1, \ldots, m-1$, *then the p.d.f. of* $(\mathbf{W}_1, \mathbf{W}_2, \ldots, \mathbf{W}_{m-1})$ *is given by*

$$p(\mathbf{W}_1, \mathbf{W}_2, \ldots, \mathbf{W}_{m-1}) = \frac{\Gamma\left(\frac{pn}{2}\right)|\Sigma|^{-\frac{n-n_m}{2}}}{\Gamma\left(\frac{pn_m}{2}\right)\prod_{i=1}^{m-1}\Gamma_p\left(\frac{pn_i}{2}\right)} \prod_{i=1}^{m-1} |\mathbf{W}_i|^{\frac{n_i-p-1}{2}}$$
$$\times \int_{\left(tr\left(\Sigma^{-1}\sum_{i=1}^{m-1}\mathbf{W}_i\right)\right)^{\frac{1}{2}}}^{\infty} r^{2-pn}\left(r^2 - tr\left(\Sigma^{-1}\sum_{i=1}^{m-1}\mathbf{W}_i\right)\right)^{\frac{pn_m}{2}-1} dF(r), \quad (5.4)$$

where $\mathbf{W}_i > \mathbf{0}$, $i = 1, \ldots, m$, *and* $F(r)$ *is the distribution function of* r.

PROOF: Let $\mathbf{Y} = \Sigma^{-\frac{1}{2}}\mathbf{X}$, $\mathbf{Y}_i = \Sigma^{-\frac{1}{2}}\mathbf{X}_i$, $i = 1, 2, \ldots, m$, and $\mathbf{V}_i = \mathbf{Y}_i\mathbf{Y}_i'$, $i = 1, 2, \ldots, m-1$. Then $\mathbf{Y} \sim E_{p,n}(\mathbf{0}, \mathbf{I}_p \otimes \mathbf{I}_n, \psi)$. From Theorem 3.6, it follows that the density of $\mathbf{Y} = (\mathbf{Y}_1, \mathbf{Y}_2, \ldots, \mathbf{Y}_{m-1})$ is

$$f(\mathbf{Y}_1, \mathbf{Y}_2, \ldots, \mathbf{Y}_{m-1}) = \frac{\Gamma\left(\frac{pn}{2}\right)}{\pi^{\frac{p(n-n_m)}{2}}\Gamma\left(\frac{pn_m}{2}\right)}$$
$$\times \int_{\left(tr\left(\sum_{i=1}^{m-1}\mathbf{Y}_i\mathbf{Y}_i'\right)\right)^{\frac{1}{2}}}^{\infty} r^{2-pn}\left(r^2 - tr\left(\sum_{i=1}^{m-1}\mathbf{Y}_i\mathbf{Y}_i'\right)\right)^{\frac{pn_m}{2}-1} dF(r).$$

Then, Lemma 5.1 gives the p.d.f. of $(\mathbf{V}_1, \mathbf{V}_2, \ldots, \mathbf{V}_{m-1})$ as

$$p(\mathbf{V}_1, \mathbf{V}_2, \ldots, \mathbf{V}_{m-1}) = \frac{\Gamma\left(\frac{pn}{2}\right)}{\Gamma\left(\frac{pn_m}{2}\right) \prod_{i=1}^{m-1} \Gamma_p\left(\frac{pn_i}{2}\right)} \prod_{i=1}^{m-1} |\mathbf{V}_i|^{\frac{n_i-p-1}{2}} \tag{5.5}$$

$$\times \int_{\left(tr\left(\sum_{i=1}^{m-1} \mathbf{V}_i\right)\right)^{\frac{1}{2}}}^{\infty} r^{2-pn} \left(r^2 - tr\left(\sum_{i=1}^{m-1} \mathbf{V}_i\right)\right)^{\frac{pn_m}{2}-1} dF(r).$$

Since, $(\mathbf{V}_1, \mathbf{V}_2, \ldots, \mathbf{V}_{m-1}) = \boldsymbol{\Sigma}^{-\frac{1}{2}}(\mathbf{W}_1, \mathbf{W}_2, \ldots, \mathbf{W}_{m-1})\boldsymbol{\Sigma}^{-\frac{1}{2}}$ and

$$J((\mathbf{V}_1, \mathbf{V}_2, \ldots, \mathbf{V}_{m-1}) \to (\mathbf{W}_1, \mathbf{W}_2, \ldots, \mathbf{W}_{m-1})) = |\boldsymbol{\Sigma}|^{-\frac{(p+1)(m-1)}{2}},$$

from (5.5) we get

$$p(\mathbf{W}_1, \mathbf{W}_2, \ldots, \mathbf{W}_{m-1}) = \frac{\Gamma\left(\frac{pn}{2}\right)}{\Gamma\left(\frac{pn_m}{2}\right) \prod_{i=1}^{m-1} \Gamma_p\left(\frac{pn_i}{2}\right)} \prod_{i=1}^{m-1} |\boldsymbol{\Sigma}|^{-\frac{n_i-p-1}{2}} \prod_{i=1}^{m-1} |\mathbf{W}_i|^{\frac{n_i-p-1}{2}}$$

$$\times \int_{\left(tr\left(\boldsymbol{\Sigma}^{-1} \sum_{i=1}^{m-1} \mathbf{W}_i\right)\right)^{\frac{1}{2}}}^{\infty} r^{2-pn} \left(r^2 - tr\left(\boldsymbol{\Sigma}^{-1} \sum_{i=1}^{m-1} \mathbf{W}_i\right)\right)^{\frac{pn_m}{2}-1} |\boldsymbol{\Sigma}|^{-\frac{(p+1)(m-1)}{2}} dF(r),$$

and since $\left(\prod_{i=1}^{m-1} |\boldsymbol{\Sigma}|^{-\frac{n_i-p-1}{2}}\right) |\boldsymbol{\Sigma}|^{-\frac{(p+1)(m-1)}{2}} = |\boldsymbol{\Sigma}|^{-\frac{1}{2}\sum_{i=1}^{m-1} n_i}$, we obtain (5.6). ■

If $\mathbf{X}$ is absolutely continuous, we obtain the following result.

Theorem 5.5. *Let* $\mathbf{X} \sim E_{p,n}(\mathbf{0}, \boldsymbol{\Sigma} \otimes \mathbf{I}_n, \psi)$ *have the p.d.f.*

$$f(\mathbf{X}) = \frac{1}{|\boldsymbol{\Sigma}|^{\frac{n}{2}}} h(tr(\mathbf{X}'\boldsymbol{\Sigma}^{-1}\mathbf{X})).$$

Partition $\mathbf{X}$ *as* $\mathbf{X} = (\mathbf{X}_1, \mathbf{X}_2, \ldots, \mathbf{X}_m)$, *where* $\mathbf{X}_i$ *is* $p \times n_i$, $i = 1, \ldots, m$, $1 \le m \le p$. *Let* $\mathbf{W}_i = \mathbf{X}_i\mathbf{X}_i'$, $i = 1, \ldots, m$, *then the p.d.f. of* $(\mathbf{W}_1, \mathbf{W}_2, \ldots, \mathbf{W}_m)$

$$p(\mathbf{W}_1, \mathbf{W}_2, \ldots, \mathbf{W}_m) = \frac{\pi^{\frac{pn}{2}} |\boldsymbol{\Sigma}|^{-\frac{n}{2}}}{\prod_{i=1}^{m} \Gamma_p\left(\frac{n_i}{2}\right)} \prod_{i=1}^{m} |\mathbf{W}_i|^{\frac{n_i-p-1}{2}} h\left(tr\left(\boldsymbol{\Sigma}^{-1} \sum_{i=1}^{m} \mathbf{W}_i\right)\right),$$

$$\mathbf{W}_i > \mathbf{0}, \quad i = 1, \ldots, m. \tag{5.6}$$

PROOF: Since the p.d.f. of $\mathbf{X} = (\mathbf{X}_1, \mathbf{X}_2, \ldots, \mathbf{X}_m)$ is $|\boldsymbol{\Sigma}|^{-\frac{n}{2}} h\left(tr\boldsymbol{\Sigma}^{-1} \sum_{i=1}^{m} \mathbf{X}_i\mathbf{X}_i'\right)$, from Lemma 5.2, we obtain (5.6). ■

Corollary 5.1. *Let* $\mathbf{X} \sim E_{p,n}(\mathbf{0}, \Sigma \otimes \mathbf{I}_n, \psi)$ *with the p.d.f.*

$$f(\mathbf{X}) = \frac{1}{|\Sigma|^{\frac{n}{2}}} h(tr(\mathbf{X}'\Sigma^{-1}\mathbf{X})).$$

Then, the p.d.f. of $\mathbf{A} = \mathbf{X}\mathbf{X}'$ *is given by*

$$p(\mathbf{A}) = \frac{\pi^{\frac{pn}{2}} |\Sigma|^{-\frac{n}{2}}}{\Gamma_p\left(\frac{n}{2}\right)} |\mathbf{A}|^{\frac{n-p-1}{2}} h(tr(\Sigma^{-1}\mathbf{A})), \quad \mathbf{A} > \mathbf{0}.$$

Lemma 5.3. *Let x and z be independent, and y and z be independent one-dimensional random variables, such that $P(x > 0) = P(y > 0) = P(z > 0) = 1$ and $xz \approx yz$. Assume one of the following conditions holds:*

(i) $\phi_{log z}(t) \neq 0$ *almost everywhere*
(ii) $P(x < 1) = 1$.

Then, $x \approx y$.

PROOF: Note that $xz \approx yz$ is equivalent to $log x + log z \approx log y + log z$, which is again equivalent to

$$\phi_{log x}(t)\phi_{log z}(t) = \phi_{log y}(t)\phi_{log z}(t), \tag{5.7}$$

where $\phi_x(t)$ denotes the characteristic function of x.

(i) Since $\phi_{log z}(t) \neq 0$ almost everywhere, and the characteristic functions are continuous, we get

$$\phi_{log x}(t) = \phi_{log y}(t).$$

Hence, $x \approx y$.

(ii) Since $\phi_{log z}(0) = 1$ and the characteristic functions are continuous, there exists $\delta > 0$, such that $\phi_{log z}(t) \neq 0$ for $t \in (-\delta, \delta)$.
Then, from (5.7) we get

$$\phi_{log y}(t) = \phi_{log y}(t) \quad \text{for} \quad t \in (-\delta, \delta) \tag{5.8}$$

Since $P(x < 1) = 1$, we have $P(log x < 0) = 1$. However, $P(log x < 0) = 1$, together with (5.8), implies that $log x \approx log y$ (see Marcinkiewicz, 1938). Thus, we get $x \approx y$. ∎

Theorem 5.6. *Let* $(\mathbf{W}_1, \mathbf{W}_2, \ldots, \mathbf{W}_m) \sim G_{p,m}\left(\Sigma, \frac{n_1}{2}, \frac{n_2}{2}, \ldots, \frac{n_m}{2}, \psi\right)$, *where n_i is positive integer and $\mathbf{W}_i$ is $p \times p$, $i = 1, 2, \ldots, m$ and $\Sigma > \mathbf{0}$. Let $\mathbf{v}$ be a p-dimensional constant nonzero vector. Then,*

$$(\mathbf{v}'\mathbf{W}_1\mathbf{v}, \mathbf{v}'\mathbf{W}_2\mathbf{v}, \ldots, \mathbf{v}'\mathbf{W}_m\mathbf{v}) \sim G_{1,m}\left(\frac{n_1}{2}, \frac{n_2}{2}, \ldots, \frac{n_m}{2}, \psi^*\right), \tag{5.9}$$

where $\psi^*(z) = \psi(\mathbf{v}'\Sigma\mathbf{v}z)$.

PROOF: Let $(\mathbf{X}_1, \mathbf{X}_2, \ldots, \mathbf{X}_m) \sim E_{p,n}(\mathbf{0}, \Sigma \otimes \mathbf{I}_n, \psi)$, with $n = \sum_{i=1}^m n_i$ and $\mathbf{X}_i$ is $p \times n_i$, $i = 1, 2, \ldots, m$. Then, by definition,

$$(\mathbf{X}_1\mathbf{X}_1', \mathbf{X}_2\mathbf{X}_2', \ldots, \mathbf{X}_m\mathbf{X}_m') \approx (\mathbf{W}_1, \mathbf{W}_2, \ldots, \mathbf{W}_m).$$

Hence,

$$(\mathbf{v}'\mathbf{X}_1\mathbf{X}_1'\mathbf{v}, \mathbf{v}'\mathbf{X}_2\mathbf{X}_2'\mathbf{v}, \ldots, \mathbf{v}'\mathbf{X}_m\mathbf{X}_m'\mathbf{v}) \approx (\mathbf{v}'\mathbf{W}_1\mathbf{v}, \mathbf{v}'\mathbf{W}_2\mathbf{v}, \ldots, \mathbf{v}'\mathbf{W}_m\mathbf{v}) \tag{5.10}$$

Define $\mathbf{y}_i' = \mathbf{v}'\mathbf{X}_i$, $i = 1, 2, \ldots, m$. Then, $\mathbf{y}_i$ is $n_i \times 1$ and

$$(\mathbf{y}_1', \mathbf{y}_2', \ldots, \mathbf{y}_m') = \mathbf{v}'(\mathbf{X}_1, \mathbf{X}_2, \ldots, \mathbf{X}_m),$$

and hence,

$$(\mathbf{y}_1', \mathbf{y}_2', \ldots, \mathbf{y}_m') \sim E_{1,n}(\mathbf{0}, \mathbf{v}'\Sigma\mathbf{v} \otimes \mathbf{I}_n, \psi) = E_{1,n}(\mathbf{0}, \mathbf{I}_n, \psi^*)$$

with $\psi^*(z) = \psi(\mathbf{v}'\Sigma\mathbf{v}z)$. Therefore,

$$(\mathbf{y}_1'\mathbf{y}_1, \mathbf{y}_2'\mathbf{y}_2, \ldots, \mathbf{y}_m'\mathbf{y}_m) \sim G_{1,m}\left(\frac{n_1}{2}, \frac{n_2}{2}, \ldots, \frac{n_m}{2}, \psi^*\right). \tag{5.11}$$

Since, $\mathbf{y}_1'\mathbf{y}_1 = \mathbf{v}'\mathbf{X}_i\mathbf{X}_i'\mathbf{v}$, $i = 1, 2, \ldots, m$, (5.10) and (5.11) give (5.9). ■

Now, we derive some result which can be regarded as the generalizations of Cochran's theorem for normal variables to the m.e.c. distribution.

Theorem 5.7. *Let* $\mathbf{X} \sim E_{p,n}(\mathbf{0}, \Sigma \otimes \mathbf{I}_n, \psi)$, $\Sigma > \mathbf{0}$ *and assume there exists a p-dimensional constant vector* $\mathbf{v}$ *such that* $P(\mathbf{v}'\Sigma^{-\frac{1}{2}}\mathbf{X} = \mathbf{0}) = 0$. *Let* $\mathbf{A}$ *be an* $n \times n$ *symmetric matrix and* $k \leq n$ *a positive integer. Then,*

$$\mathbf{X}\mathbf{A}\mathbf{X}' \sim G_{p,1}\left(\Sigma, \frac{k}{2}, \psi\right) \tag{5.12}$$

if and only if $\mathbf{A}^2 = \mathbf{A}$ *and* $rk(\mathbf{A}) = k$.

PROOF: It is enough to consider the case $\Sigma = \mathbf{I}_p$ because otherwise we can define $Y = \Sigma^{-\frac{1}{2}}\mathbf{X}$, and $\mathbf{X}\mathbf{A}\mathbf{X}' \sim G_{p,1}\left(\Sigma, \frac{k}{2}, \psi\right)$ is equivalent to $\mathbf{Y}\mathbf{A}\mathbf{Y}' \sim G_{p,1}\left(\frac{k}{2}, \psi\right)$.

First, assume $\mathbf{A}^2 = \mathbf{A}$ and $rk(\mathbf{A}) = k$. Then, using Theorem 1.12, we can write

$$\mathbf{A} = \mathbf{G}\begin{pmatrix} \mathbf{I}_k & \mathbf{0} \\ \mathbf{0} & \mathbf{0} \end{pmatrix}\mathbf{G}',$$

where $\mathbf{G} \in O(n)$ and $\mathbf{0}$'s denote zero matrices of appropriate dimensions.

Define $n \times k$ matrix $\mathbf{C} = \begin{pmatrix} \mathbf{I}_k \\ \mathbf{0} \end{pmatrix}$ and let $\mathbf{Y} = \mathbf{XGC}$. Then, $\mathbf{Y} \sim E_{p,k}(\mathbf{0}, \mathbf{I}_p \otimes \mathbf{I}_k, \psi)$ and

$$\mathbf{XAX}' = \mathbf{YY}'. \tag{5.13}$$

From Definition 5.1, we have

$$\mathbf{YY}' \sim G_{p,1}\left(\frac{k}{2}, \psi\right). \tag{5.14}$$

From (5.13) and (5.14), we obtain (5.12).

On the other hand, assume (5.12) holds, and define $\mathbf{y} = \mathbf{X}'\mathbf{v}$. Then,

$$\mathbf{y} \sim E_n(\mathbf{0}, \mathbf{I}_n, \psi^*) \quad \text{with} \quad \psi^*(t) = \psi(\mathbf{v}'\mathbf{v}t).$$

Moreover, $P(\mathbf{y} = \mathbf{0}) = 0$. From Theorem 5.2, we get

$$\mathbf{y}'\mathbf{A}\mathbf{y} \sim G_{1,1}\left(\frac{k}{2}, \psi^*\right). \tag{5.15}$$

Let $\mathbf{y} \approx r\mathbf{u}$ be the stochastic representation of $\mathbf{y}$. Then,

$$\mathbf{y}'\mathbf{A}\mathbf{y} \approx r^2\mathbf{u}'\mathbf{A}\mathbf{u}. \tag{5.16}$$

Let $\mathbf{y} = \begin{pmatrix} \mathbf{y}_1 \\ \mathbf{y}_2 \end{pmatrix}$, where $\mathbf{y}_1$ is k-dimensional. Then, from Corollary 2.4, we get

$$\begin{pmatrix} \mathbf{y}_1 \\ \mathbf{y}_2 \end{pmatrix} \approx \begin{pmatrix} r\sqrt{w}\mathbf{u}_1 \\ r\sqrt{1-w}\mathbf{u}_2 \end{pmatrix}, \tag{5.17}$$

where r, w, $\mathbf{u}_1$ and $\mathbf{u}_2$ are independent, $\mathbf{u}_1$ is uniformly distributed on S_k, $\mathbf{u}_2$ is uniformly distributed on S_{n-k}, and $w \sim B\left(\frac{k}{2}, \frac{n-k}{2}\right)$. Since $\mathbf{y}_1 \sim E_k(\mathbf{0}, \mathbf{I}_k, \psi^*)$, we get

$$\mathbf{y}_1'\mathbf{y}_1 \sim G_{1,1}\left(\frac{k}{2}, \psi^*\right). \tag{5.18}$$

From (5.17), we obtain

$$\mathbf{y}_1'\mathbf{y}_1 \approx r^2 w \mathbf{u}_1'\mathbf{u}_1 = r^2 w. \tag{5.19}$$

From (5.15), (5.16), (5.18), and (5.19), we get

$$r^2\mathbf{u}'\mathbf{A}\mathbf{u} \approx r^2 w. \tag{5.20}$$

Since $P(0 < w < 1) = 1$ and $P(r^2 > 0) = 1$, we have $P(r^2 w > 0) = 1$. Therefore, (5.20) implies $P(\mathbf{u}'\mathbf{A}\mathbf{u} > 0) = 1$. Using (ii) of Lemma 5.3, from (5.20) we obtain $\mathbf{u}'\mathbf{A}\mathbf{u} \approx w$. Thus,

$$\mathbf{u}'\mathbf{A}\mathbf{u} \sim B\left(\frac{k}{2}, \frac{n-k}{2}\right). \tag{5.21}$$

Define $\mathbf{z} \sim N_n(\mathbf{0}, \mathbf{I}_n)$. Then, from Theorem 2.15 it follows that $\frac{\mathbf{z}}{\|\mathbf{z}\|} \approx u$, and from (5.21) we get

$$\frac{\mathbf{z}'\mathbf{A}\mathbf{z}}{\|\mathbf{z}\|^2} \sim B\left(\frac{k}{2}, \frac{n-k}{2}\right). \tag{5.22}$$

Now, $\mathbf{A} = \mathbf{G}\mathbf{D}\mathbf{G}'$, where $\mathbf{G} \in O(n)$ and $\mathbf{D} = \begin{pmatrix} \mathbf{I}_k & \mathbf{0} \\ \mathbf{0} & \mathbf{0} \end{pmatrix}$, where $\mathbf{0}$'s denote zero matrices of appropriate dimensions. Let $\mathbf{t} = \mathbf{G}'\mathbf{z}$. Then, $\mathbf{t} \sim N_n(\mathbf{0}, \mathbf{I}_n)$ and $\frac{\mathbf{z}'\mathbf{A}\mathbf{z}}{\|\mathbf{z}\|^2} = \frac{\mathbf{t}'\mathbf{D}\mathbf{t}}{\|\mathbf{t}\|^2} \sim B\left(\frac{k}{2}, \frac{n-k}{2}\right)$. However $\mathbf{z} \approx \mathbf{t}$

$$\frac{\mathbf{z}'\mathbf{D}\mathbf{z}}{\|\mathbf{z}\|^2} \sim B\left(\frac{k}{2}, \frac{n-k}{2}\right). \tag{5.23}$$

From (5.22) and (5.23), we get

$$\frac{\mathbf{z}'\mathbf{A}\mathbf{z}}{\|\mathbf{z}\|^2} \approx \frac{\mathbf{z}'\mathbf{D}\mathbf{z}}{\|\mathbf{z}\|^2}. \tag{5.24}$$

Now, $\mathbf{z}'\mathbf{A}\mathbf{z} = \|\mathbf{z}\|^2 \frac{\mathbf{z}'\mathbf{A}\mathbf{z}}{\|\mathbf{z}\|^2}$ with $\|\mathbf{z}\|^2$ and $\frac{\mathbf{z}'\mathbf{A}\mathbf{z}}{\|\mathbf{z}\|^2}$ being independent. Moreover, $\mathbf{z}'\mathbf{D}\mathbf{z} = \|\mathbf{z}\|^2 \frac{\mathbf{z}'\mathbf{A}\mathbf{z}}{\|\mathbf{z}\|^2}$, with $\|\mathbf{z}\|^2$ and $\frac{\mathbf{z}'\mathbf{D}\mathbf{z}}{\|\mathbf{z}\|^2}$ being independent. Therefore, from (5.23), we get $\mathbf{z}'\mathbf{A}\mathbf{z} \sim \mathbf{z}'\mathbf{D}\mathbf{z}$. Since, $\mathbf{z}'\mathbf{D}\mathbf{z} \sim \chi_k^2$, we get

$$\mathbf{z}'\mathbf{A}\mathbf{z} \sim \chi_k^2. \tag{5.25}$$

Now, (5.25) implies that $\mathbf{A}^2 = \mathbf{A}$, and $rk(\mathbf{A}) = k$. ■

Theorem 5.7 can be generalized in the following way.

Theorem 5.8. *Let* $\mathbf{X} \sim E_{p,n}(\mathbf{0}, \Sigma \otimes \mathbf{I}_n, \psi)$, $\Sigma > \mathbf{0}$ *and assume there exists a p-dimensional constant vector* $\mathbf{v}$ *such that* $P(\mathbf{v}'\Sigma^{-\frac{1}{2}}\mathbf{X} = \mathbf{0}) = 0$. *Let* $\mathbf{A}_1, \mathbf{A}_2, \ldots, \mathbf{A}_m$ *be* $n \times n$ *symmetric matrices and* $k_1, k_2, \ldots, k_m$ *positive integers with* $\sum_{i=1}^n k_i \leq n$. *Then,*

$$(\mathbf{X}\mathbf{A}_1\mathbf{X}', \mathbf{X}\mathbf{A}_2\mathbf{X}', \ldots, \mathbf{X}\mathbf{A}_m\mathbf{X}') \sim G_{p,m}\left(\Sigma, \frac{k_1}{2}, \frac{k_2}{2}, \ldots, \frac{k_m}{2}, \psi\right) \tag{5.26}$$

if and only if

$$rk(\mathbf{A}_i) = k_i, \quad i = 1, \ldots, m, \tag{5.27}$$

$$\mathbf{A}_i^2 = \mathbf{A}_i, \quad i = 1, \ldots, m \tag{5.28}$$

and

$$\mathbf{A}_i \mathbf{A}_j = \mathbf{0}, \quad i \neq j,\ i, j = 1, \ldots, m. \tag{5.29}$$

PROOF: As in Theorem 5.7, it is enough to consider the case $\mathbf{\Sigma} = \mathbf{I}_p$.

First, assume (5.26)–(5.28) are satisfied. Then, from Theorem 1.14, there exists $\mathbf{G} \in O(n)$ such that

$$\mathbf{G}'\mathbf{A}_1\mathbf{G} = \begin{pmatrix} \mathbf{I}_{k_1} & \mathbf{0} \\ \mathbf{0} & \mathbf{0} \end{pmatrix}, \quad \mathbf{G}'\mathbf{A}_2\mathbf{G} = \begin{pmatrix} \mathbf{0} & \mathbf{0} & \mathbf{0} \\ \mathbf{0} & \mathbf{I}_{k_2} & \mathbf{0} \\ \mathbf{0} & \mathbf{0} & \mathbf{0} \end{pmatrix}$$

$$\ldots, \mathbf{G}'\mathbf{A}_m\mathbf{G} = \begin{pmatrix} \mathbf{0} & \mathbf{0} & \mathbf{0} \\ \mathbf{0} & \mathbf{I}_{k_m} & \mathbf{0} \\ \mathbf{0} & \mathbf{0} & \mathbf{0} \end{pmatrix}.$$

Let $k = \sum_{i=1}^m k_i$ and define the $n \times k$ matrix $\mathbf{C} = \begin{pmatrix} \mathbf{I}_k \\ \mathbf{0} \end{pmatrix}$. Moreover, define the $k \times k$ matrices $\mathbf{C}_i = \begin{pmatrix} \mathbf{0} & \mathbf{0} & \mathbf{0} \\ \mathbf{0} & \mathbf{I}_{k_i} & \mathbf{0} \\ \mathbf{0} & \mathbf{0} & \mathbf{0} \end{pmatrix}$, $i = 1, \ldots, m$. Then, $\mathbf{G}'\mathbf{A}_i\mathbf{G} = \mathbf{C}\mathbf{C}_i\mathbf{C}'$, $i = 1, \ldots, m$.

Define $\mathbf{Y} = \mathbf{XGC}$. Then, $\mathbf{Y} \sim E_{p,k}(\mathbf{0}, \mathbf{I}_p \otimes \mathbf{I}_k, \psi)$ and

$$\mathbf{X}\mathbf{A}_i\mathbf{X}' = \mathbf{Y}\mathbf{C}_i\mathbf{Y}', \quad i = 1, \ldots, m. \tag{5.30}$$

Partition $\mathbf{Y}$ into $\mathbf{Y} = (\mathbf{Y}_1, \mathbf{Y}_2, \ldots, \mathbf{Y}_m)$ where $\mathbf{Y}_i$ is $p \times k_i$, $i = 1, \ldots, m$. Then,

$$\mathbf{Y}\mathbf{C}_i\mathbf{Y}' = \mathbf{Y}_i\mathbf{Y}_i', \quad i = 1, \ldots, m \tag{5.31}$$

and by Definition 5.1, we get

$$(\mathbf{Y}_1\mathbf{Y}_1', \mathbf{Y}_2\mathbf{Y}_2', \ldots, \mathbf{Y}_m\mathbf{Y}_m') \sim G_{p,m}\left(\frac{k_1}{2}, \frac{k_2}{2}, \ldots, \frac{k_m}{2}, \psi\right). \tag{5.32}$$

From (5.30), (5.31), and (5.32) we obtain (5.26).

Next, assume (5.26) holds. Then, it follows from Theorem 5.2, that

$$\mathbf{X}\mathbf{A}_1\mathbf{X}' \sim G_{p,1}\left(\frac{k_1}{2}, \psi\right), \quad i = 1, \ldots, m.$$

Using Theorem 5.7, we get $rk(\mathbf{A}_i) = k_i$ and $\mathbf{A}_i^2 = \mathbf{A}_i$. It also follows from Theorem 5.2, that

$$(\mathbf{X}\mathbf{A}_i\mathbf{X}', \mathbf{X}\mathbf{A}_j\mathbf{X}') \sim G_{p,2}\left(\frac{k_i}{2}, \frac{k_j}{2}, \psi\right).$$

Now, using Theorem 5.3, we get

$$(\mathbf{X}(\mathbf{A}_i + \mathbf{A}_j)\mathbf{X}') \sim G_{p,1}\left(\frac{k_1 + k_2}{2}, \psi\right), i \neq j.$$

Using Theorem 5.7 again, we get $(\mathbf{A}_i + \mathbf{A}_j)^2 = \mathbf{A}_i + \mathbf{A}_j$. However, we already know $\mathbf{A}_i^2 = \mathbf{A}_i$, and $\mathbf{A}_j^2 = \mathbf{A}_j$. Hence, we get $\mathbf{A}_i\mathbf{A}_j = \mathbf{0}$. ■

Theorem 5.9. *Let $\mathbf{X} \sim E_{p,n}(\mathbf{0}, \Sigma \otimes \mathbf{I}_n, \psi)$, $\Sigma > \mathbf{0}$ and assume there exists a p-dimensional constant vector $\mathbf{v}$ such that $P(\mathbf{v}'\Sigma^{-\frac{1}{2}}\mathbf{X} = \mathbf{0}) = 0$. Let $\mathbf{A}$ and $\mathbf{B}$ be symmetric idempotent $n \times n$ matrices, with $1 \leq rk(\mathbf{A}) < n$, $1 \leq rk(\mathbf{B}) < n$, such that $\mathbf{AB} = \mathbf{0}$. Then, $\mathbf{XAX}'$ and $\mathbf{XBX}'$ are independent if and only if $\mathbf{X} \sim N_{p,n}(\mathbf{0}, \sigma^2\mathbf{I}_p \otimes \mathbf{I}_n)$, where $\sigma^2 > 0$.*

PROOF: Without loss of generality, we can assume $\Sigma = \mathbf{I}_p$.

Let $n_1 = rk(\mathbf{A})$, $n_2 = rk(\mathbf{B})$. Since $\mathbf{A}$ and $\mathbf{B}$ are symmetric, we have $\mathbf{BA} = \mathbf{AB} = \mathbf{0}$. Using Theorem 1.14, we can find $\mathbf{G} \in O(n)$, such that

$$\mathbf{G}'\mathbf{A}\mathbf{G} = \begin{pmatrix} \mathbf{I}_{n_1} & \mathbf{0} & \mathbf{0} \\ \mathbf{0} & \mathbf{0} & \mathbf{0} \\ \mathbf{0} & \mathbf{0} & \mathbf{0} \end{pmatrix}, \quad \text{and} \quad \mathbf{G}'\mathbf{B}\mathbf{G} = \begin{pmatrix} \mathbf{0} & \mathbf{0} & \mathbf{0} \\ \mathbf{0} & \mathbf{I}_{n_2} & \mathbf{0} \\ \mathbf{0} & \mathbf{0} & \mathbf{0} \end{pmatrix}.$$

Let $n_0 = n_1 + n_2$ and define the $n \times n_0$ matrix $\mathbf{C} = \begin{pmatrix} \mathbf{I}_{n_0} \\ \mathbf{0} \end{pmatrix}$. Moreover, define the $n_0 \times n_0$ matrices $\mathbf{C}_1 = \begin{pmatrix} \mathbf{I}_{n_1} & \mathbf{0} \\ \mathbf{0} & \mathbf{0} \end{pmatrix}$ and $\mathbf{C}_2 = \begin{pmatrix} \mathbf{0} & \mathbf{0} \\ \mathbf{0} & \mathbf{I}_{n_2} \end{pmatrix}$. Then, $\mathbf{G}'\mathbf{A}\mathbf{G} = \mathbf{C}\mathbf{C}_1\mathbf{C}'$ and $\mathbf{G}'\mathbf{B}\mathbf{G} = \mathbf{C}\mathbf{C}_2\mathbf{C}'$.

Define $\mathbf{Y} = \mathbf{XGC}$, then $\mathbf{XAX}' = \mathbf{Y}\mathbf{C}_1\mathbf{Y}'$ and $\mathbf{XBX}' = \mathbf{Y}\mathbf{C}_2\mathbf{Y}'$. Partition $\mathbf{Y}$ into $\mathbf{Y} = (\mathbf{Y}_1\ \mathbf{Y}_2)$, where $\mathbf{Y}_1$ is $p \times n_1$. Then, $\mathbf{Y}\mathbf{C}_1\mathbf{Y}' = \mathbf{Y}_1\mathbf{Y}_1'$ and $\mathbf{Y}\mathbf{C}_2\mathbf{Y}' = \mathbf{Y}_2\mathbf{Y}_2'$.

First, assume $\mathbf{X} \sim N_{p,n}(\mathbf{0}, \sigma^2\mathbf{I}_p \otimes \mathbf{I}_n)$. Then, $\mathbf{Y} \sim N_{p,n_0}(\mathbf{0}, \sigma^2\mathbf{I}_p \otimes \mathbf{I}_{n_0})$. Thus, the columns of $\mathbf{Y}$ are independent and so are $\mathbf{Y}_1$ and $\mathbf{Y}_2$. Hence, $\mathbf{Y}_1\mathbf{Y}_1'$ and $\mathbf{Y}_2\mathbf{Y}_2'$ are independent. Therefore, $\mathbf{XAX}'$ and $\mathbf{XBX}'$ are independent.

On the other hand, assume $\mathbf{XAX}'$ and $\mathbf{XBX}'$ are independent. Define $\mathbf{y} = \mathbf{X}'\mathbf{v}$. Then, $\mathbf{y} \sim E_n(\mathbf{0}, \mathbf{I}_n, \psi^*)$, where $\psi^*(t) = \psi(\mathbf{v}'\mathbf{v}t)$. Moreover, $P(\mathbf{y} = \mathbf{0}) = 0$. Since,

$\mathbf{XAX}'$ and $\mathbf{XBX}'$ are independent, so are $\mathbf{y}'\mathbf{Ay}$ and $\mathbf{y}'\mathbf{By}$. Define $\mathbf{w} = \mathbf{G}'\mathbf{y}$, then $\mathbf{w} \sim E_n(\mathbf{0}, \mathbf{I}_n, \psi^*)$ with $P(\mathbf{w} = \mathbf{0}) = 0$. Let $r\mathbf{u}$ be the stochastic representation of $\mathbf{w}$. Then, $P(r = 0) = 0$. Partition $\mathbf{w}$ into $\mathbf{w} = \begin{pmatrix} \mathbf{w}_1 \\ \mathbf{w}_2 \\ \mathbf{w}_3 \end{pmatrix}$ where $\mathbf{w}_1$ is n_1-dimensional, and $\mathbf{w}_2$ is n_2-dimensional.

Let $r\begin{pmatrix} \sqrt{z_1}\mathbf{u}_1 \\ \sqrt{z_2}\mathbf{u}_2 \\ \sqrt{z_3}\mathbf{u}_3 \end{pmatrix}$ be the representation of $\begin{pmatrix} \mathbf{w}_1 \\ \mathbf{w}_2 \\ \mathbf{w}_3 \end{pmatrix}$ given by Theorem 2.17. Since $P(r = 0) = 0$, we get $P(\mathbf{w}_1 = \mathbf{0}) = P(\mathbf{w}_2 = \mathbf{0}) = 0$. Now, $\mathbf{y}'\mathbf{Ay} = \mathbf{w}_1'\mathbf{w}_1$ and $\mathbf{y}'\mathbf{By} = \mathbf{w}_2'\mathbf{w}_2$. Define $\mathbf{w}_0 = \begin{pmatrix} \mathbf{w}_1 \\ \mathbf{w}_2 \end{pmatrix}$. Then, $\mathbf{w}_0 \sim E_{n_0}(\mathbf{0}, \mathbf{I}_{n^*}, \psi^*)$. Let $r_0\mathbf{u}_0$ be the stochastic representation of $\mathbf{w}_0$. Then,

$$(\mathbf{w}_1'\mathbf{w}_1, \mathbf{w}_2'\mathbf{w}_2) \approx r_0^2(s_1, s_2)$$

where r_0^2 and (s_1, s_2) are independent, $s_1 + s_2 = 1$ and

$$s_1 \sim B\left(\frac{n_1}{2}, \frac{n_2}{2}\right). \tag{5.33}$$

Moreover, $\frac{\mathbf{w}_1'\mathbf{w}_1}{\mathbf{w}_2'\mathbf{w}_2} \approx \frac{s_1}{s_2}$ and $\mathbf{w}_1'\mathbf{w}_1 + \mathbf{w}_2'\mathbf{w}_2 \approx r_0^2$ are independent. Therefore, $\mathbf{w}_1'\mathbf{w}_1 \approx \sigma_0^2\chi^2_{p_1}$ and $\mathbf{w}_2'\mathbf{w}_2 \approx \sigma_0^2\chi^2_{p_2}$ for $\sigma_0 > 0$ (see Lukacs, 1956, p. 208). Since $r_0^2 \approx \mathbf{w}_1'\mathbf{w}_1 + \mathbf{w}_2'\mathbf{w}_2$, we have

$$r_0^2 \sim \chi^2_p \quad \text{with } p = p_1 + p_2. \tag{5.34}$$

We also have $\frac{1-s_2}{s_2} \approx \frac{\mathbf{w}_1'\mathbf{w}_1}{\mathbf{w}_2'\mathbf{w}_2}$. Consequently, $\frac{1}{s_2} \approx \frac{\mathbf{w}_1'\mathbf{w}_1 + \mathbf{w}_2'\mathbf{w}_2}{\mathbf{w}_2'\mathbf{w}_2}$, and $s_2 \approx \frac{\mathbf{w}_2'\mathbf{w}_2}{\mathbf{w}_1'\mathbf{w}_1 + \mathbf{w}_2'\mathbf{w}_2} \sim B\left(\frac{p_2}{2}, \frac{p_1}{2}\right)$. Hence,

$$s_1 \sim B\left(\frac{p_1}{2}, \frac{p_2}{2}\right). \tag{5.35}$$

From (5.33) and (5.35), we get $p_1 = n_1$, $p_2 = n_2$. Thus, $p = n_0$. From (5.34), we get $r_0^2 \approx \sigma_0^2\chi^2_{n_0}$. Since, $\mathbf{u}_0$ is uniformly distributed on S_{n_0}, we get $\mathbf{w}_0 \sim N_{n_0}(\mathbf{0}, \sigma_0^2\mathbf{I}_{r_0})$. Consequently, $\psi^*(z) = \exp\left(-\frac{\sigma_0^2 z}{2}\right)$. Therefore, $\psi(z) = \exp\left(-\frac{\sigma^2 z}{2}\right)$, with $\sigma^2 = \frac{\sigma_0^2}{\mathbf{v}'\mathbf{v}}$. Hence, $\mathbf{X} \sim N_{p,n}(\mathbf{0}, \sigma^2\mathbf{I}_p \otimes \mathbf{I}_n)$. ∎

Corollary 5.2. *Let $\mathbf{X} \sim E_{p,n}(\mathbf{0}, \mathbf{I}_p \otimes \mathbf{I}_n, \psi)$ and let $\mathbf{x}_i$ denote the ith column of $\mathbf{X}$, $i = 1, 2, \ldots, n$. Define $\bar{\mathbf{x}} = \frac{1}{n}\sum_{i=1}^n \mathbf{x}_i$ and $\mathbf{S}(\mathbf{X}) = \frac{1}{n}\sum_{i=1}^n (\mathbf{x}_i - \bar{\mathbf{x}})(\mathbf{x}_i - \bar{\mathbf{x}})'$. Assume there exists a p-dimensional constant vector $\mathbf{v}$ such that $P(\mathbf{v}'\mathbf{X} = 0) = 0$. Then, $\bar{\mathbf{x}}$ and $\mathbf{S}(\mathbf{X})$ are independent if and only if $\mathbf{X} \sim N_{p,n}(\mathbf{0}, \sigma^2\mathbf{I}_p \otimes \mathbf{I}_n)$ with $\sigma^2 > 0$.*

PROOF: If $\mathbf{X} \sim N_{p,n}(\mathbf{0}, \sigma^2 \mathbf{I}_p \otimes \mathbf{I}_n)$, then $\mathbf{x}_i$'s constitute a random sample from the distribution $N_p(\mathbf{0}, \sigma^2 \mathbf{I}_p)$ and the independence of $\bar{\mathbf{x}}$ and $\mathbf{S}(\mathbf{X})$ is a well-known result (see Anderson, 2003, p. 74).

On the other hand, if $\bar{\mathbf{x}}$ and $\mathbf{S}(\mathbf{X})$ are independent, then $\bar{\mathbf{x}}\bar{\mathbf{x}}'$ and $\mathbf{S}(\mathbf{X})$ are also independent. We have $\bar{\mathbf{x}}\bar{\mathbf{x}}' = \mathbf{X}\mathbf{e}_n\mathbf{e}_n'\mathbf{X}'$ and

$$\mathbf{S}(\mathbf{X}) = \mathbf{X}\left(\mathbf{I}_n - \frac{\mathbf{e}_n\mathbf{e}_n'}{n}\right)\mathbf{X}'.$$

Let $\mathbf{A} = \mathbf{e}_n\mathbf{e}_n'$ and $\mathbf{B} = \mathbf{I}_n - \frac{\mathbf{e}_n\mathbf{e}_n'}{n}$. Then, $\mathbf{A}$ and $\mathbf{B}$ satisfy the conditions of Theorem 5.9. Therefore $\mathbf{X} \sim N_{p,n}(\mathbf{0}, \sigma^2 \mathbf{I}_p \otimes \mathbf{I}_n)$. ∎

5.2 Rank of Quadratic Forms

The main result of this section uses the following lemma proved by Okamoto (1973).

Lemma 5.4. *Let* $\mathbf{X}$ *be a* $p \times n$ *random matrix with absolute continuous distribution. Let* $\mathbf{A}$ *be an* $n \times n$ *symmetric matrix with* $rk(\mathbf{A}) = q$. *Then*

(i) $P\{rk(\mathbf{X}\mathbf{A}\mathbf{X}') = \min(p, q)\} = 1$ *and*
(ii) $P\{$*nonzero eigenvalues of* $\mathbf{X}\mathbf{A}\mathbf{X}'$ *are distinct*$\} = 1$.

PROOF: See Okamoto (1973). ∎

Matrix variate elliptically contoured distributions are not necessarily absolutely continuous. However, as the following theorem shows (see Gupta and Varga, 1991), a result similar to that of Okamoto can be derived for this class of distributions also, if we assume that the distribution is symmetric about the origin and it assumes zero with probability zero.

Theorem 5.10. *Let* $\mathbf{X} \sim E_{p,n}(\mathbf{0}, \Sigma \otimes \Phi, \psi)$ *with* $P(\mathbf{X} = \mathbf{0}) = 0$. *Let* $\mathbf{A}$ *be an* $n \times n$ *symmetric matrix. Then*

(i) $P\{rk(\mathbf{X}\mathbf{A}\mathbf{X}') = \min(rk(\Sigma), rk(\Phi\mathbf{A}\Phi))\} = 1$ *and*
(ii) $P\{$*the nonzero eigenvalues of* $\mathbf{X}\mathbf{A}\mathbf{X}'$ *are distinct*$\} = 1$.

PROOF: Let $rk(\Sigma) = q$, $rk(\Phi) = m$ and let $\mathbf{X} \approx r\mathbf{C}\mathbf{U}\mathbf{D}'$ be the stochastic representation of $\mathbf{X}$. Then, $vec(\mathbf{U}')$ is uniformly distributed on S_{qm}. Using Theorem 1.9 we can write $\mathbf{C} = \mathbf{G}\begin{pmatrix}\mathbf{B}\\ \mathbf{0}\end{pmatrix}$, where $\mathbf{G} \in O(p)$ and $\mathbf{B}$ is a $q \times q$ positive definite matrix. Then

$$\mathbf{X}\mathbf{A}\mathbf{X}' \approx r^2\mathbf{G}\begin{pmatrix}\mathbf{B}\mathbf{U}\mathbf{D}'\mathbf{A}\mathbf{D}\mathbf{U}'\mathbf{B}' & \mathbf{0}\\ \mathbf{0} & \mathbf{0}\end{pmatrix}\mathbf{G}',$$

where **0**'s denote zero matrices of appropriate dimensions. Since $P(\mathbf{X} = \mathbf{0}) = 0$, we have $P(r = 0) = 0$. Moreover, the nonzero eigenvalues of $\mathbf{G} \begin{pmatrix} \mathbf{BUD'ADU'B'} & \mathbf{0} \\ \mathbf{0} & \mathbf{0} \end{pmatrix} \mathbf{G}'$ are the same as those of $\mathbf{BUD'ADU'B'}$. Hence,

$$P\{rk(\mathbf{XAX}') = rk(\mathbf{BUD'ADU'B'})\} = 1 \tag{5.36}$$

and

$$\begin{aligned} &P\{\text{the nonzero eigenvalues of } \mathbf{XAX}' \text{ are distinct}\} \\ &= P\{\text{the nonzero eigenvalues of } \mathbf{BUD'ADU'B'} \text{ are distinct}\}. \end{aligned} \tag{5.37}$$

Let $\Sigma^* = \mathbf{BB}'$, $\mathbf{A}^* = \mathbf{D'AD}$ and define $\mathbf{Y} \sim N_{q,m}(\mathbf{0}, \Sigma^* \otimes \mathbf{I}_m)$. Since $\mathbf{B}$ is nonsingular, $\Sigma^* > \mathbf{0}$ and so $\mathbf{Y}$ is absolutely continuous. Let $\mathbf{Y} \approx r^*\mathbf{BU}^*$ be the stochastic representation of $\mathbf{Y}$. Then, $vec(\mathbf{U}^{*\prime})$ is uniformly distributed on S_{qm}. Now, $\mathbf{YA}^*\mathbf{Y}' \approx r^{*2}\mathbf{BU}^*\mathbf{D'ADU}^{*\prime}\mathbf{B}'$ and therefore,

$$P\{rk(\mathbf{YA}^*\mathbf{Y}') = rk(\mathbf{BU}^*\mathbf{D'ADU}^{*\prime}\mathbf{B}')\} = 1 \tag{5.38}$$

and

$$\begin{aligned} &P\{\text{the nonzero eigenvalues of } \mathbf{YA}^*\mathbf{Y}' \text{ are distinct}\} \\ &= P\{\text{the nonzero eigenvalues of } \mathbf{BU}^*\mathbf{D'ADU}^{*\prime}\mathbf{B}' \text{ are distinct}\}. \end{aligned} \tag{5.39}$$

However, from Lemma 5.4 we know that

$$P\{rk(\mathbf{YA}^*\mathbf{Y}') = \min(q, rk(\mathbf{A}^*))\} = 1 \tag{5.40}$$

and

$$P\{\text{the nonzero eigenvalues of } \mathbf{YA}^*\mathbf{Y}' \text{ are distinct}\} = 1. \tag{5.41}$$

Moreover,

$$rk(\mathbf{A}^*) = rk(\mathbf{D'AD}) \geq rk(\mathbf{DD'ADD'}) \geq rk(\mathbf{D^-DD'ADD'D'^-}) = rk(\mathbf{D'AD}),$$

where we used $\mathbf{D^-D} = \mathbf{I}_m$ which follows from Theorem 1.23. Hence,

$$rk(\mathbf{A}^*) = rk(\mathbf{DD'ADD'}) = rk(\boldsymbol{\Phi}\mathbf{A}\boldsymbol{\Phi}).$$

Since $vec(\mathbf{U}') \approx vec(\mathbf{U}^{*\prime})$, we have $\mathbf{U} \approx \mathbf{U}^*$ and then (i) follows from (5.36), (5.38), and (5.40) and (ii) follows from (5.37), (5.39) and (5.41). ∎

If $\Sigma > \mathbf{0}$ and $\boldsymbol{\Phi} > \mathbf{0}$ in Theorem 5.10, we obtain the following result.

Theorem 5.11. *Let* $\mathbf{X} \sim E_{p,n}(\mathbf{0}, \Sigma \otimes \Phi, \psi)$ *with* $\Sigma > \mathbf{0}$, $\Phi > \mathbf{0}$ *and* $P(\mathbf{X} = \mathbf{0}) = 0$. *Let* $\mathbf{A}$ *be an* $n \times n$ *symmetric matrix. Then*

(i) $P\{rk(\mathbf{XAX}') = \min(p, rk(\mathbf{A}))\} = 1$ *and*
(ii) $P\{$*the nonzero eigenvalues of* $\mathbf{XAX}'$ *are distinct*$\} = 1$.

PROOF: It follows, directly, from Theorem 5.10. ∎

Corollary 5.3. *Let* $\mathbf{X} \sim E_{p,n}(\mu\mathbf{e}_n', \Sigma \otimes \Phi, \psi)$, *where* $p < n$, $\mu \in I\!R^p$, $\Sigma > \mathbf{0}$ *and* $\Phi > \mathbf{0}$. *Assume* $P(\mathbf{X} = \mu\mathbf{e}_n') = 0$. *Let* $\mathbf{x}_i$ *be the columns of* $\mathbf{X}$, $i = 1, \ldots, n$ *and define* $\bar{\mathbf{x}} = \frac{1}{n}\sum_{i=1}^n \mathbf{x}_i$. *Then* $\mathbf{S}(\mathbf{X}) = \frac{1}{n}\sum_{i=1}^n (\mathbf{x}_i - \bar{\mathbf{x}})(\mathbf{x}_i - \bar{\mathbf{x}})'$ *is positive definite and its characteristic roots are distinct with probability one.*

PROOF: Here

$$\mathbf{S}(\mathbf{X}) = \mathbf{X}\left(\mathbf{I}_n - \frac{\mathbf{e}_n\mathbf{e}_n'}{n}\right)\mathbf{X}' = (\mathbf{X} - \mu\mathbf{e}_n')\left(\mathbf{I}_n - \frac{\mathbf{e}_n\mathbf{e}_n'}{n}\right)(\mathbf{X} - \mu\mathbf{e}_n')'.$$

Now let $\mathbf{Y} = \mathbf{X} - \mu\mathbf{e}_n'$ and $\mathbf{A} = \mathbf{I}_n - \frac{\mathbf{e}_n\mathbf{e}_n'}{n}$. Then, $\mathbf{Y} \sim E_{p,n}(\mathbf{0}, \Sigma \otimes \Phi, \psi)$, $P(\mathbf{Y} = \mathbf{0}) = 0$, $S(\mathbf{X}) = \mathbf{YAY}'$ and from Theorem 5.11, we obtain the desired result. ∎

5.3 Distributions of Invariant Matrix Variate Functions

In this section, we will derive the distributions of invariant functions of random matrices with m.e.c. distributions. In order to do this, we will need the following theorem (Gupta and Varga, 1994d).

Theorem 5.12. *Let* $\mathbf{X} \sim E_{p,n}(\mathbf{0}, \Sigma \otimes \Phi, \psi)$ *with* $P(\mathbf{X} = \mathbf{0}) = 0$. *Assume* $\mathbf{Y} \sim N_{p,n}(\mathbf{0}, \Sigma \otimes \Phi)$. *Let* $\mathscr{F}$ *be a subset of the* $p \times n$ *real matrices, such that if* $\mathbf{Z} \in I\!R^{p \times n}$, $\mathbf{Z} \in \mathscr{F}$, *and* $a > 0$ *then* $a\mathbf{Z} \in \mathscr{F}$ *and* $P(\mathbf{X} \notin \mathscr{F}) = P(\mathbf{Y} \notin \mathscr{F}) = 0$. *Let* $K(\mathbf{Z})$ *be a function defined on* $\mathscr{F}$, *such that if* $\mathbf{Z} \in \mathscr{F}$ *and* $a > 0$, *then* $K(\mathbf{Z}) = K(a\mathbf{Z})$. *Then,* $K(\mathbf{X})$ *and* $K(\mathbf{Y})$ *are defined with probability one and* $K(\mathbf{X})$ *and* $K(\mathbf{Y})$ *are identically distributed.*

PROOF: $K(\mathbf{X})$ and $K(\mathbf{Y})$ are defined if $\mathbf{X} \in \mathscr{F}$ and $\mathbf{Y} \in \mathscr{F}$. Since $P(\mathbf{X} \notin \mathscr{F}) = P(\mathbf{Y} \notin \mathscr{F}) = 0$ we see that $K(\mathbf{X})$ and $K(\mathbf{Y})$ are defined with probability one. Let $r_1\mathbf{AU}_1\mathbf{B}'$ be the stochastic representation of $\mathbf{X}$ and $r_2\mathbf{AU}_2\mathbf{B}'$, the stochastic representation of $\mathbf{Y}$. It follows, from the conditions of the theorem, that if $a\mathbf{Z} \in \mathscr{F}$ and $a > 0$ then $\mathbf{Z} \in \mathscr{F}$. Since $P(\mathbf{X} = \mathbf{0}) = 0$, we have $P(r_1 = 0) = 0$. Since $P(r_1\mathbf{AU}_1\mathbf{B}' \in \mathscr{F}) = 1$, we get $P(\mathbf{AU}_1\mathbf{B}' \in \mathscr{F}) = 1$. So, $K(\mathbf{AU}_1\mathbf{B}')$ is defined with probability one. Moreover, $P\{K(r_1\mathbf{AU}_1\mathbf{B}') = K(\mathbf{AU}_1\mathbf{B}')\} = 1$.

Similarly, $P\{K(r_2\mathbf{AU}_2\mathbf{B}') = K(\mathbf{AU}_2\mathbf{B}')\} = 1$. But $\mathbf{AU}_1\mathbf{B}' \approx \mathbf{AU}_2\mathbf{B}'$. Hence,

$$K(\mathbf{AU}_1\mathbf{B}') \approx K(\mathbf{AU}_2\mathbf{B}').$$

Therefore, $K(r_1\mathbf{AU}_1\mathbf{B}') \approx K(r_2\mathbf{AU}_2\mathbf{B}')$, which means $K(\mathbf{X}) \approx K(\mathbf{Y})$. ■

Remark 5.3. The significance of Theorem 5.12 is the following. Assume the conditions of the theorem are satisfied. Then, it is enough to determine the distribution of the function for the normal case, in order to get the distribution of the function, when the underlying distribution is elliptically contoured.

Now, we apply Theorem 5.12 to special cases.

Theorem 5.13. *Let* $\mathbf{X} \sim E_{p,n}(\mathbf{0}, \mathbf{I}_p \otimes \mathbf{I}_n, \psi)$, *with* $P(\mathbf{X} = \mathbf{0}) = 0$. *Let* $\mathbf{G} : n \times m$ *be such that* $n - m \geq p$ *and* $\mathbf{G}'\mathbf{G} = \mathbf{I}_m$. *Then,*

$$(\mathbf{X}(\mathbf{I}_n - \mathbf{GG}')\mathbf{X}')^{-\frac{1}{2}}\mathbf{XG} \sim T_{p,m}(n - (m+p) + 1, \mathbf{0}, \mathbf{I}_p, \mathbf{I}_m).$$

PROOF: Let $K(\mathbf{Z}) = (\mathbf{Z}(\mathbf{I}_n - \mathbf{GG}')\mathbf{Z}')^{-\frac{1}{2}}\mathbf{ZG}$. Let $\mathscr{F} = \{\mathbf{Z} | \mathbf{Z}$ is $p \times n$ matrix, such that $\mathbf{Z}(\mathbf{I}_n - \mathbf{GG}')\mathbf{Z}'$ is nonsingular$\}$. Clearly, if $\mathbf{Z} \in \mathbb{R}^{p\times n}$, $\mathbf{Z} \in \mathscr{F}$ and $a > 0$, then $a\mathbf{Z} \in \mathscr{F}$. If $\mathbf{Z} \in \mathscr{F}$ and $a > 0$, then

$$K(a\mathbf{Z}) = (a\mathbf{Z}(\mathbf{I}_n - \mathbf{GG}')(a\mathbf{Z})')^{-\frac{1}{2}}a\mathbf{ZG} = (\mathbf{Z}(\mathbf{I}_n - \mathbf{GG}')\mathbf{Z}')^{-\frac{1}{2}}\mathbf{ZG} = K(\mathbf{Z}).$$

Let $\mathbf{E} : n \times m$ be defined as $\mathbf{E} = \begin{pmatrix} \mathbf{I}_m \\ \mathbf{0} \end{pmatrix}$. Then, $\mathbf{E}'\mathbf{E} = \mathbf{G}'\mathbf{G}$. Now, Theorem 1.11 says that there exists an $n \times n$ matrix $\mathbf{H}$, such that $\mathbf{HH}' = \mathbf{I}_n$ and $\mathbf{G}'\mathbf{H} = \mathbf{E}'$. That means, $\mathbf{H}$ is orthogonal and $\mathbf{H}'\mathbf{G} = \mathbf{E}$. So, we have

$$\begin{aligned}
\mathbf{I}_n - \mathbf{GG}' &= \mathbf{I}_n - \mathbf{HEE}'\mathbf{H}' \\
&= \mathbf{H}(\mathbf{I}_n - \mathbf{EE}')\mathbf{H}' \\
&= \mathbf{H}\left(\mathbf{I}_n - \begin{pmatrix} \mathbf{I}_m & \mathbf{0} \\ \mathbf{0} & \mathbf{0} \end{pmatrix}\right)\mathbf{H}' \\
&= \mathbf{H}\begin{pmatrix} \mathbf{0} & \mathbf{0} \\ \mathbf{0} & \mathbf{I}_{n-m} \end{pmatrix}\mathbf{H}' = \mathbf{HDH}',
\end{aligned}$$

where $\mathbf{D} = \begin{pmatrix} \mathbf{0} & \mathbf{0} \\ \mathbf{0} & \mathbf{I}_{n-m} \end{pmatrix}$. Clearly, $\mathbf{D} = \mathbf{BB}'$, where $\mathbf{B}$ is an $n \times (n-m)$ matrix defined as $\mathbf{B} = \begin{pmatrix} \mathbf{0} \\ \mathbf{I}_{n-m} \end{pmatrix}$. Using Theorem 5.11, we get

$$\begin{aligned}
P\{rk(\mathbf{X}(\mathbf{I}_n - \mathbf{GG}')\mathbf{X}') &= \min(rk(\mathbf{I}_n - \mathbf{GG}'), p)\} \\
&= 1.
\end{aligned}$$

However, $rk(\mathbf{I}_n - \mathbf{GG}') = rk(\mathbf{HDH}') = rk(\mathbf{D}) = n - m$, and since $n - m \geq p$, we see that $rk(\mathbf{X}(\mathbf{I}_n - \mathbf{GG}')\mathbf{X}') = p$, with probability one. Therefore, $\mathbf{X}(\mathbf{I}_n - \mathbf{GG}')\mathbf{X}'$ is of full rank with probability one. So, $P(\mathbf{X} \notin \mathscr{F}) = 0$.

Hence, we can use Theorem 5.12. That means we can assume $\mathbf{X} \sim N_{p,n}(\mathbf{0}, \mathbf{I}_p \otimes \mathbf{I}_n)$. Let $\mathbf{V} = \mathbf{XH}$. Then, $\mathbf{V} \sim N_{p,n}(\mathbf{0}, \mathbf{I}_p \otimes \mathbf{I}_n)$. Partition $\mathbf{V}$ as $\mathbf{V} = (\mathbf{V}_1, \mathbf{V}_2)$, where $\mathbf{V}_1$ is $p \times m$. Then, $\mathbf{V}_1 \sim N_{p,m}(\mathbf{0}, \mathbf{I}_p \otimes \mathbf{I}_m)$, and $\mathbf{V}_2 \sim N_{p,n-m}(\mathbf{0}, \mathbf{I}_p \otimes \mathbf{I}_{n-m})$, where $\mathbf{V}_1$ and $\mathbf{V}_2$ are independent. Now,

$$\begin{aligned}
(\mathbf{X}(\mathbf{I}_n - \mathbf{GG}')\mathbf{X}')^{-\frac{1}{2}}\mathbf{XG} &= (\mathbf{XHDH}'\mathbf{X}')^{-\frac{1}{2}}\mathbf{XHE} \\
&= (\mathbf{VBB}'\mathbf{V}')^{-\frac{1}{2}}\mathbf{VE} \\
&= \left(\mathbf{V}\begin{pmatrix}\mathbf{0}\\ \mathbf{I}_{n-m}\end{pmatrix}(\mathbf{0}\ \mathbf{I}_{n-m})\mathbf{V}'\right)^{-\frac{1}{2}}\mathbf{V}\begin{pmatrix}\mathbf{I}_m\\ \mathbf{0}\end{pmatrix} \\
&= (\mathbf{V}_2\mathbf{V}_2')^{-\frac{1}{2}}\mathbf{V}_1.
\end{aligned}$$

Here, $\mathbf{V}_2\mathbf{V}_2' \sim W_p(n-m, \mathbf{I}_p)$, $\mathbf{V}_1 \sim N_{p,m}(\mathbf{0}, \mathbf{I}_p \otimes \mathbf{I}_m)$ and $\mathbf{V}_1$ and $\mathbf{V}_2\mathbf{V}_2'$ are independent. From Dickey (1967), we get that under these conditions, $(\mathbf{V}_2\mathbf{V}_2')^{-\frac{1}{2}}\mathbf{V}_1 \sim T_{p,m}(n-(m+p)+1, \mathbf{0}, \mathbf{I}_p, \mathbf{I}_m)$ (also see Javier and Gupta, 1985a). ■

Theorem 5.14. *Let* $\mathbf{X} \sim E_{p,n}(\mathbf{0}, \mathbf{I}_p \otimes \mathbf{I}_n, \psi)$ *with* $P(\mathbf{X} = \mathbf{0}) = 0$. *Let* $\mathbf{B}$: $n \times n$ *be a symmetric, idempotent matrix of rank* m *where* $m \geq p$ *and* $n - m \geq p$. *Then,*

$$(\mathbf{XX}')^{-\frac{1}{2}}(\mathbf{XBX}')(\mathbf{XX}')^{-\frac{1}{2}} \sim B_p^I\left(\frac{m}{2}, \frac{n-m}{2}\right).$$

PROOF: Let $K(\mathbf{Z}) = (\mathbf{ZZ}')^{-\frac{1}{2}}(\mathbf{ZBZ}')(\mathbf{ZZ}')^{-\frac{1}{2}}$. Let $\mathscr{F} = \{\mathbf{Z} | \mathbf{Z}$ is $p \times n$ matrix, such that $\mathbf{ZZ}'$ is nonsingular$\}$. Clearly, if $\mathbf{Z} \in \mathbb{R}^{p \times n}$, $\mathbf{Z} \in \mathscr{F}$ and $a > 0$, then $a\mathbf{Z} \in \mathscr{F}$. If $\mathbf{Z} \in \mathscr{F}$ and $a > 0$, then

$$\begin{aligned}
K(a\mathbf{Z}) &= (a^2\mathbf{ZZ}')^{-\frac{1}{2}}(a^2\mathbf{ZBZ}')(a^2\mathbf{ZZ}')^{-\frac{1}{2}} \\
&= (\mathbf{ZZ}')^{-\frac{1}{2}}(\mathbf{ZBZ}')(\mathbf{ZZ}')^{-\frac{1}{2}} \\
&= K(\mathbf{Z}).
\end{aligned}$$

Using Theorem 5.11, we get

$$P\{rk(\mathbf{XX}') = \min(rk(\mathbf{I}_n), p)\} = 1.$$

Since $m \geq p$ and $n - m \geq p$ we have $n \geq 2p$ and hence, $n \geq p$. Thus, $\min(rk(\mathbf{I}_n), p) = p$. Therefore, $\mathbf{XX}'$ is of full rank with probability one. So, $P(\mathbf{X} \notin \mathscr{F}) = 0$. Similarly, $P(\mathbf{Y} \notin \mathscr{F}) = 0$. Hence, we can use Theorem 5.12. That means we can assume $\mathbf{X} \sim N_{p,n}(\mathbf{0}, \mathbf{I}_p \otimes \mathbf{I}_n)$. Since $\mathbf{B}$ is a symmetric, idempotent matrix of rank m, there exists an $n \times n$ orthogonal matrix $\mathbf{H}$, such that $\mathbf{B} = \mathbf{H}\begin{pmatrix} \mathbf{I}_q & \mathbf{0} \\ \mathbf{0} & \mathbf{0} \end{pmatrix}\mathbf{H}'$. We can write $\begin{pmatrix} \mathbf{I}_q & \mathbf{0} \\ \mathbf{0} & \mathbf{0} \end{pmatrix} = \mathbf{CC}'$ where $\mathbf{C}$ is an $n \times m$ matrix defined as $\mathbf{C} = \begin{pmatrix} \mathbf{I}_q \\ \mathbf{0} \end{pmatrix}$. Let $\mathbf{V} = \mathbf{XH}$. Then, $\mathbf{V} \sim N_{p,n}(\mathbf{0}, \mathbf{I}_p \otimes \mathbf{I}_n)$. Partition $\mathbf{V}$ as $\mathbf{V} = (\mathbf{V}_1, \mathbf{V}_2)$ where $\mathbf{V}_1$ is $p \times m$. Then, $\mathbf{V}_1 \sim N_{p,m}(\mathbf{0}, \mathbf{I}_p \otimes \mathbf{I}_m)$, and $\mathbf{V}_2 \sim N_{p,n-m}(\mathbf{0}, \mathbf{I}_p \otimes \mathbf{I}_{n-m})$, where $\mathbf{V}_1$ and $\mathbf{V}_2$ are independent.

Now,

$$\begin{aligned}(\mathbf{XX}')^{-\frac{1}{2}}(\mathbf{XBX}')(\mathbf{XX}')^{-\frac{1}{2}} &= (\mathbf{XHH}'\mathbf{X}')^{-\frac{1}{2}}(\mathbf{XHCC}'\mathbf{H}'\mathbf{X}')(\mathbf{XHH}'\mathbf{X}')^{-\frac{1}{2}} \\ &= (\mathbf{VV}')^{-\frac{1}{2}}(\mathbf{VCC}'\mathbf{V}')(\mathbf{VV}')^{-\frac{1}{2}} \\ &= (\mathbf{V}_1\mathbf{V}_1' + \mathbf{V}_2\mathbf{V}_2')^{-\frac{1}{2}}(\mathbf{V}_1\mathbf{V}_1')(\mathbf{V}_1\mathbf{V}_1' + \mathbf{V}_2\mathbf{V}_2')^{-\frac{1}{2}}.\end{aligned}$$

Here $\mathbf{V}_1\mathbf{V}_1' \sim W_p(m, \mathbf{I}_p)$, $\mathbf{V}_2\mathbf{V}_2' \sim W_p(n-m, \mathbf{I}_p)$ and $\mathbf{V}_1\mathbf{V}_1'$ and $\mathbf{V}_2\mathbf{V}_2'$ are independent. Finally from Olkin and Rubin (1964), we get that under these conditions

$$(\mathbf{V}_1\mathbf{V}_1' + \mathbf{V}_2\mathbf{V}_2')^{-\frac{1}{2}}(\mathbf{V}_1\mathbf{V}_1')(\mathbf{V}_1\mathbf{V}_1' + \mathbf{V}_2\mathbf{V}_2')^{-\frac{1}{2}} \sim B_p^I\left(\frac{m}{2}, \frac{n-m}{2}\right). \qquad \blacksquare$$

Theorem 5.15. *Let* $\mathbf{X} \sim E_{p,n}(\mathbf{0}, \mathbf{I}_p \otimes \mathbf{I}_n, \psi)$ *with* $P(\mathbf{X} = \mathbf{0}) = 0$. *Let* $\mathbf{A}$ *and* $\mathbf{B}$ *be* $n \times n$ *symmetric, idempotent matrices,* $rk(\mathbf{A}) = n_1$, $rk(\mathbf{B}) = n_2$, *such that* n_1, $n_2 \geq p$ *and* $\mathbf{AB} = \mathbf{0}$. *Then,*

$$(\mathbf{XAX}')^{-\frac{1}{2}}(\mathbf{XBX}')(\mathbf{XAX}')^{-\frac{1}{2}} \sim B_p^{II}\left(\frac{n_2}{2}, \frac{n_1}{2}\right).$$

PROOF: Let $K(\mathbf{Z}) = (\mathbf{ZAZ}')^{-\frac{1}{2}}(\mathbf{ZBZ}')(\mathbf{ZAZ}')^{-\frac{1}{2}}$. Let $\mathscr{F} = \{\mathbf{Z} | \mathbf{Z}$ is $p \times n$ matrix, such that $\mathbf{ZAZ}'$ is nonsingular$\}$. Clearly, if $\mathbf{Z} \in \mathbb{R}^{p \times n}$, $\mathbf{Z} \in \mathscr{F}$ and $a > 0$, then $a\mathbf{Z} \in \mathscr{F}$. If $\mathbf{Z} \in \mathscr{F}$ and $a > 0$, then

$$\begin{aligned}K(a\mathbf{Z}) &= (a^2\mathbf{ZAZ}')^{-\frac{1}{2}}(a^2\mathbf{ZBZ}')(a^2\mathbf{ZAZ}')^{-\frac{1}{2}} \\ &= (\mathbf{ZAZ}')^{-\frac{1}{2}}(\mathbf{ZBZ}')(\mathbf{ZAZ}')^{-\frac{1}{2}} \\ &= K(\mathbf{Z}).\end{aligned}$$

Using Theorem 5.11, we get

$$P\{rk(\mathbf{XAX}') = \min(rk(\mathbf{A}), p)\} = 1.$$

However $rk(\mathbf{A}) = n_1 \geq p$, and hence $\min(rk(\mathbf{A}), p) = p$. Therefore, $\mathbf{XAX}'$ is of full rank with probability one. So, $P(\mathbf{X} \notin \mathscr{F}) = 0$. Similarly, $P(\mathbf{Y} \notin \mathscr{F}) = 0$.

Hence, we can use Theorem 5.12. That means we can assume $\mathbf{X} \sim N_{p,n}(\mathbf{0}, \mathbf{I}_p \otimes \mathbf{I}_n)$. Since $\mathbf{A}$ and $\mathbf{B}$ are symmetric, idempotent matrices and $\mathbf{AB} = \mathbf{0}$, there exists an $n \times n$ orthogonal matrix $\mathbf{H}$, such that

$$\mathbf{H}'\mathbf{A}\mathbf{H} = \begin{pmatrix} \mathbf{I}_{n_1} & \mathbf{0} & \mathbf{0} \\ \mathbf{0} & \mathbf{0} & \mathbf{0} \\ \mathbf{0} & \mathbf{0} & \mathbf{0} \end{pmatrix} \quad \text{and} \quad \mathbf{H}'\mathbf{B}\mathbf{H} = \begin{pmatrix} \mathbf{0} & \mathbf{0} & \mathbf{0} \\ \mathbf{0} & \mathbf{I}_{n_2} & \mathbf{0} \\ \mathbf{0} & \mathbf{0} & \mathbf{0} \end{pmatrix}$$

(see Hocking, 1985).

We can write $\mathbf{H}'\mathbf{A}\mathbf{H} = \mathbf{C}\mathbf{C}'$ and $\mathbf{H}'\mathbf{A}\mathbf{H} = \mathbf{D}\mathbf{D}'$, where $\mathbf{C}' = (\mathbf{I}_{n_1}, \mathbf{0}, \mathbf{0})$ and $\mathbf{D}' = (\mathbf{0}, \mathbf{I}_{n_2}, \mathbf{0})$. Let $\mathbf{V} = \mathbf{XH}$. Then, $\mathbf{V} \sim N_{p,n}(\mathbf{0}, \mathbf{I}_p \otimes \mathbf{I}_n)$. Partition $\mathbf{V}$ as $\mathbf{V} = (\mathbf{V}_1, \mathbf{V}_2, \mathbf{V}_3)$ where $\mathbf{V}_1$ is $p \times n_1$ and $\mathbf{V}_2$ is $p \times n_2$. Then, $\mathbf{V}_1 \sim N_{p,n_1}(\mathbf{0}, \mathbf{I}_p \otimes \mathbf{I}_{n_1})$ and $\mathbf{V}_2 \sim N_{p,n_2}(\mathbf{0}, \mathbf{I}_p \otimes \mathbf{I}_{n_2})$ where $\mathbf{V}_1$ and $\mathbf{V}_2$ are independent.

$$\begin{aligned}(\mathbf{XAX}')^{-\frac{1}{2}}(\mathbf{XBX}')(\mathbf{XAX}')^{-\frac{1}{2}} &= (\mathbf{XHCC}'\mathbf{H}'\mathbf{X}')^{-\frac{1}{2}}(\mathbf{XHDD}'\mathbf{H}'\mathbf{X}')(\mathbf{XHCC}'\mathbf{H}'\mathbf{X}')^{-\frac{1}{2}} \\ &= (\mathbf{VCC}'\mathbf{V}')^{-\frac{1}{2}}(\mathbf{VDD}'\mathbf{V}')(\mathbf{VCC}'\mathbf{V}')^{-\frac{1}{2}} \\ &= (\mathbf{V}_1\mathbf{V}_1')^{-\frac{1}{2}}(\mathbf{V}_2\mathbf{V}_2')(\mathbf{V}_1\mathbf{V}_1')^{-\frac{1}{2}}.\end{aligned}$$

Here $\mathbf{V}_1\mathbf{V}_1' \sim W_p(n_1, \mathbf{I}_p)$, $\mathbf{V}_2\mathbf{V}_2' \sim W_p(n_2, \mathbf{I}_p)$ and $\mathbf{V}_1\mathbf{V}_1'$ and $\mathbf{V}_2\mathbf{V}_2'$ are independent. Finally from Olkin and Rubin (1964), we get that under these conditions

$$(\mathbf{V}_1\mathbf{V}_1')^{-\frac{1}{2}}(\mathbf{V}_2\mathbf{V}_2')(\mathbf{V}_1\mathbf{V}_1')^{-\frac{1}{2}} \sim B_p^{II}\left(\frac{n_2}{2}, \frac{n_1}{2}\right). \qquad \blacksquare$$

The next theorem shows that under some general conditions, Hotelling's T^2 statistic has the same distribution in the elliptically contoured case, as in the normal case; that is, we get an F distribution.

Theorem 5.16. *Let* $\mathbf{X} \sim E_{p,n}(\mathbf{0}, \Sigma \otimes \mathbf{I}_n, \psi)$ *with* $P(\mathbf{X} = \mathbf{0}) = 0$. *Assume* $p < n$. *Let* $\mathbf{x}_i$ *be the ith column of* $\mathbf{X}$, $i = 1, \ldots, n$. *Let* $\bar{\mathbf{x}} = \frac{1}{n}\sum_{i=1}^{n} \mathbf{x}_i$. *Define* $\mathbf{S}(\mathbf{X}) = \frac{1}{n-1}\sum_{i=1}^{n}(\mathbf{x}_i - \bar{\mathbf{x}})(\mathbf{x}_i - \bar{\mathbf{x}})'$ *and* $T^2(\mathbf{X}) = n\bar{\mathbf{x}}'\mathbf{S}(\mathbf{x})^{-1}\bar{\mathbf{x}}$. *Then*

$$\frac{T^2(\mathbf{X})}{n-1}\frac{n-p}{p} \sim F_{p,n-p}.$$

PROOF: Let $\mathbf{A} = \mathbf{I}_n - \frac{\mathbf{e}_n\mathbf{e}_n'}{n}$. Then, $\mathbf{S}(\mathbf{X}) = \mathbf{XAX}'$. We also have $\bar{\mathbf{x}} = \frac{1}{n}\mathbf{X}\mathbf{e}_n$. Thus, $\mathbf{T}^2(\mathbf{X}) = \mathbf{e}_n'\mathbf{X}'(\mathbf{XAX}')^{-1}\mathbf{X}\mathbf{e}_n$.

Let $K(\mathbf{Z}) = \mathbf{e}_n'\mathbf{Z}'(\mathbf{ZAZ}')^{-1}\mathbf{Z}\mathbf{e}_n$. Let $\mathscr{F} = \{\mathbf{Z}|\mathbf{Z}$ is $p \times n$ matrix, such that $\mathbf{ZAZ}'$ is nonsingular$\}$. Clearly, if $\mathbf{Z} \in \mathbb{R}^{p\times n}$, $\mathbf{Z} \in \mathscr{F}$ and $a > 0$, then $a\mathbf{Z} \in \mathscr{F}$. If $\mathbf{Z} \in \mathscr{F}$ and $a > 0$, then

$$\begin{aligned} K(a\mathbf{Z}) &= \mathbf{e}_n' a\mathbf{Z}'(a^2\mathbf{ZAZ}')^{-1} a\mathbf{Z}\mathbf{e}_n \\ &= \mathbf{e}_n'\mathbf{Z}'(\mathbf{ZAZ}')^{-1}\mathbf{Z}\mathbf{e}_n \\ &= K(\mathbf{Z}). \end{aligned}$$

From Corollary 5.3, we see that $\mathbf{XAX}'$ is of full rank with probability one. So, $P(\mathbf{X} \notin \mathscr{F}) = 0$. Similarly, $P(\mathbf{Y} \notin \mathscr{F}) = 0$.

Hence, we can use Theorem 5.12. That means we can assume $\mathbf{X} \sim N_{p,n}(\mathbf{0}, \mathbf{I}_p \otimes \mathbf{I}_n)$. However, for the normal case this is a well known result (see Corollary 5.2.1 of Anderson, 2003, p. 176). ■

Theorem 5.17. *Let* $\mathbf{X} \sim E_{p,n}(\mu\mathbf{e}_n', \Sigma \otimes \mathbf{I}_n, \psi)$ *with* $P(\mathbf{X} = \mu\mathbf{e}_n') = 0$ *and* $\mu \in I\!R^p$. *Assume* $n > p$. *Let* $\mathbf{Y} \sim N_{p,n}(\mu\mathbf{e}_n', \Sigma \otimes \mathbf{I}_n)$, *and* $\mathbf{S}(\mathbf{X}) = \mathbf{X}\left(\mathbf{I}_n - \frac{\mathbf{e}_n\mathbf{e}_n'}{n}\right)\mathbf{X}'$. *Then, the principal components of* $\mathbf{S}(\mathbf{X})$ *have the same joint distribution as the principal components of* $\mathbf{S}(\mathbf{Y})$.

PROOF: Let $\mathbf{A} = \mathbf{I}_n - \frac{\mathbf{e}_n\mathbf{e}_n'}{n}$ and $\mathbf{S}(\mathbf{Z}) = \mathbf{ZAZ}'$ for $\mathbf{Z} \in I\!R^{p\times n}$. First, note that $\mathbf{S}(\mathbf{Z}) = \mathbf{S}(\mathbf{Z} - \mu\mathbf{e}_n')$ therefore, without loss of generality, we can assume $\mu = \mathbf{0}$.

Let $K(\mathbf{Z}) = \{$ normalized characteristic vectors of $\mathbf{ZAZ}'\}$. Let $\mathscr{F} = \{\mathbf{Z}|\mathbf{Z}$ is $p \times n$ matrix, such that $\mathbf{ZAZ}'$ is nonsingular$\}$. Clearly, if $\mathbf{Z} \in I\!R^{p\times n}$, $\mathbf{Z} \in \mathscr{F}$ and $a > 0$, then $a\mathbf{Z} \in \mathscr{F}$. If $\mathbf{Z} \in \mathscr{F}$ and $a > 0$, then obviously, $K(a\mathbf{Z}) = K(\mathbf{Z})$. Using Corollary 5.3, we find that the characteristic roots of $\mathbf{XAX}'$ are nonzero and distinct with probability one. So, $P(\mathbf{X} \notin \mathscr{F}) = 0$. Similarly, $P(\mathbf{Y} \notin \mathscr{F}) = 0$.

Now applying Theorem 5.12, we obtain the desired result. ■

Chapter 6
Characterization Results

6.1 Characterization Based on Invariance

In this section, we characterize the parameters of m.e.c. distributions which are invariant under certain linear transformations. First we prove the following lemma.

Lemma 6.1. *The $p \times p$ matrix Σ defined by*

$$\Sigma = a\mathbf{I}_p + b\mathbf{e}_p\mathbf{e}_p',$$

is positive semidefinite if and only if $a \geq 0$ and $a \geq -pb$.

PROOF: From part (vi) of Theorem 1.2, we have to show that the characteristic roots of Σ are nonnegative. From Theorem 1.5 we obtain

$$\begin{aligned}|\Sigma - \lambda\mathbf{I}_p| &= |(a-\lambda)\mathbf{I}_p + b\mathbf{e}_p\mathbf{e}_p'| \\ &= ((a-\lambda+b)-b)^{p-1}(a-\lambda+b+(p-1)b) \\ &= (a-\lambda)^{p-1}(a+pb-\lambda).\end{aligned}$$

Hence, the characteristic roots of Σ are $\lambda_1 = a$ and $\lambda_2 = a+pb$. Therefore, the characteristic roots of Σ are nonnegative if and only if $a \geq 0$ and $a+pb \geq 0$. ■

Theorem 6.1. *Let $\mathbf{X} \sim E_{p,n}(\mathbf{M}, \Sigma \otimes \Phi, \psi)$ with $p > 1$ and $\Phi \neq \mathbf{0}$. Define*

$$\begin{aligned}\mathscr{P} &= \{\mathbf{P} : \mathbf{P} \textit{ is } p \times p \textit{ permutation matrix}\}, \\ \mathscr{R} &= \{\mathbf{R} : \mathbf{R} \textit{ is } p \times p \textit{ signed permutation matrix}\}, \quad \textit{and} \\ \mathscr{G} &= \{\mathbf{G} : \mathbf{G} \textit{ is } p \times p \textit{ orthogonal matrix}\}.\end{aligned}$$

A.K. Gupta et al., *Elliptically Contoured Models in Statistics and Portfolio Theory*,
DOI 10.1007/978-1-4614-8154-6_6, © Springer Science+Business Media New York 2013

Then,

(a) *For every* $\mathbf{P} \in \mathscr{P}$, $\mathbf{PX} \approx \mathbf{X}$ *if and only if* $\mathbf{M} = \mathbf{e}_p\mu'$ *where* $\mu \in I\!R^n$, $\Sigma = a\mathbf{I}_p + b\mathbf{e}_p\mathbf{e}_p'$, $a, b \in I\!R$, $a \geq 0$, *and* $a \geq -pb$,
(b) *For every* $\mathbf{R} \in \mathscr{R}$, $\mathbf{RX} \approx \mathbf{X}$ *if and only if* $\mathbf{M} = \mathbf{0}$ *and* $\Sigma = a\mathbf{I}_p$, *where* $a \geq 0$,
(c) *For every* $\mathbf{G} \in \mathscr{G}$, $\mathbf{GX} \approx \mathbf{X}$ *if and only if* $\mathbf{M} = \mathbf{0}$ *and* $\Sigma = a\mathbf{I}_p$, *where* $a \geq 0$.

PROOF:

(a) First, assume that $\mathbf{X} \sim E_{p,n}(\mathbf{e}_p\mu', (a\mathbf{I}_p + b\mathbf{e}_p\mathbf{e}_p') \otimes \Phi, \psi)$ with $a \geq 0$, and $a \geq -pb$. Then, from Lemma 6.1, Σ is positive semidefinite.
Let $\mathbf{P} \in \mathscr{P}$. Then, $\mathbf{PX} \sim E_{p,n}(\mathbf{Pe}_p\mu', (\mathbf{P}(a\mathbf{I}_p + b\mathbf{e}_p\mathbf{e}_p')\mathbf{P}') \otimes \Phi, \psi)$. Since $\mathbf{Pe}_p = \mathbf{e}_p$ and $\mathbf{PP}' = \mathbf{I}_p$, we get $\mathbf{PX} \sim E_{p,n}(\mathbf{e}_p\mu', (a\mathbf{I}_p + b\mathbf{e}_p\mathbf{e}_p') \otimes \Phi, \psi)$, which proves that $\mathbf{PX} \approx \mathbf{X}$.

Next, assume $\mathbf{X} \sim E_{p,n}(\mathbf{M}, \Sigma \otimes \Phi, \psi)$ and $\mathbf{PX} \approx \mathbf{X}$ for every $\mathbf{P} \in \mathscr{P}$. Then, $\mathbf{PX} \sim E_{p,n}(\mathbf{PM}, \mathbf{P}\Sigma\mathbf{P}' \otimes \Phi, \psi)$ and hence, $\mathbf{PM} = \mathbf{M}$ and $\mathbf{P}\Sigma\mathbf{P}' = \Sigma$ for every $\mathbf{P} \in \mathscr{P}$. Now, we introduce the following notation. Let $\mathbf{P}(k,l)$, $1 \leq k,l \leq p$ denote a $p \times p$ symmetric matrix, whose (i,j)th element is

$$\begin{cases} 1 \text{ if } & i = j, \quad i \neq k, i \neq l \\ 1 \text{ if } & i = l, \quad j = k \\ 1 \text{ if } & i = k, \quad j = l \\ 0 \text{ otherwise} & \end{cases}$$

Then, it is easy to see that $\mathbf{P}(k,l) \in \mathscr{P}$.

From $\mathbf{P}(1,i)\mathbf{M} = \mathbf{M}$, $i = 2, \ldots, p$, we get $\mathbf{M} = \mathbf{e}_p\mu'$, where $\mu \in I\!R^n$. From $\mathbf{P}(1,i)\Sigma\mathbf{P}(1,i) = \Sigma$, $i = 2, \ldots, p$ we get

$$\sigma_{ii} = \sigma_{11} \tag{6.1}$$

and

$$\sigma_{ij} = \sigma_{1j} \quad \text{if} \quad j > i. \tag{6.2}$$

If $p \geq 3$, then from $\mathbf{P}(2,i)\Sigma\mathbf{P}(2,i) = \Sigma$, $i = 3, \ldots, p$, we get

$$\sigma_{1j} = \sigma_{12} \quad \text{if} \quad j \geq 3. \tag{6.3}$$

From (6.1), it is clear that the diagonal elements of Σ are equal, whereas (6.2) and (6.3) show that the off-diagonal elements of Σ are also equal. Therefore, $\Sigma = a\mathbf{I}_p + b\mathbf{e}_p\mathbf{e}_p'$, and $|\Sigma| = (a + pb)a^{p-1}$. From Lemma 6.1, in order for $a\mathbf{I}_p + b\mathbf{e}_p\mathbf{e}_p'$ to be positive semidefinite, we must have $a \geq 0$ and $a \geq -pb$.

(b) First assume that $\mathbf{X} \sim E_{p,n}(\mathbf{0}, a\mathbf{I}_p \otimes \Phi, \psi)$ with $a \geq 0$. Let $\mathbf{R} \in \mathscr{R}$, then $\mathbf{RX} \sim E_{p,n}(\mathbf{0}, a\mathbf{RI}_p\mathbf{R}' \otimes \Phi, \psi)$. Since $\mathbf{RR}' = \mathbf{I}_p$, we have $\mathbf{RX} \approx \mathbf{X}$.

Next, assume that $\mathbf{X} \sim E_{p,n}(\mathbf{M}, \Sigma \otimes \Phi, \psi)$ and $\mathbf{RX} \approx \mathbf{X}$ for every $\mathbf{R} \in \mathscr{R}$. Let $\mathbf{R} = -\mathbf{I}_p$, then $\mathbf{RX} \sim E_{p,n}(-\mathbf{M}, \Sigma \otimes \Phi, \psi)$. Therefore $\mathbf{M} = -\mathbf{M}$ i.e. $\mathbf{M} = \mathbf{0}$. Since $\mathscr{P} \subset \mathscr{R}$, Σ must be of the form $a\mathbf{I}_p + b\mathbf{e}_p\mathbf{e}_p'$, $a \geq 0$ and $a \geq -pb$. So, $\mathbf{X} \sim E_{p,n}(\mathbf{0}, (a\mathbf{I}_p + b\mathbf{e}_p\mathbf{e}_p') \otimes \Phi, \psi)$.

Let $\mathbf{R}$ be a $p \times p$ symmetric matrix, whose (i, j)th element is

$$\begin{cases} -1 \text{ if } & i = j = 1 \\ 1 \quad \text{if } & i = j > l \\ 0 \quad \text{otherwise} \end{cases}$$

Then, $\mathbf{R} \in \mathscr{R}$ and $\mathbf{RX} \sim E_{p,n}(\mathbf{0}, \mathbf{R}(a\mathbf{I}_p + b\mathbf{e}_p\mathbf{e}_p')\mathbf{R}' \otimes \Phi, \psi)$. Since, $\mathbf{R}(a\mathbf{I}_p + b\mathbf{e}_p\mathbf{e}_p')\mathbf{R}' = a\mathbf{I}_p + b\mathbf{R}\mathbf{e}_p\mathbf{e}_p'\mathbf{R}'$, we must have $b\mathbf{R}\mathbf{e}_p\mathbf{e}_p' = b\mathbf{e}_p\mathbf{e}_p'$, which can also be written as

$$b(\mathbf{e}_p\mathbf{e}_p' - \mathbf{R}\mathbf{e}_p\mathbf{e}_p'\mathbf{R}) = \mathbf{0}. \tag{6.4}$$

Now, $\mathbf{e}_p\mathbf{e}_p' - \mathbf{R}\mathbf{e}_p\mathbf{e}_p'\mathbf{R}$ is a matrix whose (i, j)th element is

$$\begin{cases} 2 \text{ if } & i = 1, \quad j \geq 2 \\ 2 \text{ if } & i \geq j, \quad j = 1 \\ 0 \text{ otherwise} \end{cases}$$

Hence, $\mathbf{e}_p\mathbf{e}_p' - \mathbf{R}\mathbf{e}_p\mathbf{e}_p'\mathbf{R} \neq \mathbf{0}$ and thus, from (6.4), we conclude that $b = 0$. Therefore, $\mathbf{X} \sim E_{p,n}(\mathbf{0}, a\mathbf{I}_p \otimes \Phi, \psi)$ with $a \geq 0$.

(c) First assume $\mathbf{X} \sim E_{p,n}(\mathbf{0}, a\mathbf{I}_p \otimes \Phi, \psi)$ with $a \geq 0$. Let $\mathbf{G} \in \mathscr{G}$, then $\mathbf{GX} \sim E_{p,n}(\mathbf{0}, a\mathbf{GG}' \otimes \Phi, \psi)$ and since $\mathbf{GG}' = \mathbf{I}_p$, we have $\mathbf{GX} \approx \mathbf{X}$.
On the other hand, if $\mathbf{X} \sim E_{p,n}(\mathbf{0}, a\mathbf{I}_p \otimes \Phi, \psi)$ and $\mathbf{GX} \approx \mathbf{X}$ for every $\mathbf{G} \in \mathscr{G}$, then $\mathbf{RX} \approx \mathbf{X}$ for $\mathbf{R} \in \mathscr{R}$ must also hold, since $\mathscr{R} \subset \mathscr{G}$. Then, using part (b), we obtain $\mathbf{X} \sim E_{p,n}(\mathbf{0}, a\mathbf{I}_p \otimes \Phi, \psi)$ with $a \geq 0$. ■

Definition 6.1. Let $\mathbf{X} \sim E_{p,n}(\mathbf{M}, \Sigma \otimes \Phi, \psi)$. Then, $\mathbf{X}$ is called left-spherical, if $\mathbf{GX} \approx \mathbf{X}$, for every $\mathbf{G} \in O(p)$, right-spherical if $\mathbf{XH} \approx \mathbf{X}$ for every $\mathbf{H} \in O(n)$, and spherical if it is both left- and right-spherical.

Theorem 6.2. *Let $\mathbf{X} \sim E_{p,n}(\mathbf{M}, \Sigma \otimes \Phi, \psi)$. Then, $\mathbf{X}$ is left-spherical if and only if $\mathbf{M} = \mathbf{0}$ and $\Sigma = a\mathbf{I}_p$, with $a \geq 0$; right-spherical if and only if $\mathbf{M} = \mathbf{0}$ and $\Phi = b\mathbf{I}_n$; and spherical if and only if $\mathbf{M} = \mathbf{0}$, $\Sigma = a\mathbf{I}_p$ and $\Phi = b\mathbf{I}_n$, with $a \geq 0$.*

PROOF: It follows, from Theorem 6.1. ■

6.2 Characterization of Normality

In this section, it is shown that if m.e.c. distributions possess certain properties, they must be normal. The first result shows that the normality of one element of a random matrix with m.e.c. distribution implies the normality of the whole random matrix.

Theorem 6.3. *Let* $\mathbf{X} \sim E_{p,n}(\mathbf{M}, \Sigma \otimes \Phi, \psi)$ *and assume that there exist i and j such that* x_{ij} *is nondegenerate normal. Then,* $\mathbf{X} \sim N_{p,n}(\mathbf{M}, \Sigma \otimes \Phi)$.

PROOF: It follows, from Theorems 2.10, that $x_{ij} \sim E_1(m_{ij}, \sigma_{ii}\phi_{jj}, \psi)$. Since x_{ij} is normal, we have $\psi(z) = \exp\left(-\frac{z}{2}\right)$. Then $E_{p,n}(\mathbf{M}, \Sigma \otimes \Phi, \psi)$ becomes $N_{p,n}(\mathbf{M}, \Sigma \otimes \Phi, \psi)$. ∎

The following characterization results are based on independence.

Theorem 6.4. *Let* $\mathbf{X} \sim E_{p,n}(\mathbf{M}, \Sigma \otimes \Phi, \psi)$. *If* $\mathbf{X}$ *has two elements that are nondegenerate and independent, then* $\mathbf{X} \sim N_{p,n}(\mathbf{M}, \Sigma \otimes \Phi)$.

PROOF: Without loss of generality we can assume $\mathbf{M} = \mathbf{0}$. Let us denote the two independent elements of $\mathbf{X}$ by y and z. Then, we get

$$\begin{pmatrix} y \\ z \end{pmatrix} \sim E_2\left(\begin{pmatrix} 0 \\ 0 \end{pmatrix}, \begin{pmatrix} a & c \\ c & b \end{pmatrix}, \psi\right).$$

From Theorem 2.8, we have $y \sim E_1(0, a, \psi)$ and from Theorem 2.19, we obtain

$$y|z \sim E_1\left(\frac{c}{b}z, a - \frac{c^2}{b}, \psi^*\right).$$

Since, y and z are independent, $y|z \approx y$. Hence, $y \sim E_1\left(\frac{c}{b}z, a - \frac{c^2}{b}, \psi^*\right)$ and in view of Theorem 2.4 we must have $\frac{c}{b}z = 0$ for every z real number. This is possible, only if $c = 0$. Therefore,

$$\begin{pmatrix} y \\ z \end{pmatrix} \sim E_2\left(\begin{pmatrix} 0 \\ 0 \end{pmatrix}, \begin{pmatrix} a & 0 \\ 0 & b \end{pmatrix}, \psi\right).$$

Now let $y_1 = \sqrt{a}y$, $z_1 = \sqrt{b}z$, and $\mathbf{w} = \begin{pmatrix} y_1 \\ z_1 \end{pmatrix}$. Then, $\mathbf{w} \sim E_2\left(\begin{pmatrix} 0 \\ 0 \end{pmatrix}, \begin{pmatrix} 1 & 0 \\ 0 & 1 \end{pmatrix}, \psi\right)$, and its characteristic function at $\mathbf{t} = \begin{pmatrix} t_1 \\ t_2 \end{pmatrix}$ is

$$\phi_{\mathbf{w}}(\mathbf{t}) = \psi\left((t_1\ t_2)\begin{pmatrix} 1 & 0 \\ 0 & 1 \end{pmatrix}\begin{pmatrix} t_1 \\ t_2 \end{pmatrix}\right) = \psi(t_1^2 + t_2^2).$$

The characteristic functions of y_1 and z_1 are $\phi_{y_1}(t_1) = \psi(t_1^2)$ and $\phi_{z_1}(t_2) = \psi(t_2^2)$. Since y and z are independent, so are y_1 and z_1. Therefore, $\phi_{\mathbf{w}}(\mathbf{t}) = \phi_{y_1}(t_1)\phi_{z_1}(t_2)$; that is $\psi(t_1^2+t_2^2) = \psi(t_1^2)\psi(t_2^2)$, or equivalently

$$\psi(t_1+t_2) = \psi(t_1)\psi(t_2), \qquad t_1 \geq 0,\ t_2 \geq 0. \tag{6.5}$$

Now, using Corollary 1.1 we see that $\psi(t) = e^{-kt}$, $k \geq 0$, $t \geq 0$. Moreover, $k = 0$ is impossible since this would make y degenerate. Therefore, $k > 0$ and hence $\mathbf{X}$ is normal. ■

Corollary 6.1. *Let* $\mathbf{X} \sim E_{p,n}(\mathbf{M}, \Sigma \otimes \Phi, \psi)$ *and* $\mathbf{x}_i$, $i = 1, \ldots, n$ *denote the columns of* $\mathbf{X}$. *If* $\mathbf{x}_1, \mathbf{x}_2, \ldots, \mathbf{x}_n$ *are all nondegenerate and independent, then* $\mathbf{X} \sim N_{p,n}(\mathbf{M}, \Sigma \otimes \Phi)$, *where* Φ *is diagonal.*

PROOF: This follows, from Theorem 6.4. Since if two columns are independent, then any two elements, picked one from each of these columns will also be independent. The structure of Φ is implied by the fact that $\mathbf{x}_1, \mathbf{x}_2, \ldots, \mathbf{x}_n$ are independent and normal. ■

Remark 6.1. For the case $p = 1$, a result similar to Corollary 6.1, was given by Kelker (1970). However, he had stronger conditions since he made the diagonality of Φ an assumption of the theorem. The following theorem, although not a characterization result gives the idea for a further characterization of normality.

Theorem 6.5. *Let* $\mathbf{X} \sim E_{p,n}(\mathbf{M}, \Sigma \otimes \Phi, \psi)$ *be nondegenerate, with finite second order moment. Assume* $\mathbf{A} : q \times p$, $\mathbf{B} : n \times k$, $\mathbf{C} : r \times p$, *and* $\mathbf{D} : n \times l$ *are constant matrices. Then,* $\mathbf{AXB}$ *and* $\mathbf{CXD}$ *are uncorrelated if and only if either* $\mathbf{A}\Sigma\mathbf{C}' = \mathbf{0}$ *or* $\mathbf{B}'\Phi\mathbf{D} = \mathbf{0}$.

PROOF: Without loss of generality, we can assume $\mathbf{M} = \mathbf{0}$. Then, using Theorem 1.17, we can write

$$\begin{aligned} Cov(vec(\mathbf{AXB})', vec(\mathbf{CXD})') &= E(vec(\mathbf{AXB})'(vec(\mathbf{CXD})')') \\ &= E((\mathbf{A} \otimes \mathbf{B}')vec(\mathbf{X}')(vec(\mathbf{X}'))'(\mathbf{C}' \otimes \mathbf{D})) \\ &= -2\psi'(0)(\mathbf{A} \otimes \mathbf{B}')(\Sigma \otimes \Phi)(\mathbf{C}' \otimes \mathbf{D}) \\ &= -2\psi'(0)(\mathbf{A}\Sigma\mathbf{C}') \otimes (\mathbf{B}'\Phi\mathbf{D}). \end{aligned}$$

Here, $\psi'(0) \neq 0$, since $\mathbf{X}$ is nondegenerate, so we must have $(\mathbf{A}\Sigma\mathbf{C}') \otimes (\mathbf{B}'\Phi\mathbf{D}) = \mathbf{0}$. This holds iff $\mathbf{A}\Sigma\mathbf{C}' = \mathbf{0}$ or $\mathbf{B}'\Phi\mathbf{D} = \mathbf{0}$. ■

Remark 6.2. Since, in the normal case, uncorrelatedness and independence are equivalent, Theorem 6.5 implies that for $\mathbf{X} \sim N_{p,n}(\mathbf{M}, \Sigma \otimes \Phi)$, $\mathbf{AXB}$ and $\mathbf{CXD}$ are independent iff $\mathbf{A}\Sigma\mathbf{C}' = \mathbf{0}$ or $\mathbf{B}'\Phi\mathbf{D} = \mathbf{0}$. This property of the matrix variate normal distribution was obtained by Nel (1977).

Theorem 6.5 shows that under certain conditions, two linear expressions in a random matrix with m.e.c. distribution are uncorrelated, whereas Remark 6.2 says that if the underlying distribution is normal the two linear transforms are independent. The question arises whether the independence of the linear transforms characterizes normality in the class of m.e.c. distributions. The answer is yes, as the next theorem shows.

Theorem 6.6. *Let* $\mathbf{X} \sim E_{p,n}(\mathbf{M}, \Sigma \otimes \Phi, \psi)$ *and* $\mathbf{A} : q \times p$, $\mathbf{B} : n \times k$, $\mathbf{C} : r \times p$, *and* $\mathbf{D} : n \times l$ *be constant nonzero matrices. If* $\mathbf{AXB}$ *and* $\mathbf{CXD}$ *are nondegenerate and independent, then* $\mathbf{X}$ *is normal.*

PROOF: Without loss of generality, we can assume $\mathbf{M} = \mathbf{0}$. Since $\mathbf{AXB}$ and $\mathbf{CXD}$ are independent, so are $vec(\mathbf{AXB})' = (\mathbf{A} \otimes \mathbf{B}')vec(\mathbf{X}')$ and $vec(\mathbf{CXD})' = (\mathbf{C} \otimes \mathbf{D}')vec(\mathbf{X}')$. Let $\mathbf{x} = vec(\mathbf{X})$, then $\mathbf{x} \sim E_{pn}(\mathbf{0}, \Sigma \otimes \Phi, \psi)$. Let $\mathbf{v}'$ be a nonzero row of $\mathbf{A} \otimes \mathbf{B}'$ and $\mathbf{w}'$ be a nonzero row of $\mathbf{C} \otimes \mathbf{D}'$. Then, $\mathbf{v}'\mathbf{x}$ and $\mathbf{w}'\mathbf{x}$ are independent. Let $\mathbf{H} = \begin{pmatrix} \mathbf{v}' \\ \mathbf{w}' \end{pmatrix}$. Then $\mathbf{Hx} = \begin{pmatrix} \mathbf{v}'\mathbf{x} \\ \mathbf{w}'\mathbf{x} \end{pmatrix} \sim E_2(\mathbf{0}, \mathbf{H}(\Sigma \otimes \Phi)\mathbf{H}', \psi)$. Since $\mathbf{v}'\mathbf{x}$ and $\mathbf{w}'\mathbf{x}$ are independent and their joint distribution is elliptically contoured, from Theorem 6.4 we conclude that $\mathbf{Hx}$ is normal. Therefore, $\mathbf{X}$ is normal. ■

The following characterization results are based on conditional distributions (see Gupta and Varga, 1990, 1992, 1994a, 1997).

Theorem 6.7. *Let* $\mathbf{X} \sim E_{p,n}(\mathbf{M}, \Sigma \otimes \Phi, \psi)$ *be nondegenerate. Let* $\mathbf{X}$, $\mathbf{M}$, *and* Σ *be partitioned as* $\mathbf{X} = \begin{pmatrix} \mathbf{X}_1 \\ \mathbf{X}_2 \end{pmatrix}$, $\mathbf{M} = \begin{pmatrix} \mathbf{M}_1 \\ \mathbf{M}_2 \end{pmatrix}$, $\Sigma = \begin{pmatrix} \Sigma_{11} & \Sigma_{12} \\ \Sigma_{21} & \Sigma_{22} \end{pmatrix}$, *where* $\mathbf{X}_1$ *and* $\mathbf{M}_1$ *are* $q \times n$ *and* Σ_{11} *is* $q \times q$. *Assume* $rk(\Sigma_{22}) \geq 1$ *and* $rk(\Sigma) - rk(\Sigma_{22}) \geq 1$. *Let* $\mathbf{X}_1|\mathbf{X}_2 \sim E_{q,n}(\mathbf{M}_1 + \Sigma_{12}\Sigma_{22}^{-1}(\mathbf{X}_2 - \mathbf{M}_2), \Sigma_{11\cdot 2} \otimes \Phi, \psi_{q(\mathbf{X}_2)})$ *with* $q(\mathbf{X}_2) = tr((\mathbf{X}_2 - \mathbf{M}_2)'\Sigma_{11}^{-}(\mathbf{X}_2 - \mathbf{M}_2)\Phi^{-})$. *Then,* $\psi_{q(\mathbf{X}_2)}$ *does not depend on* $\mathbf{X}_2$, *with probability one if and only if* $\mathbf{X} \sim N_{p,n}(\mathbf{M}, \Sigma \otimes \Phi)$.

PROOF: It is known that if $\mathbf{X} \sim N_{p,n}(\mathbf{M}, \Sigma \otimes \Phi)$, then $\psi_{q(\mathbf{X}_2)} = \exp\left(-\frac{z}{2}\right)$ and hence, $\psi_{q(\mathbf{X}_2)}$ does not depend on $\mathbf{X}_2$ (see, e.g. Anderson, 2003, p. 33).

Conversely, assume $\psi_{q(\mathbf{X}_2)}$ does not depend on $\mathbf{X}_2$ for $\mathbf{X}_2 \in A$, where $P(\mathbf{X}_2 \in A) = 1$. Thus, for $\mathbf{X}_2 \in A$, we can write $\psi_{q(\mathbf{X}_2)} = \psi_0(z)$, where ψ_0 does not depend on $\mathbf{X}_2$. It follows from the definition of $q(\mathbf{X}_2)$, that it suffices to consider the case $\mathbf{M} = \mathbf{0}$, $\Sigma = \mathbf{I}_p$, $\Phi = \mathbf{I}_n$ (see Theorem 2.22). Let $\mathbf{T}$ be a $p \times n$ matrix and partition it as $\mathbf{T} = \begin{pmatrix} \mathbf{T}_1 \\ \mathbf{T}_2 \end{pmatrix}$, where $\mathbf{T}_1$ is $q \times n$. Then, the characteristic function of $\mathbf{X}$ is

$$\begin{aligned} \phi_{\mathbf{X}}(\mathbf{T}) &= \psi(tr(\mathbf{T}'\mathbf{T})) \\ &= \psi(tr(\mathbf{T}_1'\mathbf{T}_1 + \mathbf{T}_2'\mathbf{T}_2)) \\ &= \psi(tr(\mathbf{T}_1'\mathbf{T}_1) + tr(\mathbf{T}_2'\mathbf{T}_2)). \end{aligned} \tag{6.6}$$

On the other hand,

$$\begin{aligned}
\phi_{\mathbf{X}}(\mathbf{T}) &= E(etr(i\mathbf{T}'\mathbf{X})) \\
&= E(etr(i\mathbf{X}\mathbf{T}')) \\
&= E(etr(i(\mathbf{X}_1\mathbf{T}_1' + \mathbf{X}_2\mathbf{T}_2'))) \\
&= E(E(etr(i(\mathbf{X}_1\mathbf{T}_1' + \mathbf{X}_2\mathbf{T}_2'))|\mathbf{X}_2)) \\
&= E(etr(i\mathbf{X}_2\mathbf{T}_2'))E(E(etr(i\mathbf{X}_1\mathbf{T}_1')|\mathbf{X}_2)) \\
&= E(etr(i\mathbf{X}_2\mathbf{T}_2'))\psi_0(tr(\mathbf{T}_1'\mathbf{T}_1)) \\
&= \psi(tr(\mathbf{T}_2'\mathbf{T}_2))\psi_0(tr(\mathbf{T}_1'\mathbf{T}_1)). \qquad (6.7)
\end{aligned}$$

Let $u = tr(\mathbf{T}_1'\mathbf{T}_1)$ and $v = tr(\mathbf{T}_2'\mathbf{T}_2)$. Then, from (6.6) and (6.7), we obtain

$$\psi(u+v) = \psi(u)\psi_0(v) \qquad (6.8)$$

for $u, v \geq 0$. Taking $u = 0$ in (6.8), we get $\psi(v) = \psi_0(v)$. Hence, (6.8) gives

$$\psi(u+v) = \psi(u)\psi(v) \qquad (6.9)$$

for $u, v \geq 0$. Now from Corollary 1.1 we see that $\psi(z) = e^{-kz}$, $k \geq 0, t \geq 0$. Moreover, k cannot be zero, since this would make $\mathbf{X}$ degenerate. Therefore, $k > 0$ and hence $\mathbf{X}$ is normal. ∎

The next theorem shows that the normality of the conditional distributional characterizes the normal distribution in the class of m.e.c. distributions.

Theorem 6.8. *Let* $\mathbf{X} \sim E_{p,n}(\mathbf{M}, \Sigma \otimes \Phi, \psi)$ *be nondegenerate. Let* $\mathbf{X}$ *and* Σ *be partitioned as* $\mathbf{X} = \begin{pmatrix} \mathbf{X}_1 \\ \mathbf{X}_2 \end{pmatrix}$, $\Sigma = \begin{pmatrix} \Sigma_{11} & \Sigma_{12} \\ \Sigma_{21} & \Sigma_{22} \end{pmatrix}$, *where* $\mathbf{X}_1$ *is* $q \times n$ *and* Σ_{11} *is* $q \times q$. *Assume* $rk(\Sigma_{22}) \geq 1$ *and* $rk(\Sigma) - rk(\Sigma_{22}) \geq 1$. *Then,* $P(\mathbf{X}_1|\mathbf{X}_2$ *is nondegenerate normal*$) = 1$ *if and only if* $\mathbf{X} \sim N_{p,n}(\mathbf{M}, \Sigma \otimes \Phi)$.

PROOF: It has already been mentioned in the proof of Theorem 6.7 that if $\mathbf{X} \sim N_{p,n}(\mathbf{M}, \Sigma \otimes \Phi)$, then $\mathbf{X}_1|\mathbf{X}_2$ is nondegenerate normal with probability one.

Conversely, assume $\mathbf{X}_1|\mathbf{X}_2$ is nondegenerate normal with probability one. Then, from the definition of $q(\mathbf{X}_2)$, it follows that it suffices to consider the case $\mathbf{M} = \mathbf{0}$, $\Sigma = \mathbf{I}_p$, $\Phi = \mathbf{I}_n$, where now $q(\mathbf{X}_2) = tr(\mathbf{X}_2'\mathbf{X}_2)$. So $q(\mathbf{X}_2) = 0$ if and only if $\mathbf{X}_2 = \mathbf{0}$. Thus, $P(q(\mathbf{X}_2) = 0) = P(\mathbf{X}_2 = \mathbf{0})$.

On the other hand, from Corollary 3.2 it follows that $P(\mathbf{X}_2 = \mathbf{0}) = P(\mathbf{X} = \mathbf{0})$. Hence, $P(q(\mathbf{X}_2) = 0) = P(\mathbf{X} = \mathbf{0})$. However $P(\mathbf{X} = \mathbf{0}) > 0$ is not possible since this would imply that $\mathbf{X}_1|\mathbf{X}_2$ is degenerate with positive probability. Therefore, $P(q(\mathbf{X}_2) = 0) = 0$ must hold. Hence, there exists a set $A \subset I\!R^{q \times n}$ such that $P(\mathbf{X}_2 \in A) = 1$. If $\mathbf{X}_2 \in A$, then $q(\mathbf{X}_2) > 0$ and $\mathbf{X}_1|\mathbf{X}_2$ is nondegenerate normal. So if $\mathbf{X}_2 \in A$, then we get

$$\mathbf{X}_1|\mathbf{X}_2 \sim E_{q,n}(\mathbf{0}, \mathbf{I}_q \otimes \mathbf{I}_n, \psi_{q(\mathbf{X}_2)}) \quad \text{with} \quad \psi_{q(\mathbf{X}_2)}(z) = \exp\left(-\frac{c(q(\mathbf{X}_2))z}{2}\right). \tag{6.10}$$

Here, c denotes a function of $q(\mathbf{X}_2)$, such that $c(q(\mathbf{X}_2)) > 0$ for $\mathbf{X}_2 \in A$. Let $r_{q(\mathbf{X}_2)}\mathbf{U}_1$ be the stochastic representation of $\mathbf{X}_1|\mathbf{X}_2$. Since

$$r^2_{q(\mathbf{X}_2)} \approx tr((\mathbf{X}_1|\mathbf{X}_2)'(\mathbf{X}_1|\mathbf{X}_2)),$$

from (6.10), we get

$$r^2_{q(\mathbf{X}_2)} \sim c(q(\mathbf{X}_2))\chi^2_{qn}. \tag{6.11}$$

The p.d.f. of χ^2_{qn} is

$$g(y) = \frac{1}{2^{\frac{qn}{2}}\Gamma\left(\frac{qn}{2}\right)} y^{\frac{qn}{2}-1} e^{-\frac{y}{2}}, \quad y > 0. \tag{6.12}$$

Let $v^2 = c(q(\mathbf{X}_2))y$ for fixed $\mathbf{X}_2$ with $v \geq 0$ and $J(y \to v) = \frac{1}{c(q(\mathbf{X}_2))}2v$. Then, $v \approx r_{q(\mathbf{X}_2)}$, hence the p.d.f. of v, say $p(v)$, is the same as the p.d.f. of $r_{q(\mathbf{X}_2)}$. Therefore, from (6.12) we get

$$\begin{aligned} p(v) &= \frac{2v}{2^{\frac{qn}{2}}\Gamma\left(\frac{qn}{2}\right)c(q(\mathbf{X}_2))} \left(\frac{v^2}{c(q(\mathbf{X}_2))}\right)^{\frac{qn}{2}-1} e^{-\frac{v^2}{2c(q(\mathbf{X}_2))}} \\ &= \frac{1}{\Gamma\left(\frac{qn}{2}\right)2^{\frac{qn}{2}-1}(c(q(\mathbf{X}_2)))^{\frac{qn}{2}}} v^{qn-1} e^{-\frac{v^2}{2c(q(\mathbf{X}_2))}}. \end{aligned} \tag{6.13}$$

Let $r\mathbf{U}$ be the stochastic representation of $\mathbf{X}$ and F be the distribution function of r. By appealing to Corollary 2.5, from (6.13), we obtain

$$\begin{aligned} 1 - F(z) &= K_{q(\mathbf{X}_2)} \int_{\sqrt{z^2 - q(\mathbf{X}_2)}}^{\infty} (v^2 + q(\mathbf{X}_2))^{\frac{pn}{2}-1} v^{-(qn-2)} p(v) dv \\ &= L_{\mathbf{X}_2} \int_{\sqrt{z^2 - q(\mathbf{X}_2)}}^{\infty} (v^2 + q(\mathbf{X}_2))^{\frac{pn}{2}-1} v e^{-\frac{v^2}{2c(q(\mathbf{X}_2))}} dv, \end{aligned} \tag{6.14}$$

for $z \geq q(\mathbf{X}_2)$ where

$$L_{\mathbf{X}_2} = \frac{K_{q(\mathbf{X}_2)}}{\Gamma\left(\frac{qn}{2}\right)2^{\frac{qn}{2}-1}(c(q(\mathbf{X}_2)))^{\frac{qn}{2}}}.$$

Substituting $t^2 = v^2 + q(\mathbf{X}_2)$ and $J(v \to t) = \frac{t}{v}$ in (6.14), we get

$$1 - F(z) = L_{\mathbf{X}_2} \int_z^\infty t^{pn-2} t e^{-\frac{t^2 - q(\mathbf{X}_2)}{2c(q(\mathbf{X}_2))}} dt$$
$$= L_{\mathbf{X}_2} e^{\frac{q(\mathbf{X}_2)}{2c(q(\mathbf{X}_2))}} \int_z^\infty t^{pn-1} e^{-\frac{t^2}{2c(q(\mathbf{X}_2))}} dt. \tag{6.15}$$

Now, (6.15) holds for $z \geq q(\mathbf{X}_2)$. Differentiating (6.15), with respect to z, we get

$$F'(z) = J_{\mathbf{X}_2} z^{pn-1} e^{-\frac{z^2}{2c(q(\mathbf{X}_2))}} \quad \text{for} \quad z \geq q(\mathbf{X}_2), \tag{6.16}$$

where $J_{\mathbf{X}_2} = -L_{\mathbf{X}_2} e^{\frac{q(\mathbf{X}_2)}{2c(q(\mathbf{X}_2))}}$.

Let $\mathbf{X}_{2,1} \in A$ and $\mathbf{X}_{2,2} \in A$ and let $z \in [\max(q(\mathbf{X}_{2,1}), q(\mathbf{X}_{2,2})), \infty)$. Then, from (6.16), it follows that $c(q(\mathbf{X}_{2,1})) = c(q(\mathbf{X}_{2,2}))$. Therefore, (6.10) shows that $\phi_{q(\mathbf{X}_{2,1})} = \phi_{q(\mathbf{X}_{2,2})}$. Since $P(A) = 1$, this means that $\phi_{q(\mathbf{X}_2)}$ does not depend on $\mathbf{X}_2$, with probability one. Hence, from Theorem 6.7, it is easily seen that $\mathbf{X}$ is normally distributed. ■

The next characterization result shows that if a m.e.c. distribution is absolutely continuous and one of its marginals has the p.d.f. whose functional form coincides with that of the p.d.f. of the original distribution up to a constant multiplier, then the distribution must be normal. More precisely, we can prove the following theorem.

Theorem 6.9. *Let $\mathbf{X} \sim E_{p,n}(\mathbf{M}, \Sigma \otimes \Phi, \psi)$ be absolutely continuous with p.d.f.*

$$|\Sigma|^{-\frac{n}{2}} |\Phi|^{-\frac{p}{2}} h(tr((\mathbf{X}-\mathbf{M})'\Sigma^{-1}(\mathbf{X}-\mathbf{M})\Phi^{-1})).$$

Let $\mathbf{X}$, $\mathbf{M}$, and Σ be partitioned as $\mathbf{X} = \begin{pmatrix} \mathbf{X}_1 \\ \mathbf{X}_2 \end{pmatrix}$, $\mathbf{M} = \begin{pmatrix} \mathbf{M}_1 \\ \mathbf{M}_2 \end{pmatrix}$, $\Sigma = \begin{pmatrix} \Sigma_{11} & \Sigma_{12} \\ \Sigma_{21} & \Sigma_{22} \end{pmatrix}$, where $\mathbf{X}_1$ and $\mathbf{M}_1$ are $q \times n$ and Σ_{11} is $q \times q$ with $1 \leq q < p$. Let the p.d.f. of $\mathbf{X}_1$ be

$$|\Sigma_{11}|^{-\frac{n}{2}} |\Phi|^{-\frac{p}{2}} h_1(tr((\mathbf{X}_1-\mathbf{M}_1)'\Sigma_{11}^{-1}(\mathbf{X}_1-\mathbf{M}_1)\Phi^{-1})).$$

Then, h and h_1 agree up to a constant multiplier; that is,

$$h(z) = c h_1(z) \tag{6.17}$$

if and only if $\mathbf{X}$ is normal.

PROOF: If $\mathbf{X}$ is normal, then $h(z) = (2\pi)^{-\frac{pn}{2}} \exp\left(-\frac{z}{2}\right)$. Moreover, $\mathbf{X}_1$ is also normal with $h_1(z) = (2\pi)^{-\frac{qn}{2}} \exp\left(-\frac{z}{2}\right)$. Thus, $h(z) = (2\pi)^{\frac{(q-p)n}{2}} h_1(z)$, so (6.17) is satisfied with $c = (2\pi)^{\frac{(q-p)n}{2}}$.

Conversely, assume (6.17) holds. Without loss of generality, we can assume $\mathbf{M} = \mathbf{0}$, $\Sigma = \mathbf{I}_p$, $\Phi = \mathbf{I}_n$. From (6.17), we get

$$h(tr(\mathbf{X}\mathbf{X}')) = c h_1(tr(\mathbf{X}_1\mathbf{X}_1')). \tag{6.18}$$

We also use the fact that $h_1(tr(\mathbf{X}_1\mathbf{X}_1'))$, being marginal p.d.f. of $\mathbf{X}$, can be obtained from $h(tr(\mathbf{XX}'))$ through integration.

Let $\mathbf{X} = (\mathbf{X}_1, \mathbf{X}_2)$, then $h(tr(\mathbf{XX}')) = h(tr(\mathbf{X}_1\mathbf{X}_1' + \mathbf{X}_2\mathbf{X}_2')) = h(tr(\mathbf{X}_1\mathbf{X}_1') + tr(\mathbf{X}_2\mathbf{X}_2'))$. Hence we have

$$h_1(tr(\mathbf{X}_1\mathbf{X}_1')) = \int_{\mathbb{R}^{(p-q)\times n}} h(tr(\mathbf{X}_1\mathbf{X}_1') + tr(\mathbf{X}_2\mathbf{X}_2'))d\mathbf{X}_2. \tag{6.19}$$

From (6.18) and (6.19), we get

$$h_1(tr(\mathbf{X}_1\mathbf{X}_1')) = c\int_{\mathbb{R}^{(p-q)\times n}} h_1(tr(\mathbf{X}_1\mathbf{X}_1') + tr(\mathbf{X}_2\mathbf{X}_2'))d\mathbf{X}_2.$$

Hence

$$h_1(z) = c\int_{\mathbb{R}^{(p-q)\times n}} h_1(z + tr(\mathbf{X}_2\mathbf{X}_2'))d\mathbf{X}_2, \quad z \geq 0.$$

Using (6.17) again, we have

$$h(z) = c^2\int_{\mathbb{R}^{(p-q)\times n}} h_1(z + tr(\mathbf{X}_2\mathbf{X}_2'))d\mathbf{X}_2, \quad z \geq 0.$$

which can also be written as

$$h(z) = c^2\int_{\mathbb{R}^{(p-q)\times n}} h_1(z + tr(\mathbf{Y}_2\mathbf{Y}_2'))d\mathbf{Y}_2. \tag{6.20}$$

Define the $(p+q)\times n$ matrix $\mathbf{Y} = \begin{pmatrix}\mathbf{Y}_1\\ \mathbf{Y}_2\end{pmatrix}$, with $\mathbf{Y}_1 : p\times n$. Let $z = tr(\mathbf{Y}_1\mathbf{Y}_1')$. Then, from (6.20), we have

$$h(tr(\mathbf{Y}_1\mathbf{Y}_1')) = c^2\int_{\mathbb{R}^{(p-q)\times n}} h_1(tr(\mathbf{Y}_1\mathbf{Y}_1') + tr(\mathbf{Y}_2\mathbf{Y}_2'))d\mathbf{Y}_2. \tag{6.21}$$

Now, the left-hand side of (6.21) is a p.d.f., since $\mathbf{Y}_1$ is $p\times n$. Hence,

$$\int_{\mathbb{R}^{p\times n}} h(tr(\mathbf{Y}_1\mathbf{Y}_1'))d\mathbf{Y}_1 = 1.$$

Therefore, integrating the right-hand side of (6.21), with respect to $\mathbf{Y}_1$ we get

$$c^2\int_{\mathbb{R}^{p\times n}}\int_{\mathbb{R}^{(p-q)\times n}} h_1(tr(\mathbf{Y}_1\mathbf{Y}_1') + tr(\mathbf{Y}_2\mathbf{Y}_2'))d\mathbf{Y}_2 d\mathbf{Y}_1 = 1,$$

which can be rewritten as

$$\int_{\mathbb{R}^{(p+(p-q))\times n}} c^2 h_1(tr(\mathbf{YY}'))d\mathbf{Y} = 1.$$

Therefore, $c^2h_1(tr(\mathbf{YY}'))$ is the p.d.f. of a $(2p-q)\times n$ random matrix $\mathbf{Y}$ with m.e.c. distribution. Moreover, it follows from (6.21) and (6.17), that $ch_1(tr(\mathbf{Y}_1\mathbf{Y}_1'))$ is the p.d.f. of the $p\times n$ dimensional marginal of $\mathbf{Y}$. Since $\mathbf{Y}_1 \sim E_{p,n}(\mathbf{0},\mathbf{I}_p\otimes\mathbf{I}_n,\psi)$, we must have

$$\mathbf{Y}\sim E_{2p-q,n}(\mathbf{0},\mathbf{I}_{2p-q}\otimes\mathbf{I}_n,\psi).$$

By iterating the above procedure we see that for any $j\geq 1$, there exists a $(p+j(p-q))\times n$ random matrix $\mathbf{Y}_j$ with m.e.c. distribution, such that

$$\mathbf{Y}\sim E_{p+j(p-q),n}(\mathbf{0},\mathbf{I}_{p+j(p-q)}\otimes\mathbf{I}_n,\psi).$$

Then it follows from the Definition 4.1, Theorem 4.3, and Remark 4.2, that there exists a distribution function $G(u)$ on $(0,\infty)$, such that

$$\psi(s)=\int_0^\infty \exp\left(-\frac{su}{2}\right)dG(u). \tag{6.22}$$

Therefore,

$$h(z)=\int_0^\infty \frac{1}{(2\pi z)^{\frac{pn}{2}}}\exp\left(-\frac{zu}{2}\right)dG(u)$$

and

$$h_1(z)=\int_0^\infty \frac{1}{(2\pi z)^{\frac{qn}{2}}}\exp\left(-\frac{zu}{2}\right)dG(u)$$

Using (6.17), we get

$$\int_0^\infty\left(\frac{1}{(2\pi z)^{\frac{pn}{2}}}-\frac{c}{(2\pi z)^{\frac{qn}{2}}}\right)\exp\left(-\frac{zu}{2}\right)dG(u)=0. \tag{6.23}$$

Using the inverse Laplace transform in (6.23), we obtain

$$\left(\frac{1}{(2\pi z)^{\frac{pn}{2}}}-\frac{c}{(2\pi z)^{\frac{qn}{2}}}\right)dG(z)=0.$$

Hence,

$$(2\pi z)^{-\frac{pn}{2}}\left(1-(2\pi z)^{\frac{(p-q)n}{2}}c\right)dG(z)=0.$$

However, this is possible only if G is degenerate at $z_0=\frac{1}{2\pi}c^{\frac{2}{(p-q)n}}$; that is,

$$G(z) = \begin{cases} 0 \text{ if } z < z_0 \\ 1 \text{ if } z \geq z_0. \end{cases}$$

Now, let $z_0 = \sigma^2$. Then, (6.22) gives $\psi(s) = \exp\left(-\frac{s\sigma^2}{2}\right)$, which implies that $\mathbf{X}$ is normal. ■

Since in the normal case $\psi_{q(\mathbf{X}_2)}$ does not depend on $\mathbf{X}_2$, the conditional distribution $r_{q(\mathbf{X}_2)}|\mathbf{X}_2$ is also independent of $\mathbf{X}_2$. Hence, for every k positive integer the conditional moment $E(r^k_{q(\mathbf{X}_2)}|\mathbf{X}_2)$, is independent of $\mathbf{X}_2$. The next theorem shows that normal distribution is the only m.e.c. distribution which possesses this property.

Theorem 6.10. *Let $\mathbf{X} \sim E_{p,n}(\mathbf{M}, \Sigma \otimes \Phi, \psi)$ be nondegenerate. Let $\mathbf{X}$ and Σ be partitioned as $\mathbf{X} = \begin{pmatrix} \mathbf{X}_1 \\ \mathbf{X}_2 \end{pmatrix}$, $\Sigma = \begin{pmatrix} \Sigma_{11} & \Sigma_{12} \\ \Sigma_{21} & \Sigma_{22} \end{pmatrix}$, where $\mathbf{X}_1$ is $q \times n$ and Σ_{11} is $q \times q$ with $1 \leq q < p$. Assume $rk(\Sigma_{22}) \geq 1$ and $rk(\Sigma) - rk(\Sigma_{22}) \geq 1$. Then, there exists positive integer k such that $E(r^k_{q(\mathbf{X}_2)}|\mathbf{X}_2)$ is finite and does not depend on $\mathbf{X}_2$, with probability one if and only if $\mathbf{X}$ is normal. Here, $r_{q(\mathbf{X}_2)}$ is the one-dimensional random variable, appearing in the stochastic representation of $\mathbf{X}_1|\mathbf{X}_2$:*

$$\mathbf{X}_1|\mathbf{X}_2 \approx r_{q(\mathbf{X}_2)}\mathbf{A}\mathbf{U}_1\mathbf{B}'.$$

PROOF: If $\mathbf{X}$ is normal, then $\psi_{q(\mathbf{X}_2)}(z) = \psi\left(-\frac{z}{2}\right)$. Hence, $r^2_{q(\mathbf{X}_2)} \sim \chi^2_{q_1 n}$, with $q_1 = rk(\Sigma) - rk(\Sigma_{22})$. Hence, $E(r^k_{q(\mathbf{X}_2)}|\mathbf{X}_2)$ is the $\frac{k}{2}$th moment of $\chi^2_{q_1 n}$ which is finite and independent of $\mathbf{X}_2$.

Conversely, assume $E(r^k_{q(\mathbf{X}_2)}|\mathbf{X}_2)$ is finite and does not depend on $\mathbf{X}_2$, with probability one. Without loss of generality, we can assume $\mathbf{M} = \mathbf{0}$, $\Sigma = \mathbf{I}_p$, $\Phi = \mathbf{I}_n$. Then, we have $q(\mathbf{X}_2) = ||\mathbf{X}_2||$ and $\mathbf{X}_1|\mathbf{X}_2 \approx r_{||\mathbf{X}_2||^2}\mathbf{U}_1$. Hence we get $tr(\mathbf{X}_1'\mathbf{X}_1)|\mathbf{X}_2 \approx r^2_{||\mathbf{X}_2||^2} tr(\mathbf{U}_1'\mathbf{U}_1)$, and since $tr(\mathbf{U}_1'\mathbf{U}_1) = 1$, we get $||\mathbf{X}_1|||\mathbf{X}_2 \approx r_{||\mathbf{X}_2||^2}$. Therefore,

$$E(||\mathbf{X}_1||^k|\mathbf{X}_2) = E(r^k_{||\mathbf{X}_2||^2}|\mathbf{X}_2).$$

Hence, $E(||\mathbf{X}_1||^k|\mathbf{X}_2)$ is finite (with probability one) and independent of $\mathbf{X}_2$.

Next, we show that $P(\mathbf{X} = \mathbf{0}) = 0$. Assume this is not the case. Let $0 < P_0 = P(\mathbf{X} = \mathbf{0}) = P\left(\begin{pmatrix} \mathbf{X}_1 \\ \mathbf{X}_2 \end{pmatrix} = \mathbf{0}\right)$. Then $P((\mathbf{X}_1|\mathbf{X}_2) = \mathbf{0}) \geq P_0$ and $P(E(||\mathbf{X}_1||^k|\mathbf{X}_2) = 0) \geq P_0$. Since $E(||\mathbf{X}_1||^k|\mathbf{X}_2)$ does not depend on $\mathbf{X}_2$ with probability one, we have $P(E(||\mathbf{X}_1||^k|\mathbf{X}_2) = 0) = 1$. Hence, $P((\mathbf{X}_1|\mathbf{X}_2) = \mathbf{0}) = 1$.

Since,

$$P(\mathbf{X}_1 \in B) = \int_{\mathbb{R}^{q \times n}} P((\mathbf{X}_1|\mathbf{X}_2) \in B) dF_{\mathbf{X}_2}(\mathbf{X}_2),$$

where $F_{\mathbf{X}_2}$ is the distribution function of $\mathbf{X}_2$ and $B \in \mathscr{B}(\mathbb{R}^{q_1 \times n})$, we get $P(\mathbf{X}_1 = \mathbf{0}) = 1$. Then, from Corollary 3.2, we get $P(\mathbf{X} = \mathbf{0}) = P(\mathbf{X}_1 = \mathbf{0}) = 1$. That means $\mathbf{X}$ is degenerate, which contradicts the assumptions of the theorem. So, $P(\mathbf{X} = \mathbf{0}) = 0$.

From Corollary 3.2, it follows that $P(\mathbf{X}_2 = \mathbf{0}) = 0$. Let $\mathbf{X} \approx r\mathbf{U}$ be the stochastic representation of $\mathbf{X}$ and F be the distribution function of r. Then, using Theorem 2.21, we obtain

$$P((\mathbf{X}_1|\mathbf{X}_2) = \mathbf{0}) = 1 \quad \text{if} \quad F(||\mathbf{X}_2||) = 1$$

and

$$F_{||\mathbf{X}_2||^2}(z) = \frac{1}{K_{||\mathbf{X}_2||^2}} \int_{\left(||\mathbf{X}_2||, \sqrt{z^2 + ||\mathbf{X}_2||^2}\right)} (s^2 - ||\mathbf{X}_2||^2)^{\frac{qn}{2}-1} s^{-(pn-2)} dF(s) \tag{6.24}$$

if $z \geq 0$ and $F(||\mathbf{X}_2||) < 1$, where $F_{||\mathbf{X}_2||^2}(z)$ denotes the distribution function of $r_{||\mathbf{X}_2||^2}$ and

$$K_{||\mathbf{X}_2||^2} = \int_{(||\mathbf{X}_2||, \infty)} (s^2 - ||\mathbf{X}_2||^2)^{\frac{qn}{2}-1} s^{-(pn-2)} dF(s).$$

From (6.24), we get

$$dF_{||\mathbf{X}_2||^2}(z) = \frac{1}{K_{||\mathbf{X}_2||^2}} (z^2)^{\frac{qn}{2}-1} (z^2 + ||\mathbf{X}_2||^2)^{-\frac{pn-2}{2}} \frac{ds}{dz} dF(s) \tag{6.25}$$

where $z^2 + ||\mathbf{X}_2||^2 = s^2$. Using (6.25), we obtain

$$\begin{aligned} E(r^k_{||\mathbf{X}_2||^2}|\mathbf{X}_2) &= \int_0^\infty z^k dF_{||\mathbf{X}_2||^2}(z) \\ &= \frac{1}{K_{||\mathbf{X}_2||^2}} \int_{(||\mathbf{X}_2||, \infty)} (s^2 - ||\mathbf{X}_2||^2)^{\frac{qn+k}{2}-1} s^{-(pn-2)} dF(s). \end{aligned}$$

Since, $E(r^k_{||\mathbf{X}_2||^2}|\mathbf{X}_2)$ does not depend on $\mathbf{X}_2$, it follows that there exists a constant $c(k)$ which does not depend on $\mathbf{X}_2$, such that

$$\begin{aligned} &\int_{(||\mathbf{X}_2||, \infty)} (s^2 - ||\mathbf{X}_2||^2)^{\frac{qn+k}{2}-1} s^{-(pn-2)} dF(s) \\ &= c(k) \int_{(||\mathbf{X}_2||, \infty)} (s^2 - ||\mathbf{X}_2||^2)^{\frac{qn}{2}-1} s^{-(pn-2)} dF(s), \end{aligned} \tag{6.26}$$

almost everywhere. Since $P(\mathbf{X} = \mathbf{0}) = 0$, Theorem 3.5 shows that $\mathbf{X}_2$ is absolutely continuous. Therefore, $||\mathbf{X}_2||$ is also absolutely continuous. Hence, (6.26) implies that

$$\int_{(y,\infty)} (s^2-y^2)^{\frac{qn+k}{2}-1} s^{-(pn-2)} dF(s) = c(k) \int_{(y,\infty)} (s^2-y^2)^{\frac{qn}{2}-1} s^{-(pn-2)} dF(s) \tag{6.27}$$

for almost every y (with respect to the Lebesgue measure) in the interval $(0, y_0)$, where $y_0 = \inf\{y : F(y) = 1\}$. Furthermore, (6.27) is also true if $y > y_0$, because in that case both sides become zero. Therefore, (6.27) holds for almost every $y > 0$.

Define the following distribution function on $[0, \infty)$

$$H(z) = \frac{\int_0^z s^k dF(s)}{\int_0^\infty s^k dF(s)}. \tag{6.28}$$

In order to do this, we have to prove that $\int_0^\infty s^k dF(s)$ is finite and positive. Now, F is the distribution function of r, where $\mathbf{X} \approx r\mathbf{U}$ is the stochastic representation of $\mathbf{X}$. Since, $P(\mathbf{X} = \mathbf{0}) = 0$, we have $P(r > 0) = 1$. Hence, $\int_0^\infty s^k dF(s) = E(r^k) > 0$. On the other hand, let $\mathbf{X}_1 \approx r_1\mathbf{U}_1$ be the stochastic representation of $\mathbf{X}_1$. It follows from Corollary 2.4, that

$$r_1 \approx rt, \tag{6.29}$$

where r and t are independent and $t^2 \sim B\left(\frac{qn}{2}, \frac{(p-q)n}{2}\right)$. Now,

$$E(\|\mathbf{X}_1\|^k) = E(E(\|\mathbf{X}_1\|^k|\mathbf{X}_2))$$

is finite, since $E(\|\mathbf{X}_1\|^k|\mathbf{X}_2)$ is finite and does not depend on $\mathbf{X}_2$, with probability one. Since $r_1 \approx \|\mathbf{X}_1\|$, we see that $E(r_1^k)$ is finite. From (6.29), we get $E(r_1^k) = E(r^k)E(t^k)$ and $E(r_1^k)$ is finite implies that $E(r^k)$ is finite. Since $P(\mathbf{X} = \mathbf{0}) = 0$, we have $F(0) = 0$ and so $H(0) = 0$. Now, (6.27) can be rewritten in terms of H as

$$\int_{(y,\infty)} (s^2-y^2)^{\frac{qn+k}{2}-1} s^{-(pn+k-2)} dH(s) = c \int_{(y,\infty)} (s^2-y^2)^{\frac{qn}{2}-1} s^{-(pn+k-2)} dH(s). \tag{6.30}$$

Let r_0 be a random variable with distribution function H. Further, let $\mathbf{u}_0$ be uniformly distributed over S_{pn+k}, independent of r_0. Define $\mathbf{y} = r_0\mathbf{u}_0$. Then, $\mathbf{y} \sim E_{pn+k}(\mathbf{0}, \mathbf{I}_{pn+k}, \psi^*)$. Since $H(0) = 0$, we get $P(r_0 = 0) = 0$. Thus, $P(\mathbf{y} = \mathbf{0}) = 0$. Let $\mathbf{y}_1$ be a $(p-q)n+k$-dimensional subvector of $\mathbf{y}$ and $\mathbf{y}_2$, a $(p-q)n$-dimensional subvector of $\mathbf{y}_1$. Then, it follows from Theorem 3.1, that both $\mathbf{y}_1$ and $\mathbf{y}_2$ are absolutely continuous. Let $h_1(\mathbf{y}_1'\mathbf{y}_1)$ be the p.d.f. of $\mathbf{y}_1$ and $h_2(\mathbf{y}_2'\mathbf{y}_2)$ that of $\mathbf{y}_2$. Let $\mathbf{y}_1 \approx r_{0,1}\mathbf{U}_{0,1}$ and $\mathbf{y}_2 \approx r_{0,2}\mathbf{U}_{0,2}$ be the stochastic representations of $\mathbf{y}_1$ and $\mathbf{y}_2$ respectively. Moreover, Theorem 3.1 shows that the p.d.f. of $r_{0,1}$ is

$$g_1(y) = c_1 y^{(p-q)n+k-1} \int_{(y,\infty)} (s^2-y^2)^{\frac{qn}{2}-1} s^{-(pn+k-2)} dH(s)$$

and that of $r_{0,2}$ is

$$g_2(y) = c_2 y^{(p-q)n-1} \int_{(y,\infty)} (s^2 - y^2)^{\frac{qn+k}{2}-1} s^{-(pn+k-2)} dH(s).$$

Here c_i denotes a positive constant, $i = 1, 2, \ldots$. It follows, from Theorem 2.16, that

$$\begin{aligned} h_1(y^2) &= c_3 y^{-((p-q)n+k)+1} g_1(y) \\ &= c_4 \int_{(y,\infty)} (s^2 - y^2)^{\frac{qn}{2}-1} s^{-(pn+k-2)} dH(s) \end{aligned}$$

and

$$\begin{aligned} h_2(y^2) &= c_5 y^{-(p-q)n+1} g_2(y) \\ &= c_6 \int_{(y,\infty)} (s^2 - y^2)^{\frac{qn+k}{2}-1} s^{-(pn+k-2)} dH(s). \end{aligned}$$

Hence,

$$h_1(y^2) = c_4 \int_{(y,\infty)} (s^2 - y^2)^{\frac{qn}{2}-1} s^{-(pn-2)} dF(s) \tag{6.31}$$

and

$$h_2(y^2) = c_6 \int_{(y,\infty)} (s^2 - y^2)^{\frac{qn+k}{2}-1} s^{-(pn-2)} dF(s). \tag{6.32}$$

Then, (6.27), (6.30), and (6.31) imply that $h_2(y^2) = c_7 h_1(y^2)$. Therefore, from Theorem 6.9 we conclude that $\mathbf{y}_1$ is normal. Since, $\mathbf{y}_1$ is a subvector of $\mathbf{y}$, it is also normal. Hence, $H(z)$ is the distribution of $c_8 \chi_{pn+k}$. Then,

$$dH(z) = l(z)dz, \tag{6.33}$$

where

$$l(z) = c_9 \left(\frac{z}{c_8}\right)^{pn+k-1} e^{-\frac{z^2}{2c_8^2}}, \quad z \geq 0. \tag{6.34}$$

From (6.28), it follows that $dH(z) = c_{10} z^k dF(z)$. Using (6.33) and (6.34), we obtain $c_{11} z^{pn+k-1} e^{-\frac{z^2}{2c_8^2}} dz = z^k dF(z)$. Hence,

$$dF(z) = c_{12} \left(\frac{z}{c_8}\right)^{pn-1} e^{-\frac{z^2}{2c_8^2}}, \quad z \geq 0.$$

Since F is the distribution of r, we obtain that $r \approx c_{12}\chi_{pn}$. Therefore, $\mathbf{X}$ is normal.■

The following theorem gives a characterization of normality based on conditional central moments.

Theorem 6.11. *Let* $\mathbf{X} \sim E_{p,n}(\mathbf{M}, \Sigma \otimes \Phi, \psi)$ *be nondegenerate. Let* $\mathbf{X}$*,* $\mathbf{M}$*, and* Σ *be partitioned as* $\mathbf{X} = \begin{pmatrix} \mathbf{X}_1 \\ \mathbf{X}_2 \end{pmatrix}$, $\mathbf{M} = \begin{pmatrix} \mathbf{M}_1 \\ \mathbf{M}_2 \end{pmatrix}$, $\Sigma = \begin{pmatrix} \Sigma_{11} & \Sigma_{12} \\ \Sigma_{21} & \Sigma_{22} \end{pmatrix}$*, where* $\mathbf{X}_1$ *and* $\mathbf{M}_1$ *are* $q \times n$ *and* Σ_{11} *is* $q \times q$*. Assume* $rk(\Sigma_{22}) \geq 1$ *and* $rk(\Sigma) - rk(\Sigma_{22}) \geq 1$*. Assume also that there exist nonnegative integers,* k_{ij}*,* $i = 1, 2, \ldots, q$*;* $j = 1, 2, \ldots, n$*, satisfying* $k = \sum_{i=1}^{p} \sum_{j=1}^{n} k_{ij} \geq 1$*, and such that* $K(\mathbf{X}_2) = E\left(\prod_{i=1}^{q} \prod_{j=1}^{n} (x_{ij} - E(x_{ij}|\mathbf{X}_2))^{k_{ij}} |\mathbf{X}_2\right)$ *is nonzero with positive probability. Then* $K(\mathbf{X}_2)$ *is finite and does not depend on* $\mathbf{X}_2$*, with probability one if and only if* $\mathbf{X}$ *is normally distributed.*

PROOF: If $\mathbf{X}$ is normal, then

$$\mathbf{X}_1|\mathbf{X}_2 \sim N_{q,n}(\mathbf{M}_1 + \Sigma_{12}\Sigma_{22}^{-}(\mathbf{X}_2 - \mathbf{M}_2), \Sigma_{11\cdot 2} \otimes \Phi).$$

Hence

$$(\mathbf{X}_1 - E(\mathbf{X}_1|\mathbf{X}_2))|\mathbf{X}_2 \sim N_{q,n}(\mathbf{0}, \Sigma_{11\cdot 2} \otimes \Phi).$$

Therefore, $E\left(\prod_{i=1}^{q} \prod_{j=1}^{n} (x_{ij} - E(x_{ij}|\mathbf{X}_2))^{k_{ij}} |\mathbf{X}_2\right)$ is finite and does not depend on $\mathbf{X}_2$.

Conversely, assume $K(\mathbf{X}_2)$ is finite and does not depend on $\mathbf{X}_2$. Now,

$$\mathbf{X}_1|\mathbf{X}_2 \sim E_{q,n}(\mathbf{M}_1 + \Sigma_{12}\Sigma_{22}^{-}(\mathbf{X}_2 - \mathbf{M}_2), \Sigma_{11\cdot 2} \otimes \Phi, \psi_{q(\mathbf{X}_2)}),$$

and hence

$$(\mathbf{X}_1 - E(\mathbf{X}_1|\mathbf{X}_2))|\mathbf{X}_2 \sim E_{q,n}(\mathbf{0}, \Sigma_{11\cdot 2} \otimes \Phi, \psi_{q(\mathbf{X}_2)}).$$

Let $r_{q(\mathbf{X}_2)}\mathbf{AUB}'$ be the stochastic representation of $(\mathbf{X}_1 - E(\mathbf{X}_1|\mathbf{X}_2))|\mathbf{X}_2$. Then,

$$\prod_{i=1}^{q} \prod_{j=1}^{n} ((x_{ij} - E(x_{ij}|\mathbf{X}_2))^{k_{ij}} |\mathbf{X}_2) \approx r_{q(\mathbf{X}_2)}^{k} \prod_{i=1}^{q} \prod_{j=1}^{n} (\mathbf{AUB}')_{ij}^{k_{ij}}. \tag{6.35}$$

The expected value of the left-hand side of (6.35) is finite and does not depend on $\mathbf{X}_2$. Since $r_{q(\mathbf{X}_2)}$ and $\mathbf{U}$ are independent, taking expectation on both sides of (6.35), we obtain

$$K(\mathbf{X}_2) = E(r_{q(\mathbf{X}_2)}^{k}|\mathbf{X}_2) E\left(\prod_{i=1}^{q} \prod_{j=1}^{n} (\mathbf{A}^*\mathbf{UB}')_{ij}^{k_{ij}}\right). \tag{6.36}$$

Now $P(0 \neq |K(\mathbf{X}_2)| < \infty) > 0$, therefore it follows, from (6.36), that

$$P\left(0 \neq \left|E\left(\prod_{i=1}^{q}\prod_{j=1}^{n}(\mathbf{A}^*\mathbf{U}\mathbf{B}')_{ij}^{k_{ij}}\right)\right|\right) > 0. \tag{6.37}$$

However $E\left(\prod_{i=1}^{q}\prod_{j=1}^{n}(\mathbf{A}^*\mathbf{U}\mathbf{B}')_{ij}^{k_{ij}}\right)$ is a constant (say c) that does not depend $\mathbf{X}_2$, therefore (6.37) implies that

$$E\left(\prod_{i=1}^{q}\prod_{j=1}^{n}(\mathbf{A}^*\mathbf{U}\mathbf{B}')_{ij}^{k_{ij}}\right) \neq 0. \tag{6.38}$$

From (6.36) and (6.38), we get

$$E(r_{q(\mathbf{X}_2)}^k|\mathbf{X}_2) = \frac{1}{c}K(\mathbf{X}_2).$$

Therefore, $E(r_{q(\mathbf{X}_2)}^k|\mathbf{X}_2)$ is finite and independent of $\mathbf{X}_2$ with probability one. Then, using Theorem 6.10, we conclude that $\mathbf{X}$ is normally distributed. ■

Theorems 6.7–6.11 are due to Cambanis, Huang, and Simons (1981). In Theorem 6.11, however, the assumption that the conditional central moments are nonzero, with probability one is missing. Without this assumption, the theorem is not correct. To see this, take $k_{11} = 1$ and $k_{ij} = 0$, $(i, j) \neq (1,1)$. Then

$$\begin{aligned} K(\mathbf{X}_2) &= E((\mathbf{X}_{11} - E(\mathbf{X}_{11}|\mathbf{X}_2))|\mathbf{X}_2) \\ &= E(\mathbf{X}_{11}|\mathbf{X}_2) - E(\mathbf{X}_{11}|\mathbf{X}_2) \\ &= O. \end{aligned}$$

Thus $K(\mathbf{X}_2)$ is finite and does not depend on $\mathbf{X}_2$, but $\mathbf{X}$ does not have to be normal.

In order to derive further characterization results, we need the following lemma.

Lemma 6.2. *Let* $\mathbf{X} \sim E_{p,n}(\mathbf{0}, \Sigma \otimes \Phi, \psi)$ *with* $P(\mathbf{X} = \mathbf{0}) = P_0$ *where* $0 \leq P_0 < 1$. *Then, there exists a one-dimensional random variable* s *and a* $p \times n$ *random matrix* $\mathbf{L}$, *such that* s *and* $\mathbf{L}$ *are independent,* $P(s = 0) = P_0$, $P(s = 1) = 1 - P_0$, $\mathbf{L} \sim E_{p,n}\left(\mathbf{0}, \Sigma \otimes \Phi, \frac{\psi - P_0}{1 - P_0}\right)$ *and* $\mathbf{X} \approx s\mathbf{L}$. *Moreover,* $P(\mathbf{L} = \mathbf{0}) = 0$.

PROOF: If $P_0 = 0$, choose s, such that $P(s = 1) = 1$ and the theorem is trivial. If $P_0 > 0$, then let us define $\mathbf{L}$ in the following way. Define a measure $P_{\mathbf{L}}$ on $\mathbb{R}^{p\times n}$ in the following way: $P_{\mathbf{L}}(B) = \frac{1}{1-P_0}P(\mathbf{X} \in (B - \{\mathbf{0}\}))$, where B is a Borel set in $\mathbb{R}^{p\times n}$. Now, $P_{\mathbf{L}}(\mathbb{R}^{p\times n}) = P(\mathbf{X} \in (\mathbb{R}^{p\times n} - \{\mathbf{0}\})) = \frac{1-P_0}{1-P_0} = 1$. Therefore, $P_{\mathbf{L}}$ defines a probability measure on $\mathbb{R}^{p\times n}$. Let $\mathbf{L}$ be a random matrix, whose distribution is defined by $P_{\mathbf{L}}$; that is, $P(\mathbf{L} \in B) = P_{\mathbf{L}}(B)$ for every Borel set B in $\mathbb{R}^{p\times n}$.

Let s be a one-dimensional random variable, with $P(s=0)=P_0$, and $P(s=1)=1-P_0$ such that s is independent of $\mathbf{L}$. Then we show $\mathbf{X} \approx s\mathbf{L}$.

Let B be a Borel set in $\mathbb{R}^{p\times n}$. Then

$$\begin{aligned}
P(s\mathbf{L} \in B) &= P(s\mathbf{L} \in B|s=1)P(s=1)+P(s\mathbf{L} \in B|s=0)P(s=0) \\
&= P(\mathbf{L} \in B|s=1)P(s=1)+P(\mathbf{0} \in B|s=0)P(s=0) \\
&= P(\mathbf{L} \in B)(1-P_0)+P(\mathbf{0} \in B)P_0 \\
&= \frac{1}{1-P_0}P(\mathbf{X} \in (B-\{\mathbf{0}\}))(1-P_0)+\chi_B(\mathbf{0})P_0 \\
&= P(\mathbf{X} \in (B-\{\mathbf{0}\}))+\chi_B(\mathbf{0})P_0.
\end{aligned}$$

If $\mathbf{0} \in B$, then $B=(B-\{\mathbf{0}\})\cup\{\mathbf{0}\}$ and $\chi_B(\mathbf{0})=1$. Therefore

$$\begin{aligned}
P(\mathbf{X} \in (B-\{\mathbf{0}\}))+\chi_B(\mathbf{0})P_0 &= P(\mathbf{X} \in (B-\{\mathbf{0}\}))+P(\mathbf{X} \in \{\mathbf{0}\}) \\
&= P(\mathbf{X} \in B).
\end{aligned}$$

If $\mathbf{0} \notin B$, then $B=B-\{\mathbf{0}\}$ and $\chi_B(\mathbf{0})=0$. So,

$$P(\mathbf{X} \in (B-\{\mathbf{0}\}))+\chi_B(\mathbf{0})P_0 = P(\mathbf{X} \in B).$$

Hence, $\mathbf{X} \approx s\mathbf{L}$. Therefore, the characteristic functions of $\mathbf{X}$ and $s\mathbf{L}$ are equal, i.e. $\phi_{\mathbf{X}}(\mathbf{T})=\phi_{s\mathbf{L}}(\mathbf{T})$. Moreover,

$$\begin{aligned}
\phi_{s\mathbf{L}}(\mathbf{T}) &= E(etr(i\mathbf{T}'s\mathbf{L})) \\
&= E(etr(i\mathbf{T}'s\mathbf{L})|s=1)P(s=1)+E(etr(i\mathbf{T}'s\mathbf{L})|s=0)P(s=0) \\
&= E(etr(i\mathbf{T}'\mathbf{L})|s=1)(1-P_0)+E(1|s=0)P_0 \\
&= E(etr(i\mathbf{T}'\mathbf{L}))(1-P_0)+1\cdot P_0 \\
&= \phi_{\mathbf{L}}(\mathbf{T})(1-P_0)+P_0.
\end{aligned}$$

Hence, $\phi_{\mathbf{X}}(\mathbf{T})=\phi_{\mathbf{L}}(\mathbf{T})(1-P_0)+P_0$, and therefore

$$\phi_{\mathbf{L}}(\mathbf{T}) = \frac{\phi_{\mathbf{X}}(\mathbf{T})-P_0}{1-P_0} = \frac{\psi(tr(\mathbf{T}'\boldsymbol{\Sigma}\mathbf{T}\boldsymbol{\Phi}))-P_0}{1-P_0}.$$

Furthermore, $P(\mathbf{L}=\mathbf{0})=\frac{1}{1-P_0}P(\mathbf{X} \in (\{\mathbf{0}\}-\{\mathbf{0}\}))=0$. ■

Theorem 6.12. *Let $\mathbf{X} \sim E_{p,n}(\mathbf{0},\boldsymbol{\Sigma}_1 \otimes \mathbf{I}_n,\psi_1)$ with $p \leq n$. Assume*

$$\mathbf{X}\mathbf{X}' \sim G_{p,1}\left(\boldsymbol{\Sigma}_2,\frac{n}{2},\psi_2\right).$$

Then there exists $c>0$ such that $\boldsymbol{\Sigma}_2=c\boldsymbol{\Sigma}_1$ and $\psi_2(z)=\psi_1\left(\frac{z}{c}\right)$.

PROOF: It follows, from Definition 5.1, that there exists a $p \times n$ random matrix $\mathbf{Y}$, such that $\mathbf{Y} \sim E_{p,n}(\mathbf{0}, \Sigma_2 \otimes \mathbf{I}_n, \psi_2)$ and $\mathbf{YY}' \sim G_{p,1}\left(\Sigma_2, \frac{n}{2}, \psi_2\right)$. So, we have $\mathbf{XX}' \approx \mathbf{YY}'$. From Theorem 2.4, it suffices to show that $\mathbf{X} \approx \mathbf{Y}$.

First, note that if $\mathbf{Z}$ is a $p \times n$ matrix, then $\mathbf{Z} = \mathbf{0}$ iff $\mathbf{ZZ}' = \mathbf{0}$. Therefore, $P(\mathbf{X} = \mathbf{0}) = P(\mathbf{XX}' = \mathbf{0})$ and $P(\mathbf{Y} = \mathbf{0}) = P(\mathbf{YY}' = \mathbf{0})$. Since, $\mathbf{XX}' \approx \mathbf{YY}'$, we have $P(\mathbf{XX}' = \mathbf{0}) = P(\mathbf{YY}' = \mathbf{0})$ and so, $P(\mathbf{X} = \mathbf{0}) = P(\mathbf{Y} = \mathbf{0})$.

Let us denote $P(\mathbf{X} = \mathbf{0})$ by P_0. If $P_0 = 1$, then $P(\mathbf{X} = \mathbf{0}) = P(\mathbf{Y} = \mathbf{0}) = 1$, and hence, $\mathbf{X} \approx \mathbf{Y}$.

If $P_0 < 1$, then from Theorem 1.8, there exists a $p \times p$ nonsingular matrix $\mathbf{H}$, such that $\mathbf{H}\Sigma_2\mathbf{H}' = \mathbf{I}_p$ and $\mathbf{H}\Sigma_1\mathbf{H}' = \mathbf{D}$, where $\mathbf{D}$ is diagonal.

Let $\mathbf{V}_1 = \mathbf{HX}$ and $\mathbf{V}_2 = \mathbf{HY}$. Then, $\mathbf{V}_1 \sim E_{p,n}(\mathbf{0}, \mathbf{D} \otimes \mathbf{I}_n, \psi_1)$ and $\mathbf{V}_2 \sim E_{p,n}(\mathbf{0}, \mathbf{I}_p \otimes \mathbf{I}_n, \psi_2)$. Moreover, $\mathbf{V}_1\mathbf{V}_1' = \mathbf{HXX}'\mathbf{H}'$ and $\mathbf{V}_2\mathbf{V}_2' = \mathbf{HYY}'\mathbf{H}'$. So, we have $\mathbf{V}_1\mathbf{V}_1' \approx \mathbf{V}_2\mathbf{V}_2'$. It suffices to prove that $\mathbf{V}_1 \approx \mathbf{V}_2$, because if this is true, then $\mathbf{H}^{-1}\mathbf{V}_1 \approx \mathbf{H}^{-1}\mathbf{V}_2$ and so, $\mathbf{X} \approx \mathbf{Y}$.

Using Lemma 6.2, we can write $\mathbf{V}_i \approx s_i\mathbf{L}_i$, where s_i and $\mathbf{L}_i$ are independent, s_i is a one-dimensional random variable, with $P(s_i = 0) = P_0$, $P(s_i = 1) = 1 - P_0$, $P(\mathbf{L}_i = \mathbf{0}) = 0$ $(i = 1, 2)$. So $s_1 \approx s_2$. Moreover, we have

$$\mathbf{L}_1 \sim E_{p,n}\left(\mathbf{0}, \mathbf{D} \otimes \mathbf{I}_n, \frac{\psi_1 - P_0}{1 - P_0}\right) \quad \text{and} \quad \mathbf{L}_2 \sim E_{p,n}\left(\mathbf{0}, \mathbf{I}_p \otimes \mathbf{I}_n, \frac{\psi_2 - P_0}{1 - P_0}\right).$$

If $P_0 = 0$, then $P(s_1 = 1) = P(s_2 = 1) = 1$ and so, $\mathbf{L}_1\mathbf{L}_1' \approx s_1^2\mathbf{L}_1\mathbf{L}_1' \approx \mathbf{V}_1\mathbf{V}_1' \approx \mathbf{V}_2\mathbf{V}_2' \approx s_2^2\mathbf{L}_2\mathbf{L}_2' \approx \mathbf{L}_1\mathbf{L}_1'$. If $0 < P_0 < 1$, then for any Borel set B in $\mathbb{R}^{p \times n}$ we have

$$\begin{aligned} P(\mathbf{V}_i\mathbf{V}_i' \in B) &= P(s_i^2\mathbf{L}_i\mathbf{L}_i' \in B) \\ &= P(s_i^2\mathbf{L}_i\mathbf{L}_i' \in B | s_i = 1)P(s_i = 1) + P(s_i^2\mathbf{L}_i\mathbf{L}_i' \in B | s_i = 0)P(s_i = 0) \\ &= P(\mathbf{L}_i\mathbf{L}_i' \in B)(1 - P_0) + \chi_B(\mathbf{0})P_0. \end{aligned}$$

Therefore,

$$P(\mathbf{L}_i\mathbf{L}_i' \in B) = \frac{P(\mathbf{V}_i\mathbf{V}_i' \in B) - \chi_B(\mathbf{0})P_0}{1 - P_0} \quad (i = 1, 2).$$

Since, $\mathbf{V}_1\mathbf{V}_1' \approx \mathbf{V}_2\mathbf{V}_2'$, we have $P(\mathbf{V}_1\mathbf{V}_1' \in B) = P(\mathbf{V}_2\mathbf{V}_2' \in B)$ and so, $P(\mathbf{L}_1\mathbf{L}_1' \in B) = P(\mathbf{L}_2\mathbf{L}_2' \in B)$. Hence, $\mathbf{L}_1\mathbf{L}_1' \approx \mathbf{L}_2\mathbf{L}_2'$. Let $r_1\mathbf{D}^{\frac{1}{2}}\mathbf{U}_1$ and $r_2\mathbf{U}_2$ be the stochastic representations of $\mathbf{L}_1$ and $\mathbf{L}_2$, respectively. Then we have $\mathbf{L}_1\mathbf{L}_1' \approx r_1^2\mathbf{D}^{\frac{1}{2}}\mathbf{U}_1\mathbf{U}_1'\mathbf{D}^{\frac{1}{2}}$ and $\mathbf{L}_2\mathbf{L}_2' \approx r_2^2\mathbf{U}_2\mathbf{U}_2'$. Thus,

$$r_1^2\mathbf{D}^{\frac{1}{2}}\mathbf{U}_1\mathbf{U}_1'\mathbf{D}^{\frac{1}{2}} \approx r_2^2\mathbf{U}_2\mathbf{U}_2'. \tag{6.39}$$

Let $\mathbf{W}_1 = \mathbf{U}_1\mathbf{U}_1'$ and $\mathbf{W}_2 = \mathbf{U}_2\mathbf{U}_2'$. Since, $\mathbf{U}_1 \approx \mathbf{U}_2$, we have $\mathbf{W}_1 \approx \mathbf{W}_2$. Note that (6.39) can be rewritten as

$$r_1^2\mathbf{D}^{\frac{1}{2}}\mathbf{W}_1\mathbf{D}^{\frac{1}{2}} \approx r_2^2\mathbf{W}_2. \tag{6.40}$$

From Theorem 5.11, it follows that $P(rk(\mathbf{L}_1\mathbf{L}_1') = p) = 1$. Since, $\mathbf{L}_1\mathbf{L}_1'$ is a positive semidefinite $p \times p$ matrix, we get $P(\mathbf{L}_1\mathbf{L}_1' > \mathbf{0}) = 1$. Now $\mathbf{L}_1\mathbf{L}_1' \approx \mathbf{L}_2\mathbf{L}_2'$, and hence we have $P(\mathbf{L}_2\mathbf{L}_2' > \mathbf{0}) = 1$. Therefore, $P(r_2^2\mathbf{W}_2 > \mathbf{0}) = 1$ and $P(r_1^2\mathbf{D}^{\frac{1}{2}}\mathbf{W}_1\mathbf{D}^{\frac{1}{2}} > \mathbf{0}) = 1$. Thus, the diagonal elements of $r_2^2\mathbf{W}_2$ and $r_1^2\mathbf{D}^{\frac{1}{2}}\mathbf{W}_1\mathbf{D}^{\frac{1}{2}}$ are positive, with probability one. If $p = 1$, then $\mathbf{D}$ is a scalar; $\mathbf{D} = c$. If $p > 1$, then it follows from (6.40) that

$$\frac{(r_2^2\mathbf{W}_2)_{11}}{(r_2^2\mathbf{W}_2)_{ii}} \approx \frac{\left(r_1^2\mathbf{D}^{\frac{1}{2}}\mathbf{W}_1\mathbf{D}^{\frac{1}{2}}\right)_{11}}{\left(r_1^2\mathbf{D}^{\frac{1}{2}}\mathbf{W}_1\mathbf{D}^{\frac{1}{2}}\right)_{ii}}, \quad i = 2,\ldots,p,$$

or equivalently,

$$\frac{(\mathbf{W}_2)_{11}}{(\mathbf{W}_2)_{ii}} \approx \frac{\left(\mathbf{D}^{\frac{1}{2}}\mathbf{W}_1\mathbf{D}^{\frac{1}{2}}\right)_{11}}{\left(\mathbf{D}^{\frac{1}{2}}\mathbf{W}_1\mathbf{D}^{\frac{1}{2}}\right)_{ii}}. \tag{6.41}$$

However, $\mathbf{D}^{\frac{1}{2}}$ is diagonal so, $\left(\mathbf{D}^{\frac{1}{2}}\mathbf{W}_1\mathbf{D}^{\frac{1}{2}}\right)_{jj} = (\mathbf{W}_1)_{jj}d_{jj}$, $j = 1,\ldots,p$ and (6.41) becomes

$$\frac{(\mathbf{W}_2)_{11}}{(\mathbf{W}_2)_{ii}} \approx \frac{(\mathbf{W}_1)_{11}}{(\mathbf{W}_1)_{ii}}\frac{d_{11}}{d_{ii}}.$$

Since $\mathbf{W}_1 \approx \mathbf{W}_2$, we have $\frac{(\mathbf{W}_2)_{11}}{(\mathbf{W}_2)_{ii}} \approx \frac{(\mathbf{W}_1)_{11}}{(\mathbf{W}_1)_{ii}}$ and so, $\frac{(\mathbf{W}_1)_{11}}{(\mathbf{W}_1)_{ii}} \approx \frac{(\mathbf{W}_1)_{11}}{(\mathbf{W}_1)_{ii}}\frac{d_{11}}{d_{ii}}$. Since $P\left(\frac{(\mathbf{W}_1)_{11}}{(\mathbf{W}_1)_{ii}} > 0\right) = 1$, this is possible only if $\frac{d_{11}}{d_{ii}} = 1$, $i = 2,\ldots,p$. So, we get $\mathbf{D} = c\mathbf{I}_p$ where c is a scalar constant. From (6.39), we get

$$cr_1^2\mathbf{U}_1\mathbf{U}_1' \approx r_2^2\mathbf{U}_2\mathbf{U}_2'.$$

Taking trace on both sides, we get

$$tr(cr_1^2\mathbf{U}_1\mathbf{U}_1') \approx tr(r_2^2\mathbf{U}_2\mathbf{U}_2')$$

and hence,

$$cr_1^2 tr(\mathbf{U}_1\mathbf{U}_1') \approx r_2^2 tr(\mathbf{U}_2\mathbf{U}_2').$$

Now, $tr(\mathbf{U}_1\mathbf{U}_1') = tr(\mathbf{U}_2\mathbf{U}_2') = 1$ and therefore, $cr_1^2 \approx r_2^2$ and $r_2 \approx \sqrt{c}r_1$. Let $r_3 \approx r_2$, such that r_3 is independent of $\mathbf{U}_1$ and $\mathbf{U}_2$. Then, we have $\mathbf{L}_1 \approx r_1\mathbf{D}^{\frac{1}{2}}\mathbf{U}_1 \approx \frac{1}{\sqrt{c}}r_3(c\mathbf{I}_p)^{\frac{1}{2}}\mathbf{U}_1 = r_3\mathbf{U}_1 \approx r_3\mathbf{U}_2$. Since $\mathbf{L}_2 \approx r_2\mathbf{U}_2 \approx r_3\mathbf{U}_2$, we have $\mathbf{L}_1 \approx \mathbf{L}_2$. Since $s_1 \approx s_2$, we get $s_1\mathbf{L}_1 \approx s_2\mathbf{L}_2$. Therefore $\mathbf{V}_1 \approx \mathbf{V}_2$. ■

Now we can prove a result on the characterization of normality.

Theorem 6.13. *Let* $\mathbf{X} \sim E_{p,n}(\mathbf{0}, \Sigma_1 \otimes \mathbf{I}_n, \psi_1)$, *with* $p \leq n$. *Assume* $\mathbf{XX}' \sim W_p(\Sigma_2, n)$. *Then,* $\mathbf{X} \sim N_{p,n}(\mathbf{0}, \Sigma_1 \otimes \mathbf{I}_n)$ *and* $\Sigma_1 = \Sigma_2$.

PROOF: The result follows immediately by taking $\psi_2(z) = \exp\left(-\frac{z}{2}\right)$, in Theorem 6.12.

The following theorem is an extension of Theorem 6.12.

Theorem 6.14. *Let* $\mathbf{X} \sim E_{p,n}(\mathbf{0}, \Sigma_1 \otimes \Phi, \psi_1)$ *and* $\mathbf{Y} \sim E_{p,n}(\mathbf{0}, \Sigma_2 \otimes \Phi, \psi_2)$, *with* $\Sigma_1 > \mathbf{0}$, $\Sigma_2 \geq \mathbf{0}$, *and* $\Phi > \mathbf{0}$. *Assume* $\mathbf{XAX}' \approx \mathbf{YAY}'$ *where* $\mathbf{A}$ *is an* $n \times n$ *positive semidefinite matrix, with* $rk(\mathbf{A}) \geq p$. *Suppose that all moments of* $\mathbf{X}$ *exist. Let* $x = tr(\mathbf{X}'\Sigma_1^{-1}\mathbf{X}\Phi^{-1})$ *and define* $m_k = E(x^k)$, $k = 1, 2, \ldots$. *If* $\sum_{k=1}^{\infty} \left(\frac{1}{m_{2k}}\right)^{\frac{1}{2k}} = \infty$, *then* $\mathbf{X} \approx \mathbf{Y}$.

PROOF: Without loss of generality, we can assume that $\Phi = \mathbf{I}_n$. Indeed, if $\Phi \neq \mathbf{I}_n$ then, define $\mathbf{X}_1 = \mathbf{X}\Phi^{-\frac{1}{2}}$, $\mathbf{Y}_1 = \mathbf{Y}\Phi^{-\frac{1}{2}}$, and $\mathbf{A}_1 = \Phi^{\frac{1}{2}}\mathbf{A}\Phi^{\frac{1}{2}}$. Then, $\mathbf{X}_1 \sim E_{p,n}(\mathbf{0}, \Sigma_1 \otimes \mathbf{I}_n, \psi_1)$, $\mathbf{Y}_1 \sim E_{p,n}(\mathbf{0}, \Sigma_2 \otimes \mathbf{I}_n, \psi_2)$, $rk(\mathbf{A}_1) = rk(\mathbf{A}) \geq p$, $\mathbf{XAX}' = \mathbf{X}_1\mathbf{A}_1\mathbf{X}_1'$, and $\mathbf{YAY}' = \mathbf{Y}_1\mathbf{A}_1\mathbf{Y}_1'$. So $\mathbf{XAX}' \approx \mathbf{YAY}'$ if and only if $\mathbf{X}_1\mathbf{A}_1\mathbf{X}_1' \approx \mathbf{Y}_1\mathbf{A}_1\mathbf{Y}_1'$ and $\mathbf{X} \approx \mathbf{Y}$ if and only if $\mathbf{X}_1 \approx \mathbf{Y}_1$.

Hence, we can assume $\Phi = \mathbf{I}_n$. From Theorem 1.8, there exists a $p \times p$ nonsingular matrix $\mathbf{H}$ such that $\mathbf{H}\Sigma_1\mathbf{H}' = \mathbf{I}_p$ and $\mathbf{H}\Sigma_2\mathbf{H}' = \mathbf{D}$, where $\mathbf{D}$ is diagonal. Let $\mathbf{Z}_1 = \mathbf{HX}$ and $\mathbf{Z}_2 = \mathbf{HY}$. Then, $\mathbf{Z}_1 \sim E_{p,n}(\mathbf{0}, \mathbf{I}_p \otimes \mathbf{I}_n, \psi_1)$ and $\mathbf{Z}_2 \sim E_{p,n}(\mathbf{0}, \mathbf{D} \otimes \mathbf{I}_n, \psi_2)$. Moreover, $\mathbf{Z}_1\mathbf{AZ}_1' = \mathbf{HXAX}'\mathbf{H}'$ and $\mathbf{Z}_2\mathbf{AZ}_2' = \mathbf{HYAY}'\mathbf{H}'$. Since $\mathbf{XAX}' \approx \mathbf{YAY}'$, we have $\mathbf{Z}_1\mathbf{AZ}_1' \approx \mathbf{Z}_2\mathbf{AZ}_2'$. It suffices to prove that $\mathbf{Z}_1 \approx \mathbf{Z}_2$, because if this is true, then $\mathbf{H}^{-1}\mathbf{Z}_1 \approx \mathbf{H}^{-1}\mathbf{Z}_2$ and therefore $\mathbf{X} \approx \mathbf{Y}$. Since $\mathbf{Z}_1\mathbf{AZ}_1' \approx \mathbf{Z}_2\mathbf{AZ}_2'$, we get $P(\mathbf{Z}_1\mathbf{AZ}_1' = \mathbf{0}) = P(\mathbf{Z}_2\mathbf{AZ}_2' = \mathbf{0})$.

If $P(\mathbf{Z}_1 = \mathbf{0}) < 1$ then, using Lemma 6.2, we can write $\mathbf{Z}_1 = s_1\mathbf{L}_1$ where s_1 and $\mathbf{L}_1$ are independent, s_1 is a one-dimensional random variable with $P(s_1 = 0) = P(\mathbf{Z}_1 = \mathbf{0})$, $P(s_1 = 1) = 1 - P(\mathbf{Z}_1 = \mathbf{0})$, $\mathbf{L}_1 \sim E_{p,n}\left(\mathbf{0}, \mathbf{I}_p \otimes \mathbf{I}_n, \frac{\psi_1 - P(\mathbf{Z}_1 = \mathbf{0})}{1 - P(\mathbf{Z}_1 = \mathbf{0})}\right)$ and $P(\mathbf{L}_1 = \mathbf{0}) = 0$. Then, from Theorem 5.11, it follows that $P(rk(\mathbf{L}_1\mathbf{AL}_1') = p) = 1$. Since, $\mathbf{A}$ is positive definite, this implies that

$$P(\mathbf{L}_1\mathbf{AL}_1' > \mathbf{0}) = 1. \tag{6.42}$$

Consequently $P(\mathbf{L}_1\mathbf{AL}_1' = \mathbf{0}) = 0$. Moreover,

$$\begin{aligned} P(\mathbf{Z}_1\mathbf{AZ}_1' = \mathbf{0}) &= P(s_1^2\mathbf{L}_1\mathbf{AL}_1' = \mathbf{0}) \\ &= P(s_1^2 = 0) \\ &= P(s_1 = 0) \\ &= P(\mathbf{Z}_1 = \mathbf{0}). \end{aligned}$$

Hence, $P(\mathbf{Z}_1\mathbf{AZ}_1' = \mathbf{0}) < 1$.

If $P(\mathbf{Z}_2 = \mathbf{0}) < 1$ then, using Lemma 6.2, we can write $\mathbf{Z}_2 = s_2\mathbf{L}_2$ where s_2 and $\mathbf{L}_2$ are independent, s_2 is a one-dimensional random variable with $P(s_2 = 0) = P(\mathbf{Z}_2 = \mathbf{0})$, $P(s_2 = 1) = 1 - P(\mathbf{Z}_2 = \mathbf{0})$, $\mathbf{L}_2 \sim E_{p,n}\left(\mathbf{0}, \mathbf{D} \otimes \mathbf{I}_n, \frac{\psi_2 - P(\mathbf{Z}_2=\mathbf{0})}{1-P(\mathbf{Z}_2=\mathbf{0})}\right)$ and $P(\mathbf{L}_2 = \mathbf{0}) = 0$. Then, from Theorem 5.10, it follows that $P(rk(\mathbf{L}_2\mathbf{A}\mathbf{L}_2') = \min(rk(\mathbf{D}), p)) = 1$. From $P(\mathbf{Z}_2 = \mathbf{0}) < 1$, it follows that $rk(\mathbf{D}) \geq 1$. Hence,

$$P(rk(\mathbf{L}_2\mathbf{A}\mathbf{L}_2') \geq 1) = 1.$$

Thus, $P(\mathbf{L}_2\mathbf{A}\mathbf{L}_2' = \mathbf{0}) = 0$. Therefore,

$$\begin{aligned} P(\mathbf{Z}_2\mathbf{A}\mathbf{Z}_2' = \mathbf{0}) &= P(s_2^2\mathbf{L}_2\mathbf{A}\mathbf{L}_2' = \mathbf{0}) \\ &= P(s_2^2 = 0) \\ &= P(s_2 = 0) \\ &= P(\mathbf{Z}_2 = \mathbf{0}). \end{aligned}$$

Hence, $P(\mathbf{Z}_2\mathbf{A}\mathbf{Z}_2' = \mathbf{0}) < 1$. Therefore, if either $P(\mathbf{Z}_1 = \mathbf{0}) < 1$ or $P(\mathbf{Z}_2 = \mathbf{0}) < 1$, then we get $P(\mathbf{Z}_1\mathbf{A}\mathbf{Z}_1' = \mathbf{0}) = P(\mathbf{Z}_2\mathbf{A}\mathbf{Z}_2' = \mathbf{0}) < 1$ and hence, $P(\mathbf{Z}_i = \mathbf{0}) < 1$ must hold for $i = 1, 2$. However, then we get

$$\begin{aligned} P(\mathbf{Z}_1 = \mathbf{0}) &= P(\mathbf{Z}_1\mathbf{A}\mathbf{Z}_1' = \mathbf{0}) \\ &= P(\mathbf{Z}_2\mathbf{A}\mathbf{Z}_2' = \mathbf{0}) \\ &= P(\mathbf{Z}_2 = \mathbf{0}) \end{aligned}$$

and $P(s_1 = 0) = P(s_2 = 0)$. Hence, $s_1 \approx s_2$.

If either $P(\mathbf{Z}_1 = \mathbf{0}) = 1$ or $P(\mathbf{Z}_2 = \mathbf{0}) = 1$, then we must have $P(\mathbf{Z}_i = \mathbf{0}) = 1$ for $i = 1, 2$. Therefore, $P(\mathbf{Z}_1 = \mathbf{0}) = P(\mathbf{Z}_2 = \mathbf{0})$ is always true. Let $P_0 = P(\mathbf{Z}_1 = \mathbf{0})$.

If $P_0 = 1$, then $\mathbf{Z}_1 = \mathbf{0} = \mathbf{Z}_2$ and the theorem is proved.

If $P_0 < 1$, then we first prove that

$$\mathbf{L}_1\mathbf{A}\mathbf{L}_1' \approx \mathbf{L}_2\mathbf{A}\mathbf{L}_2'. \tag{6.43}$$

If $P_0 = 0$, then $P(s_1 = 1) = P(s_2 = 1) = 1$ and so,

$$\mathbf{L}_1\mathbf{A}\mathbf{L}_1' \approx s_1\mathbf{L}_1\mathbf{A}\mathbf{L}_1' \approx \mathbf{Z}_1\mathbf{A}\mathbf{Z}_1' \approx \mathbf{Z}_2\mathbf{A}\mathbf{Z}_2' \approx s_2\mathbf{L}_2\mathbf{A}\mathbf{L}_2' \approx \mathbf{L}_2\mathbf{A}\mathbf{L}_2'.$$

If $0 < P_0 < 1$, then for any Borel set B in $\mathbb{R}^{p\times n}$, we have

$$\begin{aligned} P(\mathbf{Z}_1\mathbf{A}\mathbf{Z}_1' \in B) &= P(s_1\mathbf{L}_1\mathbf{A}\mathbf{L}_1' \in B) \\ &= P(s_1\mathbf{L}_1\mathbf{A}\mathbf{L}_1' \in B | s_1 = 1)P(s_1 = 1) \\ &+ P(s_1\mathbf{L}_1\mathbf{A}\mathbf{L}_1' \in B | s_1 = 0)P(s_1 = 0) \\ &= P(\mathbf{L}_1\mathbf{A}\mathbf{L}_1' \in B)(1 - P_0) + \chi_B(\mathbf{0})P_0. \end{aligned}$$

Therefore,

$$P(\mathbf{L}_1\mathbf{A}\mathbf{L}_1' \in B) = \frac{P(\mathbf{Z}_1\mathbf{A}\mathbf{Z}_1' \in B) - \chi_B(\mathbf{0})P_0}{1-P_0}.$$

Similarly,

$$P(\mathbf{L}_2\mathbf{A}\mathbf{L}_2' \in B) = \frac{P(\mathbf{Z}_2\mathbf{A}\mathbf{Z}_2' \in B) - \chi_B(\mathbf{0})P_0}{1-P_0}.$$

Since, $\mathbf{Z}_1\mathbf{A}\mathbf{Z}_1' \approx \mathbf{Z}_2\mathbf{A}\mathbf{Z}_2'$, we get $P(\mathbf{Z}_1\mathbf{A}\mathbf{Z}_1' \in B) = P(\mathbf{Z}_2\mathbf{A}\mathbf{Z}_2' \in B)$ and so, $P(\mathbf{L}_1\mathbf{A}\mathbf{L}_1' \in B) = P(\mathbf{L}_2\mathbf{A}\mathbf{L}_2' \in B)$. Therefore, $\mathbf{L}_1\mathbf{A}\mathbf{L}_1' \approx \mathbf{L}_2\mathbf{A}\mathbf{L}_2'$ which establishes (6.43).

Let $r_1\mathbf{U}_1$ and $r_2\mathbf{D}^{\frac{1}{2}}\mathbf{U}_2$ be the stochastic representations of $\mathbf{L}_1$ and $\mathbf{L}_2$. Here $\mathbf{D}$ is diagonal. Then, we have $\mathbf{L}_1\mathbf{A}\mathbf{L}_1' \approx r_1^2\mathbf{U}_1\mathbf{A}\mathbf{U}_1'$ and $\mathbf{L}_2\mathbf{A}\mathbf{L}_2' \approx r_2^2\mathbf{D}^{\frac{1}{2}}\mathbf{U}_2\mathbf{A}\mathbf{U}_2'\mathbf{D}^{\frac{1}{2}}$. Hence,

$$r_1^2\mathbf{U}_1\mathbf{A}\mathbf{U}_1' \approx r_2^2\mathbf{D}^{\frac{1}{2}}\mathbf{U}_2\mathbf{A}\mathbf{U}_2'\mathbf{D}^{\frac{1}{2}}. \tag{6.44}$$

Let $\mathbf{W}_1 = \mathbf{U}_1\mathbf{A}\mathbf{U}_1'$ and $\mathbf{W}_2 = \mathbf{U}_2\mathbf{A}\mathbf{U}_2'$. Since $\mathbf{U}_1 \approx \mathbf{U}_2$, we have $\mathbf{W}_1 \approx \mathbf{W}_2$. So, (6.44) can be rewritten as

$$r_1^2\mathbf{W}_1 \approx r_2^2\mathbf{D}^{\frac{1}{2}}\mathbf{W}_2\mathbf{D}^{\frac{1}{2}}. \tag{6.45}$$

If $p = 1$, then $\mathbf{D}$ is a scalar; $\mathbf{D} = c\mathbf{I}_1$. If $p > 1$, then from (6.42) it follows that the diagonal elements of $\mathbf{L}_1\mathbf{A}\mathbf{L}_1'$ are positive, with probability one. From (6.43), it is seen that $P(\mathbf{L}_2\mathbf{A}\mathbf{L}_2' > \mathbf{0}) = 1$ and the diagonal elements of $\mathbf{L}_2\mathbf{A}\mathbf{L}_2'$ are also positive with probability one. Using (6.45), we obtain

$$\frac{(r_1^2\mathbf{W}_1)_{11}}{(r_1^2\mathbf{W}_1)_{ii}} \approx \frac{\left(r_2^2\mathbf{D}^{\frac{1}{2}}\mathbf{W}_2\mathbf{D}^{\frac{1}{2}}\right)_{11}}{\left(r_2^2\mathbf{D}^{\frac{1}{2}}\mathbf{W}_2\mathbf{D}^{\frac{1}{2}}\right)_{ii}}, \quad i = 2,\ldots,p,$$

or equivalently,

$$\frac{(\mathbf{W}_1)_{11}}{(\mathbf{W}_1)_{ii}} \approx \frac{\left(\mathbf{D}^{\frac{1}{2}}\mathbf{W}_2\mathbf{D}^{\frac{1}{2}}\right)_{11}}{\left(\mathbf{D}^{\frac{1}{2}}\mathbf{W}_2\mathbf{D}^{\frac{1}{2}}\right)_{ii}}. \tag{6.46}$$

However $\mathbf{D}^{\frac{1}{2}}$ is diagonal and hence, $\left(\mathbf{D}^{\frac{1}{2}}\mathbf{W}_2\mathbf{D}^{\frac{1}{2}}\right)_{jj} = (\mathbf{W}_2)_{jj}d_{jj}$, $j = 1,\ldots,p$. Therefore, (6.46) becomes

$$\frac{(\mathbf{W}_1)_{11}}{(\mathbf{W}_1)_{ii}} \approx \frac{(\mathbf{W}_2)_{11}}{(\mathbf{W}_2)_{ii}}\frac{d_{11}}{d_{ii}}.$$

Since $\mathbf{W}_1 \approx \mathbf{W}_2$, we have $\frac{(\mathbf{W}_1)_{11}}{(\mathbf{W}_1)_{ii}} \approx \frac{(\mathbf{W}_2)_{11}}{(\mathbf{W}_2)_{ii}}$ and so, $\frac{(\mathbf{W}_2)_{11}}{(\mathbf{W}_2)_{ii}} \approx \frac{(\mathbf{W}_2)_{11}}{(\mathbf{W}_2)_{ii}} \frac{d_{11}}{d_{ii}}$. Now $P\left(\frac{(\mathbf{W}_2)_{11}}{(\mathbf{W}_2)_{ii}} > 0\right) = 1$, which is possible only if $\frac{d_{11}}{d_{ii}} = 1$, $i = 2, \ldots, p$. Consequently we get $\mathbf{D} = c\mathbf{I}_p$, where c is a scalar constant. From (6.45), we get

$$r_1^2 \mathbf{W}_1 \approx r_2^2 c \mathbf{W}_2.$$

Taking trace on both sides, we have $tr(r_1^2\mathbf{W}_1) \approx tr(r_2^2 c\mathbf{W}_2)$, and hence,

$$r_1^2 tr(\mathbf{W}_1) \approx r_2^2 c tr(\mathbf{W}_2). \tag{6.47}$$

Since, $tr(\mathbf{U}_1\mathbf{U}_1') = 1$, all the elements of $\mathbf{U}_1$ are less than 1. Therefore, there exists a positive constant K such that $tr(\mathbf{U}_1\mathbf{U}_1') < K$. From Theorem 5.10, it follows that $P(rk(\mathbf{U}_1\mathbf{A}\mathbf{U}_1') = p) = 1$. Consequently $\mathbf{U}_1\mathbf{A}\mathbf{U}_1' > \mathbf{0}$ with probability one. Therefore, $E((tr(\mathbf{W}_1))^k)$ is a finite positive number for $k = 1, 2, \ldots$. From (6.47), it follows that

$$E(r_1^{2k})E((tr(\mathbf{W}_1))^k) = E((cr_2^2)^k)E((tr(\mathbf{W}_2))^k), \quad k = 1, 2, \ldots.$$

Hence,

$$E(r_1^{2k}) = E((cr_2^2)^k), \quad k = 1, 2, \ldots. \tag{6.48}$$

Since, $\mathbf{Z}_1 \approx s_1\mathbf{L}_1$ and $\mathbf{L}_1 \approx r_1\mathbf{U}_1$, we can write $\mathbf{Z}_1 \approx s_1 r_1 \mathbf{U}_1$, where s_1, r_1, and $\mathbf{U}_1$ are independent.

Similarly, $\mathbf{Z}_2 \approx s_2\sqrt{c} r_2 \mathbf{U}_2$, with s_2, r_2, $\mathbf{U}_2$ independent. Since $s_1 \approx s_2$, we have

$$E(s_1^{2k}) = E(s_2^{2k}) = (1 - P_0), \quad k = 0, 1, 2, \ldots. \tag{6.49}$$

From (6.48) and (6.49), it follows that

$$E\left(\left(\frac{1}{c} s_1^2 r_1^2\right)^k\right) = E\left((s_2^2 r_2^2)^k\right).$$

Now,

$$\begin{aligned} x &= tr(\mathbf{X}'\Sigma_1^{-1}\mathbf{X}) \\ &= tr(\mathbf{Z}_1'\mathbf{H}'^{-1}\Sigma_1^{-1}\mathbf{H}^{-1}\mathbf{Z}_1) \\ &= tr(\mathbf{Z}_1'\mathbf{Z}_1) \\ &= s_1^2 r_1^2 tr(\mathbf{U}_1'\mathbf{U}_1) \\ &= s_1^2 r_1^2. \end{aligned}$$

Thus, $m_k = E\left((s_1^2r_1^2)^k\right) = E\left((cs_2^2r_2^2)^k\right)$. However, if m_k is the kth moment of a random variable and $\sum_{k=1}^{\infty}\left(\frac{1}{m_{2k}}\right)^{\frac{1}{2k}} = \infty$, then the distribution of the random variable is uniquely determined (see Rao, 1973, p. 106). Thus we have $s_1^2r_1^2 \approx cs_2^2r_2^2$. Therefore, $s_1r_1 \approx \sqrt{c}s_2r_2$, and hence $\mathbf{Z}_1 \approx \mathbf{Z}_2$. ■

Part III
Estimation and Hypothesis Testing

Chapter 7
Estimation

7.1 Maximum Likelihood Estimators of the Parameters

Let $\mathbf{X} \sim E_{p,n}(\mathbf{M}, \Sigma \otimes \Phi, \psi)$, with p.d.f.

$$f(\mathbf{X}) = \frac{1}{|\Sigma|^{\frac{n}{2}}|\Phi|^{\frac{p}{2}}} h(tr(\mathbf{X}-\mathbf{M})'\Sigma^{-1}(\mathbf{X}-\mathbf{M})\Phi^{-1}).$$

Let $n \geq p$ and assume h is known. We want to estimate $\mathbf{M}$, Σ, and Φ based on a single observation from $\mathbf{X}$.

First, we show that, without imposing some restrictions on $\mathbf{M}$, Σ, and Φ, indeed the maximum likelihood estimators (MLE's) do not exist.

Let a be a positive number for which $h(a) \neq 0$, and Φ be any $n \times n$ positive definite matrix. Let s be a positive number, and define $k = \sqrt{\frac{sa}{p}}$. Let $\mathbf{V}$ be a $p \times n$ matrix such that $\mathbf{V} = k(\mathbf{I}_p, \mathbf{0})$, $\mathbf{M} = \mathbf{X} - \mathbf{V}\Phi^{\frac{1}{2}}$, and $\Sigma = s\mathbf{I}_p$. Then, $f(\mathbf{X})$ can be expressed as

$$\begin{aligned} f(\mathbf{X}) &= \frac{1}{|s^p|^{\frac{n}{2}}|\Phi|^{\frac{p}{2}}} h(tr(\Phi^{\frac{1}{2}}\mathbf{V}'\Sigma^{-1}\mathbf{V}\Phi^{\frac{1}{2}}\Phi^{-1})) \\ &= \frac{1}{s^{\frac{pn}{2}}|\Phi|^{\frac{p}{2}}} h(tr(k^2(s\mathbf{I}_p)^{-1}\mathbf{I}_p)) \\ &= \frac{1}{s^{\frac{pn}{2}}|\Phi|^{\frac{p}{2}}} h\left(\frac{k^2}{s}p\right) \\ &= \frac{h(a)}{s^{\frac{pn}{2}}|\Phi|^{\frac{p}{2}}}. \end{aligned}$$

A.K. Gupta et al., *Elliptically Contoured Models in Statistics and Portfolio Theory*,
DOI 10.1007/978-1-4614-8154-6_7, © Springer Science+Business Media New York 2013

Therefore, if $s \to 0$, then $f(\mathbf{X}) \to \infty$. Hence, the MLE's do not exist. This example shows that even if Φ is known, there are no MLE's for $\mathbf{M}$ and Σ. Thus, we have to restrict the parameter space, in order to get the MLE's. The following lemma will be needed to obtain the estimation results.

Lemma 7.1. *Let* $\mathbf{X}$ *be a* $p \times n$ *matrix,* μ *a* p*-dimensional vector,* $\mathbf{v}$ *an* n*-dimensional vector and* $\mathbf{A}$ *an* $n \times n$ *symmetric matrix, such that* $\mathbf{v}'\mathbf{A}\mathbf{v} \neq 0$. *Then,*

$$(\mathbf{X}-\mu\mathbf{v}')\mathbf{A}(\mathbf{X}-\mu\mathbf{v}')' = \mathbf{X}\left(\mathbf{A}-\frac{\mathbf{A}\mathbf{v}\mathbf{v}'\mathbf{A}}{\mathbf{v}'\mathbf{A}\mathbf{v}}\right)\mathbf{X}'$$
$$+ (\mathbf{v}'\mathbf{A}\mathbf{v})\left(\mathbf{X}\frac{\mathbf{A}\mathbf{v}}{\mathbf{v}'\mathbf{A}\mathbf{v}}-\mu\right)\left(\mathbf{X}\frac{\mathbf{A}\mathbf{v}}{\mathbf{v}'\mathbf{A}\mathbf{v}}-\mu\right)'. \tag{7.1}$$

PROOF: The right hand side of (7.1) equals

$$\begin{aligned}
&\mathbf{X}\mathbf{A}\mathbf{X}' - \mathbf{X}\frac{\mathbf{A}\mathbf{v}\mathbf{v}'\mathbf{A}}{\mathbf{v}'\mathbf{A}\mathbf{v}}\mathbf{X}' \\
&+ (\mathbf{v}'\mathbf{A}\mathbf{v})\left(\mathbf{X}\frac{\mathbf{A}\mathbf{v}\mathbf{v}'\mathbf{A}}{(\mathbf{v}'\mathbf{A}\mathbf{v})^2}\mathbf{X}' - \mathbf{X}\frac{\mathbf{A}\mathbf{v}}{\mathbf{v}'\mathbf{A}\mathbf{v}}\mu' - \mu\frac{\mathbf{v}'\mathbf{A}}{\mathbf{v}'\mathbf{A}\mathbf{v}} + \mu\mu'\right)' \\
&= \mathbf{X}\mathbf{A}\mathbf{X}' - \mathbf{X}\mathbf{A}\mathbf{v}\mu' - \mu\mathbf{v}'\mathbf{A}\mathbf{X}' + (\mathbf{v}'\mathbf{A}\mathbf{v})\mu\mu' \\
&= (\mathbf{X}-\mu\mathbf{v}')\mathbf{A}(\mathbf{X}-\mu\mathbf{v}')',
\end{aligned}$$

which is the left hand side of (7.1). ■

Now denote by $\mathbf{x}_i$ the ith column of the matrix $\mathbf{X}$, then

$$\sum_{i=1}^{n}(\mathbf{x}_i-\mu)(\mathbf{x}_i-\mu)' = (\mathbf{X}-\mu\mathbf{e}_n')(\mathbf{X}-\mu\mathbf{e}_n')'$$

and since $\bar{\mathbf{x}} = \frac{1}{n}\sum_{i=1}^{n}\mathbf{x}_i = \frac{1}{n}\mathbf{X}\mathbf{e}_n$, we have

$$\begin{aligned}
\sum_{i=1}^{n}(\mathbf{x}_i-\bar{\mathbf{x}})(\mathbf{x}_i-\bar{\mathbf{x}})' &= (\mathbf{X}-\bar{\mathbf{x}}\mathbf{e}_n')(\mathbf{X}-\bar{\mathbf{x}}\mathbf{e}_n')' \\
&= \left(\mathbf{X}-\frac{1}{n}\mathbf{X}\mathbf{e}_n\mathbf{e}_n'\right)\left(\mathbf{X}-\frac{1}{n}\mathbf{X}\mathbf{e}_n\mathbf{e}_n'\right)' \\
&= \mathbf{X}\left(\mathbf{I}_n-\frac{1}{n}\mathbf{e}_n\mathbf{e}_n'\right)\left(\mathbf{I}_n-\frac{1}{n}\mathbf{e}_n\mathbf{e}_n'\right)'\mathbf{X}' \\
&= \mathbf{X}\left(\mathbf{I}_n-\frac{1}{n}\mathbf{e}_n\mathbf{e}_n'\right)\mathbf{X}'.
\end{aligned}$$

Choosing $\mathbf{v} = \mathbf{e}_n$, and $\mathbf{A} = \mathbf{I}_n$, we have $\mathbf{v}'\mathbf{A}\mathbf{v} = n$, $\mathbf{A} - \frac{\mathbf{A}\mathbf{v}\mathbf{v}'\mathbf{A}}{\mathbf{v}'\mathbf{A}\mathbf{v}} = \mathbf{I}_n - \frac{1}{n}\mathbf{e}_n\mathbf{e}_n'$, and $\mathbf{X}\frac{\mathbf{A}\mathbf{v}}{\mathbf{v}'\mathbf{A}\mathbf{v}} = \frac{1}{n}\mathbf{e}_n$. Then, from Lemma 7.1, we get

$$\sum_{i=1}^{n}(\mathbf{x}_i - \mu)(\mathbf{x}_i - \mu)' = \sum_{i=1}^{n}(\mathbf{x}_i - \bar{\mathbf{x}})(\mathbf{x}_i - \bar{\mathbf{x}})' + n(\bar{\mathbf{x}} - \mu)(\bar{\mathbf{x}} - \mu)',$$

which is a well-known identity.

Theorem 7.1. *Let* $\mathbf{X} \sim E_{p,n}(\mathbf{M}, \Sigma \otimes \Phi, \psi)$, *with p.d.f.*

$$f(\mathbf{X}) = \frac{1}{|\Sigma|^{\frac{n}{2}}|\Phi|^{\frac{p}{2}}} h(tr((\mathbf{X}-\mathbf{M})'\Sigma^{-1}(\mathbf{X}-\mathbf{M})\Phi^{-1})),$$

where $h(z)$ *is monotone decreasing on* $[0,\infty)$. *Suppose* h, Σ, *and* Φ *are known and we want to find the MLE of* $\mathbf{M}$ *(say* $\hat{\mathbf{M}}$*), based on a single observation* $\mathbf{X}$. *Then,*

(a) $\hat{\mathbf{M}} = \mathbf{X}$,
(b) If $\mathbf{M} = \mu\mathbf{v}'$, *where* μ *is p-dimensional,* $\mathbf{v}$ *is n-dimensional vector and* $\mathbf{v} \neq \mathbf{0}$ *is known, the MLE of* μ *is* $\hat{\mu} = \mathbf{X}\frac{\Phi^{-1}\mathbf{v}}{\mathbf{v}'\Phi^{-1}\mathbf{v}}$, *and*
(c) If $\mathbf{M}$ *is of the form* $\mathbf{M} = \mu\mathbf{e}_n'$, *the MLE of* μ *is* $\hat{\mu} = \mathbf{X}\frac{\Phi^{-1}\mathbf{e}_n}{\mathbf{e}_n'\Phi^{-1}\mathbf{e}_n}$.

PROOF:

(a) It holds that

$$\begin{aligned} f(\mathbf{X}) &= \frac{1}{|\Sigma|^{\frac{n}{2}}|\Phi|^{\frac{p}{2}}} h(tr((\mathbf{X}-\mathbf{M})'\Sigma^{-1}(\mathbf{X}-\mathbf{M})\Phi^{-1})) \\ &= \frac{1}{|\Sigma|^{\frac{n}{2}}|\Phi|^{\frac{p}{2}}} h\left(tr\left(\left(\Sigma^{-\frac{1}{2}}(\mathbf{X}-\mathbf{M})\Phi^{-\frac{1}{2}}\right)'\left(\Sigma^{-\frac{1}{2}}(\mathbf{X}-\mathbf{M})\Phi^{-\frac{1}{2}}\right)\right)\right). \end{aligned}$$

Let $\mathbf{y} = vec\left(\Sigma^{-\frac{1}{2}}(\mathbf{X}-\mathbf{M})\Phi^{-\frac{1}{2}}\right)'$. Then, we have

$$tr\left(\left(\Sigma^{-\frac{1}{2}}(\mathbf{X}-\mathbf{M})\Phi^{-\frac{1}{2}}\right)'\left(\Sigma^{-\frac{1}{2}}(\mathbf{X}-\mathbf{M})\Phi^{-\frac{1}{2}}\right)\right) = \mathbf{y}'\mathbf{y}.$$

Since h is monotone decreasing in $[0,\infty)$ the last expression attains its minimum when $\mathbf{y}'\mathbf{y}$ is minimum. Now $\mathbf{y}'\mathbf{y} \geq \mathbf{0}$ and $\mathbf{y}'\mathbf{y} = \mathbf{0}$ iff $\mathbf{y} = \mathbf{0}$, therefore $f(\mathbf{X})$ is minimized for $\mathbf{y} = \mathbf{0}$. This means that $\mathbf{X} = \mathbf{M}$. Hence, $\hat{\mathbf{M}} = \mathbf{X}$.

(b) We have

$$f(\mathbf{X}) = \frac{1}{|\Sigma|^{\frac{n}{2}}|\Phi|^{\frac{p}{2}}} h(tr(\Sigma^{-1}(\mathbf{X}-\mu\mathbf{v}')\Phi^{-1}(\mathbf{X}-\mu\mathbf{v}')')).$$

Using Lemma 7.1, we can write

$$
\begin{aligned}
f(\mathbf{X}) &= \frac{1}{|\Sigma|^{\frac{n}{2}}|\Phi|^{\frac{p}{2}}} h\left(tr\left(\Sigma^{-1}\left[\mathbf{X}\left(\Phi^{-1} - \frac{\Phi^{-1}\mathbf{v}\mathbf{v}'\Phi^{-1}}{\mathbf{v}'\Phi^{-1}\mathbf{v}}\right)\mathbf{X}'\right.\right.\right. \\
&+ \left.\left.\left.(\mathbf{v}'\Phi^{-1}\mathbf{v})\left(\mathbf{X}\frac{\Phi^{-1}\mathbf{v}}{\mathbf{v}'\Phi^{-1}\mathbf{v}} - \mu\right)\left(\mathbf{X}\frac{\Phi^{-1}\mathbf{v}}{\mathbf{v}'\Phi^{-1}\mathbf{v}} - \mu\right)'\right]\right)\right) \\
&= \frac{1}{|\Sigma|^{\frac{n}{2}}|\Phi|^{\frac{p}{2}}} h\left(tr\left(\Sigma^{-1}\mathbf{X}\left(\Phi^{-1} - \frac{\Phi^{-1}\mathbf{v}\mathbf{v}'\Phi^{-1}}{\mathbf{v}'\Phi^{-1}\mathbf{v}}\right)\mathbf{X}'\right)\right. \\
&+ \left.(\mathbf{v}'\Phi^{-1}\mathbf{v})tr\left(\left(\mathbf{X}\frac{\Phi^{-1}\mathbf{v}}{\mathbf{v}'\Phi^{-1}\mathbf{v}} - \mu\right)'\Sigma^{-1}\left(\mathbf{X}\frac{\Phi^{-1}\mathbf{v}}{\mathbf{v}'\Phi^{-1}\mathbf{v}} - \mu\right)\right)\right).
\end{aligned}
$$

Again, since h is monotone decreasing in $[0,\infty)$ the last expression attains its minimum when $tr\left(\left(\mathbf{X}\frac{\Phi^{-1}\mathbf{v}}{\mathbf{v}'\Phi^{-1}\mathbf{v}} - \mu\right)'\Sigma^{-1}\left(\mathbf{X}\frac{\Phi^{-1}\mathbf{v}}{\mathbf{v}'\Phi^{-1}\mathbf{v}} - \mu\right)\right)$ is minimum. Writing $\mathbf{y} = vec\left(\Sigma^{-\frac{1}{2}}\left(\mathbf{X}\frac{\Phi^{-1}\mathbf{v}}{\mathbf{v}'\Phi^{-1}\mathbf{v}} - \mu\right)\right)'$, we have to minimize $\mathbf{y}\mathbf{y}'$. Therefore, minimum is attained at $\mathbf{y} = \mathbf{0}$. So we must have $\mathbf{X}\frac{\Phi^{-1}\mathbf{v}}{\mathbf{v}'\Phi^{-1}\mathbf{v}} = \mu$. Hence,

$$
\hat{\mu} = \mathbf{X}\frac{\Phi^{-1}\mathbf{v}}{\mathbf{v}'\Phi^{-1}\mathbf{v}}.
$$

(c) This is a special case of (b), with $\mathbf{v} = \mathbf{e}_n$. ■

The next result is based on a theorem due to Anderson, Fang and Hsu (1986).

Theorem 7.2. *Assume we have an observation* $\mathbf{X}$ *from the distribution* $E_{p,n}(\mathbf{M}, \Sigma \otimes \Phi, \psi)$, *where* $(\mathbf{M}, \Sigma \otimes \Phi) \in \Omega \subset I\!R^{p\times n} \times I\!R^{pn\times pn}$. *Suppose* Ω *has the property that if* $(\mathbf{Q}, \mathbf{S}) \in \Omega$ $(\mathbf{Q} \in I\!R^{p\times n}$, $\mathbf{S} \in I\!R^{pn\times pn})$, *then* $(\mathbf{Q}, c\mathbf{S}) \in \Omega$ *for any* $c > 0$ *scalar. Moreover, let* $\mathbf{X}$ *have the p.d.f.*

$$
f(\mathbf{X}) = \frac{1}{|\Sigma|^{\frac{n}{2}}|\Phi|^{\frac{p}{2}}} h(tr((\mathbf{X}-\mathbf{M})'\Sigma^{-1}(\mathbf{X}-\mathbf{M})\Phi^{-1})),
$$

where $l(z) = z^{\frac{pn}{2}}h(z)$, $z \geq 0$ *has a finite maximum at* $z = z_h > 0$. *Furthermore, suppose that under the assumption that* $\mathbf{X}$ *has the distribution* $N_{p,n}(\mathbf{M}, \Sigma \otimes \Phi)$, $(\mathbf{M}, \Sigma \otimes \Phi) \in \Omega$, *the MLE's of* $\mathbf{M}$ *and* $\Sigma \otimes \Phi$ *are* $\mathbf{M}^*$ *and* $(\Sigma \otimes \Phi)^*$, *which are unique and* $P((\Sigma \otimes \Phi)^* > \mathbf{0}) = 1$. *Then, under the condition* $\mathbf{X} \sim E_{p,n}(\mathbf{M}, \Sigma \otimes \Phi, \psi)$, $(\mathbf{M}, \Sigma \otimes \Phi) \in \Omega$, *the MLE's of* $\mathbf{M}$ *and* $\Sigma \otimes \Phi$ *are* $\hat{\mathbf{M}} = \mathbf{M}^*$ *and* $\widehat{(\Sigma \otimes \Phi)} = \frac{pn}{z_h}(\Sigma \otimes \Phi)^*$ *and the maximum of the likelihood is*

$$
|\widehat{(\Sigma \otimes \Phi)}|^{-\frac{1}{2}} h(z_h).
$$

PROOF: Define

$$\Sigma_1 = \frac{\Sigma}{|\Sigma|^{\frac{1}{p}}}, \quad \Phi_1 = \frac{\Phi}{|\Phi|^{\frac{1}{n}}} \tag{7.2}$$

and $z = tr((\mathbf{X}-\mathbf{M})'\Sigma^{-1}(\mathbf{X}-\mathbf{M})\Phi^{-1})$. Then, we have

$$z = \frac{tr((\mathbf{X}-\mathbf{M})'\Sigma_1^{-1}(\mathbf{X}-\mathbf{M})\Phi_1^{-1})}{|\Sigma|^{\frac{1}{p}}|\Phi|^{\frac{1}{n}}}. \tag{7.3}$$

Therefore, we can write

$$\begin{aligned} f(\mathbf{X}) &= \frac{1}{|\Sigma|^{\frac{n}{2}}|\Phi|^{\frac{p}{2}}} h(z) \\ &= z^{\frac{pn}{2}} h(z)(tr((\mathbf{X}-\mathbf{M})'\Sigma_1^{-1}(\mathbf{X}-\mathbf{M})\Phi_1^{-1}))^{-\frac{pn}{2}}. \end{aligned} \tag{7.4}$$

Hence, maximizing $f(\mathbf{X})$ is equivalent to maximizing $l(z)$ and $(tr((\mathbf{X}-\mathbf{M})'\Sigma_1^{-1}(\mathbf{X}-\mathbf{M})\Phi_1^{-1}))^{-\frac{pn}{2}}$. If $\mathbf{X} \sim N_{p,n}(\mathbf{M}, \Sigma \otimes \Phi)$, then $h(z) = (2\pi)^{-\frac{pn}{2}} e^{-\frac{z}{2}}$, and

$$l(z) = (2\pi)^{-\frac{pn}{2}} z^{\frac{pn}{2}} e^{-\frac{z}{2}}.$$

Therefore,

$$\begin{aligned} \frac{dl(z)}{dz} &= (2\pi)^{-\frac{pn}{2}} \left(\frac{pn}{2} z^{\frac{pn}{2}-1} e^{-\frac{z}{2}} + z^{\frac{pn}{2}} \left(-\frac{1}{2}\right) e^{-\frac{z}{2}} \right) \\ &= \frac{1}{2}(2\pi)^{-\frac{pn}{2}} z^{\frac{pn}{2}-1} e^{-\frac{z}{2}} (pn - z). \end{aligned}$$

Consequently, $l(z)$ attains its maximum at $z = pn$. From the conditions of the theorem it follows that under normality, $(tr((\mathbf{X}-\mathbf{M})'\Sigma_1^{-1}(\mathbf{X}-\mathbf{M})\Phi_1^{-1}))^{-\frac{pn}{2}}$ is maximized for $\mathbf{M} = \mathbf{M}^*$ and $\Sigma_1 \otimes \Phi_1 = (\Sigma_1 \otimes \Phi_1)^* = \frac{(\Sigma \otimes \Phi)^*}{|(\Sigma \otimes \Phi)^*|^{\frac{1}{pn}}}$. Since, $(tr((\mathbf{X}-\mathbf{M})'\Sigma_1^{-1}(\mathbf{X}-\mathbf{M})\Phi_1^{-1}))^{-\frac{pn}{2}}$ does not depend on h, in the case of $\mathbf{X} \sim E_{p,n}(\mathbf{M}, \Sigma \otimes \Phi, \psi)$, it also attains its maximum for $\mathbf{M} = \hat{\mathbf{M}} = \mathbf{M}^*$ and $\Sigma_1 \otimes \Phi_1 = (\widehat{\Sigma_1 \otimes \Phi_1}) = (\Sigma_1 \otimes \Phi_1)^*$. On the other hand, $l(z)$ is maximized for $z = z_h$. Then, using (7.2) and (7.3) we get

$$\begin{aligned} (\widehat{\Sigma \otimes \Phi}) &= |(\widehat{\Sigma \otimes \Phi})|^{\frac{1}{pn}} (\widehat{\Sigma_1 \otimes \Phi_1}) \\ &= |\hat{\Sigma}|^{\frac{1}{p}} |\hat{\Phi}|^{\frac{1}{n}} (\widehat{\Sigma_1 \otimes \Phi_1}) \end{aligned}$$

$$= \frac{tr((\mathbf{X}-\mathbf{M})'\hat{\Sigma}_1^{-1}(\mathbf{X}-\mathbf{M})\hat{\Phi}_1^{-1})}{z_h}(\widehat{\Sigma_1 \otimes \Phi_1})$$

$$= \frac{pn}{z_h}\frac{tr((\mathbf{X}-\mathbf{M}^*)'\Sigma_1^{*-1}(\mathbf{X}-\mathbf{M}^*)\Phi_1^{*-1})}{pn}(\Sigma_1 \otimes \Phi_1)^*$$

$$= \frac{pn}{z_h}(\Sigma \otimes \Phi)^*.$$

The maximum of the likelihood is

$$\frac{1}{|(\widehat{\Sigma \otimes \Phi})|^{\frac{1}{2}}}h(z_h) = |(\widehat{\Sigma \otimes \Phi})|^{-\frac{1}{2}}h(z_h). \qquad \blacksquare$$

Remark 7.1. Assume $\mathbf{X} \sim N_{p,n}(\mathbf{M}, \Sigma \otimes \Phi)$. Then

$$h(z) = (2\pi)^{-\frac{pn}{2}}e^{-\frac{z}{2}}, \quad l(z) = \left(\frac{z}{2\pi}\right)^{\frac{pn}{2}}e^{-\frac{z}{2}},$$

and $l(z)$ attains its maximum at $z_h = pn$. Moreover, $h(z_h) = (2\pi e)^{-\frac{pn}{2}}$.

It is natural to ask whether z_h, as defined in Theorem 7.2, exists in a large class of m.e.c. distributions. The following lemma, essentially due to Anderson, Fang, and Hsu (1986), gives a sufficient condition for the existence of z_h.

Lemma 7.2. *Let $\mathbf{X} \sim E_{p,n}(\mathbf{M}, \Sigma \otimes \Phi, \psi)$ have the p.d.f.*

$$f(\mathbf{X}) = \frac{1}{|\Sigma|^{\frac{n}{2}}|\Phi|^{\frac{p}{2}}}h(tr((\mathbf{X}-\mathbf{M})'\Sigma^{-1}(\mathbf{X}-\mathbf{M})\Phi^{-1})).$$

Assume $h(z)$, $(z \geq 0)$ is continuous and monotone decreasing, if z is sufficiently large. Then, there exists $z_h > 0$, such that $l(z) = z^{\frac{pn}{2}}h(z)$ attains its maximum at $z = z_h$.

PROOF: From Theorem 2.16, $r = (tr((\mathbf{X}-\mathbf{M})'\Sigma^{-1}(\mathbf{X}-\mathbf{M})\Phi^{-1}))^{\frac{1}{2}}$ has the p.d.f.

$$p_1(r) = \frac{2\pi^{\frac{pn}{2}}}{\Gamma\left(\frac{pn}{2}\right)}r^{pn-1}h(r^2), \qquad r \geq 0.$$

Let $y = r^2$. Then, $J(r \to y) = \frac{1}{2r}$, and hence the p.d.f. of y is

$$p_2(y) = \frac{\pi^{\frac{pn}{2}}}{\Gamma\left(\frac{pn}{2}\right)}y^{\frac{pn}{2}-1}h(y), \qquad y \geq 0.$$

Consequently,

$$\int_0^\infty \frac{\pi^{\frac{pn}{2}}}{\Gamma\left(\frac{pn}{2}\right)} y^{\frac{pn}{2}-1} h(y)dy = 1.$$

Therefore, $\int_z^\infty y^{\frac{pn}{2}-1}h(y)dy \to 0$ as $z \to \infty$.

Now, let $t = \frac{z}{2}$. Then we prove that for sufficiently large z, $g(z) \leq c\int_t^\infty y^{\frac{pn}{2}-1}h(y)dy$, where c is a constant.

If $np > 1$ then

$$\begin{aligned}
l(z) &= l(2t) \\
&= 2^{\frac{pn}{2}} t^{\frac{pn}{2}} h(2t) \\
&= 2^{\frac{pn}{2}} t^{\frac{pn}{2}-1} t h(2t) \\
&\leq 2^{\frac{pn}{2}} t^{\frac{pn}{2}-1} \int_t^{2t} h(y)dy \\
&\leq 2^{\frac{pn}{2}} \int_t^{2t} y^{\frac{pn}{2}-1} h(y)dy \\
&\leq 2^{\frac{pn}{2}} \int_t^{\infty} y^{\frac{pn}{2}-1} h(y)dy.
\end{aligned}$$

If $np = 1$, then

$$\begin{aligned}
l(z) &= l(2t) \\
&= 2^{\frac{1}{2}} t^{\frac{1}{2}} h(2t) \\
&= 2^{\frac{1}{2}} t^{\frac{1}{2}-1} t h(2t) \\
&\leq 2^{\frac{1}{2}} t^{\frac{1}{2}-1} \int_t^{2t} h(y)dy \\
&= 2(2t)^{\frac{1}{2}-1} \int_t^{2t} h(y)dy \\
&\leq 2\int_t^{2t} y^{\frac{1}{2}-1} h(y)dy \\
&\leq 2\int_t^{\infty} y^{\frac{1}{2}-1} h(y)dy.
\end{aligned}$$

However, $\int_t^\infty y^{\frac{pn}{2}-1}h(y)dy \to 0$ as $t \to \infty$, thus $l(z) \to 0$ as $z \to \infty$. Moreover, $l(0) = 0h(0) = 0$. Since, $l(z)$ is continuous, nonnegative and $\lim_{z\to 0} l(z) = \lim_{z\to\infty} l(z) = 0$, $l(z)$ attains its minimum at a positive number z_h. ■

For the next estimation result we need the following lemma, given in Anderson (2003, p. 69).

Lemma 7.3. *Let* $\mathbf{A}$ *be a* $p \times p$ *positive definite matrix and define a function* g *on the set of* $p \times p$ *positive definite matrices as*

$$g(\mathbf{B}) = -nlog|\mathbf{B}| - tr(\mathbf{B}^{-1}\mathbf{A}).$$

Then, $g(\mathbf{B})$ *attains its maximum at* $\mathbf{B} = \frac{\mathbf{A}}{n}$ *and its maximum value is*

$$g\left(\frac{\mathbf{A}}{n}\right) = pn(logn - 1) - nlog|\mathbf{A}|.$$

PROOF: We can write

$$\begin{aligned} g(\mathbf{B}) &= -nlog|\mathbf{B}| - tr(\mathbf{B}^{-1}\mathbf{A}) \\ &= n(log|\mathbf{A}| - log|\mathbf{B}|) - tr(\mathbf{B}^{-1}\mathbf{A}) - nlog|\mathbf{A}| \\ &= nlog|\mathbf{B}^{-1}\mathbf{A}| - tr(\mathbf{B}^{-1}\mathbf{A}) - nlog|\mathbf{A}|. \end{aligned}$$

Now, $g(\mathbf{B})$ is maximized for the same $\mathbf{B}$ as $h(\mathbf{B}) = nlog|\mathbf{B}^{-1}\mathbf{A}| - tr(\mathbf{B}^{-1}\mathbf{A})$. We have

$$\begin{aligned} h(\mathbf{B}) &= nlog\left|\mathbf{B}^{-1}\mathbf{A}^{\frac{1}{2}}\mathbf{A}^{\frac{1}{2}}\right| - tr\left(\mathbf{B}^{-1}\mathbf{A}^{\frac{1}{2}}\mathbf{A}^{\frac{1}{2}}\right) \\ &= nlog\left|\mathbf{A}^{\frac{1}{2}}\mathbf{B}^{-1}\mathbf{A}^{\frac{1}{2}}\right| - tr\left(\mathbf{A}^{\frac{1}{2}}\mathbf{B}^{-1}\mathbf{A}^{\frac{1}{2}}\right). \end{aligned}$$

Now, from Theorem 1.1, it follows that $\mathbf{A}^{\frac{1}{2}}\mathbf{B}^{-1}\mathbf{A}^{\frac{1}{2}}$ is also positive definite and from Theorem 1.7, it can be written as

$$\mathbf{A}^{\frac{1}{2}}\mathbf{B}^{-1}\mathbf{A}^{\frac{1}{2}} = \mathbf{GDG}',$$

where $\mathbf{G}$ is $p \times p$ orthogonal and $\mathbf{D}$ is $p \times p$ diagonal with positive diagonal elements $\lambda_1, \lambda_2, \ldots, \lambda_p$. We obtain

$$\begin{aligned} h(\mathbf{B}) &= nlog\left(\prod_{i=1}^{p}\lambda_i\right) - \sum_{i=1}^{p}\lambda_i \\ &= \sum_{i=1}^{p}(nlog\lambda_i - \lambda_i). \end{aligned}$$

Now, $\frac{\partial h(\mathbf{B})}{\partial \lambda_i} = \frac{n}{\lambda_i} - 1$. Thus from $\frac{\partial h(\mathbf{B})}{\partial \lambda_i} = 0$, we get $\lambda_i = n$, $i = 1, 2, \ldots, p$. Hence, $h(\mathbf{B})$ attains its maximum when $\mathbf{D} = n\mathbf{I}_p$ and so

$$\mathbf{A}^{\frac{1}{2}}\mathbf{B}^{-1}\mathbf{A}^{\frac{1}{2}} = \mathbf{G}n\mathbf{I}_p\mathbf{G}' = n\mathbf{I}_p.$$

Therefore, $\mathbf{B}^{-1} = n\mathbf{A}^{-1}$, and then $\mathbf{B} = \frac{\mathbf{A}}{n}$. Moreover,

$$g\left(\frac{\mathbf{A}}{n}\right) = -nlog\left(\left|\frac{\mathbf{A}}{n}\right|\right) - tr(n\mathbf{I}_p)$$
$$= pn(log n - 1) - nlog|\mathbf{A}|. \qquad \blacksquare$$

Theorem 7.3. *Let* $\mathbf{X} \sim E_{p,n}(\mathbf{M}, \Sigma \otimes \Phi, \psi)$ *have the p.d.f.*

$$f(\mathbf{X}) = \frac{1}{|\Sigma|^{\frac{n}{2}}|\Phi|^{\frac{p}{2}}} h(tr((\mathbf{X}-\mathbf{M})'\Sigma^{-1}(\mathbf{X}-\mathbf{M})\Phi^{-1})),$$

where $n \geq p$ *and the function* $l(z) = z^{\frac{pn}{2}} h(z)$, $z \in [0, \infty)$, *attains its maximum for a positive* z *(say* z_h*). Suppose* h, $\mathbf{M}$, *and* Φ *are known, and we want to find the MLE of* Σ *(say* $\hat{\Sigma}$*), based on a single observation* $\mathbf{X}$. *Then,*

$$\hat{\Sigma} = \frac{p}{z_h}(\mathbf{X}-\mathbf{M})\Phi^{-1}(\mathbf{X}-\mathbf{M})'.$$

PROOF: Step 1. First, we prove the result for normal distribution; that is, when $\mathbf{X} \sim N_{p,n}(\mathbf{M}, \Sigma \otimes \Phi)$. Here, $h(z) = \exp\left(-\frac{z}{2}\right)$, and

$$f(\mathbf{X}) = \frac{1}{(2\pi)^{\frac{pn}{2}}|\Sigma|^{\frac{n}{2}}|\Phi|^{\frac{p}{2}}} etr\left\{-\frac{(\Sigma^{-1}(\mathbf{X}-\mathbf{M})\Phi^{-1}(\mathbf{X}-\mathbf{M})')}{2}\right\}.$$

Taking logarithm of both sides of the last equation and applying Lemma 7.3, we obtain that $f(\mathbf{X})$ attains its maximum in Σ, if

$$\hat{\Sigma} = \frac{1}{n}(\mathbf{X}-\mathbf{M})\Phi^{-1}(\mathbf{X}-\mathbf{M})'.$$

Step 2. Let $\mathbf{X} \sim E_{p,n}(\mathbf{M}, \Sigma \otimes \Phi, \psi)$. Since we proved, in Step 1, that for the normal case, the MLE of Σ is $\frac{1}{n}(\mathbf{X}-\mathbf{M})\Phi^{-1}(\mathbf{X}-\mathbf{M})'$, by using Theorem 7.2 we get

$$\hat{\Sigma} = \frac{pn}{z_h}\frac{1}{n}(\mathbf{X}-\mathbf{M})\Phi^{-1}(\mathbf{X}-\mathbf{M})' = \frac{p}{z_h}(\mathbf{X}-\mathbf{M})\Phi^{-1}(\mathbf{X}-\mathbf{M})'.$$

It follows, from Theorem 5.11, that $rk((\mathbf{X}-\mathbf{M})\Phi^{-1}(\mathbf{X}-\mathbf{M})') = p$ with probability one. Hence, $P(\hat{\Sigma} > \mathbf{0}) = 1$. $\blacksquare$

The next result is an extension of a result of Anderson, Fang, and Hsu (1986).

Theorem 7.4. *Let* $\mathbf{X} \sim E_{p,n}(\mu\mathbf{v}', \Sigma \otimes \Phi, \psi)$ *have the p.d.f.*

$$f(\mathbf{X}) = \frac{1}{|\Sigma|^{\frac{n}{2}} |\Phi|^{\frac{p}{2}}} h(tr((\mathbf{X} - \mu\mathbf{v}')'\Sigma^{-1}(\mathbf{X} - \mu\mathbf{v}')\Phi^{-1})),$$

where $n > p$, μ *is a p-dimensional vector,* $\mathbf{v}$ *is an n-dimensional nonzero vector, and the function* $l(z) = z^{\frac{pn}{2}} h(z)$, $z \in [0, \infty)$, *attains its maximum for a positive z (say* z_h*). Suppose h,* $\mathbf{v}$*, and* Φ *are known and we want to find the MLEs of* μ *and* Σ *(say* $\hat{\mu}$ *and* $\hat{\Sigma}$*) based on a single observation* $\mathbf{X}$*. Then*

$$\hat{\mu} = \mathbf{X}\frac{\Phi^{-1}\mathbf{v}}{\mathbf{v}'\Phi^{-1}\mathbf{v}} \quad \text{and} \quad \hat{\Sigma} = \frac{p}{z_h}\mathbf{X}\left(\Phi^{-1} - \frac{\Phi^{-1}\mathbf{v}\mathbf{v}'\Phi^{-1}}{\mathbf{v}'\Phi^{-1}\mathbf{v}}\right)\mathbf{X}'.$$

In the special case, when $\mathbf{v} = \mathbf{e}_n$*, we have*

$$\hat{\mu} = \mathbf{X}\frac{\Phi^{-1}\mathbf{e}_n}{\mathbf{e}_n'\Phi^{-1}\mathbf{e}_n} \quad \text{and} \quad \hat{\Sigma} = \frac{p}{z_h}\mathbf{X}\left(\Phi^{-1} - \frac{\Phi^{-1}\mathbf{e}_n\mathbf{e}_n'\Phi^{-1}}{\mathbf{e}_n'\Phi^{-1}\mathbf{e}_n}\right)\mathbf{X}'.$$

PROOF: Step 1. First we prove the result for normal distribution; that is, when $\mathbf{X} \sim N_{p,n}(\mu\mathbf{v}', \Sigma \otimes \Phi)$. Here, $h(z) = \exp\left(-\frac{z}{2}\right)$.

Using Lemma 7.1, the p.d.f. of $\mathbf{X}$ can be written as

$$f(\mathbf{X}) = \frac{1}{(2\pi)^{\frac{pn}{2}} |\Sigma|^{\frac{n}{2}} |\Phi|^{\frac{p}{2}}} \exp\left\{-\frac{1}{2}\left(tr\Sigma^{-1}\left[\mathbf{X}\left(\Phi^{-1} - \frac{\Phi^{-1}\mathbf{v}\mathbf{v}'\Phi^{-1}}{\mathbf{v}'\Phi^{-1}\mathbf{v}}\right)\mathbf{X}'\right.\right.\right.$$
$$\left.\left.\left. + \ (\mathbf{v}'\Phi^{-1}\mathbf{v})\left(\mathbf{X}\frac{\Phi^{-1}\mathbf{v}}{\mathbf{v}'\Phi^{-1}\mathbf{v}} - \mu\right)\left(\mathbf{X}\frac{\Phi^{-1}\mathbf{v}}{\mathbf{v}'\Phi^{-1}\mathbf{v}} - \mu\right)'\right]\right)\right\}.$$

Minimizing the expression on the right-hand side, we get

$$\hat{\mu} = \mathbf{X}\frac{\Phi^{-1}\mathbf{v}}{\mathbf{v}'\Phi^{-1}\mathbf{v}} \quad \text{and} \quad \hat{\Sigma} = \frac{\mathbf{X}\left(\Phi^{-1} - \frac{\Phi^{-1}\mathbf{v}\mathbf{v}'\Phi^{-1}}{\mathbf{v}'\Phi^{-1}\mathbf{v}}\right)\mathbf{X}'}{n}.$$

As noted in Remark 7.1, $l(z)$ is maximized for $z_h = pn$. Thus,

$$\hat{\Sigma} = \frac{p}{z_h}\mathbf{X}\left(\Phi^{-1} - \frac{\Phi^{-1}\mathbf{v}\mathbf{v}'\Phi^{-1}}{\mathbf{v}'\Phi^{-1}\mathbf{v}}\right)\mathbf{X}'.$$

Step 2. Let $\mathbf{X} \sim E_{p,n}(\mu\mathbf{v}', \Sigma \otimes \Phi, \psi)$. We found the MLE's of μ and Σ for the normal case in Step 1. Now using Theorem 7.2, we get

$$\hat{\mu} = \mathbf{X}\frac{\Phi^{-1}\mathbf{v}}{\mathbf{v}'\Phi^{-1}\mathbf{v}}$$

and

$$\hat{\Sigma} = \frac{pn}{z_h} \frac{\mathbf{X}\left(\Phi^{-1} - \frac{\Phi^{-1}\mathbf{v}\mathbf{v}'\Phi^{-1}}{\mathbf{v}'\Phi^{-1}\mathbf{v}}\right)\mathbf{X}'}{n} = \frac{p}{z_h}\mathbf{X}\left(\Phi^{-1} - \frac{\Phi^{-1}\mathbf{v}\mathbf{v}'\Phi^{-1}}{\mathbf{v}'\Phi^{-1}\mathbf{v}}\right)\mathbf{X}'.$$

From Theorem 5.11 it follows that with probability one,

$$rk\left(\mathbf{X}\left(\Phi^{-1} - \frac{\Phi^{-1}\mathbf{v}\mathbf{v}'\Phi^{-1}}{\mathbf{v}'\Phi^{-1}\mathbf{v}}\right)\mathbf{X}'\right) = \min\left(rk\left(\Phi^{-1} - \frac{\Phi^{-1}\mathbf{v}\mathbf{v}'\Phi^{-1}}{\mathbf{v}'\Phi^{-1}\mathbf{v}}\right), p\right).$$

Since, $\Phi^{-1} - \frac{\Phi^{-1}\mathbf{v}\mathbf{v}'\Phi^{-1}}{\mathbf{v}'\Phi^{-1}\mathbf{v}} = \Phi^{-\frac{1}{2}}\left(\mathbf{I}_n - \frac{\Phi^{-\frac{1}{2}}\mathbf{v}\mathbf{v}'\Phi^{-\frac{1}{2}}}{\mathbf{v}'\Phi^{-1}\mathbf{v}}\right)\Phi^{-\frac{1}{2}}$ and Φ is of full rank, we have

$$\begin{aligned} rk\left(\Phi^{-1} - \frac{\Phi^{-1}\mathbf{v}\mathbf{v}'\Phi^{-1}}{\mathbf{v}'\Phi^{-1}\mathbf{v}}\right) &= rk\left(\mathbf{I}_n - \frac{\Phi^{-\frac{1}{2}}\mathbf{v}\mathbf{v}'\Phi^{-\frac{1}{2}}}{\mathbf{v}'\Phi^{-1}\mathbf{v}}\right) \\ &\geq rk(\mathbf{I}_n) - rk\left(\frac{\Phi^{-\frac{1}{2}}\mathbf{v}\mathbf{v}'\Phi^{-\frac{1}{2}}}{\mathbf{v}'\Phi^{-1}\mathbf{v}}\right) \\ &= n - rk\left(\frac{\mathbf{v}'\Phi^{-1}\mathbf{v}}{\mathbf{v}'\Phi^{-1}\mathbf{v}}\right) \\ &= n - 1 \end{aligned}$$

where we used part (ii) of Theorem 1.3. Hence, $P(\hat{\Sigma} > \mathbf{0}) = 1$. ∎

7.2 Properties of Estimators

Now, we derive the distributions of the estimators of μ and Σ. These theorems are based on Anderson and Fang (1982a).

Theorem 7.5. $\mathbf{X} \sim E_{p,n}(\mu\mathbf{v}', \Sigma \otimes \Phi, \psi)$ *have the p.d.f.*

$$f(\mathbf{X}) = \frac{1}{|\Sigma|^{\frac{n}{2}}|\Phi|^{\frac{p}{2}}} h(tr((\mathbf{X} - \mu\mathbf{v}')'\Sigma^{-1}(\mathbf{X} - \mu\mathbf{v}')\Phi^{-1})),$$

where $n > p$, μ is a p-dimensional vector and $\mathbf{v}$ *is an n-dimensional nonzero vector,* $\mathbf{v} \neq \mathbf{0}$. *Let*

$$\hat{\mu} = \mathbf{X}\frac{\Phi^{-1}\mathbf{v}}{\mathbf{v}'\Phi^{-1}\mathbf{v}} \quad \textit{and} \quad \mathbf{A} = \mathbf{X}\left(\Phi^{-1} - \frac{\Phi^{-1}\mathbf{v}\mathbf{v}'\Phi^{-1}}{\mathbf{v}'\Phi^{-1}\mathbf{v}}\right)\mathbf{X}'.$$

(a) Then, the joint density of $\hat{\mu}$ *and* $\mathbf{A}$ *is*

$$p(\hat{\mu},\mathbf{A}) = \frac{(\mathbf{v}'\boldsymbol{\Phi}^{-1}\mathbf{v})^{\frac{p}{2}}|\mathbf{A}|^{\frac{n-p}{2}-1}\pi^{\frac{p(n-1)}{2}}}{\Gamma_p\left(\frac{n-1}{2}\right)|\Sigma|^{\frac{n}{2}}} h(\mathbf{v}'\boldsymbol{\Phi}^{-1}\mathbf{v}(\hat{\mu}-\mu)'\Sigma^{-1}(\hat{\mu}-\mu) + tr(\Sigma^{-1}\mathbf{A})). \tag{7.5}$$

(b) $\hat{\mu} \sim E_p\left(\mu, \frac{1}{\mathbf{v}'\boldsymbol{\Phi}^{-1}\mathbf{v}}\Sigma, \psi\right)$ *and* $\hat{\mu}$ *has the p.d.f.*

$$p_1(\hat{\mu}) = \frac{2(\mathbf{v}'\boldsymbol{\Phi}^{-1}\mathbf{v})^{\frac{p}{2}}\pi^{\frac{p(n-1)}{2}}}{\Gamma\left(\frac{p(n-1)}{2}\right)|\Sigma|^{\frac{1}{2}}} \int_0^\infty r^{p(n-1)-1} h(r^2 + \mathbf{v}'\boldsymbol{\Phi}^{-1}\mathbf{v}(\hat{\mu}-\mu)'\Sigma^{-1}(\hat{\mu}-\mu))dr. \tag{7.6}$$

(c) $\mathbf{A} \sim G_{p,1}\left(\Sigma, \frac{n-1}{2}, \psi\right)$ *and its p.d.f. is given by*

$$p_2(\mathbf{A}) = \frac{2\pi^{\frac{pn}{2}}|\mathbf{A}|^{\frac{n-p}{2}-1}}{\Gamma\left(\frac{p}{2}\right)\Gamma_p\left(\frac{n-1}{2}\right)|\Sigma|^{\frac{n-1}{2}}} \int_0^\infty r^{p-1} h(r^2 + tr(\Sigma^{-1}\mathbf{A}))dr. \tag{7.7}$$

PROOF:

(a) First, we derive the result for the case $\boldsymbol{\Phi} = \mathbf{I}_n$.
Let $\mathbf{G} \in O(n)$, whose first column is $\frac{\mathbf{v}}{\sqrt{\mathbf{v}'\mathbf{v}}}$ and let us use the notation $\mathbf{G} = \left(\frac{\mathbf{v}}{\sqrt{\mathbf{v}'\mathbf{v}}}, \mathbf{G}_1\right)$. Since $\mathbf{G}$ is orthogonal, we have $\mathbf{G}\mathbf{G}' = \mathbf{I}_n$; that is, $\frac{\mathbf{v}}{\sqrt{\mathbf{v}'\mathbf{v}}}\frac{\mathbf{v}'}{\sqrt{\mathbf{v}'\mathbf{v}}} + \mathbf{G}_1\mathbf{G}_1' = \mathbf{I}_n$. Thus,

$$\mathbf{G}_1\mathbf{G}_1' = \mathbf{I}_n - \frac{\mathbf{v}\mathbf{v}'}{\mathbf{v}'\mathbf{v}}. \tag{7.8}$$

Define $\mathbf{Y} = \mathbf{X}\mathbf{G}$. Then,

$$\mathbf{Y} \sim E_{p,n}(\mu\mathbf{v}'\mathbf{G}, \Sigma \otimes \mathbf{I}_n, \psi). \tag{7.9}$$

Partition $\mathbf{Y}$ as $\mathbf{Y} = (\mathbf{y}_1, \mathbf{Y}_2)$, where $\mathbf{y}_1$ is a p-dimensional vector. Since, $\mathbf{G}$ is orthogonal, $\mathbf{v}'\mathbf{G}_1 = \mathbf{0}$. Therefore,

$$\mathbf{v}'\mathbf{G} = \mathbf{v}'\left(\frac{\mathbf{v}}{\sqrt{\mathbf{v}'\mathbf{v}}}, \mathbf{G}_1\right) = (\sqrt{\mathbf{v}'\mathbf{v}}, \mathbf{0}).$$

Now, (7.9) can be written as

$$(\mathbf{y}_1, \mathbf{Y}_2) \sim E_{p,n}((\sqrt{\mathbf{v}'\mathbf{v}}\mu, \mathbf{0}), \Sigma \otimes \mathbf{I}_n, \psi). \tag{7.10}$$

Moreover, $(\mathbf{y}_1, \mathbf{Y}_2) = \mathbf{X}\left(\frac{\mathbf{v}}{\sqrt{\mathbf{v}'\mathbf{v}}}, \mathbf{G}_1\right) = \left(\mathbf{X}\frac{\mathbf{v}}{\sqrt{\mathbf{v}'\mathbf{v}}}, \mathbf{X}\mathbf{G}_1\right)$, hence $\hat{\mu} = \frac{\mathbf{y}_1}{\sqrt{\mathbf{v}'\mathbf{v}}}$ and using (7.8) we get $\mathbf{A} = \mathbf{Y}_2\mathbf{Y}_2'$.
Now, the density of $\mathbf{Y}$; that is, the joint density of $\mathbf{y}_1$ and $\mathbf{Y}_2$ is

$$\begin{aligned} p_3(\mathbf{y}_1, \mathbf{Y}_2) &= \frac{1}{|\Sigma|^{\frac{n}{2}}} h((\mathbf{y}_1 - \sqrt{\mathbf{v}'\mathbf{v}}\mu)'\Sigma^{-1}(\mathbf{y}_1 - \sqrt{\mathbf{v}'\mathbf{v}}\mu) + tr(\mathbf{Y}_2'\Sigma^{-1}\mathbf{Y}_2)) \\ &= \frac{1}{|\Sigma|^{\frac{n}{2}}} h((\mathbf{y}_1 - \sqrt{\mathbf{v}'\mathbf{v}}\mu)'\Sigma^{-1}(\mathbf{y}_1 - \sqrt{\mathbf{v}'\mathbf{v}}\mu) + tr(\Sigma^{-1}\mathbf{Y}_2\mathbf{Y}_2')). \quad (7.11) \end{aligned}$$

Using Lemma 5.2, we can write the joint density of $\mathbf{y}_1$ and $\mathbf{A}$ as

$$\begin{aligned} p_4(\mathbf{y}_1, \mathbf{A}) &= \frac{\pi^{\frac{p(n-1)}{2}}}{\Gamma_p\left(\frac{n-1}{2}\right)} |\mathbf{A}|^{\frac{n-1-p-1}{2}} \frac{1}{|\Sigma|^{\frac{n}{2}}} \\ &\times h((\mathbf{y}_1 - \sqrt{\mathbf{v}'\mathbf{v}}\mu)'\Sigma^{-1}(\mathbf{y}_1 - \sqrt{\mathbf{v}'\mathbf{v}}\mu) + tr(\Sigma^{-1}\mathbf{A})). \quad (7.12) \end{aligned}$$

We have $\mathbf{y}_1 = \sqrt{\mathbf{v}'\mathbf{v}}\hat{\mu}$. Hence, $J(\mathbf{y}_1 \to \hat{\mu}) = (\mathbf{v}'\mathbf{v})^{\frac{p}{2}}$. Therefore, the joint p.d.f. of $\hat{\mu}$ and $\mathbf{A}$ is

$$p(\hat{\mu}, \mathbf{A}) = \frac{(\mathbf{v}'\mathbf{v})^{\frac{p}{2}} |\mathbf{A}|^{\frac{n-p}{2}-1} \pi^{\frac{p(n-1)}{2}}}{\Gamma_p\left(\frac{n-1}{2}\right) |\Sigma|^{\frac{n}{2}}} h(\mathbf{v}'\mathbf{v}(\hat{\mu} - \mu)'\Sigma^{-1}(\hat{\mu} - \mu) + tr(\Sigma^{-1}\mathbf{A})). \quad (7.13)$$

Now, for $\Phi \neq \mathbf{I}_n$, define $\mathbf{X}^* = \mathbf{X}\Phi^{-\frac{1}{2}}$. Then, $\mathbf{X}^* \sim E_{p,n}(\mu\mathbf{v}^{*\prime}, \Sigma \otimes \mathbf{I}_n, \psi)$, with $\mathbf{v}^* = \Phi^{-\frac{1}{2}}\mathbf{v}$. Thus, we get $\mathbf{v}^{*\prime}\mathbf{v}^* = \mathbf{v}'\Phi^{-1}\mathbf{v}$,

$$\hat{\mu} = \mathbf{X}^* \frac{\mathbf{v}^*}{\mathbf{v}^{*\prime}\mathbf{v}^*} \quad \text{and} \quad \mathbf{A} = \mathbf{X}^* \left(\mathbf{I}_n - \frac{\mathbf{v}^*\mathbf{v}^{*\prime}}{\mathbf{v}^{*\prime}\mathbf{v}^*}\right) \mathbf{X}^{*\prime}.$$

So, using the first part of the proof, from (7.13), we obtain (7.5).

(b) Since, $\hat{\mu} = \mathbf{X}\frac{\Phi^{-1}\mathbf{v}}{\mathbf{v}'\Phi^{-1}\mathbf{v}}$, we get

$$\hat{\mu} \sim E_p\left(\mu\frac{\mathbf{v}'\Phi^{-1}\mathbf{v}}{\mathbf{v}'\Phi^{-1}\mathbf{v}}, \Sigma \otimes \frac{\mathbf{v}'\Phi^{-1}\Phi\Phi^{-1}\mathbf{v}}{(\mathbf{v}'\Phi^{-1}\mathbf{v})^2}, \psi\right) = E_p\left(\mu, \frac{1}{\mathbf{v}'\Phi^{-1}\mathbf{v}}\Sigma, \psi\right).$$

Now, assume $\Phi = \mathbf{I}_n$. Then, from (7.11), we derive the p.d.f. of $\mathbf{y}_1 = \sqrt{\mathbf{v}'\mathbf{v}}\hat{\mu}$, as

$$p_5(\mathbf{y}_1) = \int_{\mathbb{R}^{p\times(n-1)}} p_3(\mathbf{y}_1, \mathbf{Y}_2) d\mathbf{Y}_2. \quad (7.14)$$

Let $\mathbf{W} = \Sigma^{-\frac{1}{2}}\mathbf{Y}_2$. Then, $J(\mathbf{Y}_2 \to \mathbf{W}) = |\Sigma|^{\frac{n-1}{2}}$, and from (7.11) and (7.14) we get

$$p_5(\mathbf{y}_1) = \frac{1}{|\Sigma|^{\frac{n}{2}}} |\Sigma|^{\frac{n-1}{2}} \int_{I\!R^{p\times(n-1)}} h((\mathbf{y}_1 - \sqrt{\mathbf{v}'\mathbf{v}}\mu)'\Sigma^{-1}(\mathbf{y}_1 - \sqrt{\mathbf{v}'\mathbf{v}}\mu) + tr(\mathbf{W}\mathbf{W}'))d\mathbf{W}. \tag{7.15}$$

Writing $\mathbf{w} = vec(\mathbf{W}')$, (7.15) becomes

$$p_5(\mathbf{y}_1) = \frac{1}{|\Sigma|^{\frac{1}{2}}} \int_{I\!R^{p(n-1)}} h((\mathbf{y}_1 - \sqrt{\mathbf{v}'\mathbf{v}}\mu)'\Sigma^{-1}(\mathbf{y}_1 - \sqrt{\mathbf{v}'\mathbf{v}}\mu) + \mathbf{w}'\mathbf{w})d\mathbf{w}.$$

Using Lemma 2.1, we get

$$p_5(\mathbf{y}_1) = \frac{1}{|\Sigma|^{\frac{1}{2}}} \frac{2\pi^{\frac{p(n-1)}{2}}}{\Gamma\left(\frac{p(n-1)}{2}\right)} \int_0^\infty r^{p(n-1)-1} h(r^2 + (\mathbf{y}_1 - \sqrt{\mathbf{v}'\mathbf{v}}\mu)' \times \Sigma^{-1}(\mathbf{y}_1 - \sqrt{\mathbf{v}'\mathbf{v}}\mu))dr.$$

Since $\hat{\mu} = (\mathbf{v}'\mathbf{v})^{-\frac{1}{2}}\mathbf{y}_1$ and $J(\mathbf{y}_1 \to \hat{\mu}) = (\mathbf{v}'\mathbf{v})^{\frac{p}{2}}$, the p.d.f. of $\hat{\mu}$ is given by

$$p_1(\hat{\mu}) = \frac{2(\mathbf{v}'\mathbf{v})^{\frac{p}{2}}\pi^{\frac{p(n-1)}{2}}}{\Gamma\left(\frac{p(n-1)}{2}\right)|\Sigma|^{\frac{1}{2}}} \int_0^\infty r^{p(n-1)-1} h(r^2 + \mathbf{v}'\mathbf{v}(\hat{\mu}-\mu)'\Sigma^{-1}(\hat{\mu}-\mu))dr. \tag{7.16}$$

For $\Phi \neq \mathbf{I}_n$, define $\mathbf{X}^* = \mathbf{X}\Phi^{-\frac{1}{2}}$. Then, $\mathbf{X}^* \sim E_{p,n}(\mu\mathbf{v}^{*\prime}, \Sigma \otimes \mathbf{I}_n, \psi)$, with $\mathbf{v}^* = \Phi^{-\frac{1}{2}}\mathbf{v}$. Thus, we get $\mathbf{v}^{*\prime}\mathbf{v}^* = \mathbf{v}'\Phi^{-1}\mathbf{v}$ and $\hat{\mu} = \mathbf{X}^*\frac{\mathbf{v}^*}{\mathbf{v}^{*\prime}\mathbf{v}^*}$ and from (7.16) we obtain (7.6).

(c) First, assume $\Phi = \mathbf{I}_n$. Then from (7.10) we get

$$\mathbf{Y}_2 \sim E_{p,n-1}(\mathbf{0}, \Sigma \otimes \mathbf{I}_{n-1}, \psi).$$

Hence, by Definition 5.1, $\mathbf{A} = \mathbf{Y}_2\mathbf{Y}_2' \sim G_p\left(\Sigma, \frac{n-1}{2}, \psi\right)$, and its p.d.f., using (7.12), is given by

$$p_2(\mathbf{A}) = \int_{I\!R^p} p_4(\mathbf{y}_1, \mathbf{A})d\mathbf{y}_1. \tag{7.17}$$

Let $\mathbf{w} = \Sigma^{-\frac{1}{2}}(\mathbf{y}_1 - \sqrt{\mathbf{v}'\mathbf{v}}\mu)$. Then, $J(\mathbf{y}_1 \to \hat{\mu}) = |\Sigma|^{\frac{1}{2}}$, and using (7.12) and (7.17) we can write

$$p_2(\mathbf{A}) = \frac{\pi^{\frac{p(n-1)}{2}}}{\Gamma_p\left(\frac{n-1}{2}\right)|\Sigma|^{\frac{n}{2}}} |\Sigma|^{\frac{1}{2}} |\mathbf{A}|^{\frac{n-p}{2}-1} \int_{I\!R^p} h(\mathbf{w}'\mathbf{w} + tr(\Sigma^{-1}\mathbf{A}))d\mathbf{w}.$$

Using polar coordinates, we get

$$p_2(\mathbf{A}) = \frac{\pi^{\frac{p(n-1)}{2}}}{\Gamma_p\left(\frac{n-1}{2}\right)|\Sigma|^{\frac{n-1}{2}}}|\mathbf{A}|^{\frac{n-p}{2}-1}\frac{2\pi^{\frac{p}{2}}}{\Gamma\left(\frac{p}{2}\right)}\int_0^\infty r^{p-1}h(r^2+tr(\Sigma^{-1}\mathbf{A}))dr$$
$$= \frac{2\pi^{\frac{pn}{2}}|\mathbf{A}|^{\frac{n-p}{2}-1}}{\Gamma\left(\frac{p}{2}\right)\Gamma_p\left(\frac{n-1}{2}\right)|\Sigma|^{\frac{n-1}{2}}}\int_0^\infty r^{p-1}h(r^2+tr(\Sigma^{-1}\mathbf{A}))dr.$$

For $\boldsymbol{\Phi} \neq \mathbf{I}_n$, define $\mathbf{X}^* = \mathbf{X}\boldsymbol{\Phi}^{-\frac{1}{2}}$. Then, $\mathbf{X}^* \sim E_{p,n}(\mu\mathbf{v}^{*\prime}, \Sigma\otimes\mathbf{I}_n, \psi)$, with $\mathbf{v}^* = \boldsymbol{\Phi}^{-\frac{1}{2}}\mathbf{v}$. Since, $\mathbf{A} = \mathbf{X}^*\left(\mathbf{I}_n - \frac{\mathbf{v}^*\mathbf{v}^{*\prime}}{\mathbf{v}^{*\prime}\mathbf{v}^*}\right)\mathbf{X}^{*\prime}$ and $\mathbf{A}$ does not depend on $\mathbf{v}^*$, it has the same distribution under the m.e.c. distribution with $\boldsymbol{\Phi} = \mathbf{I}_n$ as under the m.e.c. distribution with $\boldsymbol{\Phi} \neq \mathbf{I}_n$. ∎

Theorem 7.6. *Let $\mathbf{X} \sim E_{p,n}(\mathbf{0}, \Sigma\otimes\boldsymbol{\Phi}, \psi)$ have the p.d.f.*

$$f(\mathbf{X}) = \frac{1}{|\Sigma|^{\frac{n}{2}}|\boldsymbol{\Phi}|^{\frac{p}{2}}}h(tr(\mathbf{X}'\Sigma^{-1}\mathbf{X}\boldsymbol{\Phi}^{-1})),$$

where $n \geq p$. Let $\mathbf{B} = \mathbf{X}\boldsymbol{\Phi}^{-1}\mathbf{X}'$, then $\mathbf{B} \sim G_{p,1}\left(\Sigma, \frac{n}{2}, \psi\right)$, and the p.d.f. of $\mathbf{B}$ is

$$p(\mathbf{B}) = \frac{\pi^{\frac{pn}{2}}|\mathbf{B}|^{\frac{n-p-1}{2}}}{\Gamma_p\left(\frac{n}{2}\right)|\Sigma|^{\frac{n}{2}}}h(tr(\Sigma^{-1}\mathbf{W})).$$

PROOF: First, assume $\boldsymbol{\Phi} = \mathbf{I}_n$. Then, by Definition 5.1, we have $\mathbf{B} = \mathbf{X}\mathbf{X}' \sim G_{p,1}\left(\Sigma, \frac{n}{2}, \psi\right)$. Moreover, we have $f(\mathbf{X}) = \frac{1}{|\Sigma|^{\frac{n}{2}}}h(tr(\Sigma^{-1}\mathbf{X}\mathbf{X}'))$. Using Lemma 5.1, we obtain the p.d.f. of $\mathbf{B}$ as

$$p(\mathbf{B}) = \frac{\pi^{\frac{pn}{2}}}{\Gamma_p\left(\frac{n}{2}\right)|\Sigma|^{\frac{n}{2}}}|\mathbf{B}|^{\frac{n-p-1}{2}}h(tr(\Sigma^{-1}\mathbf{B})). \tag{7.18}$$

For $\boldsymbol{\Phi} \neq \mathbf{I}_n$, define $\mathbf{X}^* = \mathbf{X}\boldsymbol{\Phi}^{-\frac{1}{2}}$. Then, $\mathbf{X}^* \sim E_{p,n}(\mathbf{0}, \Sigma\otimes\mathbf{I}_n, \psi)$, with $\mathbf{v}^* = \boldsymbol{\Phi}^{-\frac{1}{2}}\mathbf{v}$. Since, $\mathbf{B} = \mathbf{X}^*\mathbf{X}^{*\prime}$ and the distribution of $\mathbf{B}$ is the same under the m.e.c. distribution with $\boldsymbol{\Phi} = \mathbf{I}_n$ as under the m.e.c. distribution with $\boldsymbol{\Phi} \neq \mathbf{I}_n$. ∎

Before we derive the joint density of the characteristic roots of the estimators of Σ, we give a lemma taken from Anderson (2003).

Lemma 7.4. *Let $\mathbf{A}$ be a symmetric random matrix. Let $\lambda_1 > \lambda_2 > \ldots > \lambda_p$ be the characteristic roots of $\mathbf{A}$, and assume that the density of $\mathbf{A}$ is a function of $(\lambda_1, \lambda_2, \ldots, \lambda_p)$, i.e. $f(\mathbf{A}) = g(\lambda_1, \lambda_2, \ldots, \lambda_p)$. Then, the p.d.f. of $(\lambda_1, \lambda_2, \ldots, \lambda_p)$ is*

$$p(\lambda_1, \lambda_2, \ldots, \lambda_p) = \frac{\pi^{\frac{p^2}{2}}g(\lambda_1, \lambda_2, \ldots, \lambda_p)\prod_{i<j}(\lambda_i - \lambda_j)}{\Gamma_p\left(\frac{p}{2}\right)}.$$

PROOF: See Anderson (2003), p. 538. ∎

Theorem 7.7. *Let* $\mathbf{X} \sim E_{p,n}(\mu\mathbf{v}', \mathbf{I}_p \otimes \Phi, \psi)$ *have the p.d.f.*

$$f(\mathbf{X}) = \frac{1}{|\Phi|^{\frac{p}{2}}} h(tr((\mathbf{X} - \mu\mathbf{v}')'(\mathbf{X} - \mu\mathbf{v}')\Phi^{-1})),$$

where $n > p$, μ *is a p-dimensional vector and* $\mathbf{v}$ *is an n-dimensional nonzero vector,* $\mathbf{v} \neq \mathbf{0}$. *Let* $\mathbf{A} = \mathbf{X}\left(\Phi^{-1} - \frac{\Phi^{-1}\mathbf{v}\mathbf{v}'\Phi^{-1}}{\mathbf{v}'\Phi^{-1}\mathbf{v}}\right)\mathbf{X}'$.

Further let $\lambda_1 > \lambda_2 > \ldots > \lambda_p$ *be the characteristic roots of* $\mathbf{A}$. *Then, the p.d.f. of* $(\lambda_1, \lambda_2, \ldots, \lambda_p)$ *is*

$$p(\lambda_1, \lambda_2, \ldots, \lambda_p) = \frac{2\pi^{\frac{p(p+n)}{2}}}{\Gamma\left(\frac{p}{2}\right)\Gamma_p\left(\frac{p}{2}\right)\Gamma_p\left(\frac{n-1}{2}\right)} \left(\prod_{i=1}^{p} \lambda_i\right)^{\frac{n-p}{2}-1} \times \prod_{i<j}(\lambda_i - \lambda_j) \int_0^\infty r^{p-1} h\left(r^2 + \sum_{i=1}^{p} \lambda_i\right) dr. \tag{7.19}$$

PROOF: From (7.7), the p.d.f. of $\mathbf{A}$ is

$$p_2(\mathbf{A}) = \frac{2\pi^{\frac{pn}{2}}}{\Gamma\left(\frac{p}{2}\right)\Gamma_p\left(\frac{n-1}{2}\right)} \left(\prod_{i=1}^{p} \lambda_i\right)^{\frac{n-p}{2}-1} \int_0^\infty r^{p-1} h\left(r^2 + \sum_{i=1}^{p} \lambda_i\right) dr$$

and then, using Lemma 7.4, we obtain (7.19). ■

Theorem 7.8. *Let* $\mathbf{X} \sim E_{p,n}(\mathbf{0}, \mathbf{I}_p \otimes \Phi, \psi)$ *have the p.d.f.*

$$f(\mathbf{X}) = \frac{1}{|\Phi|^{\frac{p}{2}}} h(tr(\mathbf{X}'\mathbf{X}\Phi^{-1})),$$

where $n \geq p$. *Let* $\mathbf{B} = \mathbf{X}\Phi^{-1}\mathbf{X}'$. *Further let* $\lambda_1 > \lambda_2 > \ldots > \lambda_p$ *be the characteristic roots of* $\mathbf{B}$. *Then, the p.d.f. of* $(\lambda_1, \lambda_2, \ldots, \lambda_p)$ *is*

$$p(\lambda_1, \lambda_2, \ldots, \lambda_p) = \frac{\pi^{\frac{p(p+n)}{2}}}{\Gamma_p\left(\frac{p}{2}\right)\Gamma_p\left(\frac{n}{2}\right)} \left(\prod_{i=1}^{p} \lambda_i\right)^{\frac{n-p-1}{2}} \prod_{i<j}(\lambda_i - \lambda_j) h\left(\sum_{i=1}^{p} \lambda_i\right). \tag{7.20}$$

PROOF: From (7.18), the p.d.f. of $\mathbf{B}$ is

$$p_2(\mathbf{A}) = \frac{\pi^{\frac{pn}{2}} \left(\prod_{i=1}^{p} \lambda_i\right)^{\frac{n-p-1}{2}}}{\Gamma_p\left(\frac{n-1}{2}\right)} h\left(\sum_{i=1}^{p} \lambda_i\right)$$

and using Lemma 7.4, we obtain (7.20). ■

Next, we want to give a representation of the generalized variance of the estimator of Σ, for which we need the following result.

Lemma 7.5. *Let* $\mathbf{X} \sim N_{p,n}(\mathbf{0}, \mathbf{I}_p \otimes \mathbf{I}_n)$, *with* $n \geq p$. *Then, there exists a lower triangular random matrix* $\mathbf{T}$ *(that is,* $t_{ij} = 0$ *if* $i < j$*), such that*

$$t_{ii}^2 \sim \chi^2_{n-i+1}, \quad i = 1, \ldots, p, \tag{7.21}$$

$$t_{ij} \sim N(0,1), \quad i > j, \tag{7.22}$$

$$t_{ij}, \quad i \geq j \quad \text{are independent}, \tag{7.23}$$

and $\mathbf{XX}' \approx \mathbf{TT}'$.

PROOF: See Anderson (2003), p. 253. ■

Theorem 7.9. *Let* $\mathbf{X} \sim E_{p,n}(\mathbf{0}, \Sigma \otimes \Phi, \psi)$ *have the p.d.f.*

$$f(\mathbf{X}) = \frac{1}{|\Sigma|^{\frac{n}{2}} |\Phi|^{\frac{p}{2}}} h(tr(\mathbf{X}'\Sigma^{-1}\mathbf{X}\Phi^{-1})),$$

where $n \geq p$. *Let* $\mathbf{X} \approx r\Sigma^{\frac{1}{2}}\mathbf{U}\Phi^{\frac{1}{2}}$ *be the stochastic representation of* $\mathbf{X}$. *Let* $\mathbf{B} = \mathbf{X}\Phi^{-1}\mathbf{X}'$, *then*

$$|\mathbf{B}| \approx r^{2p}|\Sigma| \frac{\prod_{i=1}^{p} y_i}{\left(\sum_{i=1}^{p+1} y_i\right)^p}, \tag{7.24}$$

where

$$y_i \sim \chi^2_{n-i+1}, \quad i = 1, \ldots, p, \tag{7.25}$$

$$y_{p+1} \sim \chi^2_{\frac{p(p-1)}{2}}, \quad \text{and} \tag{7.26}$$

$$y_i, \quad i = 1, \ldots, p+1 \quad \text{are independent}. \tag{7.27}$$

PROOF: It follows, from Theorem 7.6, that the distribution of $\mathbf{B}$ does not depend on Φ. So, we can assume $\Phi = \mathbf{I}_n$.

Let $\mathbf{V} \sim N_{p,n}(\mathbf{0}, \mathbf{I}_p \otimes \mathbf{I}_n)$. Then, from Theorem 2.15, it follows that $\mathbf{U} \approx \frac{\mathbf{V}}{\sqrt{tr(\mathbf{VV}')}}$. Assume $\mathbf{X}$ and $\mathbf{V}$ are independent. Then, we get

$$\mathbf{X} \approx r\Sigma^{\frac{1}{2}} \frac{\mathbf{V}}{\sqrt{tr(\mathbf{VV}')}},$$

and hence,

$$|\mathbf{B}| = |\mathbf{XX}'| \approx \left| r^2 \Sigma^{\frac{1}{2}} \frac{\mathbf{VV}'}{tr(\mathbf{VV}')} \Sigma^{\frac{1}{2}} \right| = r^{2p}|\Sigma| \frac{|\mathbf{VV}'|}{(tr(\mathbf{VV}'))^p}.$$

From Lemma 7.5, we can find a lower triangular matrix $\mathbf{T}$, such that

$$\mathbf{VV}' \approx \mathbf{TT}',$$

and $\mathbf{T}$ satisfies (7.21), (7.22) and (7.23). Then,

$$|\mathbf{VV}'| \approx \prod_{i=1}^{p} t_{ii}^2$$

and

$$tr(\mathbf{VV}') \approx \sum_{i=1}^{p} t_{ii}^2 t + \sum_{i>j} t_{ij}.$$

Define $\mathbf{y}_i = t_{ii}^2$, $i = 1, \ldots, p$ and $y_{p+1} = \Sigma_{i>j} t_{ij}$. Then, (7.24)–(7.27) are satisfied. ■

Theorems 7.8 and 7.9 are adapted from Anderson and Fang (1982b). Now, we study the question of unbiasedness of the estimators of μ and Σ.

Theorem 7.10. *Let* $\mathbf{X} \sim E_{p,n}(\mathbf{M}, \Sigma \otimes \Phi, \psi)$.

(a) Then, $\hat{\mathbf{M}} = \mathbf{X}$ *is an unbiased estimator of* $\mathbf{M}$.

(b) If $\mathbf{M}$, Φ, *and* ψ *are known then*

$$\hat{\Sigma}_1 = \frac{-1}{2n\psi'(0)} (\mathbf{X} - \mathbf{M}) \Phi^{-1} (\mathbf{X} - \mathbf{M})'$$

is an unbiased estimator of Σ.

(c) If $\mathbf{M} = \mu \mathbf{v}'$, μ *is a p-dimensional vector,* $\mathbf{v}$ *is an n-dimensional nonzero vector,* $\mathbf{v} \neq \mathbf{0}$, $\mathbf{v}$ *and* Φ *are known, then* $\hat{\mu} = \mathbf{X}\frac{\Phi^{-1}\mathbf{v}}{\mathbf{v}'\Phi^{-1}\mathbf{v}}$ *is an unbiased estimator of* μ. *Moreover, if* ψ *is also known, then*

$$\hat{\Sigma}_2 = \frac{-1}{2(n-1)\psi'(0)} \mathbf{X} \left(\Phi^{-1} - \frac{\Phi^{-1}\mathbf{vv}'\Phi^{-1}}{\mathbf{v}'\Phi^{-1}\mathbf{v}} \right) \mathbf{X}'$$

is an unbiased estimator of Σ.

We assume that the first-order moment of $\mathbf{X}$ *exists when we state the unbiasedness of the estimators of* μ *and that the second-order moment exists when we consider the unbiasedness of the estimators of* Σ.

PROOF:

(a) $E(\hat{\mathbf{M}}) = E(\mathbf{X}) = \mathbf{M}$.

(b) Let $\mathbf{Y} = (\mathbf{X} - \mathbf{M})\Phi^{-\frac{1}{2}}$. Then, $\mathbf{Y} \sim E_{p,n}(\mathbf{0}, \Sigma \otimes \mathbf{I}_n, \psi)$. So $\hat{\Sigma}_1 = \frac{-1}{2n\psi'(0)} \mathbf{YY}'$ and it follows, from Theorem 3.18 that

$$E(\hat{\Sigma}_1) = -\frac{-1}{2n\psi'(0)} (-2\psi'(0)) \Sigma tr(\mathbf{I}_n) = \Sigma.$$

(c) Now we have

$$E(\hat{\mu}) = E\left(\mathbf{X}\frac{\Phi^{-1}\mathbf{v}}{\mathbf{v}'\Phi^{-1}\mathbf{v}}\right) = \mathbf{X}\frac{\mu\mathbf{v}'\Phi^{-1}\mathbf{v}}{\mathbf{v}'\Phi^{-1}\mathbf{v}} = \mu.$$

From Theorem 3.18, we obtain

$$\begin{aligned}
E(\hat{\Sigma}_2) &= -\frac{-1}{2(n-1)\psi'(0)}(-2\psi'(0))\Sigma tr\left(\left(\Phi^{-1} - \frac{\Phi^{-1}\mathbf{v}\mathbf{v}'\Phi^{-1}}{\mathbf{v}'\Phi^{-1}\mathbf{v}}\right)\Phi\right) \\
&+ \mu\mathbf{v}'\left(\Phi^{-1} - \frac{\Phi^{-1}\mathbf{v}\mathbf{v}'\Phi^{-1}}{\mathbf{v}'\Phi^{-1}\mathbf{v}}\right)\mathbf{v}\mu' \\
&= \frac{1}{n-1}\Sigma\left(tr(\mathbf{I}_n) - tr\left(\frac{\Phi^{-1}\mathbf{v}\mathbf{v}'}{\mathbf{v}'\Phi^{-1}\mathbf{v}}\right)\right) + \mu\left(\mathbf{v}'\Phi^{-1}\mathbf{v} - \frac{\mathbf{v}'\Phi^{-1}\mathbf{v}\mathbf{v}'\Phi^{-1}\mathbf{v}}{\mathbf{v}'\Phi^{-1}\mathbf{v}}\right)\mu'
\end{aligned}$$

and since $tr\left(\frac{\Phi^{-1}\mathbf{v}\mathbf{v}'}{\mathbf{v}'\Phi^{-1}\mathbf{v}}\right) = tr\left(\frac{\mathbf{v}'\Phi^{-1}\mathbf{v}}{\mathbf{v}'\Phi^{-1}\mathbf{v}}\right) = \frac{\mathbf{v}'\Phi^{-1}\mathbf{v}}{\mathbf{v}'\Phi^{-1}\mathbf{v}} = 1$, we get

$$E(\hat{\Sigma}_2) = \frac{1}{n-1}\Sigma(n-1) + \mathbf{0} = \Sigma.$$

■

The next theorem focuses on the sufficiency of the estimators.

Theorem 7.11. *Let $\mathbf{X} \sim E_{p,n}(\mathbf{M}, \Sigma \otimes \Phi, \psi)$, with the p.d.f.*

$$f(\mathbf{X}) = \frac{1}{|\Sigma|^{\frac{n}{2}}|\Phi|^{\frac{p}{2}}}h(tr((\mathbf{X}-\mathbf{M})'\Sigma^{-1}(\mathbf{X}-\mathbf{M})\Phi^{-1})).$$

and assume Φ is known.

(a) If Σ is known, then $\hat{\mathbf{M}} = \mathbf{X}$ is sufficient for $\mathbf{M}$.
(b) If $\mathbf{M}$ is known, then $\mathbf{A} = (\mathbf{X}-\mathbf{M})\Phi^{-1}(\mathbf{X}-\mathbf{M})'$ is sufficient for Σ.
(c) If $\mathbf{M} = \mu\mathbf{v}'$, where μ is a p-dimensional vector, and $\mathbf{v} \neq \mathbf{0}$ is an n-dimensional known vector, then $(\hat{\mu}, \mathbf{B})$ is sufficient for (μ, Σ), where

$$\hat{\mu} = \mathbf{X}\frac{\Phi^{-1}\mathbf{v}}{\mathbf{v}'\Phi^{-1}\mathbf{v}} \quad \text{and} \quad \mathbf{B} = \mathbf{X}\left(\Phi^{-1} - \frac{\Phi^{-1}\mathbf{v}\mathbf{v}'\Phi^{-1}}{\mathbf{v}'\Phi^{-1}\mathbf{v}}\right)\mathbf{X}'.$$

PROOF:

(a) Trivial.
(b) This statement follows, if we write

$$\begin{aligned}
f(\mathbf{X}) &= \frac{1}{|\Sigma|^{\frac{n}{2}}|\Phi|^{\frac{p}{2}}}h(tr(\Sigma^{-1}[(\mathbf{X}-\mathbf{M})\Phi^{-1}(\mathbf{X}-\mathbf{M})'])) \\
&= \frac{1}{|\Sigma|^{\frac{n}{2}}|\Phi|^{\frac{p}{2}}}h(tr(\Sigma^{-1}\mathbf{A})).
\end{aligned}$$

(c) Using Lemma 7.1, we can write

$$f(\mathbf{X}) = \frac{1}{|\Sigma|^{\frac{n}{2}}|\Phi|^{\frac{p}{2}}} h\left(tr\left(\Sigma^{-1}\left[\mathbf{X}\left(\Phi^{-1} - \frac{\Phi^{-1}\mathbf{v}\mathbf{v}'\Phi^{-1}}{\mathbf{v}'\Phi^{-1}\mathbf{v}}\right)\mathbf{X}'\right.\right.\right.$$

$$\left.\left.\left. + \ (\mathbf{v}'\Phi^{-1}\mathbf{v})\left(\mathbf{X}\frac{\Phi^{-1}\mathbf{v}}{\mathbf{v}'\Phi^{-1}\mathbf{v}} - \mu\right)\left(\mathbf{X}\frac{\Phi^{-1}\mathbf{v}}{\mathbf{v}'\Phi^{-1}\mathbf{v}} - \mu\right)'\right]\right)\right)$$

$$= \frac{1}{|\Sigma|^{\frac{n}{2}}|\Phi|^{\frac{p}{2}}} h(tr(\Sigma^{-1}[\mathbf{B} + (\mathbf{v}'\Phi^{-1}\mathbf{v})(\hat{\mu} - \mu)(\hat{\mu} - \mu)'])).$$

This proves that $(\hat{\mu}, \mathbf{B})$ is sufficient for (μ, Σ). ■

Remark 7.2. Let $\mathbf{X} \sim E_{p,n}(\mu\mathbf{e}_n', \Sigma \otimes \mathbf{I}_n, \psi)$ have the p.d.f.

$$f(\mathbf{X}) = \frac{1}{|\Sigma|^{\frac{n}{2}}|\Phi|^{\frac{p}{2}}} h(tr((\mathbf{X} - \mu\mathbf{e}_n')'\Sigma^{-1}(\mathbf{X} - \mu\mathbf{e}_n'))),$$

where μ is a p-dimensional vector and $h : \mathbb{R}_0^+ \to \mathbb{R}_0^+$ a decreasing function. Let $g : \mathbb{R}_0^+ \to \mathbb{R}_0^+$ be an increasing function. Assume we have an observation from $\mathbf{X}$ and want to estimate μ. Then, $\hat{\mu} = \mathbf{X}\frac{\mathbf{e}_n'}{n} = \bar{\mathbf{x}}$ is a minimax estimator of μ under the loss function $l(\mathbf{z}) = g((\mathbf{z} - \mu)'\Sigma^{-1}(\mathbf{z} - \mu))$. This result has been proved by Fan and Fang (1985a). In another paper, Fan and Fang (1985b) showed that if in addition to the above mentioned conditions, we assume that $n > p \geq 4$, and g is a convex function whose second derivative exists almost everywhere with respect to the Lebesgue measure and $P(\mathbf{X} = \mu\mathbf{e}_n') = 0$, then $\bar{\mathbf{x}}$ is an inadmissible estimator of μ. More precisely, they showed that the estimator $\hat{\mu}_c = \left(1 - \frac{c}{\bar{\mathbf{x}}'\hat{\Sigma}^{-1}\bar{\mathbf{x}}}\right)\bar{\mathbf{x}}$, where $\hat{\Sigma} = \mathbf{X}\left(\mathbf{I}_n - \frac{\mathbf{e}_n\mathbf{e}_n'}{n}\right)\mathbf{X}'$ and

$$0 < c \leq \frac{2(pn - p + 2)(n-1)(p-2)(p-3)}{n(n-p+2)(p-1)(n^2 - 2pn + 2n + p - 2)}$$

dominates $\bar{\mathbf{x}}$. As a consequence, $\hat{\mu}_c$ is also a minimax estimator of μ.

Chapter 8
Hypothesis Testing

8.1 General Results

Before studying concrete hypotheses, we derive some general theorems. These results are based on Anderson, Fang, and Hsu (1986) and Hsu (1985b).

Theorem 8.1. *Assume we have an observation* $\mathbf{X}$ *from the distribution* $E_{p,n}(\mathbf{M}, \Sigma \otimes \Phi, \psi)$*, where* $(\mathbf{M}, \Sigma \otimes \Phi) \in \Omega \subset \mathbb{R}^{p\times n} \times \mathbb{R}^{pn\times pn}$*, and we want to test*

$$H_0 : (\mathbf{M}, \Sigma \otimes \Phi) \in \omega \quad \textit{against} \quad H_1 : (\mathbf{M}, \Sigma \otimes \Phi) \in \Omega - \omega, \tag{8.1}$$

where $\omega \subset \Omega$*. Suppose* Ω *and* ω *have the properties that if* $\mathbf{Q} \in \mathbb{R}^{p\times n}$*,* $\mathbf{S} \in \mathbb{R}^{pn\times pn}$*, then* $(\mathbf{Q}, \mathbf{S}) \in \Omega$ *implies* $(\mathbf{Q}, c\mathbf{S}) \in \Omega$ *and* $(\mathbf{Q}, \mathbf{S}) \in \omega$ *implies* $(\mathbf{Q}, c\mathbf{S}) \in \omega$ *for any positive scalar c. Moreover, let* $\mathbf{X}$ *have the p.d.f.*

$$f(\mathbf{X}) = \frac{1}{|\Sigma|^{\frac{n}{2}} |\Phi|^{\frac{p}{2}}} h(tr((\mathbf{X}-\mathbf{M})'\Sigma^{-1}(\mathbf{X}-\mathbf{M})\Phi^{-1})),$$

where $l(z) = z^{\frac{pn}{2}} h(z)$ $(z \geq 0)$ *has a finite maximum at* $z = z_h > 0$*. Furthermore, suppose that under the assumption that* $\mathbf{X} \sim N_{p,n}(\mathbf{M}, \Sigma \otimes \Phi)$*,* $(\mathbf{M}, \Sigma \otimes \Phi) \in \Omega$*, the MLE's of* $\mathbf{M}$ *and* $\Sigma \otimes \Phi$ *are* $\mathbf{M}^*$ *and* $(\Sigma \otimes \Phi)^*$*, which are unique and* $P((\Sigma \otimes \Phi)^* > \mathbf{0}) = 1$*.*

Assume also that under the assumption that $\mathbf{X} \sim N_{p,n}(\mathbf{M}, \Sigma \otimes \Phi)$*,* $(\mathbf{M}, \Sigma \otimes \Phi) \in \omega$*, the MLE's of* $\mathbf{M}$ *and* $\Sigma \otimes \Phi$ *are* $\mathbf{M}_0^*$ *and* $(\Sigma \otimes \Phi)_0^*$*, which are unique and* $P((\Sigma \otimes \Phi)_0^* > \mathbf{0}) = 1$*.*

Then, the likelihood ratio test (LRT) statistic for testing (8.1) under the assumption that $\mathbf{X} \sim E_{p,n}(\mathbf{M}, \Sigma \otimes \Phi, \psi)$*, is the same as under the assumption that* $\mathbf{X} \sim N_{p,n}(\mathbf{M}, \Sigma \otimes \Phi)$*, namely* $\frac{|(\Sigma \otimes \Phi)^*|}{|(\Sigma \otimes \Phi)_0^*|}$*.*

PROOF: From Theorem 7.2 it follows that under the condition $E_{p,n}(\mathbf{M}, \Sigma \otimes \Phi, \psi)$, $(\mathbf{M}, \Sigma \otimes \Phi) \in \Omega$, the MLE's of $\mathbf{M}$ and $\Sigma \otimes \Phi$ are $\hat{\mathbf{M}} = \mathbf{M}^*$ and $\widehat{(\Sigma \otimes \Phi)} = \frac{np}{z_h}(\Sigma \otimes$

A.K. Gupta et al., *Elliptically Contoured Models in Statistics and Portfolio Theory*,
DOI 10.1007/978-1-4614-8154-6_8, © Springer Science+Business Media New York 2013

$\boldsymbol{\Phi})^*$ and the maximum of the likelihood is

$$|(\widehat{\Sigma \otimes \boldsymbol{\Phi}})|^{-\frac{1}{2}} h(z_h).$$

Similarly, under the condition $E_{p,n}(\mathbf{M}, \Sigma \otimes \boldsymbol{\Phi}, \psi)$, $(\mathbf{M}, \Sigma \otimes \boldsymbol{\Phi}) \in \omega$, the MLE's of $\mathbf{M}$ and $\Sigma \otimes \boldsymbol{\Phi}$ are $\hat{\mathbf{M}}_0 = \mathbf{M}_0^*$ and $(\widehat{\Sigma \otimes \boldsymbol{\Phi}})_0 = \frac{np}{z_h}(\Sigma \otimes \boldsymbol{\Phi})_0^*$ and the maximum of the likelihood is

$$|(\widehat{\Sigma \otimes \boldsymbol{\Phi}})_0|^{-\frac{1}{2}} h(z_h).$$

Hence, the LRT statistic is

$$\frac{|(\widehat{\Sigma \otimes \boldsymbol{\Phi}})_0|^{-\frac{1}{2}} h(z_h)}{|(\widehat{\Sigma \otimes \boldsymbol{\Phi}})|^{-\frac{1}{2}} h(z_h)} = \frac{\left|\frac{np}{z_h}(\Sigma \otimes \boldsymbol{\Phi})_0^*\right|^{-\frac{1}{2}}}{\left|\frac{np}{z_h}(\Sigma \otimes \boldsymbol{\Phi})^*\right|^{-\frac{1}{2}}} = \left(\frac{|(\Sigma \otimes \boldsymbol{\Phi})^*|}{|(\Sigma \otimes \boldsymbol{\Phi})_0^*|}\right)^{\frac{1}{2}}$$

which is equivalent to the test statistic $\frac{|(\Sigma \otimes \boldsymbol{\Phi})^*|}{|(\Sigma \otimes \boldsymbol{\Phi})_0^*|}$. ■

Theorem 8.2. *Assume we have an observation* $\mathbf{X}$ *from the absolutely continuous distribution* $E_{p,n}(\mathbf{M}, \Sigma \otimes \boldsymbol{\Phi}, \psi)$, *where* $(\mathbf{M}, \Sigma \otimes \boldsymbol{\Phi}) \in \Omega = \Omega_1 \times \Omega_2$ *with* $\Omega_1 \in \mathbb{R}^{p \times n}$, $\Omega_2 \in \mathbb{R}^{pn \times pn}$. *We want to test*

$$H_0 : (\mathbf{M}, \Sigma \otimes \boldsymbol{\Phi}) \in \omega \quad \textit{against} \quad H_1 : (\mathbf{M}, \Sigma \otimes \boldsymbol{\Phi}) \in \Omega - \omega,$$

where $\omega = \omega_1 \times \omega_2$, $\omega_1 \subset \Omega_1$, $\omega_2 \subset \Omega_2$. *Assume that* $\mathbf{0} \in \omega_1$. *Let* $f(\mathbf{Z})$ *be a test statistic, such that* $f(c\mathbf{Z}) = f(\mathbf{Z})$ *for any scalar* $c > 0$. *Then, we have the following*

(a) If

$$f(\mathbf{Z}) = f(\mathbf{Z} - \mathbf{M}) \tag{8.2}$$

for every $\mathbf{M} \in \omega_1$, *then the null distribution of* $f(\mathbf{X})$ *is the same under* $\mathbf{X} \sim E_{p,n}(\mathbf{M}, \Sigma \otimes \boldsymbol{\Phi}, \psi)$, *as under* $\mathbf{X} \sim N_{p,n}(\mathbf{M}, \Sigma \otimes \boldsymbol{\Phi})$.

(b) If $f(\mathbf{Z}) = f(\mathbf{Z} - \mathbf{M})$ *for every* $\mathbf{M} \in \Omega_1$, *then the distribution of* $f(\mathbf{X})$, *null as well as the nonnull, is the same under* $\mathbf{X} \sim E_{p,n}(\mathbf{M}, \Sigma \otimes \boldsymbol{\Phi}, \psi)$, *as under* $\mathbf{X} \sim N_{p,n}(\mathbf{M}, \Sigma \otimes \boldsymbol{\Phi})$.

PROOF:

(a) Let $\mathbf{X} \sim E_{p,n}(\mathbf{M}, \Sigma \otimes \boldsymbol{\Phi}, \psi)$, $\mathbf{M} \in \omega_1$. Define $\mathbf{Y} = \mathbf{X} - \mathbf{M}$. Then $\mathbf{Y} \sim E_{p,n}(\mathbf{0}, \Sigma \otimes \boldsymbol{\Phi}, \psi)$ and $f(\mathbf{X}) = f(\mathbf{X} - \mathbf{M}) = f(\mathbf{Y})$. Thus, the distribution of $f(\mathbf{X})$ is the same as the distribution of $f(\mathbf{Y})$.
From Theorem 5.12, it follows that the distribution of $f(\mathbf{Y})$ is the same under $\mathbf{Y} \sim E_{p,n}(\mathbf{0}, \Sigma \otimes \boldsymbol{\Phi}, \psi)$, as under $\mathbf{Y} \sim N_{p,n}(\mathbf{0}, \Sigma \otimes \boldsymbol{\Phi})$. However, $f(\mathbf{Y}) = f(\mathbf{Y} + \mathbf{M})$, therefore, the distribution of $f(\mathbf{Y})$ is the same under $\mathbf{Y} \sim N_{p,n}(\mathbf{0}, \Sigma \otimes \boldsymbol{\Phi})$, as under $\mathbf{Y} \sim N_{p,n}(\mathbf{M}, \Sigma \otimes \boldsymbol{\Phi})$.

(b) Let $\mathbf{X} \sim E_{p,n}(\mathbf{M}, \Sigma \otimes \Phi, \psi)$, $\mathbf{M} \in \Omega_1$. Define $\mathbf{Y} = \mathbf{X} - \mathbf{M}$. Then $\mathbf{Y} \sim E_{p,n}(\mathbf{0}, \Sigma \otimes \Phi, \psi)$ and the proof can be completed in exactly the same way as the proof of part (a). ■

Corollary 8.1. *Assume that the conditions of Theorem 8.2 are satisfied, including the condition of part (b). Assume that a test based on* $f(\mathbf{X})$ *is unbiased under* $\mathbf{X} \sim N_{p,n}(\mathbf{M}, \Sigma \otimes \Phi)$. *Then the same test is also unbiased under* $\mathbf{X} \sim E_{p,n}(\mathbf{M}, \Sigma \otimes \Phi, \psi)$.

If a test based on $f(\mathbf{X})$ *is strictly unbiased under* $\mathbf{X} \sim N_{p,n}(\mathbf{M}, \Sigma \otimes \Phi)$*; that is, the power function is greater under* H_1 *than under* H_0*; then the same test is also strictly unbiased under* $\mathbf{X} \sim E_{p,n}(\mathbf{M}, \Sigma \otimes \Phi, \psi)$.

PROOF: Since the distribution of the test statistic does not depend on ψ, neither does the power function. Now, the unbiasedness is determined by the power function. So, if the test is unbiased when $\psi(z) = \exp\left(-\frac{z}{2}\right)$, then it is also unbiased for the other ψ's. The other part of the statement follows similarly. ■

Next, we look at the hypothesis testing problem from the point of view of invariance.

Theorem 8.3. *Assume we have an observation* $\mathbf{X}$ *from the absolutely continuous distribution* $E_{p,n}(\mathbf{M}, \Sigma \otimes \Phi, \psi)$, *where* $(\mathbf{M}, \Sigma \otimes \Phi) \in \Omega = \Omega_1 \times \Omega_2$ *with* $\Omega_1 \in \mathbb{R}^{p \times n}$, $\Omega_2 \in \mathbb{R}^{pn \times pn}$, *and we want to test*

$$H_0 : (\mathbf{M}, \Sigma \otimes \Phi) \in \omega \quad \textit{against} \quad H_1 : (\mathbf{M}, \Sigma \otimes \Phi) \in \Omega - \omega,$$

where $\omega = \omega_1 \times \omega_2$, $\omega_1 \subset \Omega_1$, $\omega_2 \subset \Omega_2$. *Let* $\mathbf{0} \in \omega_1$. *Assume the hypotheses are invariant under a group G of transformations* $g : \mathbb{R}^{p \times n} \to \mathbb{R}^{p \times n}$, *such that* $g(\mathbf{X}) = c\mathbf{X}$, $c > 0$ *scalar. Let* $f(\mathbf{X})$ *be a test statistic invariant under G. Then, we have the following.*

(a) *If* $g(\mathbf{X}) = \mathbf{X} - \mathbf{M}$, $\mathbf{M} \in \omega_1$ *are all elements of G, then the null distribution of* $f(\mathbf{X})$ *is the same under* $\mathbf{X} \sim E_{p,n}(\mathbf{M}, \Sigma \otimes \Phi, \psi)$, *as under* $\mathbf{X} \sim N_{p,n}(\mathbf{M}, \Sigma \otimes \Phi)$.
(b) *If* $g(\mathbf{X}) = \mathbf{X} - \mathbf{M}$, $\mathbf{M} \in \Omega_1$ *are all elements of G, then the distribution of* $f(\mathbf{X})$, *the null as well as the nonnull is the same under* $\mathbf{X} \sim E_{p,n}(\mathbf{M}, \Sigma \otimes \Phi, \psi)$, *as under* $\mathbf{X} \sim N_{p,n}(\mathbf{M}, \Sigma \otimes \Phi)$.

PROOF: The results follow from Theorem 8.2, since the invariance of $f(\mathbf{X})$, under the transformation $g(\mathbf{X}) = c\mathbf{X}$, $c > 0$ implies $f(\mathbf{X}) = f(c\mathbf{X})$, $c > 0$; the invariance under the transformation $g(\mathbf{X}) = \mathbf{X} - \mathbf{M}$, $\mathbf{M} \in \omega_1$, implies $f(\mathbf{X}) = f(\mathbf{X} - \mathbf{M})$, $\mathbf{M} \in \omega_1$, and the invariance under the transformation $g(\mathbf{X}) = \mathbf{X} - \mathbf{M}$, $\mathbf{M} \in \Omega_1$, implies $f(\mathbf{X}) = f(\mathbf{X} - \mathbf{M})$, $\mathbf{M} \in \Omega_1$. ■

Corollary 8.2. *Assume that the conditions of Theorem 8.3 are satisfied, including the condition of part (b). Assume that a test based on* $f(\mathbf{X})$ *is unbiased under* $\mathbf{X} \sim N_{p,n}(\mathbf{M}, \Sigma \otimes \Phi)$. *Then, the same test is also unbiased under* $\mathbf{X} \sim E_{p,n}(\mathbf{M}, \Sigma \otimes \Phi, \psi)$.

If a test based on $f(\mathbf{X})$ *is strictly unbiased under* $\mathbf{X} \sim N_{p,n}(\mathbf{M}, \Sigma \otimes \Phi)$, *then the same test is also strictly unbiased under* $\mathbf{X} \sim E_{p,n}(\mathbf{M}, \Sigma \otimes \Phi, \psi)$.

PROOF: It follows from the proof of Theorem 8.3 that the conditions of Theorem 8.2 are satisfied including the conditions of part (b). Thus Corollary 8.1 can be applied, which completes the proof. ■

Further aspects of invariant statistics are studied in the following theorems.

Theorem 8.4. *Assume we have an observation* $\mathbf{X}$, *from the absolutely continuous distribution* $\mathbf{X} \sim E_{p,n}(\mathbf{M}, \Sigma \otimes \Phi, \psi)$, *where* $(\mathbf{M}, \Sigma \otimes \Phi) \in \Omega \subset \mathbb{R}^{p \times n} \times \mathbb{R}^{pn \times pn}$, *and we want to test*

$$H_0 : (\mathbf{M}, \Sigma \otimes \Phi) \in \omega \quad \text{against} \quad H_1 : (\mathbf{M}, \Sigma \otimes \Phi) \in \Omega - \omega,$$

where $\omega \subset \Omega$. *Let* G *be a group of the linear transformations* $g : \mathbb{R}^{p \times n} \to \mathbb{R}^{p \times n}$, *where* $g(\mathbf{X}) = \mathbf{C}_1\mathbf{X}\mathbf{C}_2 + \mathbf{C}_3$ *with* $\mathbf{C}_1 : p \times p$, $\mathbf{C}_2 : n \times n$, *and* $\mathbf{C}_3 : p \times n$ *matrices. Then, the hypotheses are invariant under* G *when* $\mathbf{X} \sim E_{p,n}(\mathbf{M}, \Sigma \otimes \Phi, \psi)$, *if and only if the hypotheses are invariant under* G *when* $\mathbf{X} \sim N_{p,n}(\mathbf{M}, \Sigma \otimes \Phi)$.

Also suppose that the sufficient statistic $T(\mathbf{X})$ *for* $(\mathbf{M}, \Sigma \otimes \Phi) \in \Omega$ *is the same under* $\mathbf{X} \sim E_{p,n}(\mathbf{M}, \Sigma \otimes \Phi, \psi)$ *as under* $\mathbf{X} \sim N_{p,n}(\mathbf{M}, \Sigma \otimes \Phi)$. *Then,* $f(\mathbf{X})$ *is an invariant of the sufficient statistic under* G *when* $\mathbf{X} \sim E_{p,n}(\mathbf{M}, \Sigma \otimes \Phi, \psi)$, *if and only if* $f(\mathbf{X})$ *is an invariant of the sufficient statistic under* G *when* $\mathbf{X} \sim N_{p,n}(\mathbf{M}, \Sigma \otimes \Phi)$.

PROOF: Assume the hypotheses are invariant under G when $\mathbf{X} \sim E_{p,n}(\mathbf{M}, \Sigma \otimes \Phi, \psi)$. Let $g \in G$. Then, $g(\mathbf{X}) = \mathbf{C}_1\mathbf{X}\mathbf{C}_2 + \mathbf{C}_3$. Now,

$$g(\mathbf{X}) \sim E_{p,n}(\mathbf{C}_1\mathbf{M}\mathbf{C}_2 + \mathbf{C}_3, (\mathbf{C}_1\Sigma\mathbf{C}_1') \otimes (\mathbf{C}_2'\Phi\mathbf{C}_2), \psi) \tag{8.3}$$

Thus, we have

(i) If $(\mathbf{M}, \Sigma \otimes \Phi) \in \omega$ then $(\mathbf{C}_1\mathbf{M}\mathbf{C}_2 + \mathbf{C}_3, (\mathbf{C}_1\Sigma\mathbf{C}_1') \otimes (\mathbf{C}_2'\Phi\mathbf{C}_2)) \in \omega$ and
(ii) If $(\mathbf{M}, \Sigma \otimes \Phi) \in \Omega - \omega$ then $(\mathbf{C}_1\mathbf{M}\mathbf{C}_2 + \mathbf{C}_3, (\mathbf{C}_1\Sigma\mathbf{C}_1') \otimes (\mathbf{C}_2'\Phi\mathbf{C}_2)) \in \Omega - \omega$.

Now, assume $\mathbf{X} \sim N_{p,n}(\mathbf{M}, \Sigma \otimes \Phi)$. Then

$$g(\mathbf{X}) \sim N_{p,n}(\mathbf{C}_1\mathbf{M}\mathbf{C}_2 + \mathbf{C}_3, (\mathbf{C}_1\Sigma\mathbf{C}_1') \otimes (\mathbf{C}_2'\Phi\mathbf{C}_2)) \tag{8.4}$$

and it follows from (i) and (ii) that the hypotheses are invariant under G when $\mathbf{X}$ is normal.

Conversely, assume the hypotheses are invariant under G when $\mathbf{X} \sim N_{p,n}(\mathbf{M}, \Sigma \otimes \Phi)$ and $g \in G$. Let $g(\mathbf{X}) = \mathbf{C}_1\mathbf{X}\mathbf{C}_2 + \mathbf{C}_3$. Then, we have (8.4), and (i) and (ii) follow. Since in the case of $\mathbf{X} \sim E_{p,n}(\mathbf{M}, \Sigma \otimes \Phi, \psi)$, (8.3) holds, (i) and (ii) imply that the hypotheses are invariant under G when $\mathbf{X} \sim E_{p,n}(\mathbf{M}, \Sigma \otimes \Phi, \psi)$. Therefore, the invariant test statistics are the same for the normal case as for

$\mathbf{X} \sim E_{p,n}(\mathbf{M}, \Sigma \otimes \Phi, \psi)$. Since, by assumption, the sufficient statistic is the same for $\mathbf{X}$ normal as for $\mathbf{X} \sim E_{p,n}(\mathbf{M}, \Sigma \otimes \Phi, \psi)$, the statement about $f(\mathbf{X})$ follows. ∎

The next result follows easily from the above theorem.

Corollary 8.3. *In Theorem 8.4 if G is a group of linear transformations then the hypotheses are invariant under G without specifying what particular absolutely continuous m.e.c. distribution we are working with.*

Corollary 8.4. *Assume that the conditions of Theorem 8.4 are satisfied. Also assume the hypotheses are invariant under a group G of linear transformations and that* $s(\mathbf{X})$ *is a maximal invariant of the sufficient statistic under G when* $\mathbf{X} \sim N_{p,n}(\mathbf{M}, \Sigma \otimes \Phi)$*. Then* $s(\mathbf{X})$*, is also a maximal invariant of the sufficient statistic under G when* $\mathbf{X} \sim E_{p,n}(\mathbf{M}, \Sigma \otimes \Phi, \psi)$*.*

PROOF: From Theorem 8.4 it follows that $s(\mathbf{X})$ is an invariant of the sufficient statistic under G when $\mathbf{X} \sim E_{p,n}(\mathbf{M}, \Sigma \otimes \Phi, \psi)$. On the other hand, the maximal invariance of $s(\mathbf{X})$ means that $s(\mathbf{X}) = s(\mathbf{X}^*)$ iff there exists $g \in G$, such that $g(\mathbf{X}) = \mathbf{X}^*$ and this property does not depend on the distribution of $\mathbf{X}$. ∎

Theorem 8.5. *Assume we have an observation* $\mathbf{X}$ *from the absolutely continuous distribution* $\mathbf{X} \sim E_{p,n}(\mathbf{M}, \Sigma \otimes \Phi, \psi)$*, where* $(\mathbf{M}, \Sigma \otimes \Phi) \in \Omega = \Omega_1 \times \Omega_2$ *with* $\Omega_1 \in \mathbb{R}^{p \times n}$*,* $\Omega_2 \in \mathbb{R}^{pn \times pn}$*. We want to test*

$$H_0 : (\mathbf{M}, \Sigma \otimes \Phi) \in \omega \quad \text{against} \quad H_1 : (\mathbf{M}, \Sigma \otimes \Phi) \in \Omega - \omega,$$

where $\omega = \omega_1 \times \omega_2$*,* $\omega_1 \in \Omega_1$*,* $\omega_2 \in \Omega_2$*. Let* $\mathbf{0} \in \omega_1$*. Assume the hypotheses are invariant under a group G of linear transformations* $g : \mathbb{R}^{p \times n} \to \mathbb{R}^{p \times n}$*, where* $g(\mathbf{X}) = \mathbf{C}_1\mathbf{X}\mathbf{C}_2 + \mathbf{C}_3$ *with* $\mathbf{C}_1 : p \times p$*,* $\mathbf{C}_2 : n \times n$*, and* $\mathbf{C}_3 : p \times n$ *matrices and the transformations* $g(\mathbf{X}) = c\mathbf{X} - \mathbf{M}$*,* $c > 0$*,* $\mathbf{M} \in \Omega_1$ *are all elements of G. Also suppose that the sufficient statistic* $T(\mathbf{X})$ *for* $(\mathbf{M}, \Sigma \otimes \Phi) \in \Omega$ *is the same under* $\mathbf{X} \sim E_{p,n}(\mathbf{M}, \Sigma \otimes \Phi, \psi)$ *as under* $\mathbf{X} \sim N_{p,n}(\mathbf{M}, \Sigma \otimes \Phi)$*.*

If $s(\mathbf{X})$ *is a maximal invariant of the sufficient statistic under G and a test based on* $s(\mathbf{X})$ *is uniformly most powerful invariant (UMPI) among the tests based on the sufficient statistic in the normal case, then the same test is also UMPI among the tests based on the sufficient statistic when* $\mathbf{X} \sim E_{p,n}(\mathbf{M}, \Sigma \otimes \Phi, \psi)$*.*

PROOF: From Theorem 8.4, it follows that the invariant of the sufficient under G, in the normal case are the same as when $\mathbf{X} \sim E_{p,n}(\mathbf{M}, \Sigma \otimes \Phi, \psi)$. Corollary 8.4 implies that $s(\mathbf{X})$ is a maximal invariant statistic of the sufficient statistic when $\mathbf{X} \sim E_{p,n}(\mathbf{M}, \Sigma \otimes \Phi, \psi)$. Since $\mathbf{0} \in \omega_1$ and the transformations $g(\mathbf{X}) = c\mathbf{X} - \mathbf{M}$, $c > 0$, $\mathbf{M} \in \Omega_1$ are all elements of G, from part (b) of Theorem 8.3 it follows, that $s(\mathbf{X})$ has the same distribution under $\mathbf{X} \sim E_{p,n}(\mathbf{M}, \Sigma \otimes \Phi, \psi)$ as under $\mathbf{X} \sim N_{p,n}(\mathbf{M}, \Sigma \otimes \Phi)$. Since, $s(\mathbf{X})$ is maximal invariant, every invariant test can be expressed as a function of $s(\mathbf{X})$ (see e.g. Lehmann 1959, p. 216). Therefore, the distributions of the invariant statistics are the same under $\mathbf{X} \sim E_{p,n}(\mathbf{M}, \Sigma \otimes \Phi, \psi)$ as under $\mathbf{X} \sim N_{p,n}(\mathbf{M}, \Sigma \otimes \Phi)$. Thus, if $s(\mathbf{X})$ is uniformly most powerful among invariant tests in the normal case then it has the same property when $\mathbf{X} \sim E_{p,n}(\mathbf{M}, \Sigma \otimes \Phi, \psi)$. ∎

8.2 Two Models

Now, we describe the parameter spaces in which we want to study hypothesis testing problems.

8.2.1 *Model I*

Let $\mathbf{x}_1, \mathbf{x}_2, \ldots, \mathbf{x}_n$ be p-dimensional random vectors, such that $n > p$ and $\mathbf{x}_i \sim E_p(\mu, \Sigma, \psi)$, $i = 1, \ldots, n$. Moreover, assume that $\mathbf{x}_i$, $i = 1, \ldots, n$ are uncorrelated and their joint distribution is elliptically contoured and absolutely continuous. This model can be expressed as

$$\mathbf{X} \sim E_{p,n}(\mu \mathbf{e}_n', \Sigma \otimes \mathbf{I}_n, \psi), \tag{8.5}$$

where $\mathbf{X} = (\mathbf{x}_1, \mathbf{x}_2, \ldots, \mathbf{x}_n)$. Then the joint p.d.f. of $\mathbf{x}_1, \mathbf{x}_2, \ldots, \mathbf{x}_n$ can be written as

$$f(\mathbf{X}) = \frac{1}{|\Sigma|^n} h\left(\sum_{i=1}^{n} (\mathbf{x}_i - \mu)' \Sigma^{-1} (\mathbf{x}_i - \mu) \right). \tag{8.6}$$

Assume $l(z) = z^{\frac{pn}{2}} h(z)$, $z \geq 0$ has a finite maximum at $z = z_h > 0$. Define

$$\bar{\mathbf{x}} = \frac{1}{n} \sum_{i=1}^{n} \mathbf{x}_i \quad \text{and} \quad \mathbf{A} = \sum_{i=1}^{n} (\mathbf{x}_i - \bar{\mathbf{x}})(\mathbf{x}_i - \bar{\mathbf{x}})'.$$

Then $\bar{\mathbf{x}} = \mathbf{X} \frac{\mathbf{e}_n}{n}$, $\mathbf{A} = \mathbf{X} \left(\mathbf{I}_n - \frac{\mathbf{e}_n \mathbf{e}_n'}{n} \right) \mathbf{X}'$ and from Theorem 7.11, the statistic $T(\mathbf{X}) = (\bar{\mathbf{x}}, \mathbf{A})$ is sufficient for (μ, Σ).

If $\psi(z) = \exp\left(-\frac{z}{2}\right)$, then $\mathbf{X} \sim E_{p,n}(\mu \mathbf{e}_n', \Sigma \otimes \mathbf{I}_n)$. In this case $\mathbf{x}_1, \mathbf{x}_2, \ldots, \mathbf{x}_n$ are independent, and identically distributed random vectors each with distribution $N_p(\mu, \Sigma)$. Inference for this structure has been extensively studied in Anderson (2003).

8.2.2 Model II

Let $\mathbf{x}_1^{(i)}, \mathbf{x}_2^{(i)}, \ldots, \mathbf{x}_{n_i}^{(i)}$ be p-dimensional random vectors, such that $n_i > p$, $i = 1, \ldots, q$, and $\mathbf{x}_j^{(i)} \sim E_p(\mu_i, \Sigma_i, \psi)$, $j = 1, \ldots, n_i$, $i = 1, \ldots, q$. Moreover, assume that $\mathbf{x}_j^{(i)}$, $i = 1, \ldots, q$, $j = 1, \ldots, n_i$ are uncorrelated and their joint distribution is also elliptically contoured and absolutely continuous. This model can be expressed as

$$\mathbf{x} \sim E_{pn}\left(\begin{pmatrix} \mathbf{e}_{n_1} \otimes \mu_1 \\ \mathbf{e}_{n_2} \otimes \mu_2 \\ \vdots \\ \mathbf{e}_{n_q} \otimes \mu_q \end{pmatrix}, \begin{pmatrix} \mathbf{I}_{n_1} \otimes \Sigma_1 & & & \\ & \mathbf{I}_{n_2} \otimes \Sigma_2 & & \\ & & \ddots & \\ & & & \mathbf{I}_{n_q} \otimes \Sigma_q \end{pmatrix}, \psi \right) \tag{8.7}$$

where $n = \sum_{i=1}^{q} n_i$ and

$$\mathbf{x} = \begin{pmatrix} \mathbf{x}_1^{(1)} \\ \vdots \\ \mathbf{x}_{n_1}^{(1)} \\ \mathbf{x}_1^{(2)} \\ \vdots \\ \mathbf{x}_{n_2}^{(2)} \\ \vdots \\ \mathbf{x}_1^{(q)} \\ \vdots \\ \mathbf{x}_{n_q}^{(q)} \end{pmatrix}.$$

Then, the joint p.d.f. of $\mathbf{x}_j^{(i)}$, $i = 1, \ldots, q$, $j = 1, \ldots, n_i$ can be written as

$$f(\mathbf{x}) = \frac{1}{\prod_{i=1}^{q} |\Sigma_i|^{n_i}} h\left(\sum_{i=1}^{q} \sum_{j=1}^{n_i} (\mathbf{x}_j^{(i)} - \mu_i)' \Sigma_i^{-1} (\mathbf{x}_j^{(i)} - \mu_i) \right). \tag{8.8}$$

Assume $l(z) = z^{\frac{pn}{2}} h(z)$, $z \geq 0$ has a finite maximum at $z = z_h > 0$. Define

$$\bar{\mathbf{x}}^{(i)} = \frac{1}{n_i} \sum_{j=1}^{n_i} \mathbf{x}_j^{(i)}, \mathbf{A}_i = \sum_{j=1}^{n_i} (\mathbf{x}_j^{(i)} - \bar{\mathbf{x}}^{(i)})(\mathbf{x}_j^{(i)} - \bar{\mathbf{x}}^{(i)})'.$$

and $\mathbf{A} = \sum_{i=1}^{q} \mathbf{A}_i$. Also let $\bar{\mathbf{x}} = \sum_{i=1}^{q} \sum_{j=1}^{n_i} \mathbf{x}_j^{(i)}$ and $\mathbf{B} = \sum_{i=1}^{q} \sum_{j=1}^{n_i} (\mathbf{x}_j^{(i)} - \bar{\mathbf{x}})(\mathbf{x}_j^{(i)} - \bar{\mathbf{x}})'$. Then, we get

$$\begin{aligned}
&\sum_{j=1}^{n_i}(\mathbf{x}_j^{(i)}-\boldsymbol{\mu}_i)'\boldsymbol{\Sigma}_i^{-1}(\mathbf{x}_j^{(i)}-\boldsymbol{\mu}_i)\\
&=tr\left(\sum_{j=1}^{n_i}(\mathbf{x}_j^{(i)}-\boldsymbol{\mu}_i)'\boldsymbol{\Sigma}_i^{-1}(\mathbf{x}_j^{(i)}-\boldsymbol{\mu}_i)\right)\\
&=tr\left(\boldsymbol{\Sigma}_i^{-1}\left(\sum_{j=1}^{n_i}(\mathbf{x}_j^{(i)}-\boldsymbol{\mu}_i)(\mathbf{x}_j^{(i)}-\boldsymbol{\mu}_i)'\right)\right)\\
&=tr\left(\boldsymbol{\Sigma}_i^{-1}\left(\sum_{j=1}^{n_i}(\mathbf{x}_j^{(i)}-\bar{\mathbf{x}}^{(i)})(\mathbf{x}_j^{(i)}-\bar{\mathbf{x}}^{(i)})'+n(\bar{\mathbf{x}}^{(i)}-\boldsymbol{\mu}_i)(\bar{\mathbf{x}}^{(i)}-\boldsymbol{\mu}_i)'\right)\right)\\
&=tr(\boldsymbol{\Sigma}_i^{-1}(\mathbf{A}_i+n(\bar{\mathbf{x}}^{(i)}-\boldsymbol{\mu}_i)(\bar{\mathbf{x}}^{(i)}-\boldsymbol{\mu}_i)')).
\end{aligned}$$

Thus,

$$f(\mathbf{X})=\frac{1}{\prod_{i=1}^q|\boldsymbol{\Sigma}_i|^{n_i}}h\left(\sum_{i=1}^q tr(\boldsymbol{\Sigma}_i^{-1}(\mathbf{A}_i+n(\bar{\mathbf{x}}^{(i)}-\boldsymbol{\mu}_i)(\bar{\mathbf{x}}^{(i)}-\boldsymbol{\mu}_i)'))\right)$$

hence the statistic $(\bar{\mathbf{x}}^{(1)},\ldots,\bar{\mathbf{x}}^{(q)},\mathbf{A}_1,\ldots,\mathbf{A}_q)$ is sufficient for $(\boldsymbol{\mu}_1,\ldots,\boldsymbol{\mu}_q,\boldsymbol{\Sigma}_1,\ldots,\boldsymbol{\Sigma}_q)$.

If $\psi(z)=\exp\left(-\frac{z}{2}\right)$, then

$$\mathbf{x}\sim N_{pn}\left(\begin{pmatrix}\mathbf{e}_{n_1}\otimes\boldsymbol{\mu}_1\\ \mathbf{e}_{n_2}\otimes\boldsymbol{\mu}_2\\ \vdots\\ \mathbf{e}_{n_q}\otimes\boldsymbol{\mu}_q\end{pmatrix},\begin{pmatrix}\mathbf{I}_{n_1}\otimes\boldsymbol{\Sigma}_1&&&\\ &\mathbf{I}_{n_2}\otimes\boldsymbol{\Sigma}_2&&\\ &&\vdots&\\ &&&\mathbf{I}_{n_q}\otimes\boldsymbol{\Sigma}_q\end{pmatrix}\right).$$

In this case $\mathbf{x}_1^{(i)},\mathbf{x}_2^{(i)},\ldots,\mathbf{x}_{n_i}^{(i)}$ are independent, and identically distributed random variables each with distribution $N_p(\boldsymbol{\mu}_i,\boldsymbol{\Sigma}_i)$, $i=1,\ldots,q$. Moreover, $\mathbf{x}_j^{(i)}$, $i=1,\ldots,q$, $j=1,\ldots,n_i$ are jointly independent. Inference for this structure has been studied in Anderson (2003).

A special case of Model II is when $\boldsymbol{\Sigma}_1=\ldots=\boldsymbol{\Sigma}_q=\boldsymbol{\Sigma}$. Then the model can also be expressed as

$$\mathbf{X}\sim E_{p,n}((\boldsymbol{\mu}_1\mathbf{e}_{n_1}',\boldsymbol{\mu}_2\mathbf{e}_{n_2}',\ldots,\boldsymbol{\mu}_q\mathbf{e}_{n_q}'),\boldsymbol{\Sigma}\otimes\mathbf{I}_n,\psi),$$

where $n=\sum_{i=1}^q n_i$ and

$$\mathbf{X}=(\mathbf{x}_1^{(1)},\ldots,\mathbf{x}_{n_1}^{(1)},\mathbf{x}_1^{(2)},\ldots,\mathbf{x}_{n_2}^{(2)},\ldots,\mathbf{x}_1^{(q)},\ldots,\mathbf{x}_{n_q}^{(q)}).$$

This leads to the same joint p.d.f. of $\mathbf{x}_j^{(i)}$, $i = 1,\ldots,q$, $j = 1,\ldots,n_i$ as (8.7); that is

$$f(\mathbf{X}) = \frac{1}{|\Sigma|^n} h\left(\sum_{i=1}^{q}\sum_{j=1}^{n_i}(\mathbf{x}_j^{(i)} - \mu_i)'\Sigma^{-1}(\mathbf{x}_j^{(i)} - \mu_i)\right).$$

8.3 Testing Criteria

In this section, we give results on testing of hypotheses for the two models of Sect. 8.2. We use the notations of that section. We also use the theorems in Sect. 8.1 which show, that in certain cases, the hypothesis testing results of the theory of normal distributions can be easily extended to the theory of m.e.c. distributions. This section is based on Anderson and Fang (1982a), Hsu (1985a,b), and Gupta and Varga (1995b).

8.3.1 Testing That a Mean Vector Is Equal to a Given Vector

In Model I (see Sect. 8.2.1) we want to test

$$H_0 : \mu = \mu_0 \quad \text{against} \quad H_1 : \mu \neq \mu_0. \tag{8.9}$$

We assume that μ and Σ are unknown and μ_0 is given. Note that problem (8.9) is equivalent to testing

$$H_0 : \mu = \mathbf{0} \quad \text{against} \quad H_1 : \mu \neq \mathbf{0}. \tag{8.10}$$

Indeed, if $\mu_0 \neq \mathbf{0}$ then define $\mathbf{x}_i^* = \mathbf{x}_i - \mu_0$, $i = 1,\ldots,n$ and $\mu^* = \mu - \mu_0$. Then, problem (8.9) becomes

$$H_0 : \mu^* = \mathbf{0} \quad \text{against} \quad H_1 : \mu^* \neq \mathbf{0}.$$

Problem (8.10) remains invariant under the group G, where

$$G = \{g | g(\mathbf{X}) = \mathbf{C}\mathbf{X},\ \mathbf{C} \text{is } p \times p \text{ nonsingular}\}. \tag{8.11}$$

Now, we can prove the following theorem.

Theorem 8.6. *The likelihood ratio test (LRT) statistic for problem (8.9) is*

$$T_2 = n(n-1)(\bar{\mathbf{x}} - \mu_0)'\mathbf{A}^{-1}(\bar{\mathbf{x}} - \mu_0).$$

The critical region at level α is $T^2 \geq T^2_{p,n-1}(\alpha)$, where $T^2_{p,n-1}(\alpha) = \frac{(n-1)p}{n-p} F_{p,n-p}(\alpha)$ and $F_{p,n-p}(\alpha)$ denotes the 100 % point of the $F_{p,n-p}$ distribution.

If H_0 holds, then $\frac{n-p}{(n-1)p}T^2 \sim F_{p,n-p}$. Moreover, if $\mu_0 = \mathbf{0}$ then T^2 is the maximal invariant of the sufficient statistic under G.

PROOF: From Theorem 8.1 it follows that the LRT statistic is the same as in the normal case; that is, Hotelling's T^2 statistic. Since problems (8.9) and (8.10) are equivalent, we can focus on (8.10). Then, G satisfies the conditions of part (a) of Theorem 8.3 (here $\omega_1 = \{\mathbf{0}\}$) and T^2 is invariant under G, so the null distribution of T^2 is the same as in the normal case. Thus, the corresponding results of the normal theory can be used here (see Anderson 2003, Chap. 5). Since T^2 is the maximal invariant in the normal case, from Corollary 8.4 it follows that it is also maximal invariant for the present model. ∎

The nonnull distribution of T^2 depends on ψ. The nonnull p.d.f. of T^2 was derived by Hsu (1985a). He also showed that if $p > 1$, then the T^2-test is locally most powerful invariant (LMPI). On the other hand, Kariya (1981) proved that if h is a decreasing convex function, then the T^2-test is uniformly most powerful invariant (UMPI).

8.3.2 Testing That a Covariance Matrix Is Equal to a Given Matrix

In Model I (see Sect. 8.2.1), assume that h is decreasing. We want to test

$$H_0 : \Sigma = \Sigma_0 \quad \text{against} \quad H_1 : \Sigma \neq \Sigma_0 . \tag{8.12}$$

We assume that μ and Σ are unknown and $\Sigma_0 > \mathbf{0}$ is given. It is easy to see that, problem (8.12) is equivalent to testing

$$H_0 : \Sigma = \mathbf{I}_p \quad \text{against} \quad H_1 : \Sigma \neq \mathbf{I}_p \tag{8.13}$$

Theorem 8.7. *The LRT statistic for the problem (8.12) is*

$$\tau = |\Sigma_0^{-1}\mathbf{A}|^{\frac{n}{2}} h(tr(\Sigma_0^{-1}\mathbf{A})).$$

The critical region at level α is

$$\tau \le \tau_\psi(\alpha),$$

where $\tau_\psi(\alpha)$ depends on ψ, but not on Σ_0. The null distribution of τ does not depend on Σ_0.

PROOF: From Theorems 7.2 and 7.4 it follows that

$$\max_{\mu,\Sigma>\mathbf{0}} f(\mathbf{X}) = |\frac{p}{z_h}(\mathbf{A}\otimes\mathbf{I}_n)|^{-\frac{1}{2}}h(z_h)$$
$$= \left(\frac{p}{z_h}\right)^{-\frac{pn}{2}} h(z_h)|\mathbf{A}|^{-\frac{n}{2}}.$$

On the other hand, from Theorem 7.1, we obtain

$$\max_{\mu,\Sigma=\Sigma_0} f(\mathbf{X}) = \frac{1}{|\Sigma_0|^{\frac{n}{2}}}h(tr((\mathbf{X}-\bar{\mathbf{x}}\mathbf{e}_n')'\Sigma_0^{-1}(\mathbf{X}-\bar{\mathbf{x}}\mathbf{e}_n')))$$
$$= \frac{1}{|\Sigma_0|^{\frac{n}{2}}}h(tr(\Sigma_0^{-1}\mathbf{A})).$$

Thus, the likelihood ratio test statistic is given by

$$\frac{\max_{\mu,\Sigma=\Sigma_0} f(\mathbf{X})}{\max_{\mu,\Sigma>\mathbf{0}} f(\mathbf{X})} = \frac{\frac{1}{|\Sigma_0|^{\frac{n}{2}}}h(tr(\Sigma_0^{-1}\mathbf{A}))}{\left(\frac{p}{z_h}\right)^{-\frac{pn}{2}} h(z_h)|\mathbf{A}|^{-\frac{n}{2}}}$$
$$= |\Sigma_0^{-1}\mathbf{A}|^{\frac{n}{2}}h(tr(\Sigma_0^{-1}\mathbf{A}))\left(\frac{p}{z_h}\right)^{\frac{pn}{2}}\frac{1}{h(z_h)}.$$

Hence, the critical region is of the form $\tau \leq \tau_\psi(\alpha)$. Since, (8.13) is equivalent to (8.12), it follows that the null distribution of τ does not depend on Σ_0. Hence, $\tau_\psi(\alpha)$ does not depend on Σ_0, either. ■

In this problem, the distribution of the test statistic τ depends on ψ.

8.3.3 Testing That a Covariance Matrix Is Proportional to a Given Matrix

In Model I (see Sect. 8.2.1) we want to test

$$H_0 : \Sigma = \sigma^2\Sigma_0 \quad \text{against} \quad H_1 : \Sigma \neq \sigma^2\Sigma_0, \tag{8.14}$$

where μ, Σ, σ^2 are unknown, $\sigma^2 > 0$ is a scalar, and $\Sigma_0 > \mathbf{0}$ is given. Problem (8.13) remains invariant under the group G, where G is generated by the linear transformations

(i) $g(\mathbf{X}) = c\mathbf{X}$, $c > 0$ scalar and
(ii) $g(\mathbf{X}) = \mathbf{X} + \mathbf{v}\mathbf{e}_n'$, $\mathbf{v}$ is p-dimensional vector.

It is easy to see that, problem (8.14) is equivalent to testing

$$H_0 : \Sigma = \sigma^2 \mathbf{I}_p \quad \text{against} \quad H_1 : \Sigma \neq \sigma^2 \mathbf{I}_p \tag{8.15}$$

Theorem 8.8. *The LRT statistic for problem (8.14) is*

$$\tau = \frac{|\Sigma_0^{-1}\mathbf{A}|^{\frac{n}{2}}}{tr\left(\frac{1}{p}\Sigma_0^{-1}\mathbf{A}\right)^{\frac{pn}{2}}}.$$

The critical region at level α is

$$\tau \leq \tau(\alpha),$$

where $\tau(\alpha)$ is the same as in the normal case and it does not depend on Σ_0.

The distribution of τ is the same as in the normal case. The null distribution of τ does not depend on Σ_0. τ is an invariant of the sufficient statistic under G.

PROOF: From Theorem 8.1 it follows that the LRT statistic is the same as in the normal case. It is easy to see that τ is invariant under G. Moreover, G satisfies the conditions of part (b) of Theorem 8.3; therefore, the distribution of τ is the same as in the normal case. Hence, the corresponding results of the normal theory can be used here (see Anderson 2003, Sect. 10.7). Since (8.15) is equivalent to (8.14), it follows that the null distribution of τ does not depend on Σ_0. Hence, $\tau(\alpha)$ does not depend on Σ_0, either. ■

Remark 8.1. Since the distribution of τ is the same as in the normal case, its moments and asymptotic distribution under the null hypothesis are those given by the formulas in Anderson (2003, pp. 434–436), Gupta (1977), and Gupta and Nagar (1987, 1988), for the normal case.

Remark 8.2. Nagao's (1973) criterion,

$$\frac{n-1}{2} tr\left(\frac{p\mathbf{A}}{tr\mathbf{A}} - \mathbf{I}_p\right)^2,$$

is also an invariant test criterion under G and hence, it has the same distribution as in the normal case (see also Anderson 2003, pp. 436–437).

8.3.4 Testing That a Mean Vector and Covariance Matrix Are Equal to a Given Vector and Matrix

In Model I (see Sect. 8.2.1) we want to test

$$H_0 : \mu = \mu_0 \text{ and } \Sigma = \Sigma_0 \quad \text{against} \quad H_1 : \mu \neq \mu_0 \text{ or } \Sigma \neq \Sigma_0, \tag{8.16}$$

where μ, Σ are unknown and μ_0 and $\Sigma_0 > \mathbf{0}$ are given. It is easy to see that, problem (8.16) is equivalent to testing

$$H_0 : \mu = \mathbf{0} \text{ and } \Sigma = \mathbf{I}_p \quad \text{against} \quad H_1 : \mu \neq \mathbf{0} \text{ or } \Sigma \neq \mathbf{I}_p . \tag{8.17}$$

Theorem 8.9. *The LRT statistic for problem (8.16) is*

$$\tau = |\Sigma_0^{-1}\mathbf{A}|^{\frac{n}{2}} h\left(tr(\Sigma_0^{-1}\mathbf{A}) + n(\bar{\mathbf{x}} - \mu_0)'\Sigma_0^{-1}(\bar{\mathbf{x}} - \mu_0)\right).$$

The critical region at level α is

$$\tau \leq \tau_\psi(\alpha),$$

where $\tau_\psi(\alpha)$ depends on ψ, but not μ_0 or Σ_0. The null distribution of τ does not depend on μ_0 or Σ_0.

PROOF: From Theorems 7.2 and 7.4 it follows that

$$\max_{\mu,\Sigma>\mathbf{0}} f(\mathbf{X}) = \left(\frac{p}{z_h}\right)^{-\frac{pn}{2}} h(z_h)|\mathbf{A}|^{-\frac{n}{2}}.$$

On the other hand,

$$\begin{aligned}\max_{\mu=\mu_0,\Sigma=\Sigma_0} f(\mathbf{X}) &= \frac{1}{|\Sigma_0|^{\frac{n}{2}}} h(tr((\mathbf{X} - \mu_0\mathbf{e}_n')'\Sigma_0^{-1}(\mathbf{X} - \mu_0\mathbf{e}_n'))) \\ &= \frac{1}{|\Sigma_0|^{\frac{n}{2}}} h\left(tr(\Sigma_0^{-1}\mathbf{A}) + n(\bar{\mathbf{x}} - \mu_0)'\Sigma_0^{-1}(\bar{\mathbf{x}} - \mu_0)\right).\end{aligned}$$

Therefore, the likelihood ratio test statistic is given by

$$\begin{aligned}\frac{\max_{\mu=\mu_0,\Sigma=\Sigma_0} f(\mathbf{X})}{\max_{\mu,\Sigma>\mathbf{0}} f(\mathbf{X})} &= \frac{\frac{1}{|\Sigma_0|^{\frac{n}{2}}} h\left(tr(\Sigma_0^{-1}\mathbf{A}) + n(\bar{\mathbf{x}} - \mu_0)'\Sigma_0^{-1}(\bar{\mathbf{x}} - \mu_0)\right)}{\left(\frac{p}{z_h}\right)^{-\frac{pn}{2}} h(z_h)|\mathbf{A}|^{-\frac{n}{2}}} \\ &= \frac{|\Sigma_0^{-1}\mathbf{A}|^{\frac{n}{2}}}{h(z_h)} h\left(tr(\Sigma_0^{-1}\mathbf{A}) + n(\bar{\mathbf{x}} - \mu_0)'\Sigma_0^{-1}(\bar{\mathbf{x}} - \mu_0)\right) \left(\frac{p}{z_h}\right)^{\frac{pn}{2}}.\end{aligned}$$

Hence, the critical region is of the form $\tau \leq \tau_\psi(\alpha)$. Since, (8.17) is equivalent to (8.16), it follows that the null distribution of τ does not depend on μ_0 or Σ_0. Hence, $\tau_\psi(\alpha)$ does not depend on μ_0 or Σ_0, either. ∎

In this problem, the distribution of the test statistic τ depends on ψ. Nevertheless, Quan and Fang (1987) have proved that the test defined in Theorem 8.9 is unbiased.

8.3.5 Testing That a Mean Vector Is Equal to a Given Vector and a Covariance Matrix Is Proportional to a Given Matrix

In Model I (see Sect. 8.2.1) we want to test

$$H_0: \mu = \mu_0 \text{ and } \Sigma = \sigma^2\Sigma_0 \quad \text{against} \quad H_1: \mu \neq \mu_0 \text{ or } \Sigma \neq \sigma^2\Sigma_0, \tag{8.18}$$

where μ, Σ, σ^2 are unknown, $\sigma^2 > 0$ is a scalar, and μ_0, $\Sigma_0 > \mathbf{0}$ are given.

Note that problem (8.18) is equivalent to testing

$$H_0: \mu = \mathbf{0} \text{ and } \Sigma = \sigma^2\mathbf{I}_p \quad \text{against} \quad H_1: \mu \neq \mathbf{0} \text{ or } \Sigma \neq \sigma^2\mathbf{I}_p. \tag{8.19}$$

Problem (8.19) remains invariant under the group G, where

$$G = \{g | g(\mathbf{X}) = c\mathbf{X},\ c > 0 \text{ scalar}\}.$$

Theorem 8.10. *The LRT statistic for problem (8.18) is*

$$\tau = \frac{|\Sigma_0^{-1}\mathbf{A}|^{\frac{n}{2}}}{\left(tr\left(\frac{1}{p}\Sigma_0^{-1}\mathbf{A}\right) + \frac{n}{p}(\bar{\mathbf{x}} - \mu_0)'\Sigma_0^{-1}(\bar{\mathbf{x}} - \mu_0)\right)^{\frac{pn}{2}}}.$$

The critical region at level α is

$$\tau \leq \tau(\alpha),$$

where $\tau(\alpha)$ is the same as in the normal case and it does not depend on μ_0 or Σ_0. The null distribution of τ is the same as in the normal case and it does not depend on μ_0 or Σ_0. Moreover, if $\mu_0 = \mathbf{0}$ then τ is an invariant of the sufficient statistic under G.

PROOF: From Theorem 8.1 it follows that the LRT statistic is the same as in the normal case. However, since Anderson (2003) does not give the corresponding result for the normal case, we derive it here.

So, assume $\mathbf{X} \sim N_{p,n}(\mu\mathbf{e}_n', \Sigma \otimes \mathbf{I}_n)$ and we want to test (8.18). Then, from Theorem 7.2, Remark 7.1 and Theorem 7.4 it follows that

$$\begin{aligned}
\max_{\mu,\Sigma>\mathbf{0}} f(\mathbf{X}) &= |\frac{\mathbf{A}}{n} \otimes \mathbf{I}_n|^{-\frac{1}{2}} \frac{1}{(2\pi)^{\frac{pn}{2}}} e^{-\frac{pn}{2}} \\
&= \left(\frac{n}{2\pi}\right)^{\frac{pn}{2}} e^{-\frac{pn}{2}} |\mathbf{A}|^{-\frac{n}{2}}.
\end{aligned}$$

Next, we want to maximize $f(\mathbf{X})$ under H_0 where

$$f(\mathbf{X}) = \frac{1}{(2\pi)^{\frac{pn}{2}}|\sigma^2\Sigma_0|^{\frac{n}{2}}}\exp\left\{-\frac{1}{2}[tr((\sigma^2\Sigma_0)^{-1}\mathbf{A})+n(\bar{\mathbf{x}}-\mu_0)'(\sigma^2\Sigma_0)^{-1}(\bar{\mathbf{x}}-\mu_0)]\right\}.$$

Then,

$$\frac{\partial \log f(\mathbf{X})}{\partial \sigma^2} = -\frac{pn}{2\sigma^2} + \frac{1}{2\sigma^4}[tr(\Sigma_0^{-1}\mathbf{A}) + n(\bar{\mathbf{x}}-\mu_0)'\Sigma_0^{-1}(\bar{\mathbf{x}}-\mu_0)]$$

and from $\frac{\partial \log f(\mathbf{X})}{\partial \sigma^2} = 0$, we obtain

$$\hat{\sigma}^2 = \frac{tr(\Sigma_0^{-1}\mathbf{A}) + n(\bar{\mathbf{x}}-\mu_0)'\Sigma_0^{-1}(\bar{\mathbf{x}}-\mu_0)}{pn}.$$

Thus,

$$\max_{\mu=\mu_0,\Sigma=\sigma^2\Sigma_0>\mathbf{0}} f(\mathbf{X})$$

$$= \left(\frac{pn}{2\pi}\right)^{\frac{pn}{2}} e^{-\frac{pn}{2}}|\Sigma_0|^{-\frac{n}{2}}(tr(\Sigma_0^{-1}\mathbf{A}) + n(\bar{\mathbf{x}}-\mu_0)'\Sigma_0^{-1}(\bar{\mathbf{x}}-\mu_0))^{-\frac{pn}{2}}.$$

Therefore, the likelihood ratio test statistic is given by

$$\frac{\max_{\mu=\mu_0,\Sigma=\sigma^2\Sigma_0>\mathbf{0}} f(\mathbf{X})}{\max_{\mu,\Sigma>\mathbf{0}} f(\mathbf{X})}$$

$$= \frac{\left(\frac{pn}{2\pi}\right)^{\frac{pn}{2}} e^{-\frac{pn}{2}}|\Sigma_0|^{-\frac{n}{2}}(tr(\Sigma_0^{-1}\mathbf{A}) + n(\bar{\mathbf{x}}-\mu_0)'\Sigma_0^{-1}(\bar{\mathbf{x}}-\mu_0))^{-\frac{pn}{2}}}{\left(\frac{n}{2\pi}\right)^{\frac{pn}{2}} e^{-\frac{pn}{2}}|\mathbf{A}|^{-\frac{n}{2}}}$$

$$= \frac{|\Sigma_0^{-1}\mathbf{A}|^{\frac{n}{2}}}{\left(tr\left(\frac{1}{p}\Sigma_0^{-1}\mathbf{A}\right) + \frac{n}{p}(\bar{\mathbf{x}}-\mu_0)'\Sigma_0^{-1}(\bar{\mathbf{x}}-\mu_0)\right)^{\frac{pn}{2}}}.$$

Thus, the critical region is of the form $\tau \leq \tau(\alpha)$. Since problems (8.18) and (8.19) are equivalent, we can focus on (8.19). It is easy to see that τ is invariant under G if $\mu_0 = \mathbf{0}$. Moreover G satisfies the conditions of part (a) of Theorem 8.3, so the null distribution of τ is the same as in the normal case and, since problems (8.18) and (8.19) are equivalent, it does not depend on μ_0 or Σ_0. Therefore, $\tau(\alpha)$ is the same as in the normal case and it does not depend on μ_0 or Σ_0. ■

In this case, the nonnull distribution of τ depends on ψ. Nevertheless Quan and Fang (1987) have proved that the LRT is unbiased if h is decreasing.

8.3.6 Testing Lack of Correlation Between Sets of Variates

In Model I (see Sect. 8.2.1), we partition $\mathbf{x}_i$ into q subvectors

$$\mathbf{x}_i = \begin{pmatrix} \mathbf{x}_i^{(1)} \\ \mathbf{x}_i^{(2)} \\ \vdots \\ \mathbf{x}_i^{(q)} \end{pmatrix}$$

where $\mathbf{x}_i^{(j)}$ is p_j-dimensional, $j = 1, \ldots, q$, $i = 1, \ldots, n$. Partition Σ and $\mathbf{A}$ into

$$\Sigma = \begin{pmatrix} \Sigma_{11} & \Sigma_{12} & \ldots & \Sigma_{1q} \\ \Sigma_{21} & \Sigma_{22} & \ldots & \Sigma_{2q} \\ & \vdots & & \\ \Sigma_{q1} & \Sigma_{q2} & \ldots & \Sigma_{qq} \end{pmatrix}, \qquad \mathbf{A} = \begin{pmatrix} \mathbf{A}_{11} & \mathbf{A}_{12} & \ldots & \mathbf{A}_{1q} \\ \mathbf{A}_{21} & \mathbf{A}_{22} & \ldots & \mathbf{A}_{2q} \\ & \vdots & & \\ \mathbf{A}_{q1} & \mathbf{A}_{q2} & \ldots & \mathbf{A}_{qq} \end{pmatrix}$$

where Σ_{jj} and $\mathbf{A}_{jj}$ are $p_j \times p_j$, $j = 1, \ldots, q$. We want to test

$$\begin{aligned} &H_0 : \Sigma_{jk} = \mathbf{0} \quad \text{if } 1 \le j < k \le q \text{ against} \\ &H_1 : \text{there exists } j, k \text{ such that } \Sigma_{jk} \neq \mathbf{0}. \end{aligned} \tag{8.20}$$

Problem (8.20) remains invariant under the group G, where G is generated by the linear transformations

(i) $g(\mathbf{X}) = \mathbf{C}\mathbf{X}$ with $\mathbf{C} = \begin{pmatrix} \mathbf{C}_1 & & & \\ & \mathbf{C}_2 & & \\ & & \ddots & \\ & & & \mathbf{C}_q \end{pmatrix}$, where $\mathbf{C}_j$ is $p_j \times p_j$ nonsingular matrix, $j = 1, \ldots, q$,

(ii) $g(\mathbf{X}) = \mathbf{X} + \mathbf{v}\mathbf{e}_n'$, where $\mathbf{v}$ is p-dimensional vector.

Theorem 8.11. *The LRT statistic for problem (8.20) is*

$$\tau = \frac{|\mathbf{A}|}{\prod_{i=1}^{q} |\mathbf{A}_{ii}|}.$$

The critical region at level α is

$$\tau \le \tau(\alpha),$$

where $\tau(\alpha)$ is the same as in the normal case. The distribution of τ is the same as in the normal case. If H_0 holds, then $\tau \approx \prod_{i=2}^{q} v_i$, where $v_2, v_3, \ldots, v_q$ are independent *and* $v_i \sim U_{p_i,\bar{p}_i,n-\bar{p}_i}$, *with* $\bar{p}_i = \sum_{j=1}^{i-1} p_j$, $i = 2,\ldots,q$. *The LRT is strictly unbiased; that is, if* H_1 *holds, then the probability of rejecting* H_0 *is greater than* α. τ *is an invariant of the sufficient statistic under* G.

PROOF: From Theorem 8.1 it follows that the LRT statistic is the same as in the normal case. It is easy to see that τ is invariant under G. Moreover G satisfies the conditions of part (b) of Theorem 8.3. Therefore, the distribution of τ is the same as in the normal case. Hence, the corresponding results of the normal theory can be used here (see Anderson 2003, Chap. 9). The strict unbiasedness follows from Corollary 8.2 and the normal theory. ∎

Remark 8.3. Since the distribution of τ is the same as in the normal case, its moments and asymptotic distribution under the null hypothesis are these given by the formulas in Anderson (2003, pp. 388–390), and Nagar and Gupta (1986), for the normal case.

Remark 8.4. Nagao's criterion (see Anderson 2003, p. 391) is also invariant under G and hence, it has the same distribution as in the normal case. Thus, its asymptotic distribution under the null hypothesis is given by the formulas in Anderson (2003, p. 391).

Hsu (1985b) has proved that the LRT in Theorem 8.11 is admissible.

8.3.7 *Testing That a Correlation Coefficient Is Equal to a Given Number*

In Model I (see Sect. 8.2.1), let $p = 2$. Then Σ and $\mathbf{A}$ can be written as

$$\Sigma = \begin{pmatrix} \sigma_1^2 & \rho\sigma_1\sigma_2 \\ \rho\sigma_1\sigma_2 & \sigma_2^2 \end{pmatrix} \quad \text{and} \quad \mathbf{A} = \begin{pmatrix} a_{11} & a_{12} \\ a_{21} & a_{22} \end{pmatrix},$$

where $\sigma_1^2 = \sigma_{11}$, $\sigma_2^2 = \sigma_{22}$ and $\rho = \frac{\sigma_{12}}{\sqrt{\sigma_{11}\sigma_{22}}}$. We want to test

$$H_0 : \rho = \rho_0 \quad \text{against} \quad H_1 : \rho \neq \rho_0, \tag{8.21}$$

where μ and Σ are unknown and $|\rho_0| < 1$ is given. Problem (8.21) remains invariant under the group G, where G is generated by the linear transformations

(i) $g(\mathbf{X}) = \mathbf{CX}$, with $\mathbf{C} = \begin{pmatrix} c_1 & 0 \\ 0 & c_2 \end{pmatrix}$ where c_1 and c_2 are positive scalars and

(ii) $g(\mathbf{X}) = \mathbf{X} + \mathbf{v}\mathbf{e}_n'$, where $\mathbf{v}$ is p-dimensional vector.

Theorem 8.12. *The statistic*

$$r = \frac{a_{12}}{\sqrt{a_{11}a_{22}}},$$

is maximal invariant of the sufficient statistic under G. The distribution of r is the same as in the normal case. The LRT for problem (8.21) at level α has the critical region

$$r \leq r_1(\alpha) \quad or \quad r \geq r_2(\alpha),$$

where $r_1(\alpha) = \frac{\rho_0 c - (1-\rho_0^2)\sqrt{1-c}}{\rho_0^2 c + 1 - \rho_0^2}$, $r_2(\alpha) = \frac{\rho_0 c + (1-\rho_0^2)\sqrt{1-c}}{\rho_0^2 c + 1 - \rho_0^2}$ *and c is chosen such that under H_0*

$$P(r \leq r_1(\alpha)) + P(r \geq r_2(\alpha)) = \alpha.$$

The values of $r_1(\alpha)$ and $r_2(\alpha)$ depend on ρ_0, but they are the same as in the normal case.

PROOF: Since G satisfies the conditions of Corollary 8.4 and in the normal case, r is a maximal invariant under G (see Anderson 2003, p. 126), it follows that r is also maximal invariant under G in the present case. Moreover, G satisfies the conditions of part (b) of Theorem 8.3 and r is invariant under G. Therefore, the distribution of r is the same as in the normal case. Furthermore it follows from Theorem 8.1 that the LRT statistic is the same as in the normal case. Thus, the corresponding results of the normal theory can be used (see Anderson 2003, Sect. 4.2).

Remark 8.5. Since the distribution of r is the same as in the normal case, its asymptotic distribution, as well as the asymptotic distribution of Fisher's z statistic, are the same as those given by the formulas in Anderson (2003, pp. 131–134), and Konishi and Gupta (1989). Therefore, the tests based on Fisher's z statistic can also be used here.

Remark 8.6. It is known (e.g., see Anderson 2003, p. 127) that in the normal case, for the problem

$$H_0 : \rho = \rho_0 \quad \text{against} \quad H_1 : \rho > 0$$

the level α test, whose critical region has the form

$$r > r_3(\alpha)$$

is UMPI. Then, it follows from Theorem 8.5 that the same test is also UMPI in the present case.

8.3.8 Testing That a Partial Correlation Coefficient Is Equal to a Given Number

In Model I (see Sect. 8.2.1) partition Σ and $\mathbf{A}$ as

$$\Sigma = \begin{pmatrix} \Sigma_{11} & \Sigma_{12} \\ \Sigma_{21} & \Sigma_{22} \end{pmatrix} \quad \text{and} \quad \mathbf{A} = \begin{pmatrix} \mathbf{A}_{11} & \mathbf{A}_{12} \\ \mathbf{A}_{21} & \mathbf{A}_{22} \end{pmatrix},$$

where Σ_{11} and $\mathbf{A}_{11}$ are 2×2. Assume $p \geq 3$, and define

$$\Sigma_{11\cdot 2} = \Sigma_{11} - \Sigma_{12}\Sigma_{22}^{-1}\Sigma_{21} \quad \text{and} \quad \mathbf{A}_{11\cdot 2} = \mathbf{A}_{11} - \mathbf{A}_{12}\mathbf{A}_{22}^{-1}\mathbf{A}_{21}.$$

We use the notation

$$\Sigma_{11\cdot 2} = \begin{pmatrix} \sigma_{11\cdot 3,\dots,p} & \sigma_{12\cdot 3,\dots,p} \\ \sigma_{21\cdot 3,\dots,p} & \sigma_{22\cdot 3,\dots,p} \end{pmatrix} \quad \text{and} \quad \mathbf{A}_{11\cdot 2} = \begin{pmatrix} a_{11\cdot 3,\dots,p} & a_{12\cdot 3,\dots,p} \\ a_{21\cdot 3,\dots,p} & a_{22\cdot 3,\dots,p} \end{pmatrix},$$

Then, $\rho_{12\cdot 3,\dots,p} = \frac{\sigma_{12\cdot 3,\dots,p}}{\sqrt{\sigma_{11\cdot 3,\dots,p}\sigma_{22\cdot 3,\dots,p}}}$ is called the partial correlation between the first two variables, having the $(p-2)$ variables fixed (see Anderson 2003, p. 34). We want to test

$$H_0 : \rho_{12\cdot 3,\dots,p} = \rho_0 \quad \text{against} \quad H_1 : \rho_{12\cdot 3,\dots,p} \neq \rho_0, \tag{8.22}$$

where μ and Σ are unknown and $|\rho_0| < 1$ is given.

Problem (8.22) remains invariant under the group G, where G is generated by the linear transformations

(i) $g(\mathbf{X}) = \mathbf{CX}$, with $\mathbf{C} = \begin{pmatrix} \mathbf{C}_1 & \mathbf{C}_2 \\ \mathbf{C}_3 & \mathbf{C}_4 \end{pmatrix}$, where $\mathbf{C}_1 = \begin{pmatrix} c_1 & 0 \\ 0 & c_2 \end{pmatrix}$, c_1, c_2 positive scalars, $\mathbf{C}_2$ is any $2 \times (p-2)$ matrix, $\mathbf{C}_3 = \mathbf{0}$ is $(p-2) \times 2$ matrix, and $\mathbf{C}_4 = \mathbf{I}_{p-2}$.

(ii) $g(\mathbf{X}) = \mathbf{CX}$, with $\mathbf{C} = \begin{pmatrix} \mathbf{C}_1 & \mathbf{C}_2 \\ \mathbf{C}_3 & \mathbf{C}_4 \end{pmatrix}$, where $\mathbf{C}_1 = \mathbf{I}_2$, $\mathbf{C}_2$ is $2 \times (p-2)$ matrix, $\mathbf{C}_2 = \mathbf{0}$, $\mathbf{C}_3 = \mathbf{0}$ is $(p-2) \times 2$ matrix, and $\mathbf{C}_4$ is a $(p-2) \times (p-2)$ nonsingular matrix.

(iii) $g(\mathbf{X}) = \mathbf{X} + \mathbf{v}\mathbf{e}_n'$, where $\mathbf{v}$ is p-dimensional vector.

Theorem 8.13. *The statistic*

$$r_{12\cdot 3,\dots,p} = \frac{a_{12\cdot 3,\dots,p}}{\sqrt{a_{11\cdot 3,\dots,p}a_{22\cdot 3,\dots,p}}}$$

is a maximal invariant of the sufficient statistic under G. The distribution of $r_{12\cdot 3,\dots,p}$ is the same as in the normal case and it is the same as the distribution of r in Sect. 8.3.7, where we replace n by $n-p+2$.

PROOF: Since G satisfies the conditions of Corollary 8.4 and in the normal case, $r_{12\cdot 3,\ldots,p}$ is a maximal invariant under G (see Anderson 2003, p. 159), it follows that r is also a maximal invariant in the present case. Moreover, G satisfies the conditions of part (b) of Theorem 8.3 and $r_{12\cdot 3,\ldots,p}$ is invariant under G. Therefore, the distribution of $r_{12\cdot 3,\ldots,p}$ is the same as in the normal case. Furthermore, in the normal case, $r_{12\cdot 3,\ldots,p}$ has the same distribution as r when n is replaced by $n-p+2$ (see Anderson 2003, p. 143). ∎

Because of the relation between the distribution of $r_{12\cdot 3,\ldots,p}$ and r, all statistical methods mentioned in Sect. 8.3.7 can also be used here. So, for example, Fisher's z test can be applied here. Also, the test based on $r_{12\cdot 3,\ldots,p}$ for the problem

$$H_0 : \rho_{12\cdot 3,\ldots,p} = \rho_0 \quad \text{against} \quad H_1 : r_{12\cdot 3,\ldots,p} > \rho_0$$

is UMPI.

8.3.9 *Testing That a Multiple Correlation Coefficient Is Equal to a Given Number*

In Model I (see Sect. 8.2.1), let us partition Σ and $\mathbf{A}$ as

$$\Sigma = \begin{pmatrix} \sigma_{11} & \sigma_1' \\ \sigma_1 & \Sigma_{22} \end{pmatrix} \quad \text{and} \quad \mathbf{A} = \begin{pmatrix} a_{11} & \mathbf{a}_1' \\ \mathbf{a}_1 & \mathbf{A}_{22} \end{pmatrix},$$

where σ_{11} and s_{11} are one-dimensional. Then, $\bar{\rho}_{1\cdot 2,\ldots,p} = \sqrt{\frac{\sigma_1' \Sigma_{22}^{-1} \sigma_1}{\sigma_{11}}}$ is called the multiple correlation between the first variable and the other $(p-1)$ variables (see Anderson 2003, p. 38). We want to test

$$H_0 : \bar{\rho}_{1\cdot 2,\ldots,p} = 0 \quad \text{against} \quad H_1 : \bar{\rho}_{1\cdot 2,\ldots,p} \neq 0, \tag{8.23}$$

where μ and Σ are unknown. Problem (8.23) remains invariant under the group G, where G is generated by the linear transformations

(i) $g(\mathbf{X}) = \mathbf{CX}$, with $\mathbf{C} = \begin{pmatrix} c_1 & \mathbf{0}' \\ \mathbf{0} & \mathbf{C}_2 \end{pmatrix}$, where c_1 is nonnegative scalar and $\mathbf{C}_2$ is a $(p-1)\times(p-1)$ nonsingular matrix.
(ii) $g(\mathbf{X}) = \mathbf{X} + \mathbf{v}\mathbf{e}_n'$, where $\mathbf{v}$ is p-dimensional vector.

Theorem 8.14. *The statistic*

$$\bar{r}_{1\cdot 2,\ldots,p} = \sqrt{\frac{\mathbf{a}_1' \mathbf{A}_{22}^{-1} \mathbf{a}_1}{a_{11}}}$$

is maximal invariant of the sufficient statistic under G. The distribution of $\bar{r}_{1\cdot 2,\ldots,p}$ *is the same as in the normal case. If* H_0 *holds, then* $\frac{n-p}{p-1}\frac{\bar{r}^2_{1\cdot 2,\ldots,p}}{1-\bar{r}^2_{1\cdot 2,\ldots,p}} \sim F_{p-1,n-p}$. *The LRT for problem (8.23) at level* α *has the critical region*

$$\frac{n-p}{p-1}\frac{\bar{r}^2_{1\cdot 2,\ldots,p}}{1-\bar{r}^2_{1\cdot 2,\ldots,p}} \geq F_{p-1,n-p}(\alpha),$$

where $F_{p-1,n-p}(\alpha)$ *denotes the* 100 % *point of the* $F_{p-1,n-p}$ *distribution. The LRT is UMPI.*

PROOF: Since G satisfies the conditions of Corollary 8.4 and in the normal case, $\bar{r}_{1\cdot 2,\ldots,p}$ is a maximal invariant under G (see Anderson 2003, p. 157), it follows that $\bar{r}_{1\cdot 2,\ldots,p}$ is also maximal invariant under G in the present case. Moreover, G satisfies the conditions of part (b) of Theorem 8.3 and $\bar{r}_{1\cdot 2,\ldots,p}$ is invariant under G. Therefore, the distribution of $\bar{r}_{1\cdot 2,\ldots,p}$ is the same as in the normal case. Furthermore, from Theorem 8.1 it follows that the LRT statistic is the same as in the normal case. Thus, the corresponding results of the normal theory can be used (see Anderson 2003, Sect. 4.4). It follows from Anderson (2003, p. 157), that in the normal case, the LRT is UMPI. Therefore, by Theorem 8.5, the LRT is also UMPI here. ∎

Remark 8.7. Since the distribution of $\bar{r}_{1\cdot 2,\ldots,p}$ is the same as in the normal case, its moments and distribution under the nonnull hypothesis are those given by the formulas in Anderson (2003, pp. 149–157), for the normal case.

8.3.10 Testing Equality of Means

In Model II (see Sect. 8.2.2), let $\Sigma_1 = \Sigma_2 = \ldots = \Sigma_q = \Sigma$. We want to test

$$\begin{aligned} &H_0 : \mu_1 = \mu_2 = \ldots = \mu_q \quad \text{against} \\ &H_1 : \text{there exist } 1 \leq j < k \leq q, \text{ such that } \mu_j \neq \mu_k, \end{aligned} \tag{8.24}$$

where μ_i, $i = 1,2,\ldots,q$ and Σ are unknown. Problem (8.24) remains invariant under the group G, where G is generated by the linear transformations

(i) $g(\mathbf{X}) = (\mathbf{I}_n \otimes \mathbf{C})\mathbf{X}$, where $\mathbf{C}$ is $p \times p$ nonsingular matrix,
(ii) $g(\mathbf{X}) = \mathbf{X} - \mathbf{e}_n \otimes \mathbf{v}$, where $\mathbf{v}$ is p-dimensional vector.

Theorem 8.15. *The LRT statistic for problem (8.24) is*

$$\tau = \frac{|\mathbf{A}|}{|\mathbf{B}|}.$$

The critical region at level α is

$$\tau \leq U_{p,q-1,n}(\alpha),$$

where $U_{p,q-1,n}(\alpha)$ denotes the 100 % point of the $U_{p,q-1,n}$ distribution. If H_0 holds, then $\tau \approx U_{p,q-1,n}$. τ is an invariant of the sufficient statistic under G.

PROOF: From Theorem 8.1 it follows that the LRT statistic is the same as in the normal case. It is easy to see that τ is invariant under G. Moreover, G satisfies the conditions of part (a) of Theorem 8.3. Therefore, the null distribution of τ is the same as in the normal case. Hence, the corresponding results of the normal theory can be used here, see Anderson (2003, Sect. 8.8), Pillai and Gupta (1969), Gupta (1971, 1975), Gupta, Chattopadhyay, and Krishnaiah (1975), and Gupta and Javier (1986). ■

The nonnull distribution of τ depends on ψ. Nevertheless, Quan (1990) has proved that if h is decreasing then the LRT is unbiased.

8.3.11 Testing Equality of Covariance Matrices

In Model II (see Sect. 8.2.2), we want to test

$$H_0: \Sigma_1 = \Sigma_2 = \ldots = \Sigma_q \quad \text{against}$$
$$H_1: \text{there exist } 1 \leq j < k \leq q, \text{ such that } \Sigma_j \neq \Sigma_k, \tag{8.25}$$

where μ_i and Σ_i, $i = 1, 2, \ldots, q$ are unknown. Problem (8.25) remains invariant under the group G, where G is generated by the linear transformations

(i) $g(\mathbf{X}) = (\mathbf{I}_n \otimes \mathbf{C})\mathbf{X}$, where $\mathbf{C}$ is $p \times p$ nonsingular matrix,

(ii) $g(\mathbf{X}) = \mathbf{X} - \begin{pmatrix} \mathbf{e}_{n_1} \otimes \mathbf{v}_1 \\ \mathbf{e}_{n_2} \otimes \mathbf{v}_2 \vdots \\ \mathbf{e}_{n_q} \otimes \mathbf{v}_q \end{pmatrix}$, where $\mathbf{v}_i$ is p-dimensional vector, $i = 1, 2, \ldots, q$.

Theorem 8.16. *The LRT statistic for problem (8.25) is*

$$\tau = \frac{\prod_{i=1}^{q} |\mathbf{A}_i|^{\frac{n_i}{2}} \prod_{i=1}^{q} n_i^{\frac{pn_i}{2}}}{|\mathbf{A}|^{\frac{n}{2}} n^{\frac{pn}{2}}}.$$

The critical region at level α is

$$\tau \leq \tau(\alpha),$$

where $\tau(\alpha)$ is the same as in the normal case. The distribution of τ is the same as in the normal case. τ is an invariant of the sufficient statistic under G.

PROOF: From Theorem 8.1 it follows that the LRT statistic is the same as in the normal case. It is easy to see that τ is invariant under G. Moreover, G satisfies the conditions of part (b) of Theorem 8.3. Therefore, the distribution of τ is the same as in the normal case. Hence, the corresponding results of the normal theory can be used here, see Anderson (2003, Sect. 10.2), and Gupta and Tang (1984). ■

Remark 8.8. Bartlett's modified LRT statistic

$$\tau_1 = \frac{\prod_{i=1}^{q} |\mathbf{A}_i|^{\frac{n_i-1}{2}}}{|\mathbf{A}|^{\frac{n-q}{2}}}$$

is also invariant under G. So from Theorem 8.3 it follows that the distribution of τ_1 is the same as in the normal case. Quan (1990) showed that the α level test with critical region

$$\tau_1 \le \tau_1(\alpha)$$

is unbiased if h is decreasing.

Nagao's test statistic

$$\tau_2 = (n-1)^2 \sum_{i=1}^{q} \frac{tr(\mathbf{A}_i\mathbf{A}^{-1} - \mathbf{I}_p)^2}{n_i - 1}$$

is also invariant under G and has the same distribution as in the normal case.

For further details on Bartlett's modified LRT statistic and Nagao's test statistic, see Anderson (2003, pp. 413–415).

8.3.12 Testing Equality of Means and Covariance Matrices

In Model II (see Sect. 8.2.2), we want to test

$$H_0 : \mu_1 = \mu_2 = \ldots = \mu_q \quad \text{and} \quad \Sigma_1 = \Sigma_2 = \ldots = \Sigma_q \quad \text{against} \qquad (8.26)$$
$$H_1 : \text{there exist } 1 \le j < k \le q, \text{ such that } \mu_j \ne \mu_k \text{ or } \Sigma_j \ne \Sigma_k,$$

where μ_i and Σ_i, $i = 1, 2, \ldots, q$ are unknown. Problem (8.26) remains invariant under the group G, where G is generated by the linear transformations

(i) $g(\mathbf{X}) = (\mathbf{I}_n \otimes \mathbf{C})\mathbf{X}$, where $\mathbf{C}$ is $p \times p$ nonsingular matrix,
(ii) $g(\mathbf{X}) = \mathbf{X} - \mathbf{e}_n \otimes \mathbf{v}$, where $\mathbf{v}$ is p-dimensional vector.

Theorem 8.17. *The LRT statistic for problem (8.26) is*

$$\tau = \frac{\prod_{i=1}^{q} |\mathbf{A}_i|^{\frac{n_i}{2}}}{|\mathbf{B}|^{\frac{n}{2}}}$$

The critical region at level α *is*

$$\tau \leq \tau(\alpha)$$

where $\tau(\alpha)$ *is the same as in the normal case. The null distribution of* τ *is the same as in the normal case.* τ *is an invariant of the sufficient statistic under* G*.*

PROOF: From Theorem 8.1 it follows that the LRT statistic is the same as in the normal case. It is easy to see that τ is invariant under G. Moreover, G satisfies the conditions of part (a) of Theorem 8.3. Therefore, the distribution of τ is the same as in the normal case. Hence the corresponding results of the normal theory can be used here (see Anderson 2003, Sect. 10.3). ■

The nonnull distribution of τ depends on ψ. Nevertheless, Quan and Fang (1987) have proved that the LRT is unbiased if h is decreasing.

Remark 8.9. Since the null distribution of τ is the same as in the normal case, its moments, distribution, and asymptotic distribution under the null hypothesis are those given by the formulas in Anderson (2003, Sects. 10.4–5) for the normal case.

Remark 8.10. Bartlett's modified LRT statistic

$$\tau = \frac{\prod_{i=1}^{q} |\mathbf{A}_i|^{\frac{n_i-1}{2}}}{|\mathbf{B}|^{\frac{n-q}{2}}}$$

is also invariant under G. So from Theorem 8.3 it follows that its null distribution is the same as in the normal case (see Anderson 2003, pp. 412–413).

Part IV
Applications

Chapter 9
Linear Models

9.1 Estimation of the Parameters in the Multivariate Linear Regression Model

Let $\mathbf{x}_1, \mathbf{x}_2, \ldots, \mathbf{x}_n$ be p-dimensional vectors, such that $\mathbf{x}_i \sim E_p(\mathbf{B}\mathbf{z}_i, \Sigma, \psi)$, where $\mathbf{z}_i$ is a q-dimensional known vector, $i = 1, \ldots, n$, and $\mathbf{B}$ is a $p \times q$ unknown matrix. Moreover, assume that $\mathbf{x}_i$, $i = 1, \ldots, n$ are uncorrelated and their joint distribution is elliptically contoured and absolutely continuous. This model can be expressed as

$$\mathbf{X} \sim E_{p,n}(\mathbf{BZ}, \Sigma \otimes \mathbf{I}_n, \psi), \tag{9.1}$$

where $\mathbf{X} = (\mathbf{x}_1, \mathbf{x}_2, \ldots, \mathbf{x}_n)$; $\mathbf{Z} = (\mathbf{z}_1, \mathbf{z}_2, \ldots, \mathbf{z}_n)$ is a $q \times n$ known matrix; $\mathbf{B}$ $(p \times q)$ and Σ $(p \times p)$ are unknown matrices. Assume $rk(\mathbf{Z}) = q$ and $p + q \leq n$. The joint p.d.f. of $\mathbf{x}_1, \mathbf{x}_2, \ldots, \mathbf{x}_n$ can be written as

$$\begin{aligned} f(\mathbf{X}) &= \frac{1}{|\Sigma|^n} h\left(\sum_{i=1}^{n} (\mathbf{x}_i - \mathbf{B}\mathbf{z}_i)' \Sigma^{-1} (\mathbf{x}_i - \mathbf{B}\mathbf{z}_i)\right) \\ &= \frac{1}{|\Sigma|^n} h(tr(\mathbf{X} - \mathbf{BZ})' \Sigma^{-1} (\mathbf{X} - \mathbf{BZ})). \end{aligned} \tag{9.2}$$

Assume $l(z) = z^{\frac{pn}{2}} h(z)$, $z \geq 0$ has a finite maximum at $z = z_h > 0$. First, we find the MLE's of $\mathbf{B}$ and Σ.

Theorem 9.1. *The MLE's of* $\mathbf{B}$ *and* Σ *for the model (9.1) are given by*

$$\hat{\mathbf{B}} = \mathbf{XZ}'(\mathbf{ZZ}')^{-1} \tag{9.3}$$

and

$$\hat{\Sigma} = \frac{p}{z_h} \mathbf{X}(\mathbf{I}_n - \mathbf{Z}'(\mathbf{ZZ}')^{-1}\mathbf{Z})\mathbf{X}'. \tag{9.4}$$

A.K. Gupta et al., *Elliptically Contoured Models in Statistics and Portfolio Theory*, DOI 10.1007/978-1-4614-8154-6_9, © Springer Science+Business Media New York 2013

PROOF: From Theorem 1.3 it follows that $rk(\mathbf{ZZ}') = q$ and since $\mathbf{ZZ}'$ is $q \times q$, it is nonsingular.

Note that $\mathbf{B}$ and $\mathbf{BZ}$ determine each other uniquely. Indeed, from $\mathbf{B}$ we get $\mathbf{BZ}$ by postmultiplying $\mathbf{B}$ by $\mathbf{Z}$, and from $\mathbf{BZ}$ we can obtain $\mathbf{B}$ by postmultiplying $\mathbf{BZ}$ by $\mathbf{Z}'(\mathbf{ZZ}')^{-1}$. Hence finding the MLE of $\mathbf{B}$ is equivalent to finding the MLE of $\mathbf{BZ}$.

If $\mathbf{X} \sim N_{p,n}(\mathbf{BZ}, \Sigma \otimes \mathbf{I}_n)$, then the MLE's of $\mathbf{B}$ and Σ are (see Anderson 2003, Sect. 8.2),

$$\mathbf{B}^* = \mathbf{XZ}'(\mathbf{ZZ}')^{-1}$$

and

$$\Sigma^* = \frac{1}{n}(\mathbf{X} - \mathbf{B}^*\mathbf{Z})(\mathbf{X} - \mathbf{B}^*\mathbf{Z})'.$$

We can rewrite Σ^* as

$$\begin{aligned}\Sigma^* &= \frac{1}{n}\mathbf{X}(\mathbf{I}_n - \mathbf{Z}'(\mathbf{ZZ}')^{-1}\mathbf{Z})(\mathbf{I}_n - \mathbf{Z}'(\mathbf{ZZ}')^{-1}\mathbf{Z})\mathbf{X}' \\ &= \frac{1}{n}\mathbf{X}(\mathbf{I}_n - \mathbf{Z}'(\mathbf{ZZ}')^{-1}\mathbf{Z})\mathbf{X}'.\end{aligned}$$

Then, by using Theorem 7.2, we obtain the MLE's of $\mathbf{B}$ and Σ as

$$\hat{\mathbf{B}} = \mathbf{XZ}'(\mathbf{ZZ}')^{-1}$$

and

$$\begin{aligned}\hat{\Sigma} &= \frac{pn}{z_h}\frac{1}{n}\mathbf{X}(\mathbf{I}_n - \mathbf{Z}'(\mathbf{ZZ}')^{-1}\mathbf{Z})\mathbf{X}' \\ &= \frac{p}{z_h}\mathbf{X}(\mathbf{I}_n - \mathbf{Z}'(\mathbf{ZZ}')^{-1}\mathbf{Z})\mathbf{X}'.\end{aligned}$$ ■

The distributions of $\hat{\mathbf{B}}$ and $\hat{\Sigma}$ can also be obtained, and are given in the following theorem.

Theorem 9.2. *The distributions of the MLE's of* $\mathbf{B}$ *and* Σ *for the model (9.1) are given by*

$$\hat{\mathbf{B}} \sim E_{p,q}(\mathbf{B}, \Sigma \otimes (\mathbf{ZZ}')^{-1}, \psi), \tag{9.5}$$

and

$$\frac{z_h}{p}\hat{\Sigma} \sim G_{p,1}\left(\Sigma, \frac{n-q}{2}, \psi\right). \tag{9.6}$$

PROOF: From (9.1) and (9.3) we obtain

$$\begin{aligned}\hat{\mathbf{B}} &\sim E_{p,q}(\mathbf{BZZ}'(\mathbf{ZZ}')^{-1}, \Sigma \otimes (\mathbf{ZZ}')^{-1}\mathbf{ZZ}'(\mathbf{ZZ}')^{-1}, \psi) \\ &= Ep,q(\mathbf{B}, \Sigma \otimes (\mathbf{ZZ}')^{-1}, \psi).\end{aligned}$$

Then from (9.4) we get

$$\begin{aligned}\frac{z_h}{p}\hat{\Sigma} &= \mathbf{X}(\mathbf{I}_n - \mathbf{Z}'(\mathbf{ZZ}')^{-1}\mathbf{Z})\mathbf{X}' \\ &= \mathbf{X}(\mathbf{I}_n - \mathbf{Z}'(\mathbf{ZZ}')^{-1}\mathbf{Z})(\mathbf{X}(\mathbf{I}_n - \mathbf{Z}'(\mathbf{ZZ}')^{-1}\mathbf{Z}))'. \end{aligned} \tag{9.7}$$

From (9.1) we get

$$\begin{aligned}\mathbf{X}(\mathbf{I}_n - \mathbf{Z}'(\mathbf{ZZ}')^{-1}\mathbf{Z}) &\sim E_{p,n}(\mathbf{BZ}(\mathbf{I}_n - \mathbf{Z}'(\mathbf{ZZ}')^{-1}\mathbf{Z}), \Sigma \otimes (\mathbf{I}_n - \mathbf{Z}'(\mathbf{ZZ}')^{-1}\mathbf{Z}) \\ &\quad \times (\mathbf{I}_n - \mathbf{Z}'(\mathbf{ZZ}')^{-1}\mathbf{Z}), \psi) \\ &= E_{p,n}(\mathbf{0}, \Sigma \otimes (\mathbf{I}_n - \mathbf{Z}'(\mathbf{ZZ}')^{-1}\mathbf{Z}), \psi). \end{aligned} \tag{9.8}$$

Now define the $p \times n$ random matrix, $\mathbf{Y}$, by

$$\mathbf{Y} \sim E_{p,n}(\mathbf{0}, \Sigma \otimes \mathbf{I}_n, \psi) \tag{9.9}$$

Then,

$$\begin{aligned}\mathbf{Y}(\mathbf{I}_n - \mathbf{Z}'(\mathbf{ZZ}')^{-1}\mathbf{Z})\mathbf{Y}' &= \mathbf{Y}(\mathbf{I}_n - \mathbf{Z}'(\mathbf{ZZ}')^{-1}\mathbf{Z})(\mathbf{I}_n - \mathbf{Z}'(\mathbf{ZZ}')^{-1}\mathbf{Z})\mathbf{Y}' \quad (9.10) \\ &= (\mathbf{Y}(\mathbf{I}_n - \mathbf{Z}'(\mathbf{ZZ}')^{-1}\mathbf{Z}))(\mathbf{Y}(\mathbf{I}_n - \mathbf{Z}'(\mathbf{ZZ}')^{-1}\mathbf{Z}))'\end{aligned}$$

and since

$$\mathbf{Y}(\mathbf{I}_n - \mathbf{Z}'(\mathbf{ZZ}')^{-1}\mathbf{Z}) \sim E_{p,n}(\mathbf{0}, \Sigma \otimes (\mathbf{I}_n - \mathbf{Z}'(\mathbf{ZZ}')^{-1}\mathbf{Z}), \psi), \tag{9.11}$$

from (9.7), (9.8), (9.10), and (9.11) we get

$$\frac{z_h}{p}\hat{\Sigma} \approx \mathbf{Y}(\mathbf{I}_n - \mathbf{Z}'(\mathbf{ZZ}')^{-1}\mathbf{Z})\mathbf{Y}' \tag{9.12}$$

However, the matrix $(\mathbf{I}_n - \mathbf{Z}'(\mathbf{ZZ}')^{-1}\mathbf{Z})$ is idempotent of rank $n - q$. Hence, using Theorem 5.7, we get

$$\mathbf{Y}(\mathbf{I}_n - \mathbf{Z}'(\mathbf{ZZ}')^{-1}\mathbf{Z})\mathbf{Y}' \sim G_{p,1}\left(\Sigma, \frac{n-q}{2}, \psi\right).$$

■

Next, we find the unbiased estimators of $\mathbf{B}$ and Σ.

Theorem 9.3. *The statistics* $\hat{\mathbf{B}}$ *and*

$$\hat{\Sigma}_U = \frac{-1}{2\psi'(0)(n-q)}\mathbf{X}(\mathbf{I}_n - \mathbf{Z}'(\mathbf{Z}\mathbf{Z}')^{-1}\mathbf{Z})\mathbf{X}'$$

for model (9.1) are unbiased for $\mathbf{B}$ *and* Σ*, if the second order moment of* $\mathbf{X}$ *exists.*

PROOF: From (9.5) it follows that $E(\hat{\mathbf{B}}) = \mathbf{B}$. On the other hand, from Theorem 3.18, we get

$$\begin{aligned}
E(\hat{\Sigma}_U) &= \frac{-1}{2\psi'(0)(n-q)}E(\mathbf{X}(\mathbf{I}_n - \mathbf{Z}'(\mathbf{Z}\mathbf{Z}')^{-1}\mathbf{Z})\mathbf{X}') \\
&= \frac{-1}{2\psi'(0)(n-q)}(-2\psi'(0))\Sigma tr(\mathbf{I}_n - \mathbf{Z}'(\mathbf{Z}\mathbf{Z}')^{-1}\mathbf{Z}) \\
&+ \mathbf{B}\mathbf{Z}(\mathbf{I}_n - \mathbf{Z}'(\mathbf{Z}\mathbf{Z}')^{-1}\mathbf{Z})\mathbf{Z}'\mathbf{B}' \\
&= \frac{1}{n-q}\Sigma(n-q) \\
&= \Sigma.
\end{aligned}$$

■

The question of sufficiency is studied in the following theorem. First, we prove a lemma.

Lemma 9.1. *Let* $\mathbf{X}$ *be* $p \times n$*,* $\mathbf{B}$ *be* $p \times q$*, and* $\mathbf{Z}$ *be* $q \times n$ *matrices with* $rk(\mathbf{Z}) = q$*. Define* $\hat{\mathbf{B}} = \mathbf{X}\mathbf{Z}'(\mathbf{Z}\mathbf{Z}')^{-1}$*, then*

$$(\mathbf{X} - \mathbf{B}\mathbf{Z})(\mathbf{X} - \mathbf{B}\mathbf{Z})' = (\mathbf{X} - \hat{\mathbf{B}}\mathbf{Z})(\mathbf{X} - \hat{\mathbf{B}}\mathbf{Z})' + (\hat{\mathbf{B}} - \mathbf{B})\mathbf{Z}\mathbf{Z}'(\hat{\mathbf{B}} - \mathbf{B})'. \tag{9.13}$$

PROOF: We can write

$$\begin{aligned}
(\mathbf{X} - \mathbf{B}\mathbf{Z})(\mathbf{X} - \mathbf{B}\mathbf{Z})' &= (\mathbf{X} - \hat{\mathbf{B}}\mathbf{Z} + (\hat{\mathbf{B}} - \mathbf{B})\mathbf{Z})(\mathbf{X} - \hat{\mathbf{B}}\mathbf{Z} + (\hat{\mathbf{B}} - \mathbf{B})\mathbf{Z})' \\
&= (\mathbf{X} - \hat{\mathbf{B}}\mathbf{Z})(\mathbf{X} - \hat{\mathbf{B}}\mathbf{Z})' + (\hat{\mathbf{B}} - \mathbf{B})\mathbf{Z}\mathbf{Z}'(\hat{\mathbf{B}} - \mathbf{B})' \\
&+ (\mathbf{X} - \hat{\mathbf{B}}\mathbf{Z})\mathbf{Z}'(\hat{\mathbf{B}} - \mathbf{B})' + (\hat{\mathbf{B}} - \mathbf{B})\mathbf{Z}(\mathbf{X} - \hat{\mathbf{B}}\mathbf{Z})'. \qquad (9.14)
\end{aligned}$$

However,

$$(\mathbf{X} - \hat{\mathbf{B}}\mathbf{Z})\mathbf{Z}' = (\mathbf{X} - \mathbf{X}\mathbf{Z}'(\mathbf{Z}\mathbf{Z}')^{-1}\mathbf{Z})\mathbf{Z}' = \mathbf{X}\mathbf{Z}' - \mathbf{X}\mathbf{Z}' = \mathbf{0}$$

so the last two terms in (9.14) vanish. Thus, from (9.14), we get (9.13). ■

Theorem 9.4. *In model (9.1),* $\hat{\mathbf{B}}$ *and* $\hat{\Sigma}$ *are sufficient for* $\mathbf{B}$ *and* Σ.

PROOF: From (9.2) and Lemma 9.1, we get

$$\begin{aligned}
f(\mathbf{X}) &= \frac{1}{|\Sigma|^n} h(tr(\Sigma^{-1}(\mathbf{X}-\mathbf{BZ})(\mathbf{X}-\mathbf{BZ})')) \\
&= \frac{1}{|\Sigma|^n} h(tr(\Sigma^{-1}[(\mathbf{X}-\hat{\mathbf{B}}\mathbf{Z})(\mathbf{X}-\hat{\mathbf{B}}\mathbf{Z})' + (\hat{\mathbf{B}}-\mathbf{B})\mathbf{ZZ}'(\hat{\mathbf{B}}-\mathbf{B})'])).
\end{aligned}$$

Now,

$$\begin{aligned}
(\mathbf{X}-\hat{\mathbf{B}}\mathbf{Z})(\mathbf{X}-\hat{\mathbf{B}}\mathbf{Z})' &= (\mathbf{X}-\mathbf{XZ}'(\mathbf{ZZ}')^{-1}\mathbf{Z})(\mathbf{X}-\mathbf{XZ}'(\mathbf{ZZ}')^{-1}\mathbf{Z})' \\
&= \mathbf{X}(\mathbf{I}_n - \mathbf{Z}'(\mathbf{ZZ}')^{-1}\mathbf{Z})(\mathbf{I}_n - \mathbf{Z}'(\mathbf{ZZ}')^{-1}\mathbf{Z})\mathbf{X}' \\
&= \mathbf{X}(\mathbf{I}_n - \mathbf{Z}'(\mathbf{ZZ}')^{-1}\mathbf{Z})\mathbf{X}' \\
&= \frac{z_h}{p}\hat{\Sigma}.
\end{aligned}$$

Hence,

$$f(\mathbf{X}) = \frac{1}{|\Sigma|^n} h\left(tr\left(\Sigma^{-1}\left[\frac{z_h}{p}\hat{\Sigma} + (\hat{\mathbf{B}}-\mathbf{B})\mathbf{ZZ}'(\hat{\mathbf{B}}-\mathbf{B})'\right]\right)\right),$$

which proves the theorem. ■

Next, we focus on some optimality properties of the MLE's for model (9.1). Here, we assume that $\mathbf{X}$ has a finite second order moment. Let the $p \times 1$ vector $\mathbf{a}$ and the $q \times 1$ vector $\mathbf{c}$ be given and assume that we want to estimate $\xi = \mathbf{a}'\mathbf{Bc}$. We are interested in a linear unbiased estimator of ξ, that is an estimator which can be written in the form $\xi^* = \mathbf{v}'vec(\mathbf{X}')$, where $\mathbf{v}$ is a pn-dimensional vector and $E(\xi^*) = \xi$.

The MLE of ξ is $\hat{\xi} = \mathbf{a}'\hat{\mathbf{B}}\mathbf{c}$. This is a linear estimator of ξ since

$$\begin{aligned}
\mathbf{a}'\hat{\mathbf{B}}\mathbf{c} &= \mathbf{a}'\mathbf{XZ}'(\mathbf{ZZ}')^{-1}\mathbf{c} \\
&= vec(\mathbf{a}'\mathbf{XZ}'(\mathbf{ZZ}')^{-1}\mathbf{c}) \\
&= (\mathbf{a}' \otimes (\mathbf{c}'(\mathbf{ZZ}')^{-1}\mathbf{Z}))vec(\mathbf{X}').
\end{aligned}$$

Moreover, $\hat{\xi}$ is unbiased for ξ since

$$E(\hat{\xi}) = \mathbf{a}'E(\hat{\mathbf{B}})\mathbf{c} = \mathbf{a}'\mathbf{Bc} = \xi.$$

The next result, also called multivariate Gauss-Markov theorem, shows that $Var(\hat{\xi}) \leq Var(\xi^*)$ for any linear unbiased estimator ξ^*.

Theorem 9.5. *(Gauss-Markov Theorem) Let* $\mathbf{X}$ *be a* $p \times n$ *random matrix, with* $E(\mathbf{X}) = \mathbf{BZ}$ *and* $Cov(\mathbf{X}) = \Sigma \otimes \mathbf{I}_n$*, where* $\mathbf{B}$ $(p \times q)$*, and* Σ $(p \times p)$*, are unknown matrices and* $\mathbf{Z}$ $(q \times n)$ *is a known matrix. Assume* $rk(\mathbf{Z}) = q$ *and* $\mathbf{a}$ $(p \times 1)$ *and* $\mathbf{c}$

$(q\times 1)$ are known vectors. Let $\xi = \mathbf{a}'\mathbf{B}\mathbf{c}$, and define $\hat{\xi} = \mathbf{a}'\hat{\mathbf{B}}\mathbf{c}$. If $\xi^* = \mathbf{v}'vec(\mathbf{X}')$ is any linear unbiased estimator of ξ, then $Var(\hat{\xi}) \leq Var(\xi^*)$.

PROOF: See Timm (1975, p. 187). ■

Theorem 9.5 does not require that **X** have a m.e.c. distribution. Now, we show that if the distribution of **X** is elliptically contoured, we can get a stronger result. In order to do this, we need the following lemma due to Stoyan (1983).

Lemma 9.2. *Let x and y be two one-dimensional random variables. Then,*

$$E(f(x)) \leq E(f(y))$$

holds for all increasing, real function f if and only if

$$P(x \leq a) \geq P(y \leq a),$$

for all $a \in \mathbb{R}$.

PROOF: See Stoyan (1983, p. 5). ■

Now, we can derive the result on m.e.c. distributions.

Theorem 9.6. *Assume model (9.1) holds and **X** has a finite second order moment. Let **a** $(p\times 1)$, and **c** $(q\times 1)$ be known vectors, $\xi = \mathbf{a}'\mathbf{B}\mathbf{c}$, and define $\hat{\xi} = \mathbf{a}'\hat{\mathbf{B}}\mathbf{c}$. Assume $\xi^* = \mathbf{v}'vec(\mathbf{X}')$ is a linear unbiased estimator of ξ. Let $l(z)$ be a loss function, where $l: [0,\infty) \to [0,\infty)$, $l(0) = 0$, and $l(z)$ is increasing on $[0,\infty)$. Then,*

$$E(l(|\hat{\xi} - \xi|)) \leq E(l(|\xi^* - \xi|)).$$

That is, $\hat{\xi}$ is optimal in the class of linear unbiased estimators for the loss function l.

PROOF: Since $\xi^* = \mathbf{v}'vec(\mathbf{X}')$ and it is unbiased for ξ, that is, $E(\xi^*) = \xi$, we have $\hat{\xi} \sim E_1(\xi, \sigma^2_{\hat{\xi}}, \psi)$. We also have $\xi^* \sim E_1(\xi, \sigma^2_{\xi^*}, \psi)$.

Now, from Theorem 9.5, it follows that

$$Var(\hat{\xi}) \leq Var(\xi^*).$$

However, from Theorem 2.11 we get

$$Var(\hat{\xi}) = -2\psi'(0)\sigma^2_{\hat{\xi}} \quad \text{and} \quad Var(\xi^*) = -2\psi'(0)\sigma^2_{\xi^*}.$$

Thus,

$$\sigma^2_{\hat{\xi}} \leq \sigma^2_{\xi^*}. \tag{9.15}$$

Define a random variable z as $z \sim E_1(0,1,\psi)$. Then, we have $\hat{\xi} - \xi \approx \sigma_{\hat{\xi}} z$ and $\xi^* - \xi \approx \sigma_{\xi^*} z$. Consequently, for every real positive a, we obtain

$$P(|\hat{\xi} - \xi| \leq a) = P(\sigma_{\hat{\xi}} |z| \leq a) \\ = P\left(|z| \leq \frac{a}{\sigma_{\hat{\xi}}}\right) \tag{9.16}$$

and

$$P(|\xi^* - \xi| \leq a) = P(\sigma_{\xi^*} |z| \leq a) \\ = P\left(|z| \leq \frac{a}{\sigma_{\xi^*}}\right) \tag{9.17}$$

From (9.15), (9.16), and (9.17), it follows that

$$P(|\hat{\xi} - \xi| \leq a) \geq P(|\xi^* - \xi| \leq a).$$

Then, from Lemma 9.2, we obtain

$$E(l(|\hat{\xi} - \xi|)) \leq E(l(|\xi^* - \xi|)).$$

■

Next, we prove an optimality property of $\hat{\Sigma}$. First, we need some concepts and results from the theory of majorization. They are taken from Marshall and Olkin (1979). Assume $\mathbf{x}$ is an n-dimensional vector $\mathbf{x}' = (x_1, x_2, \ldots, x_n)$. Then $x_{[1]} \geq x_{[2]} \geq \ldots \geq x_{[n]}$ denote the components of $\mathbf{x}$ in decreasing order.

Now, let $\mathbf{x}$ and $\mathbf{y}$ be two n-dimensional vectors. Then, we say that $\mathbf{y}$ *majorizes* $\mathbf{x}$, and denote this by $\mathbf{x} \prec \mathbf{y}$, if $\sum_{i=1}^{j} x_{[i]} \leq \sum_{i=1}^{j} y_{[i]}$, $j = 1, \ldots, n-1$ and $\sum_{i=1}^{n} x_{[i]} = \sum_{i=1}^{n} y_{[i]}$.

Let ϕ be a real function, defined on $\mathbb{R}^n$. Then, we say that ϕ is Schur-convex, if from $\mathbf{x} \prec \mathbf{y}$, $\mathbf{x}, \mathbf{y} \in \mathbb{R}^n$, it follows that $\phi(\mathbf{x}) \leq \phi(\mathbf{y})$.

Lemma 9.3. *Let $x_1, x_2, \ldots, x_n$ be exchangeable, random variables and define $\mathbf{x} = (x_1, x_2, \ldots, x_n)'$. Assume λ is a real function defined on $\mathbb{R}^n \times \mathbb{R}^n$ and it satisfies the following conditions:*

(i) *$\lambda(\mathbf{z}, \mathbf{a})$ is convex in $\mathbf{a} \in \mathbb{R}^n$, if $\mathbf{z} \in \mathbb{R}^n$ is fixed,*
(ii) *$\lambda((z_{\pi(1)}, z_{\pi(2)}, \ldots, z_{\pi(n)}), (a_{\pi(1)}, a_{\pi(2)}, \ldots, a_{\pi(n)})) = \lambda(\mathbf{z}, \mathbf{a})$ for all permutations π of the first n positive integers, and*
(iii) *$\lambda(\mathbf{z}, \mathbf{a})$ is Borel measurable in $\mathbf{z}$, if $\mathbf{a}$ is fixed.*

Then, $\phi(\mathbf{a}) = E(\lambda(\mathbf{x}, \mathbf{a}))$ is symmetric and convex in $\mathbf{a}$.

PROOF: See Marshall and Olkin (1979, pp. 286–287). ■

Lemma 9.4. *Let ϕ be a real function defined on $I\!R^n$. If $\phi(\mathbf{a})$ is symmetric and convex in $\mathbf{a} \in I\!R^n$, then $\phi(\mathbf{a})$ is Schur-convex.*

PROOF: See Marshall and Olkin (1979, pp. 67–68). ■

Lemmas 9.3 and 9.4, together with the definition of a Schur-convex function, imply the following theorem.

Theorem 9.7. *Let $x_1, x_2, \ldots, x_n$ be exchangeable, random variables and define $\mathbf{x} = (x_1, x_2, \ldots, x_n)'$. Let $\mathbf{a}_1$ and $\mathbf{a}_2$ be n-dimensional vectors such that $\mathbf{a}_1 \prec \mathbf{a}_2$. Assume λ is a real function defined on $I\!R^n \times I\!R^n$ and it satisfies the conditions (i), (ii) and (iii) of Lemma 9.3. Then $E(\lambda(\mathbf{x}, \mathbf{a}_1)) \leq E(\lambda(\mathbf{x}, \mathbf{a}_2))$ if $\mathbf{a}_1 \prec \mathbf{a}_2$.*

PROOF: From Lemma 9.3, it follows that $\phi(\mathbf{a}) = E(\lambda(\mathbf{x}, \mathbf{a}))$ is symmetric and convex in $\mathbf{a}$. Then, from Lemma 9.4 we get that $\phi(\mathbf{a})$ is Schur-convex, and this means, by definition, that $E(\lambda(\mathbf{x}, \mathbf{a}_1)) \leq E(\lambda(\mathbf{x}, \mathbf{a}_2))$ if $\mathbf{a}_1 \prec \mathbf{a}_2$. ■

The following theorem proves another result about an estimator of Σ.

Theorem 9.8. *Assume model (9.1) holds, $\mathbf{X}$ has a finite second order moment, and ψ is known. Let*

$$\hat{\Sigma}_U = \frac{-1}{2\psi'(0)(n-q)} \mathbf{X}(\mathbf{I}_n - \mathbf{Z}'(\mathbf{Z}\mathbf{Z}')^{-1}\mathbf{Z})\mathbf{X}'.$$

Assume Σ^* is an unbiased estimator of Σ that has the form $\Sigma^* = \mathbf{X}\mathbf{C}\mathbf{X}'$, where $\mathbf{C}$ is a positive semidefinite $n \times n$ matrix depending on $\mathbf{X}$. Let $l(z)$ be a loss function, where $l : [0,\infty) \to [0,\infty)$, $l(0) = 0$, and $l(z)$ is increasing on $[0,\infty)$. Then,

$$E(l(tr(\hat{\Sigma}_U \Sigma^{-1}))) \leq E(l(tr(\Sigma^* \Sigma^{-1}))).$$

PROOF: Let Σ^* be an unbiased estimator of Σ, which can be written as $\Sigma^* = \mathbf{X}\mathbf{C}\mathbf{X}'$. From Theorem 3.18, we get

$$E(\Sigma^*) = E(\mathbf{X}\mathbf{C}\mathbf{X}') = -2\psi'(0)\Sigma tr(\mathbf{C}') + \mathbf{B}\mathbf{Z}\mathbf{C}\mathbf{Z}'\mathbf{B}'.$$

So, in order for Σ^* to be an unbiased estimator of Σ, we must have $tr(\mathbf{C}) = \frac{-1}{2\psi'(0)}$ and $\mathbf{Z}\mathbf{C}\mathbf{Z}' = \mathbf{0}$. Since $\mathbf{C}$ is positive semidefinite, we can write $\mathbf{C} = \mathbf{H}\mathbf{D}\mathbf{H}'$, where $\mathbf{H}$ is orthogonal, and $\mathbf{D}$ is diagonal with nonnegative elements. Therefore, $\mathbf{Z}\mathbf{C}\mathbf{Z}' = \mathbf{0}$ implies $\mathbf{Z}\mathbf{H}\mathbf{D}\mathbf{H}'\mathbf{Z}' = \mathbf{0}$, which can be rewritten as $(\mathbf{Z}\mathbf{H}\mathbf{D}^{\frac{1}{2}})(\mathbf{Z}\mathbf{H}\mathbf{D}^{\frac{1}{2}})' = \mathbf{0}$. Hence,

$$\begin{aligned}
&\mathbf{Z}\mathbf{H}\mathbf{D}^{\frac{1}{2}} = \mathbf{0}, \\
&\mathbf{Z}\mathbf{C} = \mathbf{Z}\mathbf{H}\mathbf{D}^{\frac{1}{2}}\mathbf{D}^{\frac{1}{2}}\mathbf{H}' = \mathbf{0}, \\
&\Sigma^* = \mathbf{X}\mathbf{C}\mathbf{X}' = (\mathbf{X} - \mathbf{B}\mathbf{Z})\mathbf{C}(\mathbf{X} - \mathbf{B}\mathbf{Z})',
\end{aligned}$$

and

$$
\begin{aligned}
tr(\Sigma^*\Sigma^{-1}) &= tr(\Sigma^{-\frac{1}{2}}\Sigma^*\Sigma^{-\frac{1}{2}}) \\
&= tr(\Sigma^{-\frac{1}{2}}(\mathbf{X}-\mathbf{BZ})\mathbf{HDH}'(\mathbf{X}-\mathbf{BZ})'\Sigma^{-\frac{1}{2}}).
\end{aligned}
$$

Now define $\mathbf{L} = \Sigma^{-\frac{1}{2}}(\mathbf{X}-\mathbf{BZ})\mathbf{H}$, where $\mathbf{L} \sim E_{p,n}(\mathbf{0}, \mathbf{I}_p \otimes \mathbf{I}_n, \psi)$, and let $\mathbf{L} \approx r\mathbf{U}$ be the stochastic representation of $\mathbf{L}$. Then,

$$
\begin{aligned}
tr(\Sigma^*\Sigma^{-1}) &= tr((r\mathbf{U})\mathbf{D}(r\mathbf{U})') \\
&= r^2 tr(\mathbf{UDU}').
\end{aligned}
$$

Since $\mathbf{ZC} = \mathbf{0}$ and $rk(\mathbf{Z}) = q$, q diagonal elements of $\mathbf{D}$ are zero, and since $tr(\mathbf{C}) = \frac{-1}{2\psi'(0)}$, the sum of the others is $\frac{-1}{2\psi'(0)}$. Define $d_i = d_{ii}$, $i = 1, \ldots, n$, and $\mathbf{d} = (d_1, d_2, \ldots, d_n)'$. Let the vector $\mathbf{d}$, corresponding to $\hat{\Sigma}_U$ be denoted by $\hat{\mathbf{d}}$. Since $(\mathbf{I}_n - \mathbf{Z}'(\mathbf{ZZ}')^{-1}\mathbf{Z})$ is an idempotent matrix with rank $n-q$, we see that $n-q$ elements of $\hat{\mathbf{d}}$ are equal to $\frac{-1}{2\psi'(0)(n-q)}$, and the rest are zeros. Since, $d_{[1]} \geq d_{[2]} \geq \ldots \geq d_{[n-q]}$, we get $\frac{1}{n-q}\sum_{i=1}^{n-q} d_{[i]} \leq \frac{1}{j}\sum_{i=1}^{j} d_{[i]}$, $j = 1, \ldots, n-q-1$, hence

$$
\begin{aligned}
\sum_{i=1}^{j} \hat{d}_{[i]} &= \left(\frac{j}{n-q}\right)\left(\frac{-1}{2\psi'(0)}\right) \\
&= \frac{j}{n-q}\sum_{i=1}^{n-q} d_{[i]} \\
&\leq \sum_{i=1}^{j} d_{[i]}, \quad j = 1, \ldots, n-q-1
\end{aligned}
$$

and $\sum_{i=1}^{j} \hat{d}_{[i]} = \frac{-1}{2\psi'(0)} = \sum_{i=1}^{j} d_{[i]}$, $j = n-q, \ldots, n$. Therefore,

$$
\hat{\mathbf{d}} \prec \mathbf{d}.
$$

Let $\mathbf{u}_i$, $i = 1, \ldots, n$ denote the ith column of the matrix $\mathbf{U}$. Then,

$$
\begin{aligned}
r^2 tr(\mathbf{UDU}') &= r^2 tr\left(\sum_{i=1}^{n} d_i \mathbf{u}_i \mathbf{u}_i'\right) \\
&= \sum_{i=1}^{n} d_i r^2 \mathbf{u}_i' \mathbf{u}_i \\
&= \sum_{i=1}^{n} d_i w_i \\
&= \mathbf{w}'\mathbf{d},
\end{aligned}
$$

where $w_i = r^2\mathbf{u}_i'\mathbf{u}_i$, $i = 1,\ldots,n$, and $\mathbf{w} = (w_1, w_2, \ldots, w_n)'$. Consequently, $w_i \geq 0$ and $w_1, w_2, \ldots, w_n$ are exchangeable random variables.

Define the real function λ on $\mathbb{R}^n \times \mathbb{R}^n$ as follows: $\lambda(\mathbf{z},\mathbf{t}) = l(\mathbf{z}'\mathbf{t})$, where $\mathbf{z},\mathbf{t} \in \mathbb{R}^n$. Then, λ satisfies the conditions of Theorem 9.7, $\hat{\mathbf{d}} \prec \mathbf{d}$ and $\mathbf{w}$ has exchangeable components. Thus, we get

$$E(l(\mathbf{w}'\hat{\mathbf{d}})) \leq E(l(\mathbf{w}'\mathbf{d})).$$

However, $\mathbf{w}'\mathbf{d} = r^2 tr(\mathbf{UDU}') = tr(\Sigma^*\Sigma^{-1})$, so we get

$$E(l(tr(\hat{\Sigma}_U\Sigma^{-1}))) \leq E(l(tr(\Sigma^*\Sigma^{-1}))). \qquad \blacksquare$$

Corollary 9.1. *Under the conditions of Theorem 9.8,*

$$E(tr(\hat{\Sigma}_U\Sigma^{-1}))^2 \leq E(tr(\Sigma^*\Sigma^{-1}))^2.$$

Theorems 9.6 and 9.8 were derived by Kuritsyn (1986) for vector variate elliptically contoured distribution. The results here are the extensions of the results of that paper to the case of matrix variate elliptically contoured distribution.

9.2 Hypothesis Testing in the Multivariate Linear Regression Model

In this section, once again, we focus on model (9.1). We use the notations of Sect. 9.1. The results are taken from Hsu (1985b).

Let the matrix $\mathbf{B}$ be partitioned as $\mathbf{B} = (\mathbf{B}_1, \mathbf{B}_2)$, where $\mathbf{B}_1$ is $p \times q_1$ $(1 \leq q_1 < q)$, and partition $\mathbf{Z}$, as $\mathbf{Z} = \begin{pmatrix} \mathbf{Z}_1 \\ \mathbf{Z}_2 \end{pmatrix}$, where $\mathbf{Z}_1$ is $q_1 \times n$. Let $q_2 = q - q_1$. Define $\mathbf{A} = \mathbf{ZZ}'$ and partition $\mathbf{A}$ as $\begin{pmatrix} \mathbf{A}_{11} & \mathbf{A}_{12} \\ \mathbf{A}_{21} & \mathbf{A}_{22} \end{pmatrix}$, where $\mathbf{A}_{11}$ is $q_1 \times q_1$. Then, $\mathbf{A}_{ij} = \mathbf{Z}_i\mathbf{Z}_j'$, $i = 1,2$, $j = 1,2$. Also, define $\mathbf{A}_{11\cdot 2} = \mathbf{A}_{11} - \mathbf{A}_{12}\mathbf{A}_{22}^{-1}\mathbf{A}_{21}$. We want to test the hypothesis

$$H_0 : B_1 = B_1^* \quad \text{against} \quad H_1 : B_1 \neq B_1^*, \tag{9.18}$$

where $\mathbf{B}$ and Σ are unknown and $\mathbf{B}_1^*$ is a $p \times q_1$ given matrix. Note that problem (9.18) is equivalent to testing

$$H_0 : B_1 = \mathbf{0} \quad \text{against} \quad H_1 : B_1 \neq \mathbf{0}. \tag{9.19}$$

Indeed, if $B_1^* \neq \mathbf{0}$, then define $\mathbf{X}^* = \mathbf{X} - \mathbf{B}_1^*\mathbf{Z}_1$. Then, we get

$$\begin{aligned} \mathbf{X}^* &\sim E_{p,n}(\mathbf{BZ} - \mathbf{B}_1^*\mathbf{Z}_1, \Sigma \otimes \mathbf{I}_n, \psi) \\ &= E_{p,n}((\mathbf{B}_1 - \mathbf{B}_1^*, \mathbf{B}_2)\mathbf{Z}, \Sigma \otimes \mathbf{I}_n, \psi) \end{aligned}$$

and $B_1 = B_1^*$ is equivalent to $B_1 - B_1^* = \mathbf{0}$.

Problem (9.19) remains invariant under the group G, where G is generated by the linear transformations

(i) $g(\mathbf{X}) = \mathbf{C}\mathbf{X}$, where $\mathbf{C}$ is $p \times p$ nonsingular matrix and
(ii) $g(\mathbf{X}) = \mathbf{X} + \mathbf{L}\mathbf{Z}_2$, where $\mathbf{L}$ is $p \times q_2$ matrix.

Now, we derive the likelihood ratio test for the problem (9.18).

Theorem 9.9. *The LRT statistic for problem (9.18) is*

$$\tau = \frac{|\mathbf{X}(\mathbf{I}_n - \mathbf{Z}'(\mathbf{Z}\mathbf{Z}')^{-1}\mathbf{Z})\mathbf{X}'|}{|(\mathbf{X} - \mathbf{B}_1^*\mathbf{Z}_1)(\mathbf{I}_n - \mathbf{Z}_2'(\mathbf{Z}_2\mathbf{Z}_2')^{-1}\mathbf{Z}_2)(\mathbf{X} - \mathbf{B}_1^*\mathbf{Z}_1)'|}.$$

The critical region at level α is

$$\tau \leq U_{p,q_1,n-q}(\alpha),$$

where $U_{p,q_1,n-q}(\alpha)$ denotes the 100 % point of the $U_{p,q_1,n-q}$ distribution. If H_0 holds, then $\tau \sim U_{p,q_1,n-q}$. Moreover, if $\mathbf{B}_1^* = \mathbf{0}$ then τ is an invariant of the sufficient statistic under G.

PROOF: From Theorem 8.1, it follows that the LRT statistic is the same as in the normal case. Since problems (9.18) and (9.19) are equivalent, we can focus on (9.19). The statistic τ is invariant under G. Moreover, G satisfies the conditions of part (a) of Theorem 8.3. Therefore, the null distribution of τ is the same as in the normal case. Thus, the corresponding results of the normal theory can be used here (see Anderson 2003, Sect. 8.4.1). ∎

For more on the null distribution of the test statistic and the asymptotic null distribution, see Anderson (2003, Sects. 8.4–8.5), Pillai and Gupta (1969), Gupta (1971), Gupta and Tang (1984, 1988), and Tang and Gupta (1984, 1986, 1987).

Next, we focus on the invariance properties of problem (9.18). The results are based on Anderson (2003). Define $\mathbf{Z}_1^* = \mathbf{Z}_1 - \mathbf{Z}_1\mathbf{Z}_2'(\mathbf{Z}_2\mathbf{Z}_2')^{-1}\mathbf{Z}_2$ and $\mathbf{B}_2^* = \mathbf{B}_2 + \mathbf{B}_1\mathbf{Z}_1\mathbf{Z}_2'(\mathbf{Z}_2\mathbf{Z}_2')^{-1}$. Then, $\mathbf{B}\mathbf{Z} = \mathbf{B}_1\mathbf{Z}_1 + \mathbf{B}_2\mathbf{Z}_2 = \mathbf{B}_1\mathbf{Z}_1^* + \mathbf{B}_2^*\mathbf{Z}_2$,

$$\begin{aligned}
\mathbf{Z}_1^*\mathbf{Z}_1^{*\prime} &= (\mathbf{Z}_1 - \mathbf{Z}_1\mathbf{Z}_2'(\mathbf{Z}_2\mathbf{Z}_2')^{-1}\mathbf{Z}_2)(\mathbf{Z}_1' - \mathbf{Z}_2'(\mathbf{Z}_2\mathbf{Z}_2')^{-1}\mathbf{Z}_2\mathbf{Z}_1') \\
&= \mathbf{Z}_1\mathbf{Z}_1' - \mathbf{Z}_1\mathbf{Z}_2'(\mathbf{Z}_2\mathbf{Z}_2')^{-1}\mathbf{Z}_2\mathbf{Z}_1' - \mathbf{Z}_1\mathbf{Z}_2'(\mathbf{Z}_2\mathbf{Z}_2')^{-1}\mathbf{Z}_2\mathbf{Z}_1' \\
&+ \mathbf{Z}_1\mathbf{Z}_2'(\mathbf{Z}_2\mathbf{Z}_2')^{-1}\mathbf{Z}_2\mathbf{Z}_2'(\mathbf{Z}_2\mathbf{Z}_2')^{-1}\mathbf{Z}_2\mathbf{Z}_1' \\
&= \mathbf{Z}_1\mathbf{Z}_1' - \mathbf{Z}_1\mathbf{Z}_2'(\mathbf{Z}_2\mathbf{Z}_2')^{-1}\mathbf{Z}_2\mathbf{Z}_1',
\end{aligned}$$

and

$$\begin{aligned}
\mathbf{Z}_1^*\mathbf{Z}_2' &= (\mathbf{Z}_1 - \mathbf{Z}_1\mathbf{Z}_2'(\mathbf{Z}_2\mathbf{Z}_2')^{-1}\mathbf{Z}_2)\mathbf{Z}_2' \\
&= \mathbf{Z}_1\mathbf{Z}_2' - \mathbf{Z}_1\mathbf{Z}_2'(\mathbf{Z}_2\mathbf{Z}_2')^{-1}\mathbf{Z}_2\mathbf{Z}_2' \\
&= \mathbf{0}.
\end{aligned}$$

Thus, (9.1) can be written in the following equivalent form:

$$\mathbf{X} \sim E_{p,n}(\mathbf{B}_1 Z_1^* + \mathbf{B}_2^* \mathbf{Z}_2, \Sigma \otimes \mathbf{I}_n, \psi), \tag{9.20}$$

where $\mathbf{Z}_1^* \mathbf{Z}_1^{*\prime} = \mathbf{A}_{11\cdot 2}$ and $\mathbf{Z}_1^* \mathbf{Z}_2' = \mathbf{0}$. We want to test

$$H_0 : B_1 = \mathbf{0} \quad \text{against} \quad H_1 : B_1 \neq \mathbf{0}. \tag{9.21}$$

Problem (9.21) remains invariant under group G, where G is generated by

(i) $g(\mathbf{Z}_1^*) = \mathbf{K}\mathbf{Z}_1^*$, where $\mathbf{K}$ is $q \times q$ nonsingular matrix, and by the transformations
(ii)

$$g(\mathbf{X}) = \mathbf{C}\mathbf{X}, \quad \text{where } \mathbf{C} \text{ is } p \times p \text{ nonsingular matrix, and} \tag{9.22}$$

(iii)

$$g(\mathbf{X}) = \mathbf{X} + \mathbf{L}\mathbf{Z}_2, \quad \text{where } \mathbf{L} \text{ is } p \times q_2 \text{ matrix.} \tag{9.23}$$

Then, we have the following theorem.

Theorem 9.10. *The maximal invariant of* $\mathbf{A}_{11\cdot 2}$, *and the sufficient statistic* $\hat{\mathbf{B}}$ *and* $\hat{\Sigma}$ *under* G *is the set of roots of*

$$|\mathbf{H} - l\mathbf{S}| = 0, \tag{9.24}$$

where $\mathbf{H} = \hat{\mathbf{B}}_1 \mathbf{A}_{11\cdot 2} \hat{\mathbf{B}}_1'$ *and* $\mathbf{S} = \frac{z_h}{p} \hat{\Sigma}$. *Here* $\hat{\mathbf{B}}_1$ *denotes the* $p \times q_1$ *matrix in the partitioning of* $\hat{\mathbf{B}}$ *into* $\hat{\mathbf{B}} = (\hat{\mathbf{B}}_1, \hat{\mathbf{B}}_2)$. *Moreover, if* H_0 *holds in (9.21), then the distribution of the roots of (9.24) are the same as in the normal case.*

PROOF: In Anderson (2003, Sect. 8.6.1), it is shown that the roots of (9.24) form a maximal invariant under the given conditions. Since the subgroup of G, which is generated by the transformations (9.22) and (9.23), satisfies the conditions of part (a) of Theorem 8.3, the null distribution of the roots of (9.24) is the same as in the normal case. ■

It is easy to see that the LRT statistic, τ, is a function of the roots of (9.24): $\tau = |\mathbf{I}_p + \mathbf{H}\mathbf{S}^{-1}|$. Other test statistics, which are also functions of the roots of (9.24) are the Lawley-Hotelling's trace criterion: $tr(\mathbf{H}\mathbf{S}^{-1})$; the Bartlett-Nanda-Pillai's trace criterion: $tr(\mathbf{H}(\mathbf{S}+\mathbf{H})^{-1})$; and the Roy's largest (smallest) root criterion, that is, the largest (smallest) characteristic root of $\mathbf{H}\mathbf{S}^{-1}$. Then, they have the same null distribution as in the normal case. For a further discussion of these test statistics, see Anderson (2003, Sect. 8.6), Pillai and Gupta (1969), and Gupta (1971). These invariant statistics were also studied by Hsu (1985b) for the case of m.e.c. distributions and it was also shown that the LRT, the Lawley-Hotelling's trace test, the Bartlett-Nanda-Pillai's trace test, and the Roy's largest root test, are all admissible.

Remark 9.1. Since the multivariate analysis of variance (MANOVA) problems can be formulated in terms of the regression model (9.1), and most hypotheses in MANOVA can be expressed as (9.19), the results of Theorem 9.9 can be used here. As a consequence, the LRT statistics are the same as those developed in the normal theory and their null distributions and critical regions are also the same as in the normal case. For the treatment of the MANOVA problems in the normal case, see Anderson (2003, Sect. 8.9).

9.3 Inference in the Random Effects Model

In Sect. 9.1, the p-dimensional vectors, $\mathbf{x}_1, \mathbf{x}_2, \ldots, \mathbf{x}_n$ have the property that $\mathbf{x}_i \sim E_p(\mathbf{B}\mathbf{z}_i, \Sigma, \psi)$, where $\mathbf{B}$ is a $p \times q$ matrix and $\mathbf{z}_i$ is a q-dimensional vector, $i = 1,2,\ldots,n$. This can also be expressed as $\mathbf{x}_i = \mathbf{B}\mathbf{z}_i + \mathbf{v}_i$, where $\mathbf{v}_i \sim E_p(\mathbf{0}, \Sigma, \psi)$, $i = 1,2,\ldots,n$. The vectors $\mathbf{z}_i$, $i = 1,2,\ldots,n$ are assumed to be known. On the other hand, the matrix $\mathbf{B}$ is unknown, but it is also constant. The random vectors $\mathbf{v}_i$, $i = 1,2,\ldots,n$ are called the error terms. Let $\mathbf{V} = (\mathbf{v}_1, \mathbf{v}_2, \ldots, \mathbf{v}_n)$. Then, the model (9.1) can be expressed as

$$\mathbf{X} = \mathbf{B}\mathbf{Z} + \mathbf{V}, \quad \text{where} \quad \mathbf{V} \sim E_{p,n}(\mathbf{0}, \Sigma \otimes \mathbf{I}_n, \psi).$$

We get a different model if we assume that the vectors $\mathbf{z}_i$, $i = 1,2,\ldots,n$ are also random. Define $\mathbf{y}_i = \begin{pmatrix} \mathbf{z}_i \\ \mathbf{v}_i \end{pmatrix}$, and assume that $\mathbf{y}_i \sim E_{p+q}\left(\begin{pmatrix} \mathbf{m}_i \\ \mathbf{0} \end{pmatrix}, \begin{pmatrix} \Sigma_1 & \mathbf{0} \\ \mathbf{0} & \Sigma_2 \end{pmatrix}, \psi\right)$, where $\mathbf{m}_i$ is a q-dimensional known vector, $i = 1,2,\ldots,n$, and Σ_1, $q \times q$, Σ_2, $(p-q) \times (p-q)$, are unknown matrices. In this case, we suppose that $\mathbf{B}$ and $\mathbf{m}_i$ are known. Moreover, let $\mathbf{y}_i$, $i = 1,2,\ldots,n$ be uncorrelated and assume that their joint distribution is elliptically contoured. Then, this model can be expressed as

$$\mathbf{X} = \mathbf{B}\mathbf{Z} + \mathbf{V}, \quad \text{where} \quad \mathbf{V} \sim E_{p,n}\left(\begin{pmatrix} \mathbf{M} \\ \mathbf{0} \end{pmatrix}, \begin{pmatrix} \Sigma_1 & \mathbf{0} \\ \mathbf{0} & \Sigma_2 \end{pmatrix}, \psi\right), \tag{9.25}$$

where $q \le p$, $\mathbf{X} = (\mathbf{x}_1, \mathbf{x}_2, \ldots, \mathbf{x}_n)$ is $p \times n$, $\mathbf{Z}$ is $q \times n$, and $\mathbf{V}$ is $p \times n$. Assume that the $p \times q$ matrix $\mathbf{B}$ and the $q \times n$ matrix $\mathbf{M}$ are known, but the $q \times q$ matrix Σ_1 and the $(p-q) \times (p-q)$ matrix Σ_2 are unknown. Also assume that $rk(\mathbf{B}) = q$, and the random matrix $\mathbf{Y} = \begin{pmatrix} \mathbf{Z} \\ \mathbf{V} \end{pmatrix}$ has finite second order moment.

We want to find the optimal mean-square estimator of $\mathbf{Z}$ given $\mathbf{X}$. This is equivalent to finding $E(\mathbf{Z}|\mathbf{X})$. We need the following result.

Lemma 9.5. *Let* $\mathbf{X} \sim E_{p,n}(\mathbf{M}, \Sigma \otimes \Phi, \psi)$, *with stochastic representation* $\mathbf{X} \approx r\mathbf{A}_0\mathbf{U}\mathbf{B}_0'$. *Let* F *be the distribution function of* r. *Define* $\mathbf{Y} = \mathbf{A}\mathbf{X}\mathbf{B}$, *with* $\mathbf{A}$ $(q \times p)$, $\mathbf{B}$ $(n \times m)$ *matrices,* $rk(\mathbf{A}) = q$, *and* $rk(\mathbf{B}) = m$. *Then,*

(a) *If* $E(\mathbf{X}|\mathbf{Y})$ *exists, we have*

$$E(\mathbf{X}|\mathbf{Y}) = \mathbf{M} + \Sigma\mathbf{A}'(\mathbf{A}\Sigma\mathbf{A}')^{-1}(\mathbf{Y} - \mathbf{AMB})(\mathbf{B}'\Phi\mathbf{B})^{-1}\mathbf{B}'\Phi.$$

(b) *If* $m = n$, $\mathbf{K} = \Phi = \mathbf{I}_n$ *and* $Cov(\mathbf{X}|\mathbf{Y})$ *exists, we have*

$$Cov(\mathbf{X}|\mathbf{Y}) = k(\Sigma - \Sigma\mathbf{A}'(\mathbf{A}\Sigma\mathbf{A}')^{-1}\mathbf{A}\Sigma) \otimes \mathbf{I}_n,$$

where $k = -2\psi'_{q(\mathbf{X}_2)}$ *and* $\psi_{q(\mathbf{X}_2)}$ *is defined by (2.30)–(2.32) with* $q(\mathbf{X}_2) = tr((\mathbf{Y} - \mathbf{AM})'(\mathbf{A}\Sigma\mathbf{A}')^{-1}(\mathbf{Y} - \mathbf{AM}))$. *Moreover, if the distribution of* $\mathbf{X}$ *is absolutely continuous and the p.d.f. of* $\mathbf{Y} = \mathbf{AX}$ *is*

$$f(\mathbf{Y}) = \frac{1}{|\mathbf{A}\Sigma\mathbf{A}'|^{\frac{n}{2}}} h_1(tr((\mathbf{Y} - \mathbf{AM})'(\mathbf{A}\Sigma\mathbf{A}')^{-1}(\mathbf{Y} - \mathbf{AM}))), \tag{9.26}$$

then

$$k = \frac{\int_r^\infty h_1(t)dt}{2h_1(r)},$$

where $r = tr((\mathbf{Y} - \mathbf{AM})'(\mathbf{A}\Sigma\mathbf{A}')^{-1}(\mathbf{Y} - \mathbf{AM}))$.

PROOF: If $q = p$, and $n = m$, the theorem is obvious. So, assume $qm < pn$.

Step 1. Assume $n = 1$, $\mathbf{x} \sim E_p(\mathbf{0}, I_p, \psi)$, and $\mathbf{B} = 1$. Using Theorem 1.9, we can write $\mathbf{A} = \mathbf{PDQ}$, where $\mathbf{P}$ is a $q \times q$ nonsingular matrix, $\mathbf{Q}$ is a $p \times p$ orthogonal matrix, and $\mathbf{D}$ is a $q \times p$ matrix with $\mathbf{D} = (\mathbf{I}_q, \mathbf{0})$. Then,

$$\begin{aligned} E(\mathbf{x}|\mathbf{y}) &= E(\mathbf{x}|\mathbf{Ax}) \\ &= E(\mathbf{x}|\mathbf{PDQx}) \\ &= \mathbf{Q}'E(\mathbf{Qx}|\mathbf{PDQx}) \\ &= \mathbf{Q}'E(\mathbf{Qx}|\mathbf{DQx}). \end{aligned} \tag{9.27}$$

Let $\mathbf{z} = \mathbf{Qx}$. Then $\mathbf{z} \sim E_p(\mathbf{0}, \mathbf{I}_p, \psi)$ and $\mathbf{DQx} = \mathbf{z}_1$, where $\mathbf{z} = \begin{pmatrix} \mathbf{z}_1 \\ \mathbf{z}_2 \end{pmatrix}$, $\mathbf{z}_1$ is q-dimensional vector. From Theorem 2.22, it follows that $E(\mathbf{z}_2|\mathbf{z}_1) = \mathbf{0}$ and $Cov(\mathbf{z}_2|\mathbf{z}_1) = k\mathbf{I}_{p-q}$ where $k = -2\psi'_{q(\mathbf{x}_2)}$ and $\psi_{q(\mathbf{x}_2)}$ is defined by (2.30)–(2.32) with $q(\mathbf{x}_2) = \mathbf{z}_1'\mathbf{z}_1$. On the other hand, $E(\mathbf{z}_1|\mathbf{z}_1) = \mathbf{z}_1$. Therefore,

$$E(\mathbf{z}|\mathbf{z}_1) = \mathbf{D}'\mathbf{Dz}. \tag{9.28}$$

From (9.27) and (9.28), it follows that

$$\begin{aligned} E(\mathbf{x}|\mathbf{y}) &= \mathbf{Q}'\mathbf{D}'\mathbf{D}\mathbf{Q}\mathbf{x} \\ &= \mathbf{Q}'\mathbf{D}'\mathbf{P}'\mathbf{P}'^{-1}\mathbf{P}^{-1}\mathbf{P}\mathbf{D}\mathbf{Q}\mathbf{x} \\ &= \mathbf{A}'(\mathbf{P}\mathbf{P}')^{-1}\mathbf{A}\mathbf{x} \\ &= \mathbf{A}'(\mathbf{P}\mathbf{D}\mathbf{Q}\mathbf{Q}'\mathbf{D}'\mathbf{P}')^{-1}\mathbf{A}\mathbf{x} \\ &= \mathbf{A}'(\mathbf{A}\mathbf{A}')^{-1}\mathbf{A}\mathbf{x} \\ &= \mathbf{A}'(\mathbf{A}\mathbf{A}')^{-1}\mathbf{y} \end{aligned}$$

and

$$\begin{aligned} Cov(\mathbf{x}|\mathbf{y}) &= E(\mathbf{x}\mathbf{x}'|\mathbf{y}) - E(\mathbf{x}|\mathbf{y})(E(\mathbf{x}|\mathbf{y}))' \\ &= E(\mathbf{Q}'\mathbf{z}\mathbf{z}'\mathbf{Q}|\mathbf{P}\mathbf{D}\mathbf{z}) - E(\mathbf{Q}'\mathbf{z}|\mathbf{P}\mathbf{D}\mathbf{z})(E(\mathbf{Q}'\mathbf{z}|\mathbf{P}\mathbf{D}\mathbf{z}))' \\ &= \mathbf{Q}'[E(\mathbf{z}\mathbf{z}'|\mathbf{D}\mathbf{z}) - E(\mathbf{z}|\mathbf{D}\mathbf{z})(E(\mathbf{z}|\mathbf{D}\mathbf{z}))']\mathbf{Q}. \end{aligned} \tag{9.29}$$

Now,

$$E(\mathbf{z}|\mathbf{D}\mathbf{z}) = E(\mathbf{z}|\mathbf{z}_1) = \mathbf{D}'\mathbf{D}\mathbf{z}, \tag{9.30}$$

$$E(\mathbf{z}\mathbf{z}'|\mathbf{D}\mathbf{z}) = E\left(\begin{pmatrix} \mathbf{z}_1\mathbf{z}_1' & \mathbf{z}_1\mathbf{z}_2' \\ \mathbf{z}_2\mathbf{z}_1' & \mathbf{z}_2\mathbf{z}_2' \end{pmatrix} |\mathbf{z}_1\right) = \begin{pmatrix} E(\mathbf{z}_1\mathbf{z}_1'|\mathbf{z}_1) & E(\mathbf{z}_1\mathbf{z}_2'|\mathbf{z}_1) \\ E(\mathbf{z}_2\mathbf{z}_1'|\mathbf{z}_1) & E(\mathbf{z}_2\mathbf{z}_2'|\mathbf{z}_1) \end{pmatrix},$$

$$E(\mathbf{z}_1\mathbf{z}_1'|\mathbf{z}_1) = \mathbf{z}_1\mathbf{z}_1' = \mathbf{D}\mathbf{z}\mathbf{z}'\mathbf{D}',$$

$$E(\mathbf{z}_1\mathbf{z}_2'|\mathbf{z}_1) = \mathbf{z}_1 E(\mathbf{z}_2'|\mathbf{z}_1) = \mathbf{0},$$

$$E(\mathbf{z}_2\mathbf{z}_1'|\mathbf{z}_1) = E(\mathbf{z}_2|\mathbf{z}_1)\mathbf{z}_1' = \mathbf{0}, \quad \text{and}$$

$$E(\mathbf{z}_2\mathbf{z}_2'|\mathbf{z}_1) = E(\mathbf{z}_2\mathbf{z}_2'|\mathbf{z}_1) - E(\mathbf{z}_2|\mathbf{z}_1)E(\mathbf{z}_2|\mathbf{z}_1)' = Cov(\mathbf{z}_2|\mathbf{z}_1) = k\mathbf{I}_{p-q}.$$

However,

$$\begin{aligned} \mathbf{z}_1'\mathbf{z}_1 &= \mathbf{x}'\mathbf{Q}'\mathbf{D}'\mathbf{D}\mathbf{Q}\mathbf{x} \\ &= \mathbf{x}'\mathbf{A}'(\mathbf{A}\mathbf{A}')^{-1}\mathbf{A}\mathbf{x} \\ &= \mathbf{y}'(\mathbf{A}\mathbf{A}')^{-1}\mathbf{y}. \end{aligned}$$

So, $k = -2\psi'_{q(\mathbf{x}_2)}$ and $\psi_{q(\mathbf{x}_2)}$ is defined by (2.30)–(2.32) with $q(\mathbf{x}_2) = \mathbf{y}'(\mathbf{A}\mathbf{A}')^{-1}\mathbf{y}$. Hence,

$$E(\mathbf{z}\mathbf{z}'|\mathbf{D}\mathbf{z}) = \mathbf{D}'\mathbf{D}\mathbf{z}\mathbf{z}'\mathbf{D}'\mathbf{D} + k(\mathbf{I}_p - \mathbf{D}\mathbf{D}'). \tag{9.31}$$

If the distribution of $\mathbf{x}$ is absolutely continuous and the p.d.f. of $\mathbf{y}=\mathbf{Hx}$ is given by (9.26), then the p.d.f. of $\mathbf{z}_1=\mathbf{DQx}$ is $h_1(\mathbf{z}_1'\mathbf{z}_1)$. Hence $k=\frac{\int_r^\infty h_1(t)dt}{2h_1(r)}$, with $r=\mathbf{z}_1'\mathbf{z}_1=\mathbf{y}'(\mathbf{AA}')^{-1}\mathbf{y}$.

It follows from (9.29), (9.30), and (9.31) that

$$\begin{aligned}Cov(\mathbf{x}|\mathbf{y}) &= \mathbf{Q}'(\mathbf{D}'\mathbf{Dzz}'\mathbf{D}'\mathbf{D}+k(\mathbf{I}_p-\mathbf{DD}')-\mathbf{D}'\mathbf{Dzz}'\mathbf{D}'\mathbf{D})\mathbf{Q}\\ &= k(\mathbf{I}_p-\mathbf{Q}'\mathbf{D}'\mathbf{DQ})\\ &= k(\mathbf{I}_p-\mathbf{A}'(\mathbf{AA}')^{-1}\mathbf{A}).\end{aligned}$$

Step 2. Let $\mathbf{X}\sim E_{p,n}(\mathbf{M},\Sigma\otimes\Phi,\psi)$. We have

$$\begin{aligned}&(vec(\mathbf{X}')|vec(\mathbf{Y}'))=((vec(\mathbf{X}-\mathbf{M})'+vec(\mathbf{M}'))|vec(\mathbf{AXB})')\\ &=(vec(\mathbf{X}-\mathbf{M})'|vec(\mathbf{A}(\mathbf{X}-\mathbf{M})\mathbf{B})')+vec(\mathbf{M}')\\ &=(((\Sigma^{\frac{1}{2}}\otimes\Phi^{\frac{1}{2}})vec(\Sigma^{-\frac{1}{2}}(\mathbf{X}-\mathbf{M})\Phi^{-\frac{1}{2}})')|(((\mathbf{A}\Sigma^{\frac{1}{2}})\otimes(\mathbf{B}'\Phi^{\frac{1}{2}}))\\ &\times vec(\Sigma^{-\frac{1}{2}}(\mathbf{X}-\mathbf{M})\Phi^{-\frac{1}{2}})'))+vec(\mathbf{M}')\\ &=(\Sigma^{\frac{1}{2}}\otimes\Phi^{\frac{1}{2}})((vec(\Sigma^{-\frac{1}{2}}(\mathbf{X}-\mathbf{M})\Phi^{-\frac{1}{2}})')|(((\mathbf{A}\Sigma^{\frac{1}{2}})\otimes(\mathbf{B}'\Phi^{\frac{1}{2}}))\\ &\times vec(\Sigma^{-\frac{1}{2}}(\mathbf{X}-\mathbf{M})\Phi^{-\frac{1}{2}})'))+vec(M').\end{aligned}\tag{9.32}$$

Now, $vec(\Sigma^{-\frac{1}{2}}(\mathbf{X}-\mathbf{M})\Phi^{-\frac{1}{2}})'\sim E_{pn}(\mathbf{0},\mathbf{I}_{pn},\psi)$ and using Step 1, we get

$$\begin{aligned}&E((vec(\Sigma^{-\frac{1}{2}}(\mathbf{X}-\mathbf{M})\Phi^{-\frac{1}{2}})')|(((\mathbf{A}\Sigma^{\frac{1}{2}})\otimes(\mathbf{B}'\Phi^{\frac{1}{2}}))vec(\Sigma^{-\frac{1}{2}}(\mathbf{X}-\mathbf{M})\Phi^{-\frac{1}{2}})'))\\ &=((\Sigma^{\frac{1}{2}}\mathbf{A}')\otimes(\Phi^{\frac{1}{2}}\mathbf{B}))((\mathbf{A}\Sigma^{\frac{1}{2}}\Sigma^{\frac{1}{2}}\mathbf{A}')\otimes(\mathbf{B}'\Phi^{\frac{1}{2}}\Phi^{\frac{1}{2}}\mathbf{B}))^{-1}((\mathbf{A}\Sigma^{\frac{1}{2}})\otimes(\mathbf{B}'\Phi^{\frac{1}{2}}))\\ &\times vec(\Sigma^{-\frac{1}{2}}(\mathbf{X}-\mathbf{M})\Phi^{-\frac{1}{2}})'\\ &=(\Sigma^{\frac{1}{2}}\mathbf{A}'(\mathbf{A}\Sigma\mathbf{A}')^{-1}\mathbf{A}\Sigma^{\frac{1}{2}})\otimes(\Phi^{\frac{1}{2}}\mathbf{B}(\mathbf{B}'\Phi\mathbf{B})^{-1}\mathbf{B}'\Phi^{\frac{1}{2}})vec(\Sigma^{-\frac{1}{2}}(\mathbf{X}-\mathbf{M})\Phi^{-\frac{1}{2}})'\\ &=vec(\Sigma^{\frac{1}{2}}\mathbf{A}'(\mathbf{A}\Sigma\mathbf{A}')^{-1}\mathbf{A}(\mathbf{X}-\mathbf{M})\mathbf{B}(\mathbf{B}'\Phi\mathbf{B})^{-1}\mathbf{B}'\Phi^{\frac{1}{2}})'.\end{aligned}\tag{9.33}$$

From (9.32) and (9.33), we get

$$\begin{aligned}E(\mathbf{X}|\mathbf{Y}) &= \mathbf{M}+\Sigma\mathbf{A}'(\mathbf{A}\Sigma\mathbf{A}')^{-1}\mathbf{A}(\mathbf{X}-\mathbf{M})\mathbf{B}(\mathbf{B}'\Phi\mathbf{B})^{-1}\mathbf{B}'\Phi\\ &= \mathbf{M}+\Sigma\mathbf{A}'(\mathbf{A}\Sigma\mathbf{A}')^{-1}(\mathbf{Y}-\mathbf{AMB})(\mathbf{B}'\Phi\mathbf{B})^{-1}\mathbf{B}'\Phi.\end{aligned}$$

If $m=n$ and $\mathbf{B}=\Phi=\mathbf{I}_n$, then from Step 1, it follows that

$$\begin{aligned}&Cov((vec(\Sigma^{-\frac{1}{2}}(\mathbf{X}-\mathbf{M}))')|(((\mathbf{A}\Sigma^{\frac{1}{2}})\otimes\mathbf{I}_n)vec(\Sigma^{-\frac{1}{2}}(\mathbf{X}-\mathbf{M}))'))\\ &=k(\mathbf{I}_{pn}-(\Sigma^{\frac{1}{2}}\mathbf{A}'(\mathbf{A}\Sigma\mathbf{A}')^{-1}\mathbf{A}\Sigma^{\frac{1}{2}})\otimes\mathbf{I}_n)\\ &=k(\mathbf{I}_p-\Sigma^{\frac{1}{2}}\mathbf{A}'(\mathbf{A}\Sigma\mathbf{A}')^{-1}\mathbf{A}\Sigma^{\frac{1}{2}})\otimes\mathbf{I}_n.\end{aligned}\tag{9.34}$$

It follows from (9.32) and (9.34), that

$$Cov(\mathbf{X}|\mathbf{Y}) = k(\Sigma - \Sigma\mathbf{A}'(\mathbf{A}\Sigma\mathbf{A}')^{-1}\mathbf{A}\Sigma) \otimes \mathbf{I}_n,$$

where $k = -2\psi'_{q(\mathbf{X}_2)}$ and $\psi_{q(\mathbf{X}_2)}$ is defined by (2.30)–(2.32) with $q(\mathbf{X}_2) = tr((\mathbf{Y} - \mathbf{AM})'(\mathbf{A}\Sigma\mathbf{A}')^{-1}(\mathbf{Y} - \mathbf{AM}))$.

If the distribution of $\mathbf{X}$ is absolutely continuous, then $k = \frac{\int_r^\infty h_1(t)dt}{2h_1(r)}$, with $r = tr((\mathbf{Y} - \mathbf{AM})'(\mathbf{A}\Sigma\mathbf{A}')^{-1}(\mathbf{Y} - \mathbf{AM}))$. ∎

Now, we find the optimal mean square estimator of $\mathbf{Z}$ given $\mathbf{X}$.

Theorem 9.11. *Assume that in model (9.25), $rk(\mathbf{B}) = q$ and the random matrix $\mathbf{Y} = \begin{pmatrix}\mathbf{Z}\\ \mathbf{V}\end{pmatrix}$ has a finite second order moment. Let $r\mathbf{A}_0\mathbf{U}\mathbf{B}_0'$ be the stochastic representation of $\mathbf{Y}$ and F be the distribution function of r. Then, the optimal mean-square estimator of $\mathbf{Z}$ given $\mathbf{X}$ is*

$$\hat{\mathbf{Z}} = E(\mathbf{Z}|\mathbf{X}) = \mathbf{M} + \Sigma_1\mathbf{B}'(\mathbf{B}\Sigma_1\mathbf{B}' + \Sigma_2)^{-1}(\mathbf{X} - \mathbf{BM}).$$

Furthermore,

$$Cov(\mathbf{Z}|\mathbf{X}) = k(\Sigma_1 - \Sigma_1\mathbf{B}'(\mathbf{B}\Sigma_1\mathbf{B}' + \Sigma_2)^{-1}\mathbf{B}\Sigma_1) \otimes \mathbf{I}_n,$$

where $k = -2\psi'_{q(\mathbf{x}_2)}$ and $\psi_{q(\mathbf{x}_2)}$ is defined by (2.30)–(2.32) with $q(\mathbf{x}_2) = tr((\mathbf{X} - \mathbf{BM})'(\mathbf{B}\Sigma_1\mathbf{B}' + \Sigma_2)^{-1}(\mathbf{X} - \mathbf{BM}))$. If $\mathbf{Y}$ is absolutely continuous and the p.d.f. of $\mathbf{X}$ is

$$f(\mathbf{X}) = \frac{1}{|\mathbf{B}\Sigma_1\mathbf{B}' + \Sigma_2|^{\frac{n}{2}}} h(tr((\mathbf{X} - \mathbf{BM})'(\mathbf{B}\Sigma_1\mathbf{B}' + \Sigma_2)^{-1}(\mathbf{X} - \mathbf{BM})))$$

then

$$k = \frac{\int_r^\infty h(t)dt}{2h(r)},$$

where $r = tr((\mathbf{X} - \mathbf{BM})'(\mathbf{B}\Sigma_1\mathbf{B}' + \Sigma_2)^{-1}(\mathbf{X} - \mathbf{BM}))$.

PROOF: We have $\mathbf{X} = (\mathbf{B}, \mathbf{I}_p)\begin{pmatrix}\mathbf{Z}\\ \mathbf{V}\end{pmatrix}$. Hence,

$$\mathbf{X} \sim E_{p,n}(\mathbf{BM}, (\mathbf{B}\Sigma_1\mathbf{B}' + \Sigma_2) \otimes \mathbf{I}_n, \psi).$$

Using Lemma 9.3, we get

$$E\left(\begin{pmatrix}\mathbf{Z}\\ \mathbf{V}\end{pmatrix}|\mathbf{X}\right)=\begin{pmatrix}\mathbf{M}\\ \mathbf{0}\end{pmatrix}+\begin{pmatrix}\Sigma_1 & \mathbf{0}\\ \mathbf{0} & \Sigma_2\end{pmatrix}\begin{pmatrix}\mathbf{B}'\\ \mathbf{I}_p\end{pmatrix}\left[(\mathbf{B},\mathbf{I}_p)\begin{pmatrix}\Sigma_1 & \mathbf{0}\\ \mathbf{0} & \Sigma_2\end{pmatrix}\begin{pmatrix}\mathbf{B}'\\ \mathbf{I}_p\end{pmatrix}\right]^{-1}$$

$$\times\ (\mathbf{X}-(\mathbf{B},\mathbf{I}_p))\begin{pmatrix}\mathbf{M}\\ \mathbf{0}\end{pmatrix}$$

$$=\begin{pmatrix}\mathbf{M}\\ \mathbf{0}\end{pmatrix}+\begin{pmatrix}\Sigma_1\mathbf{B}'\\ \Sigma_2\end{pmatrix}(\mathbf{B}\Sigma_1\mathbf{B}'+\Sigma_2)^{-1}(\mathbf{X}-\mathbf{B}\mathbf{M})$$

$$=\begin{pmatrix}\mathbf{M}+\Sigma_1\mathbf{B}'(\mathbf{B}\Sigma_1\mathbf{B}'+\Sigma_2)^{-1}(\mathbf{X}-\mathbf{B}\mathbf{M})\\ \Sigma_2(\mathbf{B}\Sigma_1\mathbf{B}'+\Sigma_2)^{-1}(\mathbf{X}-\mathbf{B}\mathbf{M})\end{pmatrix}.$$

Thus,

$$\hat{\mathbf{Z}}=E(\mathbf{Z}|\mathbf{X})=\mathbf{M}+\Sigma_1\mathbf{B}'(\mathbf{B}\Sigma_1\mathbf{B}'+\Sigma_2)^{-1}(\mathbf{X}-\mathbf{B}\mathbf{M}).$$

We also get

$$Cov\left(\begin{pmatrix}\mathbf{Z}\\ \mathbf{V}\end{pmatrix}|\mathbf{X}\right)=k\left(\begin{pmatrix}\Sigma_1 & \mathbf{0}\\ \mathbf{0} & \Sigma_2\end{pmatrix}-\begin{pmatrix}\Sigma_1\mathbf{B}'\\ \Sigma_2\end{pmatrix}(\mathbf{B}\Sigma_1\mathbf{B}'+\Sigma_2)^{-1}(\mathbf{B}\Sigma_1,\Sigma_2)\right)\otimes\mathbf{I}_n.$$

Therefore,

$$Cov(\mathbf{Z}|\mathbf{X})=k(\Sigma_1-\Sigma_1\mathbf{B}'(\mathbf{B}\Sigma_1\mathbf{B}'+\Sigma_2)^{-1}\mathbf{B}\Sigma_1)\otimes\mathbf{I}_n,$$

where $k=-2\psi'_{q(\mathbf{x}_2)}$ and $\psi_{q(\mathbf{x}_2)}$ is defined by (2.30)–(2.32) with $q(\mathbf{x}_2)=tr((\mathbf{X}-\mathbf{B}\mathbf{M})'(\mathbf{B}\Sigma_1\mathbf{B}'+\Sigma_2)^{-1}(\mathbf{X}-\mathbf{B}\mathbf{M}))$.

If $\begin{pmatrix}\mathbf{Z}\\ \mathbf{V}\end{pmatrix}$ is absolutely continuous, then

$$k=\frac{\int_r^\infty h(t)dt}{2h(r)}$$

with $r=tr((\mathbf{X}-\mathbf{B}\mathbf{M})'(\mathbf{B}\Sigma_1\mathbf{B}'+\Sigma_2)^{-1}(\mathbf{X}-\mathbf{B}\mathbf{M}))$. ■

The results of this section were derived by Chu (1973) for the vector variate case.

Chapter 10
Application in Portfolio Theory

10.1 Elliptically Contoured Distributions in Portfolio Theory

The mean-variance analysis of Markowitz (1952) is important for both practitioners and researchers in finance. This theory provides an easy access to the problem of optimal portfolio selection. However, in implementing pricing theory one is faced with a number of difficulties. The mean-variance approach seems to provide almost optimal results only if the distribution of the returns is approximately normal or the utility function looks roughly like a parabola. Kroll, Levy, and Markowitz (1984) reported that the mean-variance portfolio has a maximum expected utility or it is at least close to a maximum expected utility.

The practical pitfalls of the mean-variance analysis are mainly related to the extreme weights that often arise when the sample efficient portfolio is constructed. This phenomenon was studied by Merton (1980), who among others argued that the estimates of the variances and the covariances of the asset returns are more accurate than the estimates of the means. Best and Grauer (1991) showed that the sample efficient portfolio is extremely sensitive to changes in the asset means. Chopra and Ziemba (1993) concluded for a real data set that errors in means are over ten times as damaging as errors in variances and over 20 times as errors in covariances. For that reason many authors assume equal means for the portfolio asset returns or, in other words, the global minimum variance portfolio (GMVP). This is one reason why the GMVP is extensively discussed in literature (Chan, Karceski, and Lakonishok 1999). The GMVP has the lowest risk of any feasible portfolio. The subject of our paper are the weights of the GMVP portfolio.

Results about the distribution of the estimated optimal weights and the estimated risk measures are of great importance for evaluating the efficiency of the underlying portfolio (Barberis 1999; Fleming, Kirby, and Ostdiek 2001). Jobson and Korkie (1980) studied the weights resulting from the Sharpe ratio approach under the assumption that the returns are independent and normally distributed. They derive approximations for the mean and the variance of the estimated weights, together with the asymptotic covariance matrix. In Jobson and Korkie (1989) a test for the

A.K. Gupta et al., *Elliptically Contoured Models in Statistics and Portfolio Theory*, DOI 10.1007/978-1-4614-8154-6_10, mean-variance efficiency is derived. A performance measure is introduced which is related to the Sharpe ratio. Britten-Jones (1999) analyzed tests for the efficiency of the mean-variance optimal weights under the normality assumption of the returns. Using a regression procedure the exact distribution of the normalized weights is derived. Okhrin and Schmid (2006) proved several distributional properties for various optimal portfolio weights like, e.g., the weights based on the expected utility and the weights based on the Sharpe ratio. They considered the case of finite sample size and of infinite sample size as well.

In all of the above cited papers the stock returns are demanded to be independent and normally distributed. The assumption of normality is found appropriate due to positive theoretical features, e.g., the consistency with the mean-variance rule, the equivalence of multiperiod and single period decision rules, the consistency with the assumptions of the capital asset pricing model (Stiglitz 1989; Markowitz 1991). Fama (1976) found that monthly stock returns can be well described by a normal approach. However, in the case of daily returns the assumption of normality and independence might not be appropriate since it is very likely that the underlying distributions have heavy tails (Osborne 1959; Fama 1965, 1976; Markowitz 1991; Rachev and Mittnik 2000). For such a case the application of the multivariate t-distribution has been suggested by Zellner (1976) and Sutradhar (1988). Moreover, the assumption of independent returns turns out to be questionable, too. Numerous studies demonstrated that frequently stock returns are uncorrelated but not independent (Engle 1982, 2002; Bollerslev 1986; Nelson 1991).

In this chapter we assume that the matrix of returns follows a matrix elliptically contoured distribution. As shown in Bodnar and Schmid (2007) this family turns out to be very suitable to describe stock returns because the returns are neither assumed to be independent nor to be normally distributed. Furthermore, it is in line with the results of Andersen, Bollerslev, Diebold, and Ebens (2001) and Andersen, Bollerslev, and Diebold (2005) who showed that daily returns normalized by the realized volatility can be well approximated by the normal distribution. The family covers a wide class of distributions like, e.g., the normal distribution, the mixture of normal distributions, the multivariate t-distribution, Pearson types II and VII distributions (see Fang, Kotz, and Ng 1990). Elliptically contoured distributions have been already discussed in financial literature. For instance, Owen and Rabinovitch (1983) showed that Tobin's separation theorem, Bawa's rules of ordering certain prospects can be extended to elliptically contoured distributions. While Chamberlain (1983) showed that elliptical distributions imply mean-variance utility functions, Berk (1997) argued that one of the necessary conditions for the capital asset pricing model (CAPM) is an elliptical distribution for the asset returns. Furthermore, Zhou (1993) extended findings of Gibbons, Ross, and Shanken (1989) by applying their test of the validity of the CAPM to elliptically distributed returns. A further test for the CAPM under elliptical assumptions is proposed by Hodgson, Linton, and Vorkink (2002). The first paper dealing with the application of matrix elliptically contoured distributions in finance, however, seems to be Bodnar and Schmid (2007). They introduced a test for the global minimum variance. It is analyzed whether the lowest risk is larger than a given benchmark value or not.

10.2 Estimation of the Global Minimum Variance

In the formulation of the mean-variance portfolio problem (e.g., Markowitz 1952; Samuelson 1970; Constandinidis and Malliaris 1995) a portfolio consisting of k assets is considered. The weight of the i-th asset in the portfolio is denoted by w_i. Let $\mathbf{w} = (w_1, .., w_k)'$ and $\mathbf{w}'\mathbf{1} = 1$. Here $\mathbf{1}$ denotes a vector whose components are all equal to 1. Suppose that for the k-dimensional vector of asset returns at a fixed time point t the second moments exist. We denote the mean of this vector by μ and its covariance matrix by $\mathbf{V}$. Then the expected return of the portfolio is given by $\mathbf{w}'\mu$ and its variance is equal to $\mathbf{w}'\mathbf{V}\mathbf{w}$.

When an investor is fully risk averse optimal weights can be obtained by minimizing the portfolio variance $\mathbf{w}'\mathbf{V}\mathbf{w}$ subject to $\mathbf{w}'\mathbf{1} = 1$. If $\mathbf{V}$ is positive definite the weights are presented as

$$\mathbf{w}_M = \frac{\mathbf{V}^{-1}\mathbf{1}}{\mathbf{1}'\mathbf{V}^{-1}\mathbf{1}}. \tag{10.1}$$

The portfolio constructed using such weights is known as global minimum variance portfolio. Its variance is given by

$$\sigma_M^2 = \mathbf{w}'\mathbf{V}\mathbf{w} = \frac{1}{\mathbf{1}'\mathbf{V}^{-1}\mathbf{1}}. \tag{10.2}$$

The quantity σ_M^2 is an important measure for evaluating the portfolio because it measures its risk behavior.

Because $\mathbf{V}$ is an unknown parameter the investor cannot determine σ_M^2. He has to estimate $\mathbf{V}$ using previous observations. Given the sample of portfolio returns of k assets $\mathbf{x}_1, .., \mathbf{x}_n$ the most common estimator of $\mathbf{V}$ is its empirical counterpart, i.e.

$$\hat{\mathbf{V}} = \frac{1}{n-1}\sum_{t=1}^{n}(\mathbf{x}_t - \bar{\mathbf{x}})(\mathbf{x}_t - \bar{\mathbf{x}})' = \frac{1}{n-1}\mathbf{X}\left(\mathbf{I} - \frac{1}{n}\mathbf{1}\mathbf{1}'\right)\mathbf{X}' \tag{10.3}$$

with $\mathbf{X} = (\mathbf{x}_1, .., \mathbf{x}_n)$. Replacing $\mathbf{V}$ by $\hat{\mathbf{V}}$ in (10.2) we get the estimator $\hat{\sigma}_M^2$ of σ_M^2.

Assuming that the variables $\mathbf{x}_1, .., \mathbf{x}_n$ are independent and identically distributed with $\mathbf{x}_i \sim N_k(\mu, \mathbf{V})$ it follows from Muirhead (1982, Theorem 3.2.12) that

$$\frac{(n-1)\,\hat{\sigma}_M^2}{\sigma_M^2} \sim \chi_{n-k}^2 \quad \text{for} n > k.$$

However, the assumptions of normality and of independence are not appropriate in many situations of practical interest. Many authors have shown that, e.g., the distribution of daily stock returns is heavy tailed (cf., Osborne 1959; Fama 1965, 1976; Markowitz 1991; Mittnik and Rachev 1993). In this section we derive the exact distribution of $\hat{\sigma}_M^2 = 1/\mathbf{1}'\hat{\mathbf{V}}^{-1}\mathbf{1}$ under a weaker assumption on the underlying sample, namely we assume that $\mathbf{X}$ follows a matrix variate elliptically contoured distribution. Such a result is useful because it provides an important characteristic of the estimator $\hat{\sigma}_M^2$ and it allows to determine confidence intervals and tests for the risk measure σ_M^2 under non-normality.

10.2.1 Distribution of the Global Minimum Variance

For an arbitrary k-dimensional vector τ the density of the random variable Q defined as

$$Q = (n-1)\,\frac{\tau'\Sigma^{-1}\tau}{\tau'\hat{\mathbf{V}}^{-1}\tau} \tag{10.4}$$

is derived in this section. Because it turns out that the distribution of Q does not depend on τ it immediately leads to the distribution of the global minimum variance estimator $\hat{\sigma}_M^2$.

The stochastic representation of the random matrix $\mathbf{X}$ is essential for deriving the distribution of Q. Let Σ be positive definite. It holds that $\mathbf{X} \sim E_{k,n}(\mathbf{M}, \Sigma \otimes \mathbf{I}_n, \psi)$ if and only if $\mathbf{X}$ has the same distribution as $\mathbf{M} + R\,\Sigma^{1/2}\,\mathbf{U}$, where $\mathbf{U}$ is a $k \times n$ random matrix and $vec(\mathbf{U}')$ is uniformly distributed on the unit sphere in $I\!R^{kn}$, R is a nonnegative random variable, and R and $\mathbf{U}$ are independent (see Theorem 2.13).

The distribution of R^2 is equal to the distribution of $\sum_{i=1}^{n}(\mathbf{x}_i - \mu_i)'\Sigma^{-1}(\mathbf{x}_i - \mu_i)$. If $\mathbf{X}$ is absolutely continuous, then R is also absolutely continuous and its density is

$$f_R(r) = \frac{2\pi^{nk/2}}{\Gamma(nk/2)}\, r^{nk-1}\, h(r^2) \tag{10.5}$$

for $r \geq 0$ (cf. Theorem 2.16). Note that for the matrix variate normal distribution $R_N^2 \sim \chi_{nk}^2$, where the index N refers to the normal distribution.

Theorem 10.1. *Let $\mathbf{X} = (\mathbf{x}_1 \ldots \mathbf{x}_n) \sim E_{k,n}(\mu\,\mathbf{1}', \Sigma \otimes \mathbf{I}_n, \psi)$ and $n > k$. Let Σ be positive definite. Then it holds that*

(a) Q has a stochastic representation $R^2 b$, i.e., $Q \approx R^2 b$. R is the generating variable of $\mathbf{X}$. The random variables R and b are independent and it holds that $b \sim B(\frac{n-k}{2}, \frac{nk-n+k}{2})$. The distribution function of Q is given by

$$F_Q(y) = \frac{\Gamma(\frac{nk}{2})}{\Gamma(\frac{n-k}{2})\Gamma(\frac{nk-n+k}{2})} \int_0^1 (1-z)^{\frac{nk-n+k}{2}-1}\, z^{\frac{n-k}{2}-1} F_{R^2}\left(\frac{y}{z}\right) dz. \tag{10.6}$$

(b) Suppose that $\mathbf{X}$ is absolutely continuous. Then the density of Q is given by

$$f_Q(y) = \frac{\pi^{nk/2}}{\Gamma(\frac{n-k}{2})\Gamma(\frac{nk-n+k}{2})} y^{\frac{nk}{2}-1} \int_0^\infty t^{(nk-n+k-2)/2} h(y(t+1))\,dt\; I_{(0,\infty)}(y)\,. \tag{10.7}$$

PROOF:

(a) Using $\hat{\mathbf{V}} = \frac{1}{n-1}\mathbf{X}(\mathbf{I}-\mathbf{1}\mathbf{1}'/n)\mathbf{X}'$ and the stochastic representation of the random matrix $\mathbf{X}$ we get

$$\hat{\mathbf{V}} \approx \frac{1}{n-1} R^2 \Sigma^{1/2}\mathbf{U}(\mathbf{I}-\mathbf{1}\mathbf{1}'/n)\mathbf{U}'\Sigma^{1/2}\,.$$

Thus we obtain that

$$Q \approx R^2 \frac{\tau'\Sigma^{-1}\tau}{\tau'\left(\Sigma^{1/2}\mathbf{U}(\mathbf{I}-\mathbf{1}\mathbf{1}'/n)\mathbf{U}'\Sigma^{1/2}\right)^{-1}\tau} = R^2 Q_*$$

with

$$Q_* = \tau'\Sigma^{-1}\tau/\tau'\left(\Sigma^{1/2}\mathbf{U}(\mathbf{I}-\mathbf{1}\mathbf{1}'/n)\mathbf{U}'\Sigma^{1/2}\right)^{-1}\tau.$$

The random variables R and Q_* are independent.
The similar presentation is obtained when $\mathbf{X}$ is matrix normally distributed, i.e. $Q_N \approx R_N^2 Q_*$ with independent variables R_N and Q_*. The index N is again used to indicate on the normal case. Because $R_N^2 \sim \chi^2_{nk}$ and $R_N^2 Q_* \sim \chi^2_{n-k}$ (cf. Muirhead, 1982, Theorem 3.2.12), it follows with Fang and Zhang (1990, p. 59) that there exists a random variable $b \sim B(\frac{n-k}{2}, \frac{nk-n+k}{2})$ which is independent from R_N such that $R_N^2 Q_* \approx R_N^2 b$. We observe that $P(R_N^2 > 0) = P(b > 0) = P(Q_* > 0) = 1$. Because $P(b < 1) = 1$ it follows with Lemma 5.3 that $Q_* \approx b$.
Now we obtain the general decomposition of Theorem 10.1 by applying the results derived for normal variables. Since b can be chosen independent from R and using that R and Q_* are independent we get with Fang and Zhang (1990, p. 38) that

$$Q \approx R^2 Q_* \approx R^2 b\,.$$

Thus we have proved the stochastic representation for matrix variate elliptically contoured distributions. The representation (10.6) is directly obtained by using the well-known distribution theory for transformations of random vectors.

(b) The second part of Theorem 10.1 is an immediate consequence of part (a). ■

Note that the distribution of Q does not depend on τ.

Using (10.6) the distribution function of Q can either be calculated explicitly or at least by numerical integration. The results of Theorem 10.1 permit to derive a test that the risk of the global minimum variance portfolio $\sigma_M^2 = -2\psi'(0)/\mathbf{1}'\Sigma^{-1}\mathbf{1}$ does not exceed a certain value. For a given number $\xi > 0$ the testing problem is

$$H_0 : \sigma_M^2 \le \xi \qquad \text{against} \qquad H_1 : \sigma_M^2 > \xi \,. \tag{10.8}$$

This means that an investor is interested to know whether the risk of the global minimum variance portfolio is greater than a certain risk level ξ or not. The test statistic is given by

$$\frac{-2\psi'(0)\,(n-1)\,\hat{\sigma}_M^2}{\xi} = \frac{E(R_*^2)\,(n-1)\,\hat{\sigma}_M^2}{\xi\,k}\,. \tag{10.9}$$

R_* is the generating variable of $\mathbf{x}_1$. Equation (10.9) is valid since $\mathbf{x}_1 \sim E_k(\mu, \Sigma, \psi)$ and thus $\mathbf{x}_1$ has the stochastic representation $\mu + R_*\Sigma^{1/2}\mathbf{U}^*$. Consequently it follows that

$$\begin{aligned}\mathbf{V} = Cov(\mathbf{x}_1) &= E(R_*^2)\, E(\Sigma^{1/2}\mathbf{U}^*\mathbf{U}^{*\prime}\Sigma^{1/2})\\ &= \frac{E(R_*^2)}{E(R_{N*}^2)}\, E(R_{N*}^2\Sigma^{1/2}\mathbf{U}^*\mathbf{U}^{*\prime}\Sigma^{1/2}) \;=\; \frac{E(R_*^2)}{k}\,\Sigma\end{aligned}$$

where $R_{N*}^2 \sim \chi_k^2$. This implies (10.9). If the value of the test statistic is larger than c the hypothesis H_1 is accepted. The critical value c is determined as the solution of $F_Q(c) = 1 - \alpha$ where α is the level of significance.

Theorem 10.1 can also be used for constructing a confidence interval of σ_M^2. It is given by

$$\left[\frac{-2\psi'(0)\,(n-1)\,\hat{\sigma}_M^2}{c_1}, \frac{-2\psi'(0)\,(n-1)\,\hat{\sigma}_M^2}{c_2}\right],$$

where c_1 and c_2 are the solutions of $F_Q(c_1) = 1 - \alpha/2$ and $F_Q(c_2) = \alpha/2$.

Using the stochastic representation of the random variable Q, the moments of the estimator of the global minimum variance are obtained. It follows that for $i \in \mathbb{N}$

$$\begin{aligned}E(\hat{\sigma}_M^{2i}) &= \sigma_M^{2i}\,\frac{k^i}{E(R_*^2)^i}\,\frac{E(R^{2i})}{(n-1)^i}\,E(b^i)\\ &= \sigma_M^{2i}\,\frac{E(R^{2i})}{E(R_*^2)^i}\,\frac{k^i}{(n-1)^i}\,\frac{\Gamma(i+\frac{n-k}{2})}{\Gamma(\frac{n-k}{2})}\,\frac{\Gamma(\frac{nk}{2})}{\Gamma(i+\frac{nk}{2})}\end{aligned}$$

provided that $E(R^{2i})$ exists. This leads to

$$E(\hat{\sigma}_M^2) = \frac{n-k}{n-1}\,\sigma_M^2\,,$$

$$Var(\hat{\sigma}_M^2) = \left(\frac{n-k+2}{n(n-k)}\,\frac{k}{nk+2}\,\frac{E(R^4)}{E(R_*^2)^2} - 1\right)\frac{(n-k)^2}{(n-1)^2}\,\sigma_M^4\,.$$

Applying Theorem 3.17 (iii) and 3.22 (iii) we get that

$$E(R_*^2) = -2\psi'(0)\,k\,, \quad E(R^4) = nk\,(nk+1+4\psi''(0))$$

because $\Sigma^{-1/2}(\mathbf{X}-\mathbf{M}) \sim E_{k,n}(\mathbf{0},\mathbf{I}_k \otimes \mathbf{I}_n,\psi)$.

This shows that $\hat{\sigma}_M^2$ is an asymptotically unbiased estimator of σ_M^2. An unbiased estimator is given by $\frac{n-1}{n-k}\,\hat{\sigma}_M^2$.

10.2.2 Examples

In this section we consider some special families of matrix variate elliptical distributions. Besides the normal approach we consider several alternative models having heavier tails. We calculate the density of the global minimum variance portfolio. The moments can be easily derived and are left to the interested reader. The confidence intervals and the test statistics will be discussed in the next section.

Example 10.1. (The matrix variate normal distribution)
The density generator function for the multivariate normal distribution is

$$h(t) = \frac{1}{(2\pi)^{\frac{nk}{2}}}\exp\left(-\frac{t}{2}\right). \tag{10.10}$$

Applying Theorem 10.1 we obtain

$$\begin{aligned} f_Q(y) &= \frac{\pi^{\frac{nk}{2}}}{\Gamma(\frac{n-k}{2})\Gamma(\frac{nk-n+k}{2})}\,y^{\frac{nk}{2}-1}\exp\left(-\frac{y}{2}\right)\int_0^{\infty} t^{\frac{nk-n+k-2}{2}}\exp\left(-t\frac{y}{2}\right)dt \\ &= \frac{1}{\Gamma(\frac{n-k}{2})2^{\frac{n-k}{2}}}\,y^{\frac{n-k}{2}-1}\exp\left(-\frac{y}{2}\right). \end{aligned}$$

This is the density function of the χ^2-distribution with $n-k$ degrees of freedom.

Example 10.2. (The matrix variate symmetric Pearson type VII distribution)
If a $k\times n$ random matrix $\mathbf{X}$ follows a matrix variate symmetric multivariate Pearson Type VII distribution then $\mathbf{X}$ has a density generator h with

$$h_{kn,r,q}(t) = (\pi r)^{-nk/2}\,\frac{\Gamma(q)}{\Gamma(q-kn/2)}\,(1+\frac{t}{r})^{-q}, \qquad q>\frac{kn}{2}, \qquad r>0\,. \tag{10.11}$$

From Theorem 10.1 it holds that

$$f_Q(y) = \frac{\pi^{\frac{nk}{2}}}{\Gamma(\frac{n-k}{2})\Gamma(\frac{nk-n+k}{2})} \frac{(\pi r)^{-nk/2}\Gamma(q)}{\Gamma(q-\frac{nk}{2})} y^{\frac{nk}{2}-1} \times \int_0^\infty t^{\frac{nk-n+k-2}{2}} \left(1+\frac{y(t+1)}{r}\right)^{-q} dt$$

$$= \frac{r^{-\frac{nk}{2}}\Gamma(q)}{\Gamma(\frac{n-k}{2})\Gamma(\frac{nk-n+k}{2})\Gamma(q-\frac{nk}{2})} y^{\frac{nk}{2}-1} \times \left(1+\frac{y}{r}\right)^{-q} \int_0^\infty t^{\frac{nk-n+k-2}{2}} \left(1+\frac{\frac{y}{r}}{1+\frac{y}{r}}t\right)^{-q} dt$$

$$= \frac{r^{-(n-k)/2}}{B(\frac{n-k}{2}, q-\frac{nk}{2})} y^{\frac{n-k}{2}-1} \left(1+\frac{y}{r}\right)^{-(q-\frac{nk}{2}+\frac{n-k}{2})}$$

$$= \frac{2(q-nk/2)}{r(n-k)} f_{(n-k)/2,2(q-nk/2)}(\frac{2(q-nk/2)y}{r(n-k)}).$$

Here $f_{n,m}$ denotes the density of the $F-$ distribution with (n,m) degrees of freedom.

For the matrix variate $t-$ distribution it holds that

$$f_{R^2}(t) = \frac{1}{nk} f_{nk,r}(\frac{t}{nk}) \to \chi^2_{nk}(t) \quad \text{as} \quad r \to \infty.$$

Example 10.3. (The matrix variate symmetric Kotz type distribution)
The $k \times n$ random matrix $\mathbf{X}$ has a matrix variate symmetric Kotz type distribution if

$$h(t) = \frac{s\Gamma(\frac{nk}{2})}{\pi^{\frac{nk}{2}}\Gamma(\frac{2q+nk-2}{2s})} r^{\frac{2q+nk-2}{2s}} t^{q-1} \exp\left(-rt^s\right) \tag{10.12}$$

with $r, s > 0$, $q \in I\!R$ such that $2q + nk > 2$.

We restrict ourselves to the case $s = 1$. Then we obtain

$$f_Q(y) = \frac{\Gamma(\frac{nk}{2})}{\Gamma(\frac{n-k}{2})\Gamma(\frac{nk-n+k}{2})\Gamma(\frac{2q+nk-2}{2})} r^{q+\frac{nk}{2}-1} y^{q+\frac{nk}{2}-2} \times \exp\left(-ry\right) \int_0^\infty t^{\frac{nk-n+k-2}{2}} (1+t)^{q-1} \exp\left(-ryt\right) dt$$

$$= \frac{\Gamma(\frac{nk}{2})}{\Gamma(\frac{n-k}{2})\Gamma(\frac{2q+nk-2}{2})} r^{q+\frac{nk}{2}-1} y^{q+\frac{nk}{2}-2} \exp\left(-ry\right)$$

$$\times U(\frac{nk-n+k}{2}, \frac{2q+nk-n+k}{2}, ry)$$

where $U(a,b,x)$ is the confluent hypergeometric function.

10.2.3 Determination of the Elliptical Model

Next we want to illustrate how the obtained theoretical results of the previous sections can be applied in analyzing financial data. We consider daily data from Morgan Stanley Capital International for the equity markets returns of three developed countries (Germany, UK, and USA) for the period from January 1993 to December 1996. In our study we make use of the advantages of matrix elliptically contoured distributions. First, the returns at different time points are uncorrelated and second, the return distribution is elliptically contoured.

It has to be noted that up to now no test of goodness of fit is available for the present modeling. Our procedure is a compromise which is obtained by combining existing methods. In order to examine whether the returns of a stock are elliptically symmetric at all we use a test suggested by Heathcote, Cheng, and Rachev (1995). The type of the elliptically contoured distribution is determined by applying the moments test of Fang and Zhang (1990). The analysis will lead to a matrix variate t-distribution. Other tests on elliptical symmetry were derived by Beran (1979); Baringhaus (1991); Manzotti, Perez, and Quiroz (2002), and Zhu and Neuhaus (2003).

We consider the daily returns of the equity markets for each country separately. For each stock the sample of returns of size n is splitted into $m = [n/q]$ subsamples of size q. The i-th subsample consists of the $q(i-1)+1,\dots,qi$-th observations. In our analysis q is taken equal to 5 or 10. The test of Heathcote, Cheng, and Rachev (1995) is applied to analyze whether the 5- or 10-day vector of returns is elliptically contoured, respectively. Denoting the subsample by $\mathbf{Z}_1,..,\mathbf{Z}_m$ the null hypothesis of elliptical symmetry is rejected iff (see Heathcote, Cheng and Rachev (1995))

$$\left|\frac{\sum_{j=1}^{m} sin((\mathbf{Z}_j-\overline{\mu}_m)'\mathbf{t}_{0m})}{m\sigma_m(\mathbf{t}_{0m})}\right| > z_{\alpha/2} \tag{10.13}$$

where $z_{\alpha/2}$ is the $100(1-\alpha/2)$ percentile of the standard normal distribution. $\overline{\mu}_m$ is the vector whose components are all the same and equal to the sample mean. The function $\sigma_m(.)$ is defined by

$$\begin{aligned}\sigma_m^2(\mathbf{t}) &= \frac{1-U_m(2\mathbf{t})}{2} - 2\frac{U_m(\mathbf{t})}{m}\sum_{j=1}^{m}(\mathbf{Z}_j-\overline{\mu}_m)'\mathbf{t}\ sin((\mathbf{Z}_j-\overline{\mu}_m)'\mathbf{t})\\ &+ \left(U_m(2\mathbf{t})\right)^2\mathbf{t}'\Sigma_m\mathbf{t},\ \ \mathbf{t}\in\mathbf{S}_m\subset I\!R^q\end{aligned}$$

Table 10.1 The values of $1/\sigma_m(\mathbf{t}_{1m})$

Dimension	Germany	UK	USA
5	1.4593	1.4536	1.6101
10	1.4926	1.5195	1.5149

with

$$U_m(\mathbf{t}) = \frac{1}{m}\sum_{j=1}^{m} cos((\mathbf{Z}_j - \overline{\mu}_m)'\mathbf{t}) \,,\ \Sigma_m = \frac{1}{m}\sum_{j=1}^{m}(\mathbf{Z}_j - \overline{\mu}_m)(\mathbf{Z}_j - \overline{\mu}_m)' \,.$$

The symbol $\mathbf{t}_{0m} \in \mathbf{S}_m$ is a point where the function $\sigma_m^2(.)$ reaches its maximum value on the compact $\mathbf{S}_m$. The set $\mathbf{S}_m$ of $\mathbf{t} = (t_1, \ldots, t_q)'$ is contained in a ball of the radius $\|\mathbf{r}_m\|$, where $\mathbf{r}_m$ is the first zero of $U_m(.)$. Heathcote, Cheng and Rachev (1995) showed that $\mathbf{r}_m$ is of a greater magnitude than the first $\mathbf{t}$ (in Euclidean norm) for which $\mathbf{t}'\mathbf{A}_m\mathbf{t} = 2$ where $\mathbf{A}_m = \frac{1}{m}\sum_{j=1}^{m}\mathbf{Z}_j\mathbf{Z}_j'$. Hence, we choose $\mathbf{r}_m$ as a solution of the following minimization problem

$$\min_{\mathbf{t}\in I\!R^q} \mathbf{t}'\mathbf{t} \qquad \text{subject to} \qquad \mathbf{t}'\mathbf{A}_m\mathbf{t} = 2 \,. \tag{10.14}$$

Solving it by constructing the Lagrangian we obtain that

$$\mathbf{t} - \lambda\mathbf{A}_m\mathbf{t} = 0 \quad \text{and} \quad \mathbf{t}'\mathbf{A}_m\mathbf{t} = 2 \,.$$

The first k equations of the system have non-zero solution iff $|\mathbf{I} - \lambda\mathbf{A}_m| = 0$. Hence, $1/\lambda$ is an eigenvalue with eigenvector $\mathbf{t}$ of the matrix $\mathbf{A}_m$. It holds that $\mathbf{t}'\mathbf{t} = 2\lambda$. Thus, the radius of the ball is equal to $\sqrt{2/\lambda_{max}}$ where λ_{max} is the maximum characteristic root of the matrix $\mathbf{A}_m$.

In the statistics (10.13) the point $\mathbf{t}_{0m} \in \mathbf{S}_m$ is chosen to obtain the maximum value of the function $\sigma_m(.)$. However, it holds that

$$\left|\frac{\sum_{j=1}^{m} sin((\mathbf{Z}_j - \overline{\mu}_m)'\mathbf{t}_{0m})}{m\sigma_m(\mathbf{t}_{0m})}\right| \leq \frac{1}{\sigma_m(\mathbf{t}_{0m})} \leq \frac{1}{\sigma_m(\mathbf{t}_{1m})} \,,$$

where $\mathbf{t}_{1m}$ is a point of $\mathbf{S}_m$.

First the radius of the ball for each country and for the five and ten dimensional vector of returns are calculated. Then a vector $\mathbf{t}_{1m} \in \mathbf{S}_m$ is chosen. For the five dimensional case we fix $\mathbf{t}_{1m} = (180, 180, 180, 180, 180)'$ for German and UK and $\mathbf{t}_{1m} = (190, 190, 190, 190, 190)'$ for USA. In the ten dimensional case the vector, whose components are equal to 100 is chosen for Germany and UK, and the vector with elements equal to 150 for USA. The values of $1/\sigma_m(\mathbf{t}_{1m})$ are given in Table 10.1. The hypothesis of elliptical symmetry can be rejected in none of the cases for the significance level $\alpha = 0.1$.

Table 10.2 The values of the test statistics of the moments test proposed by Fang and Zhang (1990) (cf. Sect. 10.2.3) (here: $q = 10$, critical values for $\alpha = 0.1$: 15.99 for I moment, 68.80 for II moment)

Count.\d.f.	5		6		7		nor	
	I mom.	II mom.	I mom.	II mom.	I mom.	II mom.	I mom.	II mom.
Germany	13.67	37.53	13.67	48.92	13.67	55.75	13.67	75.71
UK	6.14	29.67	6.14	34.59	6.14	37.59	6.14	84.49
USA	4.24	12.11	4.24	16.94	4.24	19.97	4.24	178.68

In order to construct a confidence interval or to make a test for the variance of the global minimum variance portfolio it is necessary to know the type of the elliptical distribution. Many authors proposed to use the t-distribution for modeling daily returns. Blattberg and Gonedes (1974) compared two heavy tailed statistical models for stock returns, stable distributions and the t-distribution. They concluded that the student model provided a better fit for daily stock returns than the symmetric-stable model. In their study they showed that the degree of freedom for the t-model fluctuates between 2.53 and 13.26 with the mean of 4.79. Furthermore, the most estimates of the degree of freedom are located in the interval $(4,6)$. Following these proposals we fitted a multivariate t-distribution to our empirical data. Again, in the same way as described above, we examined each stock separately by building 5- and 10-day samples, respectively. The moments test of Fang and Zhang (1990, p. 185) is applied to analyze the goodness of fit. We used the testing procedure for the first and second sample moments. The test cannot be applied for higher moments since they do not exist for a multivariate t-distribution with a small degree of freedom. Table 10.2 presents the results of the test for subsamples of size $q = 10$. We fitted a 10-dimensional normal and t-distribution to the data. The degree of freedom of the t-distribution was chosen equal to 5, 6, and 7. Taking the level of significance equal to $\alpha = 0.1$ the critical value is equal to 15.99 for the test based exclusively on the first moments and 68.80 for the test using the second moments. While the null hypothesis is not rejected for the t-distribution the normal approach is excluded by the test based on the second moments.

This analysis leads us to the decision to model the data by a matrix variate t-distribution in the next section.

10.2.4 Tests for the Global Minimum Variance

Here we want to illustrate how our results can be applied. We divide our data set of daily returns in two sub-samples. The first one, that includes the data from the first 2 months of 1993, is used to estimate the mean vector and the covariance matrix. Table 10.3 shows the sample means and the sample covariance matrix of the stock returns. Because $\Sigma = \mathbf{V}/(-2\psi'(0))$ we also get an estimator of Σ. Later on these values are used as the parameters of our model. The second sub-sample consists

Table 10.3 The sample mean and the sample covariance matrix of the daily returns of Germany, UK, USA determined with the observations of January 1993 and February 1993. In the table the values of the sample covariance matrix are multiplied by 1,000, i.e., $1{,}000 Cov(\mathbf{x})$

	Germany	UK	USA
Mean	0.000530	0.000508	0.000539
Germany	0.1146	0.0244	−0.0036
UK	0.0244	0.1165	0.0218
USA	−0.0036	0.0218	0.0195

Table 10.4 Critical values $c = F_Q^{-1}(1-\alpha)$ for the test introduced in Sect. 10.2.1

d.f.	$n = 20$		$n = 21$		$n = 22$		$n = 23$	
	$\alpha = 0.05$	$\alpha = 0.1$	$\alpha = 0.05$	$\alpha = 0.1$	$\alpha = 0.05$	$\alpha = 0.1$	$\alpha = 0.05$	$\alpha = 0.1$
$d = 4$	93.51	61.62	99.14	65.59	104.8	69.36	110.4	73.12
$d = 5$	73.66	51.68	78.04	54.79	82.41	57.91	86.79	61.02
$d = 6$	62.76	45.80	66.44	48.53	70.12	51.27	73.80	54.00
$d = 7$	55.91	41.97	59.16	44.45	62.40	46.93	65.65	49.41
nor	26.30	23.54	27.59	24.77	28.87	25.99	30.14	27.20

of the values obtained within one of the following months. It is tested whether the variance of the global minimum portfolio within that month is less than or equal to the preselected value $\xi = 1.05\,\tilde{\sigma}_M^2$. The value ξ is determined with the observations of the first sample. $\tilde{\sigma}_M^2$ is the sample estimator for the global minimum variance based on the first sub-sample. Our null hypothesis says the risk within the month under consideration exceeds the risk of the comparative month more than 5 %. The critical values for the significance levels 0.05 and 0.1 are presented in Table 10.4. The last row contains the critical values for the normal distribution. Because the number of opening days of the stock exchanges may vary from 1 month to another we give the results for the sample sizes $n \in \{20,..,23\}$. In Table 10.5 the number of rejections per year of the null hypothesis in (10.8) is shown. This is done for the 5 % (10 %) level of significance for the months from March 1993 to December 1996.

Choosing a significance level of 5 % the null hypothesis is rejected only once for a t-distribution with 4 degree of freedom, twice for the t-distribution with 5, 6, and 7 degree of freedom, but 11 times for the normal distribution. For a significance level of 10 % more signals are obtained. These results are given in Table 10.5 in parenthesis. If a matrix variate t-distribution is selected we have $5 - 6$ rejections of the null hypothesis while the normal approach leads to 15 ones. The misleading assumption of normally distributed daily returns results in more frequent rejections of the null hypothesis and thus, in many cases, an analyst will unnecessarily adjust the underlying portfolio.

Table 10.5 Number of rejections per year of the null hypothesis considered in (10.8) by the test of Sect. 10.2.1 (here: $\xi = 1.05 \cdot \tilde{\sigma}_M^2$, $\alpha = 5\,\%$ in parenthesis $\alpha = 10\%$)

$D \backslash Y$	1993	1994	1995	1996
$d = 4$	0 (0)	0 (2)	0 (0)	1 (3)
$d = 5$	0 (0)	1 (2)	0 (0)	1 (3)
$d = 6$	0 (0)	1 (3)	0 (0)	1 (3)
$d = 7$	0 (0)	1 (2)	0 (0)	1 (3)
nor	2 (4)	4 (6)	1 (1)	4 (4)

10.3 Test for the Weights of the Global Minimum Variance Portfolio

The aim of the present section is to derive a test for the general linear hypothesis of the GMVP weights. This hypothesis is treated in great detail within the theory of linear models (e.g., Rao and Toutenburg 1995). It covers a large number of relevant and important testing problems. Our test statistic is derived in a similar way. Contrary to linear models its distribution under the alternative hypothesis is not a non-central F-distribution. This shows that our results cannot be obtained in a straightforward way from the theory of linear models. A great advantage of the approach suggested in this paper is that the assumptions on the distribution of the returns are very weak.

There are several possibilities how an optimal portfolio can be determined. For the expected quadratic utility the portfolio weights are chosen to maximize

$$\mathbf{w}'\mu - \frac{\alpha}{2}\mathbf{w}'\mathbf{V}\mathbf{w} \quad \text{subject to} \quad \mathbf{1}'\mathbf{w} = 1,$$

where $\alpha > 0$ describes the risk aversion of an investor. This leads to the weights

$$\mathbf{w}_{EU} = \frac{\mathbf{V}^{-1}\mathbf{1}}{\mathbf{1}'\mathbf{V}^{-1}\mathbf{1}} + \alpha^{-1}\mathbf{R}\mu \quad \text{with} \quad \mathbf{Q} = \mathbf{V}^{-1} - \frac{\mathbf{V}^{-1}\mathbf{1}\mathbf{1}'\mathbf{V}^{-1}}{\mathbf{1}'\mathbf{V}^{-1}\mathbf{1}}.$$

Another approach consists in maximizing the Sharpe ratio of a portfolio without a risk free asset. The Sharpe ratio is still one of the most popular measures for the evaluation of a portfolio and the asset performance (Cochrane 1999; MacKinley and Pastor 2000). The problem of determining optimal weights can be solved by maximizing

$$\frac{\mathbf{w}'\mu}{\sqrt{\mathbf{w}'\mathbf{V}\mathbf{w}}} \quad \text{subject to} \quad \mathbf{1}'\mathbf{w} = 1.$$

The solution is given by

$$\mathbf{w}_{SR} = \frac{\mathbf{V}^{-1}\mu}{\mathbf{1}'\mathbf{V}^{-1}\mu}$$

provided that $\mathbf{1}'\mathbf{V}^{-1}\mu \neq 0$. The portfolio with maximum Sharpe ratio can be equivalently presented as a global tangency portfolio in a classical quadratic optimization problem. In the case when an investor is fully risk averse, the weights in the sense of quadratic utility maximization and the weights of the Sharpe ratio transform to the weights

$$\mathbf{w}_M = \frac{\mathbf{V}^{-1}\mathbf{1}}{\mathbf{1}'\mathbf{V}^{-1}\mathbf{1}},$$

which are the weights of the GMVP.

10.3.1 Distribution of the Estimated Weights of the Global Minimum Variance Portfolio

The estimator $\hat{\mathbf{w}}_M$ of $\mathbf{w}_M$ is given by

$$\hat{\mathbf{w}}_M = \frac{\hat{\mathbf{V}}^{-1}\mathbf{1}}{\mathbf{1}'\hat{\mathbf{V}}^{-1}\mathbf{1}}, \tag{10.15}$$

where $\hat{\mathbf{V}}$ is given in (10.3). Note that for a normal random sample $\hat{\mathbf{V}}$ is positive definite with probability 1 if $n > k$. Okhrin and Schmid (2006) proved that in this case all marginal distributions of $\hat{\mathbf{w}}_M$ with dimension less than k follow a multivariate t-distribution.

Here we consider linear combinations of the GMVP weights. Let $\mathbf{l}_i \in I\!R^k$, $i = 1,\ldots,p$, $1 \leq p \leq k-1$, and let $\mathbf{L}' = (\mathbf{l}_1,\ldots,\mathbf{l}_p)$. We are interested in

$$\mathbf{w}_{L;p} = \mathbf{L}\mathbf{w}_M = \frac{\mathbf{L}\mathbf{V}^{-1}\mathbf{1}}{\mathbf{1}'\mathbf{V}^{-1}\mathbf{1}} = \left(\frac{\mathbf{l}_1'\mathbf{V}^{-1}\mathbf{1}}{\mathbf{1}'\mathbf{V}^{-1}\mathbf{1}},\ldots,\frac{\mathbf{l}_p'\mathbf{V}^{-1}\mathbf{1}}{\mathbf{1}'\mathbf{V}^{-1}\mathbf{1}}\right)'. \tag{10.16}$$

Applying the estimator (10.3) we get

$$\hat{\mathbf{w}}_{L;p} = \mathbf{L}\hat{\mathbf{w}}_M = \left(\frac{\mathbf{l}_1'\hat{\mathbf{V}}^{-1}\mathbf{1}}{\mathbf{1}'\hat{\mathbf{V}}^{-1}\mathbf{1}},\ldots,\frac{\mathbf{l}_p'\hat{\mathbf{V}}^{-1}\mathbf{1}}{\mathbf{1}'\hat{\mathbf{V}}^{-1}\mathbf{1}}\right)'. \tag{10.17}$$

Next we want to derive the distribution of $\hat{\mathbf{w}}_{L;p}$. The next theorem is due to Bodnar and Schmid (2008a).

Theorem 10.2. *Let $\mathbf{x}_1,\ldots,\mathbf{x}_n$ be independent and identically distributed random variables with $\mathbf{x}_1 \sim N_k(\mu,\mathbf{V})$. Let $n > k > p \geq 1$. Let $\mathbf{G} = (g_{ij})$ be a $k \times k-1$ matrix with components $g_{ii} = 1$, $g_{ki} = -1$ for $i = 1,..,k-1$, and 0 otherwise. If $rk(\mathbf{LG}) = p$ then it follows that*

$$\hat{\mathbf{w}}_{L;p} \sim t_p\left(n-k+1,\ \mathbf{w}_{L;p},\ \frac{1}{n-k+1}\frac{\mathbf{LQL}'}{\mathbf{1}'\mathbf{V}^{-1}\mathbf{1}}\right).$$

PROOF: We denote the components of $\mathbf{w}_M$ by $w_{M,1},\ldots, w_{M,k}$ and of $\hat{\mathbf{w}}_M$ by $\hat{w}_{M,1},\ldots,\hat{w}_{M,k}$. Let $\tilde{\mathbf{w}}_M = (w_{M,1},\ldots, w_{M,k-1})'$ and $\tilde{\mathbf{Q}}$ be the matrix consisting of the $k-1$ rows and columns of $\mathbf{Q}$. Okhrin and Schmid (2006) showed that the first $k-1$ components of $\hat{\mathbf{w}}_M$, which we denote by $\hat{\tilde{\mathbf{w}}}_M$, follow an $k-1$-variate t-distribution with $n-k+1$ degrees of freedom and parameters $\tilde{\mathbf{w}}_M$ and $\tilde{\mathbf{Q}}/((n-k+1)\mathbf{1}'\mathbf{V}^{-1}\mathbf{1})$. Because the sum over all components of $\hat{\mathbf{w}}_M$ is equal to 1 we can write $\hat{\mathbf{w}}_M = \mathbf{G}\hat{\tilde{\mathbf{w}}}_M$ with a $k\times k-1$ matrix $\mathbf{G}$ of rank $k-1$. If $rk(\mathbf{LG}) = p$ then by using the properties of elliptically contoured distributions it follows that

$$\mathbf{L}\hat{\mathbf{w}}_M = \mathbf{LG}\hat{\tilde{\mathbf{w}}}_M \sim t_p\left(n-k+1, \mathbf{Lw}_M, \frac{1}{n-k-1}\,\mathbf{LG}\tilde{\mathbf{Q}}\mathbf{G}'\mathbf{L}'\right).$$

Because $\mathbf{Q1} = \mathbf{0}$ it can be seen that $\mathbf{G}\tilde{\mathbf{Q}}\mathbf{G}' = \mathbf{Q}$ and thus the assertion is proved. ∎

Applying the properties of the multivariate t-distribution we get that $E(\hat{\mathbf{w}}_{L;p}) = \mathbf{w}_{L;p}$ and for $k \leq n-2$ that

$$\mathrm{Var}(\hat{\mathbf{w}}_{L;p}) = \frac{1}{n-k-1}\frac{\mathbf{LQL}'}{\mathbf{1}'\mathbf{V}^{-1}\mathbf{1}}. \tag{10.18}$$

Theorem 10.2 says that linear combinations of the components of $\hat{\mathbf{w}}_M$ are again t-distributed. However, if the matrix $\mathbf{V}$ cannot be estimated by historical data, it does not provide a test for $\mathbf{w}_{L;p}$ because the distribution of $\hat{\mathbf{w}}_{L;p}$ still depends on $\mathbf{V}$.

10.3.2 The General Linear Hypothesis for the Global Minimum Variance Portfolio Weights

Following Markowitz (1952) efficient portfolios are obtained by minimizing the variance of the portfolio return given a certain level of the expected portfolio return. The problem of testing the efficiency of a portfolio has been recently discussed in a large number of studies. In the absence of a riskless asset Gibbons (1982), Kandel (1984), Shanken (1985) and Stambaugh (1982) have analyzed multivariate testing procedures for the mean-variance efficiency of a portfolio. Jobson and Korkie (1989) and Gibbons, Ross, and Shanken (1989) derived exact F-tests for testing the efficiency of a given portfolio. More recently, Britten-Jones (1999) has given the exact F-statistics for testing the efficiency of a portfolio with respect to portfolio weights which is based on a single linear regression. In this section we introduce a test of the general linear hypothesis for the GMVP weights. First, in Sect. 10.3.2 we consider the normal case. In Sect. 10.3.3 the results are extended to matrix elliptically contoured distributions.

As above let $\mathbf{L}$ be a $p \times k$ matrix and $\mathbf{r}$ be a p-dimensional vector. $\mathbf{L}$ and $\mathbf{r}$ are assumed to be known. We consider the general linear hypothesis which is given by

$$H_0 : \mathbf{L}\mathbf{w}_M = \mathbf{r} \qquad \text{against} \qquad H_1 : \mathbf{L}\mathbf{w}_M \neq \mathbf{r} .$$

This means that the investor is interested to know whether the weights of the GMVP fulfill p linear restrictions or not. This is a very general testing problem and it includes many important special cases (cf. Greene 2003, pp. 95–96).

In Theorem 10.2 it was proved that $\hat{\mathbf{w}}_{L;p}$ follows a p-variate multivariate t-distribution with $n-k+1$ degrees of freedom, location parameter $\mathbf{w}_{L;p}$ and scale parameter $\mathbf{LQL}'/\mathbf{1}'\mathbf{V}^{-1}\mathbf{1}$. This result provides a motivation for considering the following test statistic for the present testing problem

$$T = \frac{n-k}{p}\left(\mathbf{1}'\hat{\mathbf{V}}^{-1}\mathbf{1}\right)\left(\hat{\mathbf{w}}_{L;p}-\mathbf{r}\right)'\left(\mathbf{L}\hat{\mathbf{Q}}\mathbf{L}'\right)^{-1}\left(\hat{\mathbf{w}}_{L;p}-\mathbf{r}\right) . \tag{10.19}$$

This quantity is very similar to the F statistic for testing a linear hypothesis within the linear regression model. Because the distribution of the underlying quantities is different than in the case of a linear model we cannot apply these well-known results.

Now let $F_{i,j}$ denote the F-distribution with degrees i and j. Its density is written as $f_{i,j}$. In the following we make also use of the hypergeometric function (cf. Abramowitz and Stegun 1965)

$${}_2F_1(a,b,c;x) = \frac{\Gamma(c)}{\Gamma(a)\Gamma(b)} \sum_{i=0}^{\infty} \frac{\Gamma(a+i)\Gamma(b+i)}{\Gamma(c+i)} \frac{z^i}{i!} .$$

The technical computation of a hypergeometric function is a standard routine within many mathematical software packages like, e.g., in Mathematica.

Theorem 10.3. *Let $\mathbf{x}_1,\ldots,\mathbf{x}_n$ be independent and identically distributed random variables with $\mathbf{x}_1 \sim N_k(\mu,\mathbf{V})$. Let $n > k > p \geq 1$. Let $\tilde{\mathbf{M}}' = (\mathbf{L}', \mathbf{1})$ and $rk(\tilde{\mathbf{M}}) = p+1$.*

(a) The density of T is given by

$$f_T(x) = f_{p,n-k}(x)\,(1+\lambda)^{-(n-k+p)/2} \tag{10.20}$$
$$\times\, {}_2F_1\left(\frac{n-k+p}{2}, \frac{n-k+p}{2}, \frac{p}{2}; \frac{px}{n-k+px}\,\frac{\lambda}{1+\lambda}\right)$$

with $\lambda = \mathbf{1}'\mathbf{V}^{-1}\mathbf{1}(\mathbf{r}-\mathbf{w}_{L;p})'(\mathbf{LQL}')^{-1}(\mathbf{r}-\mathbf{w}_{L;p})$.

(b) Under the null hypothesis it holds that $T \sim F_{p,n-k}$.

PROOF:

(a) Let $\tilde{\mathbf{M}}' = (\mathbf{L}', \mathbf{1})$ and $\tilde{\mathbf{M}}\mathbf{V}^{-1}\tilde{\mathbf{M}}' = \{\mathbf{H}_{ij}\}_{i,j=1,2}$, $\tilde{\mathbf{M}}\hat{\mathbf{V}}^{-1}\tilde{\mathbf{M}}' = \{\hat{\mathbf{H}}_{ij}\}_{i,j=1,2}$ with $\mathbf{H}_{22} = \mathbf{1}'\mathbf{V}^{-1}\mathbf{1}$, $\hat{\mathbf{H}}_{22} = \mathbf{1}'\hat{\mathbf{V}}^{-1}\mathbf{1}$, $\mathbf{H}_{12} = \mathbf{L}\mathbf{V}^{-1}\mathbf{1}$, $\hat{\mathbf{H}}_{12} = \mathbf{L}\hat{\mathbf{V}}^{-1}\mathbf{1}$, $\mathbf{H}_{11} = \mathbf{L}\mathbf{V}^{-1}\mathbf{L}'$, and $\hat{\mathbf{H}}_{11} = \mathbf{L}\hat{\mathbf{V}}^{-1}\mathbf{L}'$. Because $(n-1)\hat{\mathbf{V}} \sim W_k(n-1, \mathbf{V})$ and $rk(\tilde{\mathbf{M}}) = p+1$ we get with Theorem 3.2.11 of Muirhead (1982) that

$$(n-1)(\tilde{\mathbf{M}}\hat{\mathbf{V}}^{-1}\tilde{\mathbf{M}}')^{-1} \sim W_{p+1}(n-k+p, (\tilde{\mathbf{M}}\mathbf{V}^{-1}\tilde{\mathbf{M}}')^{-1}) .$$

Now it holds with $\hat{\mathbf{B}} = \mathbf{L}\hat{\mathbf{R}}\mathbf{L}'$ that

$$(\tilde{\mathbf{M}}\hat{\Sigma}^{-1}\tilde{\mathbf{M}}')^{-1} = \begin{pmatrix} \hat{\mathbf{B}}^{-1} & -\hat{\mathbf{B}}^{-1}\,\hat{\mathbf{w}}_{L;p} \\ -\hat{\mathbf{w}}_{L;p}'\,\hat{\mathbf{B}}^{-1} & (\hat{\mathbf{H}}_{22} - \hat{\mathbf{H}}_{21}\hat{\mathbf{H}}_{11}^{-1}\hat{\mathbf{H}}_{12})^{-1} \end{pmatrix}$$

because $\hat{\mathbf{w}}_{L;p} = \hat{\mathbf{H}}_{12}/\hat{\mathbf{H}}_{22}$. We obtain with Theorem 3.2.10 (ii) of Muirhead (1982)

$$\begin{aligned} & f_{\sqrt{n-1}\hat{\mathbf{B}}^{-1/2}(\hat{\mathbf{w}}_{L;p}-\mathbf{r})|(n-1)\hat{\mathbf{B}}^{-1}}(\mathbf{a}|\mathbf{C}) \\ &= f_{-(n-1)\hat{\mathbf{B}}^{-1}\hat{\mathbf{w}}_{L;p}|(n-1)\hat{\mathbf{B}}^{-1}}(-\mathbf{C}^{1/2}\mathbf{a} - \mathbf{C}\mathbf{r}|\mathbf{C})\ |\mathbf{C}|^{\frac{1}{2}} \\ &= \frac{\mathbf{H}_{22}^{p/2}}{(2\pi)^{p/2}}\ exp\left(-\frac{\mathbf{H}_{22}}{2}\ (\mathbf{a} + \mathbf{C}^{1/2}(\mathbf{r} - \mathbf{w}_{L;p}))'(\mathbf{a} + \mathbf{C}^{1/2}(\mathbf{r} - \mathbf{w}_{L;p}))\right) . \end{aligned}$$

This is the density of the p-dimensional normal distribution with mean $\mathbf{C}^{1/2}(\mathbf{r} - \mathbf{w}_{L;p})$ and covariance matrix $\mathbf{I}/\mathbf{H}_{22}$, i.e.

$$\sqrt{n-1}\ \mathbf{H}_{22}^{1/2}\ \hat{\mathbf{B}}^{-1/2}(\hat{\mathbf{w}}_{L;p} - \mathbf{r})|(n-1)\hat{\mathbf{B}}^{-1} = \mathbf{C}\ \sim\ N(\mathbf{H}_{22}^{1/2}\mathbf{C}^{1/2}(\mathbf{r} - \mathbf{w}_{L;p}), \mathbf{I}) .$$

Consequently

$$(n-1)\ \mathbf{1}'\mathbf{V}^{-1}\mathbf{1}\ (\hat{\mathbf{w}}_{L;p} - \mathbf{r})'\ \hat{\mathbf{B}}^{-1}\ (\hat{\mathbf{w}}_{L;p} - \mathbf{r})|(n-1)\hat{\mathbf{B}}^{-1} = \mathbf{C}\ \sim\ \chi^2_{p,\lambda(\mathbf{C})}$$

with $\lambda(\mathbf{C}) = \mathbf{H}_{22}(\mathbf{r} - \mathbf{w}_{L;p})'\mathbf{C}(\mathbf{r} - \mathbf{w}_{L;p})$.
From Muirhead (1982, Theorem 3.2.12) we know that

$$(n-1)\ \frac{\mathbf{1}'\mathbf{V}^{-1}\mathbf{1}}{\mathbf{1}'\hat{\mathbf{V}}^{-1}\mathbf{1}}\ \sim\ \chi^2_{n-k} .$$

Moreover, applying Theorem 3.2.10 (i) of Muirhead (1982) it follows that $\hat{\mathbf{H}}_{22}$ is independent of $\hat{\mathbf{B}}^{-1}$ and $\hat{\mathbf{B}}^{-1}\hat{\mathbf{w}}_{L;p}$ and thus $\hat{\mathbf{H}}_{22}$ is independent of $\hat{\mathbf{B}}^{-1}$ and $(\hat{\mathbf{w}}_{L;p} - \mathbf{r})'\ \hat{\mathbf{B}}^{-1}\ (\hat{\mathbf{w}}_{L;p} - \mathbf{r})$. Putting these results together we obtain

$$T|(n-1)\hat{\mathbf{B}}^{-1} = \mathbf{C}\ \sim\ F_{p,n-k,\lambda(\mathbf{C})} .$$

Because $(n-1)\hat{\mathbf{B}}^{-1} \sim W_p(n-k+p, \mathbf{B}^{-1})$ with $\mathbf{B} = \mathbf{LQL}'$ we obtain that

$$f_T(x) = \int_{\mathbf{C}>0} f_{p,n-k,\lambda(\mathbf{C})}(x) w_p(n-k+p, \mathbf{B}^{-1})(\mathbf{C}) d\mathbf{C}\,.$$

Here $f_{i,j,\lambda}$ denotes the density of the non-central F-distribution with degrees i and j and noncentrality parameter λ and w_p is the density of the Wishart distribution W_p. If $\lambda = 0$ we briefly write $f_{i,j}$. It holds that (e.g., Theorem 1.3.6 of Muirhead (1982))

$$f_{p,n-k,\lambda(\mathbf{C})}(x) = f_{p,n-k,0}(x) \exp\left(-\frac{\lambda(\mathbf{C})}{2}\right) \times \sum_{i=0}^{\infty} \frac{((n-k+p)/2)_i}{(p/2)_i} \frac{\lambda(\mathbf{C})^i}{i!} \left(\frac{px}{2(n-k+px)}\right)^i .$$

Let us denote

$$k(i) = \frac{1}{i!} \frac{((n-k+p)/2)_i}{(p/2)_i} \left(\frac{px}{2(n-k+px)}\right)^i .$$

Then it follows that

$$\begin{aligned}
f_T(x) &= f_{p,n-k}(x) \sum_{i=0}^{\infty} k(i) \int_{\mathbf{C}>0} \lambda(\mathbf{C})^i \exp\left(-\frac{\lambda(\mathbf{C})}{2}\right) \frac{1}{2^{p(n-k+p)/2}\Gamma_p(\frac{n-k+p}{2})} \\
&\quad \times |\mathbf{B}|^{\frac{n-k+p}{2}} |\mathbf{C}|^{\frac{n-k-1}{2}} etr\left(-\frac{1}{2}\mathbf{BC}\right) d\mathbf{C} \\
&= f_{p,n-k}(x) \sum_{i=0}^{\infty} k(i) \int_{\mathbf{C}>0} |\mathbf{B}|^{\frac{n-k+p}{2}} \frac{1}{2^{p(n-k+p)/2}\Gamma_p(\frac{n-k+p}{2})} \\
&\quad \times |\mathbf{C}|^{\frac{n-k-1}{2}} \left(\mathbf{1}'\mathbf{V}^{-1}\mathbf{1}(\mathbf{r}-\mathbf{w}_{L;p})'\mathbf{C}(\mathbf{r}-\mathbf{w}_{L;p})\right)^i \\
&\quad \times etr\left(-\frac{1}{2}(\mathbf{B}+\mathbf{1}'\mathbf{V}^{-1}\mathbf{1}(\mathbf{r}-\mathbf{w}_{L;p})(\mathbf{r}-\mathbf{w}_{L;p})')\mathbf{C}\right) d\mathbf{C} \\
&= f_{p,n-k}(x) |\mathbf{B}|^{\frac{n-k+p}{2}} |\mathbf{B}+\mathbf{1}'\mathbf{V}^{-1}\mathbf{1}(\mathbf{r}-\mathbf{w}_{L;p})(\mathbf{r}-\mathbf{w}_{L;p})'|^{-\frac{n-k+p}{2}} \\
&\quad \times \sum_{i=0}^{\infty} k(i) (\mathbf{1}'\mathbf{V}^{-1}\mathbf{1})^i E(((\mathbf{r}-\mathbf{w}_{L;p})'\tilde{\mathbf{C}}(\mathbf{r}-\mathbf{w}_{L;p}))^i)\,,
\end{aligned}$$

where $\tilde{\mathbf{C}}$ follows a p-dimensional Wishart distribution with $n-k+p$ degree of freedom and parameter matrix $\tilde{\mathbf{B}} = (\mathbf{B}+\mathbf{1}'\mathbf{V}^{-1}\mathbf{1}(\mathbf{r}-\mathbf{w}_{L;p})(\mathbf{r}-\mathbf{w}_{L;p})')^{-1}$. From Theorem 3.2.8 of Muirhead (1982) we obtain that

$$E(((\mathbf{r}-\mathbf{w}_{L;p})'\tilde{\mathbf{C}}(\mathbf{r}-\mathbf{w}_{L;p}))^i)$$
$$= 2^i\,((n-k+p)/2)_i\,((\mathbf{r}-\mathbf{w}_{L;p})'\tilde{\mathbf{B}}(\mathbf{r}-\mathbf{w}_{L;p}))^i$$
$$= 2^i\,((n-k+p)/2)_i\left(\frac{(\mathbf{r}-\mathbf{w}_{L;p})'\mathbf{B}^{-1}(\mathbf{r}-\mathbf{w}_{L;p})}{1+\mathbf{H}_{22}(\mathbf{r}-\mathbf{w}_{L;p})'\mathbf{B}^{-1}(\mathbf{r}-\mathbf{w}_{L;p})}\right)^i.$$

Finally

$$f_T(x) = f_{p,n-k}(x)(1+\mathbf{1}'\mathbf{V}^{-1}\mathbf{1}(\mathbf{r}-\mathbf{w}_{L;p})'\mathbf{B}^{-1}(\mathbf{r}-\mathbf{w}_{L;p}))^{-(n-k+p)/2}$$
$$\times \sum_{i=0}^{\infty}\frac{((n-k+p)/2)_i((n-k+p)/2)_i}{i!(p/2)_i}$$
$$\times \left(\frac{px\mathbf{1}'\mathbf{V}^{-1}\mathbf{1}(\mathbf{r}-\mathbf{w}_{L;p})'\mathbf{B}^{-1}(\mathbf{r}-\mathbf{w}_{L;p})}{(n-k+px)(1+\mathbf{1}'\mathbf{V}^{-1}\mathbf{1}(\mathbf{r}-\mathbf{w}_{L;p})'\mathbf{B}^{-1}(\mathbf{r}-\mathbf{w}_{L;p}))}\right)^i$$
$$= f_{p,n-k}(x)(1+\mathbf{1}'\mathbf{V}^{-1}\mathbf{1}(\mathbf{r}-\mathbf{w}_{L;p})'\mathbf{B}^{-1}(\mathbf{r}-\mathbf{w}_{L;p}))^{-(n-k+p)/2}$$
$$\times\, {}_2F_1\left(\frac{n-k+p}{2},\frac{n-k+p}{2},\frac{p}{2};\frac{px}{n-k+px}\,\frac{\mathbf{1}'\mathbf{V}^{-1}\mathbf{1}(\mathbf{r}-\mathbf{w}_{L;p})'\mathbf{B}^{-1}(\mathbf{r}-\mathbf{w}_{L;p})}{1+\mathbf{1}'\mathbf{V}^{-1}\mathbf{1}(\mathbf{r}-\mathbf{w}_{L;p})'\mathbf{B}^{-1}(\mathbf{r}-\mathbf{w}_{L;p})}\right).$$

Thus the result is proved.

(b) The statement follows by noting that $\lambda = 0$ under H_0 and

$${}_2F_1\left(\frac{n-k+p}{2},\frac{n-k+p}{2},\frac{p}{2};0\right) = 1.$$

■

It is remarkable that the distribution of T depends on the parameters μ and $\mathbf{V}$ and the matrices of the linear hypothesis $\mathbf{L}$ and $\mathbf{r}$ only via the quantity λ. The parameter λ can be interpreted as a noncentrality parameter. This fact simplifies the power study of the test. In Fig. 10.1 the power of the test, i.e. $1-F_T(c)$ with $F_{p,n-k}(c) = 0.9$, is shown as a function of λ and p. Note that the T-statistic under H_1 does not possess the non-central F-distribution which is obtained in the theory of linear models. The number of observations n is equal to 260 and k is equal to 7. The figure illustrates the good performance of the test. Even for small values of λ the test has a high performance. Moreover, it can be seen that its power decreases if p increases.

Theorem 10.3 has many important applications. If, e.g., the analyst wants to test whether the GMVP weight of the first stock in the portfolio $\hat{w}_{M,1}$ is equal to a given value r_1, perhaps a reference value from a previous time period, we choose $p = 1$ and $\mathbf{l}_1 = (1,0,\ldots,0)'$. Then we get

$$T = (n-k)\,\frac{\left(\mathbf{1}'\hat{\mathbf{V}}^{-1}\mathbf{1}\right)^2(\hat{w}_{M,1}-r_1)^2}{\mathbf{1}'\hat{\mathbf{V}}^{-1}\mathbf{1}\,\hat{v}_{11}^{(-)} - (\Sigma_{i=1}^k\hat{v}_{1i}^{(-)})^2}$$

with $\hat{\mathbf{V}}^{-1} = (\hat{v}_{ij}^{(-)})$. For the one-sided hypothesis the statistic

$$T^* = \sqrt{n-k}\,\frac{\mathbf{1}'\hat{\mathbf{V}}^{-1}\mathbf{1}(\hat{w}_{M,1}-r_1)}{\sqrt{\mathbf{1}'\hat{\mathbf{V}}^{-1}\mathbf{1}\,\hat{v}_{11}^{(-)} - (\Sigma_{i=1}^k\hat{v}_{1i}^{(-)})^2}}$$

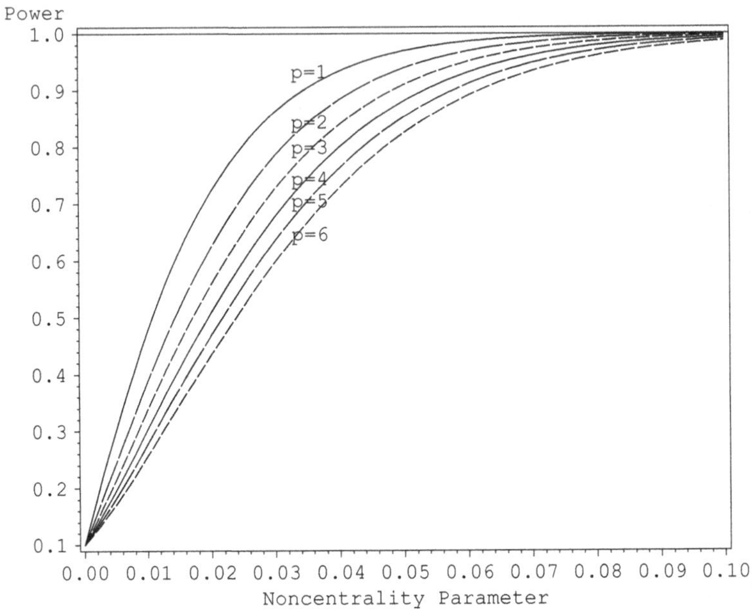

Fig. 10.1 Power function of the test for the general hypothesis (cf. Theorem 10.3) for various values of $p \in \{1,..,6\}$ and 10% level of significance. The number of portfolio assets is equal to $k = 7$ and number of observations is $n = 260$

can be used which under the null hypothesis follows a t-distribution with $n-k$ degrees of freedom. Moreover, Theorem 10.3 provides a test for the hypothesis that, e.g., two stocks have the same weights in the GMVP or that all weights are equal to reference (target) values. Consequently it can be used as a tool for monitoring the weights of the GMVP and it permits a decision whether the portfolio should be adjusted or not.

The above result can be applied to construct a $1-\alpha$ two-sided confidence interval for $\mathbf{w}_{L;p}$ as well. It is given by the set of all $\mathbf{r}$ satisfying that $T(\mathbf{r}) \leq F_{p,n-k;1-\alpha}$.

10.3.3 The General Linear Hypothesis for the Global Minimum Variance Portfolio Weights in an Elliptical Model

Using the stochastic representation of the random matrix $\mathbf{X}$ and Theorem 5.1.1 of Fang and Zhang (1990), it is proved in Theorem 10.4 that the statistics $\hat{\mathbf{w}}_{L;p}$ and T are distribution-free on the class of elliptically contoured distributions.

Theorem 10.4. *Let* $\mathbf{x} = (\mathbf{x}_1 \dots \mathbf{x}_n) \sim E_{k,n}(\mu \mathbf{1}', \Sigma \otimes \mathbf{I}_n, \psi)$. *Let* Σ *be positive definite and suppose that* $\mathbf{X}$ *is absolutely continuous. Let* $n > k > p \geq 1$. $\mathbf{G}$ *is used as in Theorem 10.2.*

(a) If $rk(\mathbf{LG}) = p$ *then*

$$\hat{\mathbf{w}}_{L;p} \sim t_p\left(n-k+1, \mathbf{w}_{L;p}, \frac{1}{n-k+1}\frac{\mathbf{LQL}'}{\mathbf{1}'\mathbf{V}^{-1}\mathbf{1}}\right).$$

(b) If $rk(\mathbf{L}', \mathbf{1}) = p+1$ *then the density of* T *is the same as in (10.20).*

PROOF: Using the stochastic representation we obtain that

$$(n-1)\hat{\mathbf{V}} \approx R^2\, \Sigma^{1/2}\mathbf{U}(\mathbf{I} - \frac{1}{n}\mathbf{1}\mathbf{1}')\mathbf{U}'\Sigma^{1/2}.$$

Consequently it holds that

$$\hat{\mathbf{w}}_{L;p} \approx \left(\frac{\mathbf{l}_i'\left(\Sigma^{1/2}\mathbf{U}(\mathbf{I} - \mathbf{1}\mathbf{1}'/n)\mathbf{U}'\Sigma^{1/2}\right)^{-1}\mathbf{1}}{\mathbf{1}'\left(\Sigma^{1/2}\mathbf{U}(\mathbf{I} - \mathbf{1}\mathbf{1}'/n)\mathbf{U}'\Sigma^{1/2}\right)^{-1}\mathbf{1}}\right)$$

$$= \left(\frac{\mathbf{l}_i'\left(R_N^2\Sigma^{1/2}\mathbf{U}(\mathbf{I} - \mathbf{1}\mathbf{1}'/n)\mathbf{U}'\Sigma^{1/2}\right)^{-1}\mathbf{1}}{\mathbf{1}'\left(R_N^2\Sigma^{1/2}\mathbf{U}(\mathbf{I} - \mathbf{1}\mathbf{1}'/n)\mathbf{U}'\Sigma^{1/2}\right)^{-1}\mathbf{1}}\right).$$

Thus $\hat{\mathbf{w}}_{L;p}$ has the same distribution as in the case of independent and normally distributed random vectors. Analogously, it can be seen that $\mathbf{L}\hat{\mathbf{Q}}\mathbf{L}'/\mathbf{1}'\hat{\mathbf{V}}^{-1}\mathbf{1}$ is distribution-free on the class of elliptically contoured distributions and thus the same property holds for T. ■

Theorem 10.4 says that the distribution of linear combinations of the estimated GMVP weights and the distribution of T are independent of the type of elliptical symmetry of the portfolio returns. Thus, in case of a fully risk averse investor neither the mean vector nor the distributional properties of asset returns have an influence on the optimal portfolio weights. In order to apply these results it is sufficient to know that the distribution of the data is a member of the family of matrix elliptical distributions but it is not necessary to have knowledge of the exact distribution. This assumption does not exclude the Pareto stable family and the multivariate t-distribution. Although the covariance matrix does not exist for the stable family (contrary to the normal case) and not always for the t-distribution, the results of Theorem 10.4 are valid for these distributional families, too.

Table 10.6 95 % and 90 % two-sided confidence intervals for the GMPV weights of France, Germany, Italy, Japan, Spain, U.K., and U.S.

Country	FR	GE	IT	JP	SP	U.K.	U.S.
95 % LCL	−0.0640	0.0292	−0.008	0.1162	−0.1751	−0.0170	0.4301
95 % UCL	0.1890	0.2107	0.0987	0.2680	0.0131	0.2615	0.6475
90 % LCL	−0.0525	0.0374	−0.0032	0.1230	−0.1666	−0.0044	0.4399
90 % UCL	0.1776	0.2025	0.0939	0.2612	0.0046	0.2489	0.6377

10.3.4 International Global Minimum Variance Portfolio

Based on the daily data from January 1, 1994 to December 31, 1994 an international global minimum variance portfolio is determined. Such a portfolio is of interest for an investor focused on international trading because it can be regarded as a benchmark portfolio for his investment. Confidence intervals for various linear restrictions are derived. Bodnar and Schmid (2007) proposed to model daily stock returns by matrix elliptically contoured distributions. Following the results of Sect. 10.2.3 the assumption of the matrix elliptical symmetry cannot be rejected for the considered data. Here we make use of this extremely useful family of matrix valued distributions.

In Table 10.6 separate confidence intervals for the GMVP weights of each country are presented. The problem of multiple comparisons (here 7) is taken into account by using the Bonferroni inequality. The null hypothesis of a weight of size 0 is not rejected for the French, Italian, Spanish, and UK returns. The GMVP weights of Germany, Japan, and USA are positive. The GMVP weight of the USA is the largest one. Its size is larger than the sum of the weights of all remaining countries. The null hypothesis of a weight of size 0.5 cannot be rejected for a 10 % level of significance. The upper bound of the 95 % one-sided confidence interval for the US weight is equal to 0.6475. This result turns out to be of interest because it does not support an investor who allocates his whole wealth into the US market. It is not in line with the result obtained by Britten-Jones (1999) for monthly price data. Finally, the lower bound for a 95 % one-sided confidence interval for the sum of the GMVP weights of the Germany, Japan, UK, and USA is given by 0.848. This shows the benefits of a portfolio allocation to the four developed markets, i.e., Germany, Japan, UK, and USA. They are able to explain nearly the whole development of the portfolio.

10.4 Inference for the Markowitz Efficient Frontier

In order to construct an optimal portfolio Markowitz (1952) proposed to choose the portfolio with the smallest risk for a given level of average portfolio return. Merton (1972) showed that the set of all of these optimal portfolios lies on a parabola in the mean-variance space, the so-called efficient frontier. This parabola is determined by three characteristics. The first two define the location of the vertex, while the third one is the slope parameter. Assuming the asset returns to be independently and normally distributed, Jobson (1991) obtained an asymptotic confidence region of the efficient frontier, while Bodnar and Schmid (2009) derived an exact joint confidence set of the three parameters of the efficient frontier. This set is used to determine a region in the mean-variance space where all optimal portfolios lie with a given probability $(1-\alpha)$.

Because the parameters of the efficient frontier are unknown quantities, the investor cannot construct the efficient frontier. Usually, the sample efficient frontier is used instead of the population efficient frontier (see, e.g. Bodnar and Schmid 2008b; Kan and Smith 2008), which is obtained by replacing the unknown parameters of the asset returns distribution with the corresponding sample counterparts. However, Basak, Jagannathan, and Ma (2005) and Siegel and Woodgate (2007) showed that the sample efficient frontier is overoptimistic and overestimates the true location of the efficient frontier in the mean-variance space. In order to correct this overoptimism Kan and Smith (2008) suggested an improved estimator of the efficient frontier, while Bodnar and Bodnar (2010) derived the unbiased estimator of the efficient frontier.

In the above cited papers, the assumption of independence and normality is maintained. However, these assumptions might not be appropriate in many situations of practical interest. Many authors have shown that the distribution of daily stock returns is heavy tailed (e.g., Fama 1965; Markowitz 1991; Mittnik and Rachev 1993; Chen, Gupta, and Troskie 2003).

10.4.1 Derivation of the Efficient Frontier

First, we derive an expression for the efficient frontier assuming that the asset returns are elliptically contoured distributed. If a random vector $\mathbf{x}$ is elliptically contoured distributed with the location parameter μ and the dispersion matrix Σ then it has the following stochastic representation (see, e.g. Fang and Zhang (1990, p. 65))

$$\mathbf{x} \approx \mu + \tilde{r}\Sigma^{1/2}\mathbf{u}, \tag{10.21}$$

where $\mathbf{u}$ is uniformly distributed on the unit sphere in $\mathbb{R}^k$ and the generating variable $\tilde{r}$ is independent of $\mathbf{u}$.

We derive an expression for the efficient frontier using the formulas for the expected return and the variance of the optimal portfolio in the sense of maximizing the expected utility function (EU portfolio). Let $\mathbf{w} = (w_1, \ldots, w_k)'$ denote the vector of portfolio weights, i.e. w_i is the part of the investor's wealth invested into the ith asset. Then the expected return of the portfolio with the weight vector $\mathbf{w}$ is given by $R_p = \mathbf{w}'E(\mathbf{x}) = \mathbf{w}'\mu$, while its variance is $V_p = \mathbf{w}'\text{Var}(\mathbf{x})\mathbf{w} = \frac{E(\tilde{r}^2)}{k}\mathbf{w}'\Sigma\mathbf{w}$. The weights of the EU portfolio are obtained by maximizing

$$R_p - \frac{\gamma}{2} V_p = \mathbf{w}'\mu - \frac{\gamma}{2} \frac{E(\tilde{r}^2)}{k} \mathbf{w}'\Sigma\mathbf{w} \tag{10.22}$$

under the constraint $\mathbf{w}'\mathbf{1} = 1$. There $\gamma > 0$ is the coefficient of the investor's risk aversion. The solution of (10.22) is given by

$$\mathbf{w}_{EU} = \frac{\Sigma^{-1}\mathbf{1}}{\mathbf{1}'\Sigma^{-1}\mathbf{1}} + \tilde{\gamma}^{-1}\mathbf{Q}\mu \quad \text{with} \quad \mathbf{Q} = \Sigma^{-1} - \frac{\Sigma^{-1}\mathbf{1}\mathbf{1}'\Sigma^{-1}}{\mathbf{1}'\Sigma^{-1}\mathbf{1}}, \tag{10.23}$$

where $\tilde{\gamma} = \gamma E(\tilde{r}^2)/k$. Using the weights (10.23), the expected return and the variance of the EU portfolio are obtained as

$$R_{EU} = \frac{\mu'\Sigma^{-1}\mathbf{1}}{\mathbf{1}'\Sigma^{-1}\mathbf{1}} + \tilde{\gamma}^{-1}\mu'\mathbf{Q}\mu = R_{GMV} + \tilde{\gamma}^{-1}s, \tag{10.24}$$

and

$$V_{EU} = \frac{E(\tilde{r}^2)}{k} \frac{1}{\mathbf{1}'\Sigma^{-1}\mathbf{1}} + \tilde{\gamma}^{-2}\frac{E(\tilde{r}^2)}{k}\mu'\mathbf{Q}\mu = \frac{E(\tilde{r}^2)}{k}V_{GMV} + \tilde{\gamma}^{-2}\frac{E(\tilde{r}^2)}{k}s, \tag{10.25}$$

where

$$R_{GMV} = \frac{\mathbf{1}'\Sigma^{-1}\mu}{\mathbf{1}'\Sigma^{-1}\mathbf{1}}, \quad \text{and} \quad V_{GMV} = \frac{1}{\mathbf{1}'\Sigma^{-1}\mathbf{1}} \tag{10.26}$$

are the expected return and the variance of the global minimum variance portfolio (GMV portfolio) and $s = \mu'\mathbf{Q}\mu$. The GMV portfolio is a special case of the EU portfolio that corresponds to the case of the fully risk averse investor, i.e. $\gamma = \infty$.

The Eqs. (10.24) and (10.25) are considered as the parametric equations of the efficient frontier. Solving (10.24) and (10.25) with respect to $\tilde{\gamma}$, the efficient frontier is expressed as

$$(R - R_{GMV})^2 = \frac{k}{E(\tilde{r}^2)} s \left(V - \frac{E(\tilde{r}^2)}{k} V_{GMV} \right). \tag{10.27}$$

From (10.27) we conclude that the efficient frontier depends on the asset return distribution. If $E(\tilde{r}^2) > k$, then the risk of the investment is higher than in the normal case. Moreover, there is a decrease in the overall market profitability since in (10.27) the slope coefficient of the parabola is multiply by $k/E(\tilde{r}^2) < 1$. When $E(\tilde{r}^2) \to \infty$, the slope coefficient of the parabola tends to zero. In this case the efficient frontier degenerates into a straight line and the only efficient portfolio is the GMV portfolio.

The same result is obtained by considering $\tilde{\gamma}$. Note, that $\tilde{\gamma}$ can be considered as a coefficient of risk inversion in elliptical models. If $\gamma = \infty$ and $E(\tilde{r}^2) < \infty$, the EU portfolio transforms to the GMV portfolio. From the other side, if $E(\tilde{r}^2) = \infty$ there is no solution of the optimization problem since in this case we get that

$$(R - R_{GMV})^2 = -sV_{GMV}.$$

10.4.2 Sample Efficient Frontier in Elliptical Models

Let $\mathbf{x}_1,\ldots,\mathbf{x}_n$ be a sample of the asset returns. We assume that the asset returns follow a matrix variate mixture of normal distributions with mean vector μ and dispersion matrix Σ. Any mixture of normal distributions belongs to the family of elliptically contoured distributions with stochastic representation of $\mathbf{x}_i$ given by

$$\mathbf{x}_i \approx \mu + r\Sigma^{1/2}\mathbf{z}_i\,, \tag{10.28}$$

where $\mathbf{z}_i \sim N_k(\mathbf{0},\mathbf{I})$ and $\mathbf{Z} = (\mathbf{z}_1,\ldots,\mathbf{z}_n)$ is independent of r. We denote this distribution by $E_{k,n}(\mu,\Sigma\otimes\mathbf{I}_n,g)$, where g is the so-called density generator which is fully determined by the distribution of r.

Note, that the asset returns are not assumed to be independently distributed in (10.28). The assumption of independence is replaced with a weaker one that the asset returns are uncorrelated. The random variable r determines the tail behavior of the asset returns. The model (10.28) is in-line with the recent modeling of the daily behavior of the asset returns. The daily asset returns are heavy-tailed distributed and they are not independent (see, e.g. Engle 1982, 2002; Bollerslev 1986; Nelson 1991).

Because μ and Σ are unknown parameters of the asset returns distribution, the investor cannot use (10.27) to construct the efficient frontier. These quantities have to be estimated from the historical values of the asset returns before the efficient frontier is determined. We consider the sample estimators of these parameters given by

$$\hat{\mu} = \frac{1}{n}\sum_{j=1}^{n}\mathbf{x}_j \quad\text{and}\quad \hat{\Sigma} = \frac{1}{n-1}\sum_{j=1}^{n}(\mathbf{x}_j-\hat{\mu})(\mathbf{x}_j-\hat{\mu})'. \tag{10.29}$$

Using $E(\tilde{r}^2)/k = E(r^2)$ and plugging (10.29) instead of μ and Σ in (10.27), the sample efficient frontier is expressed as

$$(R-\hat{R}_{GMV})^2 = \frac{1}{E(r^2)}\hat{s}(V-E(r^2)\hat{V}_{GMV})\,, \tag{10.30}$$

where

$$\hat{R}_{GMV} = \frac{\mathbf{1}'\hat{\Sigma}^{-1}\hat{\mu}}{\mathbf{1}'\hat{\Sigma}^{-1}\mathbf{1}}\,,\quad \hat{V}_{GMV} = \frac{1}{\mathbf{1}'\hat{\Sigma}^{-1}\mathbf{1}}\,,\quad\text{and}\quad \hat{s} = \hat{\mu}'\hat{\mathbf{Q}}\hat{\mu}\,, \tag{10.31}$$

with $\hat{\mathbf{Q}} = \hat{\Sigma}^{-1} - \hat{\Sigma}^{-1}\mathbf{1}\mathbf{1}'\hat{\Sigma}^{-1}/\mathbf{1}'\hat{\Sigma}^{-1}\mathbf{1}$.

Theorem 10.5 is taken from Bodnar and Gupta (2009). It presents the distribution properties of $\hat{R}_{GMV}$, $\hat{V}_{GMV}$, and $\hat{s}$.

Theorem 10.5. *Let* $\mathbf{X} = (\mathbf{x}_1, \ldots, \mathbf{x}_n) \sim E_{k,n}(\mu, \Sigma \otimes \mathbf{I}_n, g)$. *Assume that* Σ *is positive definite. Let* $k > 2$ *and* $n > k$. *Then it holds that*

(a) *Given* $r = r_0$, $\hat{V}_{GMV}$ *is independent of* $(\hat{R}_{GMV}, \hat{s})$.

(b) $(n-1)\hat{V}_{GMV}/V_{GMV}|r = r_0 \sim r_0^2 \chi^2_{n-k}$.

(c) $\frac{n(n-k+1)}{(n-1)(k-1)}\hat{s}|r = r_0 \sim F_{k-1,n-k+1,ns/r_0^2}$.

(d) $\hat{R}_{GMV}|\hat{s} = y, r = r_0 \sim \mathcal{N}\left(R_{GMV}, \frac{1+\frac{n}{n-1}y}{n} V_{GMV} r_0^2\right)$.

(e) *The joint density function is given by*

$$f_{\hat{R}_{GMV}, \hat{V}_{GMV}, \hat{s}}(x,z,y) = \int_0^\infty \frac{n(n-k+1)}{(k-1)V_{GMV} r_0^2} f_{N(R_{GMV}, \frac{1+\frac{n}{n-1}y}{n} V_{GMV} r_0^2)}(x)$$
$$\times f_{\chi^2_{n-k}}\left(\frac{n-1}{V_{GMV} r_0^2} z\right) f_{F_{k-1,n-k+1,ns/r_0^2}}\left(\frac{n(n-k+1)}{(n-1)(k-1)} y\right) f_r(r_0) dr_0 .$$

PROOF: Given $r = r_0$ it holds that the $\mathbf{x}_i$'s are independently distributed with $\mathbf{x}_i | r = r_0 \sim N_k(\mu, r_0^2 \Sigma)$. Application of Lemma 1 of Bodnar and Schmid (2009) leads to the statement of the theorem. The theorem is proved. ■

10.4.3 Confidence Region for the Efficient Frontier

A joint test for three characteristics of the efficient frontier is given by

$$H_0: R_{GMV} = R_0, V_{GMV} = V_0, s = s_0 \tag{10.32}$$

against

$$H_1: R_{GMV} = R_1 \neq R_0 \text{ or } V_{GMV} = V_1 \neq V_0 \text{ or } s = s_1 \neq s_0 .$$

For testing (10.32) we use the results of Theorem 10.5, which motivate the application of the test statistic $T = (T_R, T_V, T_S)'$ with

$$T_R = \sqrt{n} \frac{\hat{R}_{GMV} - R_0}{\sqrt{V_0}\sqrt{1 + \frac{n}{n-1}\hat{\mu}'\hat{\mathbf{Q}}\hat{\mu}}}, \tag{10.33}$$

$$T_V = (n-1)\frac{\hat{V}_{GMV}}{V_0}, \tag{10.34}$$

$$T_S = \frac{n(n-k+1)}{(k-1)(n-1)}\hat{\mu}'\hat{\mathbf{Q}}\hat{\mu} . \tag{10.35}$$

Similar test statistic was considered by Bodnar and Schmid (2009) in the normal case. The distribution of T is derived in Theorem 10.6 under the null and alternative hypotheses.

Theorem 10.6. *Let* $\mathbf{X} = (\mathbf{x}_1, \ldots, \mathbf{x}_n) \sim E_{k,n}(\mu, \Sigma \otimes \mathbf{I}_n, g)$. *Assume that* Σ *is positive definite. Let* $k > 2$ *and* $n > k$. *Then it holds that*

(a) Let $R_{GMV} = R_1$, $V_{GMV} = V_1$, *and* $s = s_1$. *The density of* T *is given by*

$$f_T(x,y,z) = \eta^{-1} \int_0^\infty r_0^{-2} f_{\chi^2_{n-k}}\left(\frac{z}{r_0^2 \eta}\right) f_{N(\sqrt{\eta}\delta(y), r_0^2 \eta)}(x) \tag{10.36}$$
$$\times f_{F_{k-1,n-k+1,ns_1/r_0^2}}(y) f_r(r_0) dr_0 .$$

with $\delta(y) = \sqrt{n}\lambda_1 / \sqrt{1 + \frac{k-1}{n-k+1} y}$, $\eta = V_1/V_0$, $\lambda_1 = (R_1 - R_0)/\sqrt{V_1}$, *and* $s_1 = \mu' \mathbf{Q} \mu$.

(b) Under the null hypothesis the density of T *under* H_0 *is given by*

$$f_T(x,y,z) = \int_0^\infty r_0^{-2} f_{\chi^2_{n-k}}(z/r_0^2) f_{N(0,r_0^2)}(x) f_{F_{k-1,n-k+1,ns_0/r_0^2}}(y) f_r(r_0) dr_0 . \tag{10.37}$$

PROOF: The proof of Theorem 10.6 follows from Proposition 3 of Bodnar and Schmid (2009) by considering first the conditional distribution of $\mathbf{X}$ given $r = r_0$ and then integrating over r_0. The theorem is proved. ∎

The results of Theorem 10.6 are used to derive the power function of the test for (10.32) which depends on μ and Σ only through the quantities η, λ_1, and s_1. The power function is equal to

$$G_{T;\alpha}(\eta, \lambda_1, s_1) = 1 - \int_0^\infty (1 - G_{T_V;\tilde{\alpha}}(\eta)) \tag{10.38}$$
$$\times \int_{z_{\tilde{\alpha}/2}}^{z_{1-\tilde{\alpha}/2}} \int_{s_{\tilde{\alpha}/2}}^{s_{1-\tilde{\alpha}/2}} f_{N(\sqrt{\eta}\delta(y), \eta r_0^2)}(x) f_{F_{k-1,n-k+1,ns_1/r_0^2}}(y) f_r(r_0) dy dx dr_0 ,$$

where

$$G_{T_V;\tilde{\alpha}}(\eta) = 1 - F_{\chi^2_{n-k}}\left(\frac{\chi^2_{n-k;1-\tilde{\alpha}/2}}{\eta r_0^2}\right) + F_{\chi^2_{n-k}}\left(\frac{\chi^2_{n-k;\tilde{\alpha}/2}}{\eta r_0^2}\right)$$

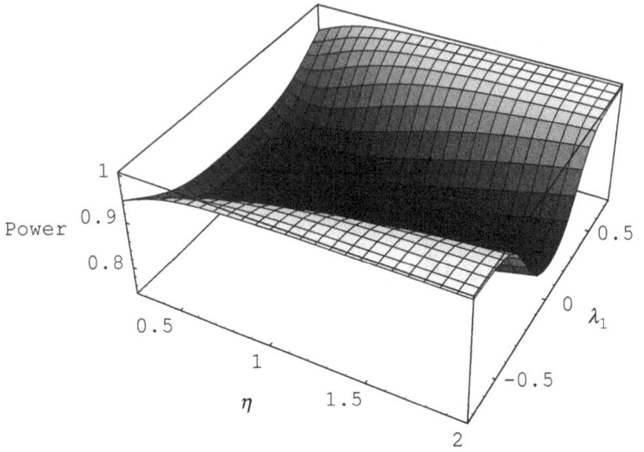

Fig. 10.2 Power of the test based on T for testing problem (10.32) as a function of $\eta = V_1/V_0$ and $\lambda_1 = (R_1 - R_0)/\sqrt{V_1}$ (data of Sect. 10.4.6, $s_1 = 0.224157$, $n = 60$, $k = 5$, and $\alpha = 5\%$). The asset returns are assumed to be matrix t-distributed with 5 degrees of freedom

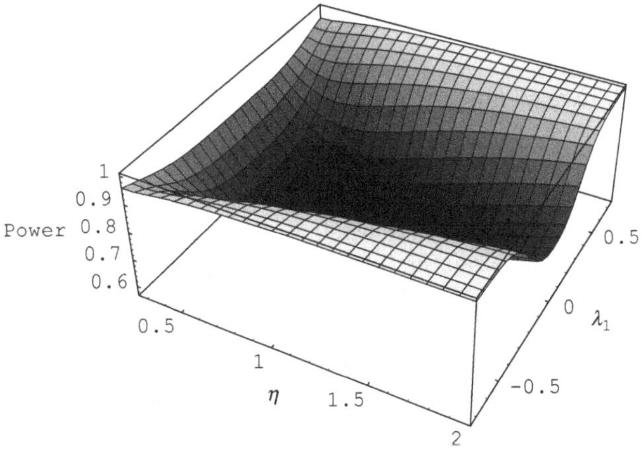

Fig. 10.3 Power of the test based on T for testing problem (10.32) as a function of $\eta = V_1/V_0$ and $\lambda_1 = (R_1 - R_0)/\sqrt{V_1}$ (data of Sect. 10.4.6, $s_1 = 0.224157$, $n = 60$, $k = 5$, and $\alpha = 5\%$). The asset returns are assumed to be matrix t-distributed with 15 degrees of freedom

and $1 - \alpha = (1 - \tilde{\alpha})^3$, i.e. $\tilde{\alpha} = 1 - \sqrt[3]{1 - \alpha}$. The quantities $s_{\tilde{\alpha}/2}$ and $s_{1-\tilde{\alpha}/2}$ are the lower and upper bounds of the $(1 - \tilde{\alpha})$-confidence interval for s.

In Figs. 10.2 and 10.3, we present (10.38) as a function of η and λ_1 in the case of the matrix t-distribution with 5 degrees of freedom and in the case of the matrix t-distribution with 15 degrees of freedom. The figures show that the test for (10.32) is more powerful when the matrix t-distribution with lager number of degrees of freedom, i.e. with the smaller tails, is considered.

Next, we construct the joint confidence set for the three characteristics of the efficient frontier. From Theorem 10.6, it follows that, given $r = r_0$, the statistics T_R, T_V, and T_S are mutually independent. This fact simplifies the construction of the confidence region. In order to account for the uncertainty of the generating variable r, we put $1-\alpha = (1-\alpha^*)^4$, e.g. $\alpha^* = 1-\sqrt[4]{1-\alpha}$. Using the fact that for given $r = r_0$ $\mathbf{x}_i$'s are independently distributed with $\mathbf{x}_i|r = r_0 \sim N_k(\mu, r_0^2\Sigma)$, the simultaneous confidence set $\mathscr{A}(r_0)$ consists of all points (R_{GMV}, V_{GMV}, s) that satisfy

$$(R_{GMV}-\hat{R}_{GMV})^2 \le z_{1-\alpha^*/2}^2\left(\frac{1}{n}+\frac{\hat{s}}{n-1}\right)V_{GMV}r_0^2,$$

$$V_{GMV} \in \left[\frac{(n-1)\hat{V}_{GMV}}{r_0^2\chi^2_{n-k;1-\alpha^*/2}}, \frac{(n-1)\hat{V}_{GMV}}{r_0^2\chi^2_{n-k;\alpha^*/2}}\right] = [\hat{V}_l, \hat{V}_u],$$

$$r_0^2\hat{s}_{\alpha^*/2} \le s \le r_0^2\hat{s}_{1-\alpha^*/2}.$$

The confidence interval for s is obtained as a confidence interval for the noncentrality parameter of the noncentral F-distribution (see Lam (1987)).

Let $F_R(r_{max}) = 1-\alpha^*/2$ and $F_R(r_{min}) = \alpha^*/2$. Then the confidence region for the efficient frontier is defined as the border of the set $\{\mathscr{A}(r) : r_{min} \le r \le r_{max}\}$, which is given by

$$(R_{GMV}-\hat{R}_{GMV})^2 \le z_{1-\alpha^*/2}^2\left(\frac{1}{n}+\frac{\hat{s}}{n-1}\right)V_{GMV}r_{max}^2 \text{ for } R_{GMV}>\hat{R}_{GMV}, \quad (10.39)$$

$$(R_{GMV}-\hat{R}_{GMV})^2 \le z_{1-\alpha^*/2}^2\left(\frac{1}{n}+\frac{\hat{s}}{n-1}\right)V_{GMV}r_{min}^2 \text{ for } R_{GMV}<\hat{R}_{GMV}, \quad (10.40)$$

$$V_{GMV} \in \left[\frac{(n-1)\hat{V}_{GMV}}{r_{max}^2\chi^2_{n-k;1-\alpha^*/2}}, \frac{(n-1)\hat{V}_{GMV}}{r_{min}^2\chi^2_{n-k;\alpha^*/2}}\right] = [\hat{V}_l, \hat{V}_u] \quad (10.41)$$

$$r_{min}^2\hat{s}_{\alpha^*/2} \le s \le r_{max}^2\hat{s}_{1-\alpha^*/2}. \quad (10.42)$$

We denote this set by $\mathscr{A}^*$. Because the efficient frontier lies in the mean-variance space, it would be interesting to derive the expression for $\mathscr{A}^*$ in the mean-variance space. Note, that the confidence region for the efficient frontier consists of all parabolas $(R-R_{GMV})^2 = s(V-V_{GMV})$, where R_{GMV} and V_{GMV} satisfy (10.39)-(10.41) and s satisfies (10.42). It can be expressed as

$$\mathscr{B} = \{(R-R_{GMV})^2 = s(V-V_{GMV}) : \quad (R_{GMV}, V_{GMV}, s) \in \mathscr{A}^*\}. \quad (10.43)$$

Let

$$g_l = z_{1-\alpha^*/2}\sqrt{1+\frac{n\hat{s}}{n-1}}\frac{\sqrt{n-1}\sqrt{\hat{V}_{GMV}}}{\sqrt{n\chi^2_{n-k;1-\alpha^*/2}}}, \quad (10.44)$$

$$g_u = z_{1-\alpha^*/2}\sqrt{1+\frac{n\hat{s}}{n-1}}\frac{\sqrt{n-1}\sqrt{\hat{V}_{GMV}}}{\sqrt{n\chi^2_{n-k;\alpha^*/2}}}\frac{r_{max}}{r_{min}}, \tag{10.45}$$

$$t = \frac{1}{1+z^{-2}_{1-\alpha^*/2}\hat{s}_{1-\alpha^*/2}\frac{n(n-1)}{n(1+\hat{s})-1}}. \tag{10.46}$$

Using the proof of Theorem 1 of Bodnar and Schmid (2009), we get the expression of $\mathscr{B}$.

Theorem 10.7. *It holds that $\mathscr{B}$ is equal to the set of all pairs (R,V) satisfying all of the following conditions*

$$\begin{cases} V \geq \frac{(n-1)\hat{V}_{GMV}}{r^2_{max}\chi^2_{n-k;1-\alpha^*/2}} & \text{for } R \in I\!R \\ V \geq z^{-2}_{1-\alpha^*/2}\frac{n(n-1)}{n(1+\hat{s})-1}r^{-2}_{max}(R-\hat{R}_{GMV})^2 & \text{for } R \in I^l_1 \\ V \leq \frac{(n-1)\hat{V}_{GMV}}{r^2_{min}\chi^2_{n-k;\alpha^*/2}} + \hat{s}^{-1}_{\alpha^*/2}r^{-2}_{min}\left(R-\hat{R}_{GMV}+g_u\right)^2 & \text{for } R \in I^l_2 \\ V \geq \frac{(n-1)\hat{V}_{GMV}}{r^2_{max}\chi^2_{n-k;1-\alpha^*/2}} + \hat{s}^{-1}_{1-\alpha^*/2}r^{-2}_{max}\left(R-\hat{R}_{GMV}-g_l\right)^2 & \text{for } R \in I^u_1 \\ V \geq \frac{z^{-2}_{1-\alpha^*/2}\frac{n(n-1)}{n(1+\hat{s})-1}r^{-2}_{max}}{1+z^{-2}_{1-\alpha^*/2}\hat{s}_{1-\alpha^*/2}\frac{n(n-1)}{n(1+\hat{s})-1}}(R-\hat{R}_{GMV})^2 & \text{for } R \in I^u_2 \\ V \geq \frac{(n-1)\hat{V}_{GMV}}{r^2_{min}\chi^2_{n-k;\alpha^*/2}} + \hat{s}^{-1}_{1-\alpha^*/2}r^{-2}_{max}\left(R-\hat{R}_{GMV}-g_u\right)^2 & \text{for } R \in I^u_3 \end{cases}, \tag{10.47}$$

where $I^u_1 = [\hat{R}_{GMV}+g_l, \hat{R}_{GMV}+g_l/t]$, $I^u_2 = [\hat{R}_{GMV}+g_l/t, \hat{R}_{GMV}+g_u/t]$, $I^u_3 = (\hat{R}_{GMV}+g_u/t, \infty)$, $I^l_1 = [\hat{R}_{GMV}-g_u, \hat{R}_{GMV}-g_l]$, *and* $I^l_2 = [\hat{R}_{GMV}-g_u, +\infty)$.

10.4.4 Unbiased Estimator of the Efficient Frontier

Basak, Jagannathan, and Ma (2005) and Siegel and Woodgate (2007) showed that the sample efficient frontier is overly optimistic and overestimates the true location of the efficient frontier in the mean-variance space. Bodnar and Bodnar (2010) corrected the overoptimism of the sample efficient frontier by deriving the unbiased estimator of the efficient frontier assuming the asset returns to be normally distributed. In Theorem 10.8, we extend this result by assuming the asset returns to follow a matrix variate mixture of normal distributions.

Theorem 10.8. *Let $\mathbf{X} = (\mathbf{x}_1, \ldots, \mathbf{x}_n) \sim E_{k,n}(\mu, \Sigma \otimes \mathbf{I}_n, g)$. Assume that Σ is positive definite. Let $k > 2$ and $n > k$. Let $E(r^2)$ and $E(r^{-2})$ exist. Then the unbiased estimator of the efficient frontier*

$$\psi(R,V) = (R - R_{GMV})^2 - \frac{1}{E(r^2)} s(V - E(r^2)V_{GMV})$$

is given by

$$\psi_u(R,V) = (R - \hat{R}_{GMV})^2 - \frac{(n-2)(n-1)}{n(n-k)(n-k-1)} \hat{V}_{GMV} - \frac{1}{E(r^2)E(r^{-2})}$$
$$\times (\frac{n-k-1}{n-1}\hat{s} - \frac{k-1}{n})(V - E(r^2)E(r^{-2})\frac{(n-k-2)(n-1)}{(n-k-1)(n-k)}\hat{V}_{GMV}). \quad (10.48)$$

PROOF: Consider

$$\begin{aligned}
E(\psi_u(R,V)) &= E(E(\psi_u(R,V)|r)) \\
&= E(E((R - \hat{R}_{GMV})^2|r)) - \frac{(n-2)(n-1)}{n(n-k)(n-k-1)} E(E(\hat{V}_{GMV}|r)) \\
&- \frac{1}{E(r^2)E(r^{-2})} E\Big((\frac{n-k-1}{n-1} E(\hat{s}|r) - \frac{k-1}{n}) \\
&\times (V - E(r^2)E(r^{-2})\frac{(n-k-2)(n-1)}{(n-k-1)(n-k)} E(\hat{V}_{GMV}|r)) \Big),
\end{aligned}$$

where in the last equality we use that $\hat{s}$ and $\hat{V}_{GMV}$ are independent given r (see Theorem 10.5).

Now from Theorem 10.5, we obtain

$$\begin{aligned}
&E(E((R - \hat{R}_{GMV})^2|r)) = (R - R_{GMV})^2 + E(\text{Var}(\hat{R}_{GMV}|r)) \\
&= (R - R_{GMV})^2 + \text{Var}(E(\hat{R}_{GMV}|\hat{s},r)) + E(\text{Var}(\hat{R}_{GMV}|\hat{s},r)) \\
&= (R - R_{GMV})^2 + \text{Var}(R_{GMV}) + E((\frac{1}{n} + \frac{1}{n-1} E(\hat{s}|r))V_{GMV} r^2) \\
&= (R - R_{GMV})^2 + E((\frac{n-2}{n(n-k-1)} + \frac{1}{n-k-1}\frac{s}{r^2})V_{GMV} r^2).
\end{aligned}$$

The last equality follows from the fact that (see Johnson, Kotz, and Balakrishnan (1995, p. 481))

$$E(\hat{s}|r) = \frac{n-1}{n-k-1}\frac{s}{r^2} + \frac{(n-1)(k-1)}{n(n-k-1)}. \quad (10.49)$$

Hence,

$$E(E((R-\hat{R}_{GMV})^2|r)) = (R-R_{GMV})^2 + \frac{(n-2)E(r^2)V_{GMV}}{n(n-k-1)} + \frac{sV_{GMV}}{n-k-1}. \tag{10.50}$$

Application of (10.49) and

$$E(\hat{V}_{GMV}|r) = \frac{n-k}{n-1} r^2 V_{GMV} \tag{10.51}$$

leads to

$$\frac{1}{E(r^2)E(r^{-2})} E\Big(\big(\frac{n-k-1}{n-1} E(\hat{s}|r) - \frac{k-1}{n}\big)$$
$$\times \big(V - E(r^2)E(r^{-2})\frac{(n-k-2)(n-1)}{(n-k-1)(n-k)} E(\hat{V}_{GMV}|r)\big)\Big)$$
$$= \frac{1}{E(r^2)E(r^{-2})} sE(r^{-2}(V - E(r^2)E(r^{-2})\frac{n-k-2}{n-k-1} V_{GMV} r^2))$$
$$= \frac{1}{E(r^2)} s(V - E(r^2)\frac{n-k-2}{n-k-1} V_{GMV}). \tag{10.52}$$

From (10.51) we get

$$\frac{(n-2)(n-1)}{n(n-k)(n-k-1)} E(E(\hat{V}_{GMV}|r)) = \frac{n-2}{n(n-k-1)} E(r^2)V_{GMV}. \tag{10.53}$$

Putting (10.50), (10.52), and (10.53) together, we obtain

$$E(\psi_u(R,V)) = (R-R_{GMV})^2 - \frac{1}{E(r^2)} s(V - E(r^2)V_{GMV}) = 0.$$

The last equality completes the proof of the theorem. ■

Although for determining the population efficient frontier $E(r^2)$ need only exist, for constructing an unbiased estimator both the moments $E(r^2)$ and $E(r^{-2})$ are used.

10.4.5 Overall F-Test

In Sect. 10.4.1, it was shown that if $\gamma = \infty$ the efficient frontier degenerates into a straight line. If $s = 0$ the efficient frontier is also a straight line without imposing any assumption on γ. In both cases, there is only one optimal portfolio, namely the GMV portfolio. Assuming the asset returns to be matrix elliptically contoured distributed Bodnar and Schmid (2008a) derived an exact test on the weight of the GMV portfolio, while Bodnar (2007, 2009) considered sequential procedures for monitoring the weights of the GMV portfolio and of the tangency portfolio, respectively.

Next, we consider a test for testing $s = 0$. The test hypothesis is given by

$$H_0 : s = 0 \quad \text{against} \quad H_1 : s > 0\,. \tag{10.54}$$

For testing (10.54) we use the results of Theorem 10.5. Because the non-central F-distribution with $s = 0$ is a central F-distribution, it holds that the null hypothesis is rejected if

$$T_S = \frac{n(n-k+1)}{(n-1)(k-1)}\,\hat{s} > F_{k-1,n-k+1;1-\alpha}\,. \tag{10.55}$$

$F_{k-1,n-k+1;1-\alpha}$ denotes the $(1-\alpha)$-quantile of the central F-distribution with $k-1$ and $n-k+1$ degrees of freedom. Note, that the distribution of the test statistic (10.55) does not depend on the distribution assumption imposed on the asset returns within the class of matrix elliptically contoured distributions. Thus, the test can easily be carried out using the the $(1-\alpha)$-quantile of the central F-distribution.

10.4.6 Empirical Illustration

In order to get a better understanding for the results presented in Sect. 10.4.3 we consider an example with real data in this section. We make use of monthly data from Morgan Stanley Capital International for the equity market returns of five developed countries (UK, Germany, USA, Canada, and Switzerland) for the period from July 1994 to June 1999. The parameters of the efficient frontier are estimated by

$$\hat{R}_{GMV} = 0.0145664, \quad \hat{V}_{GMV} = 0.0010337, \quad \text{and } \hat{s} = 0.221457\,. \tag{10.56}$$

It holds that $k = 5$ and $n = 60$. We put $\alpha = 0.15$, i.e. $\alpha^* = 0.0.0398$, $\hat{s}_{\alpha^*/2} = 0.000133$, $\hat{s}_{1-\alpha^*/2} = 0.4849$, $z_{1-\alpha^*/2} = 2.05566$, $\chi^2_{n-k;\alpha^*/2} = 35.64$, and $\chi^2_{n-k;1-\alpha^*/2} = 78.64$. Next we insert these values in Theorem 10.7 to obtain the 85 % confidence region of the efficient frontier in the mean-variance space which is bordered by five parabolas. When the matrix of the asset returns is assumed to be t-distributed with 5 degrees of freedom, it is given by

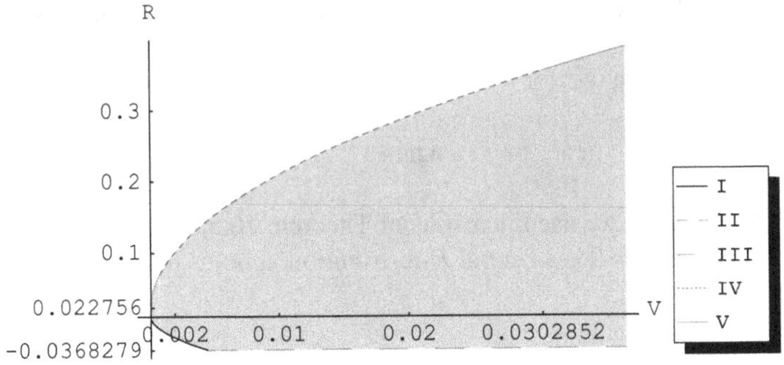

Fig. 10.4 Geometric structure of the 85 %-confidence region for the efficient frontier in the mean-variance space. The estimated parameters of the efficient frontier are given by $\hat{R}_{GMV} = 0.0145664$, $\hat{V}_{GMV} = 0.0010337$, and $\hat{s} = 0.224157$. The asset returns are assumed to be matrix t-distributed with 5 degrees of freedom

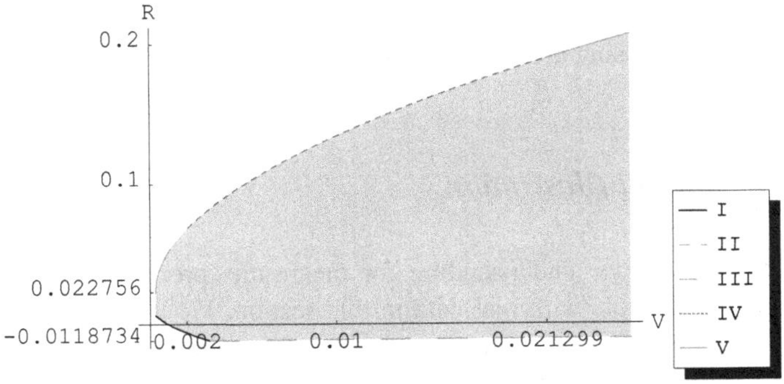

Fig. 10.5 Geometric structure of the 85 %-confidence region for the efficient frontier in the mean-variance space. The estimated parameters of the efficient frontier are given by $\hat{R}_{GMV} = 0.0145664$, $\hat{V}_{GMV} = 0.0010337$, and $\hat{s} = 0.224157$. The asset returns are assumed to be matrix t-distributed with 15 degrees of freedom

$$\begin{cases} \quad\;\; V \geq 0.000117 & \text{for } R \in \mathbb{R} \\ (I) \quad V \geq 1.74722(R-0.0145664)^2 & \text{for } R \in [-0.036644, 0.006377] \\ (II) \quad V \leq 0.004582 + 20123.7\,(R+0.036644)^2 & \text{for } R \in [-0.036644, +\infty) \\ (III)\; V \geq 0.000117 + 0.3116\,(R-0.02276)^2 & \text{for } R \in [0.02276, 0.0687) \\ (IV)\; V \geq 0.2645\,(R-0.0145664)^2 & \text{for } R \in [0.0687, 0.3529] \\ (V) \quad V \geq 0.004582 + 0.3116\,(R-0.0657)^2 & \text{for } R \in (0.3529, +\infty) \end{cases}$$

In case of the matrix t-distribution with 15 degrees of freedom, we get the following expression of the 85 % confidence region of the efficient frontier

$$\begin{cases} \quad\;\; V \geq 0.000309 & \text{for } R \in I\!R \\ (I) \quad V \geq 4.61(R - 0.0145664)^2 & \text{for } R \in [-0.01191, 0.006377] \\ (II) \quad V \leq 0.003224 + 14156.6\,(R + 0.01191)^2 & \text{for } R \in [-0.01191, +\infty) \\ (III)\; V \geq 0.000309 + 0.8222\,(R - 0.02276)^2 & \text{for } R \in [0.02276, 0.0687) \\ (IV)\; V \geq 0.69776\,(R - 0.0145664)^2 & \text{for } R \in [0.0687, 0.1895] \\ (V) \quad V \geq 0.003224 + 0.8222\,(R - 0.041)^2 & \text{for } R \in (0.0939, +\infty) \end{cases}$$

The geometrical structure of both the confidence regions are shown in Figs. 10.4 and 10.5. We observe that the area of the confidence region is smaller when the asset returns are assumed to be matrix t-distributed with 15 degrees of freedom than in the case of the matrix t-distribution with 5 degrees of freedom. There are also a number of portfolios with negative expected returns in Fig. 10.4 that belong to the confidence region. This set is much larger than the one given in Fig. 10.5.

Chapter 11
Skew Elliptically Contoured Distributions

11.1 Skew Normal Distribution

Various multivariate skew normal distributions have been proposed in the literature, with each one of them aiming to characterize a particular aspect of a given phenomenon. For example, one emphasizes invariance under quadratic forms, another one uses a general latent structure to define distributions, etc.; see Genton (2004) for an overview. Nevertheless, most of these skew normal distributions are special cases of the closed skew normal (CSN) family of distributions as defined in Domínguez-Molina, González-Farías, and Gupta (2003). The CSN class of distributions is closed under the operations of marginalization and conditioning basic to statistical modeling, includes the normal distribution, and enjoys some of the appealing properties of the latter. In particular, the expressions for its marginal and conditional densities are similar to those for the normal case. However, the distributions included in the CSN class are, in general, skewed.

Here we consider the extension of the CSN distribution from the vector to the matrix case. The distribution we propose implicitly defines the matrix variate generalizations of many other multivariate skew normal in the literature, and permits the inclusion of dependence structures, such as those for panel data, which are basic to the analysis of stochastic frontier models.

The articles by Aigner, Lovell, and Schmidt (1977) and Meeusen and van Den Broeck (1977) were seminal to the development of models capable of describing the production efficiency of companies. In them, the concept of a stochastic frontier was introduced via the model $y = f(\mathbf{x};\beta) + \varepsilon$, where the error term, $\varepsilon = \mathbf{v} - \mathbf{u}$, is composed of a symmetric disturbance term, $\mathbf{v}$, which represents measurement error, and by the non-negative, firm-specific term $\mathbf{u}$ which captures technical inefficiencies. This formulation of the error structure seeks to explain how companies with the same technical ability to manage their resources might end up with different output levels, due to the unobservable shocks $\mathbf{v}$. Developments over the last 30 years in the specification and estimation of frontier production functions are discussed in Coelli, Prasada Rao, O'Donnell, and Battese (2005).

A.K. Gupta et al., *Elliptically Contoured Models in Statistics and Portfolio Theory*,
DOI 10.1007/978-1-4614-8154-6_11,

Assuming a cross-sectional data structure, Domínguez-Molina, González-Farías, and Ramos-Quiroga (2004) proposed a stochastic frontier model based on the CSN distribution as given in González-Farías, Domínguez-Molina, and Gupta (2004a). Their proposal encompasses nested submodels with an increasing degree of complexity for the covariance structure, but within the framework of normal measurement errors and truncated normals for inefficiencies. Specifically, their model is

$$\mathbf{y} = \mathbf{f}(\mathbf{X};\beta) + \mathbf{v} + \mathbf{G}\mathbf{u} \tag{11.1}$$

where $\mathbf{y}$ is a vector consisting of the value-added values for p firms, $\mathbf{f}$ is the production function commonly based on the Cobb-Douglas model with lagged input variables, $\mathbf{v} \sim N_p(\mathbf{0},\Sigma)$ models measurement error, and $\mathbf{u} \sim N_q^{\mathbf{c}}(\nu,\Lambda)$, $q \geq p$, where $N_q^{\mathbf{c}}(\nu,\Lambda)$ denotes the $N_q(\nu,\Lambda)$ distribution truncated below at $\mathbf{c}$. The random vector $\mathbf{u}$ models technological inefficiencies in groups of firms, and is weighted by the $p \times q$ full row rank matrix $\mathbf{G}$. Also, it is assumed that $\mathbf{v}$ is independent of $\mathbf{u}$, $\mathbf{f}(\mathbf{X};\beta) = (f(\mathbf{x}_1;\beta),\ldots,f(\mathbf{x}_p;\beta))'$, $\mathbf{X} = (\mathbf{x}_1,\ldots,\mathbf{x}_p)'$ is a known matrix of covariates and β is unknown. The matrix $\mathbf{G}$ gives flexibility to the model. If it is left unspecified it can be estimated and used to validate model assumptions. On the other hand, it can be defined as $\mathbf{G} = \mathbf{I}_p$ or $\mathbf{G} = -\mathbf{I}_p$ for firm-specific cost efficiencies or technical inefficiencies, respectively.

The definition of the density of the CSN distribution, given by Domínguez-Molina, González-Farías, and Gupta (2003), is

Definition 11.1. Consider $p \geq 1$, $q \geq 1$, $\mu \in I\!R^p$, $\nu \in I\!R^q$, $\mathbf{D}$ an arbitrary $q \times p$ matrix, Σ and Δ positive definite matrices of dimensions $p \times p$ and $q \times q$, respectively. Then the density function of the CSN distribution is given by

$$g_{p,q}(\mathbf{y}) = C\phi_p(\mathbf{y};\mu,\Sigma)\Phi_q[\mathbf{D}(\mathbf{y}-\mu);\nu,\Delta], \quad \mathbf{y} \in I\!R^p,$$

with

$$C^{-1} = \Phi_q[\mathbf{0};\nu,\Delta + \mathbf{D}\Sigma\mathbf{D}'] \tag{11.2}$$

where $\phi_l(\mathbf{x};\mu,\Sigma)$ and $\Phi_l(\mathbf{x};\mu,\Sigma)$ denote the probability density function and the cumulative distribution function of the l-dimensional normal distribution with mean vector μ and covariance matrix Σ, respectively.

We will denote that the p-dimensional random vector $\mathbf{y}$ is distributed according to a CSN distribution with parameters q, μ, Σ, $\mathbf{D}$, ν, Δ by $\mathbf{y} \sim CSN_{p,q}(\mu,\Sigma,\mathbf{D},\nu,\Delta)$.

Domínguez-Molina, González-Farías, and Ramos-Quiroga (2004) show that the density of the compound error term in model (11.1), $\varepsilon = \mathbf{v} + \mathbf{G}\mathbf{u}$, is

$$g(\varepsilon) = \Phi_q^{-1}(\mathbf{0};\mathbf{c}-\nu,\Lambda)\phi_p(\varepsilon;\mathbf{G}\nu,\theta)\Phi_q[\Lambda\mathbf{G}\theta^{-1}(\varepsilon - \mathbf{G}\nu);\mathbf{c}-\nu,\Upsilon],$$

where $\theta = \Sigma + \mathbf{G}\Lambda\mathbf{G}'$ and $\Upsilon = \Lambda - \Lambda\mathbf{G}'\theta^{-1}\mathbf{G}\Lambda$. Thus,

$$\varepsilon \sim CSN_{p,q}(\mathbf{G}\nu,\theta,\Lambda\mathbf{G}'\theta^{-1},\mathbf{c}-\nu,\Upsilon)$$

The most important properties of the CSN distributions are their closure properties. For example, the joint distribution of independent CSN variables belongs to the same family as do the sums of independent CSN random variables. These closure properties allow one to study the distributional properties of random samples in a tractable way, and are very useful when considering the extension to the matrix variate case under certain types of dependencies. In what follows, we give various results which, apart from being of interest in themselves, also provide the building blocks for the matrix variate extension and the investigation of its properties.

The moment generating function of the CSN distribution, given in González-Farías, Domínguez-Molina, and Gupta (2004a), allows us to easily derive the moments of the distribution and to prove important distributional results. It is given in closed form as

$$M_{\mathbf{y}}(\mathbf{s}) = \frac{\Phi_q(\mathbf{D}\Sigma\mathbf{s};\nu,\Delta+\mathbf{D}\Sigma\mathbf{D}')}{\Phi_q(\mathbf{0};\nu,\Delta+\mathbf{D}\Sigma\mathbf{D}')}e^{\mathbf{s}'\mu+\frac{1}{2}\mathbf{s}'\Sigma\mathbf{s}}, \quad \mathbf{s}\in\mathbb{R}^p \tag{11.3}$$

The following proposition gives an alternative marginal representation of the CSN distribution which is useful, for instance, when conducting simulation or calculating moments. Moreover, the probabilistic structure defined within it can be applied directly in stochastic frontier modeling. A simpler version of this result was given in Domínguez-Molina, González-Farías, and Ramos-Quiroga (2004).

Theorem 11.1. *Let* $\mathbf{v}\sim N_p(\mathbf{0},\mathbf{I}_p)$, $\mathbf{u}\sim N_q^{\mathbf{c}}(\mathbf{0},\Delta+\mathbf{D}\Sigma\mathbf{D}')$ *and* $\mathbf{u}$ *be independent of* $\mathbf{v}$. *Then the distribution of*

$$\mathbf{y} = \mu + (\Sigma^{-1}+\mathbf{D}'\Delta^{-1}\mathbf{D})^{-\frac{1}{2}}\mathbf{v} + \Sigma\mathbf{D}'(\Delta+\mathbf{D}\Sigma\mathbf{D}')^{-1}\mathbf{u}$$

is $CSN_{p,q}(\mu,\Sigma,\mathbf{D},\nu,\Delta)$.

PROOF: In order to obtain the distribution of y we use the mgf technique. Now,

$$\begin{aligned} M_{\mathbf{y}}(\mathbf{s}) &= e^{\mathbf{s}'\mu}M_{\mathbf{v}}[(\Sigma^{-1}+\mathbf{D}'\Delta^{-1}\mathbf{D})^{-\frac{1}{2}}\mathbf{s}]M_{\mathbf{u}}[(\Delta+\mathbf{D}\Sigma\mathbf{D}')^{-1}\mathbf{D}\Sigma\mathbf{s}] \\ &= e^{\mathbf{s}'\mu}e^{\frac{1}{2}\mathbf{s}'(\Sigma^{-1}+\mathbf{D}'\Delta^{-1}\mathbf{D})^{-1}\mathbf{s}}e^{\frac{1}{2}\mathbf{s}'\Sigma\mathbf{D}'(\Delta+\mathbf{D}\Sigma\mathbf{D}')^{-1}(\Delta+\mathbf{D}\Sigma\mathbf{D}')(\Delta+\mathbf{D}\Sigma\mathbf{D}')^{-1}\mathbf{D}\Sigma\mathbf{s}} \\ &\quad\times\frac{\Phi_q(\mathbf{D}\Sigma\mathbf{s};\nu,\Delta+\mathbf{D}\Sigma\mathbf{D}')}{\Phi_q(\mathbf{0};\nu,\Delta+\mathbf{D}\Sigma\mathbf{D}')} \\ &= \frac{\Phi_q(\mathbf{D}\Sigma\mathbf{s};\nu,\Delta+\mathbf{D}\Sigma\mathbf{D}')}{\Phi_q(\mathbf{0};\nu,\Delta+\mathbf{D}\Sigma\mathbf{D}')}e^{\mathbf{s}'\mu}e^{\frac{1}{2}\mathbf{s}'[(\Sigma^{-1}+\mathbf{D}'\Delta^{-1}\mathbf{D})^{-1}+\Sigma\mathbf{D}'(\Delta+\mathbf{D}\Sigma\mathbf{D}')^{-1}\mathbf{D}\Sigma]\mathbf{s}} \end{aligned}$$

Using the Sherman-Morrison-Woodbury formula, we obtain

$$(\Sigma^{-1}+\mathbf{D}'\Delta^{-1}\mathbf{D})^{-1}+\Sigma\mathbf{D}'(\Delta+\mathbf{D}\Sigma\mathbf{D}')^{-1}\mathbf{D}\Sigma = \Sigma.$$

Thus,

$$M_{\mathbf{y}}(\mathbf{s}) = \frac{\Phi_q(\mathbf{D}\Sigma\mathbf{s}; \nu, \Delta + \mathbf{D}\Sigma\mathbf{D}')}{\Phi_q(\mathbf{0}; \nu, \Delta + \mathbf{D}\Sigma\mathbf{D}')} e^{\mathbf{s}'\mu + \frac{1}{2}\mathbf{s}'\Sigma\mathbf{s}},$$

which is the mgf of a $CSN_{p,q}(\mu, \Sigma, \mathbf{D}, \nu, \Delta)$ random vector. ∎

This representation in terms of normals and truncated normals is far more general than other representations given in the literature in terms of sums. Moreover, it proves to be very flexible when modeling different error structures for the stochastic frontier model.

An alternative way of motivating the closed skew normal distribution is via a hidden truncation process which, in many applications, will be highly plausible. For example, when the observational mechanism for measuring a variable is such that we only record a value when an external condition is satisfied, an asymmetric distribution will often be induced. The hidden truncation characterization also furnishes a useful means of establishing some of the properties of skew distributions, by so doing providing greater insight as to how they arise. For the hidden truncation process, we first condition a normal random vector on a set of latent variables subject to certain given restrictions (e.g., $Z \geq 0$), thus generating a CSN distribution. Then, if we consider operations such as marginalization, conditioning, or addition, their application results in distributions which are also members of the CSN family. However, it is important to point out that we can reverse this procedure in the following way. First, carry out the corresponding marginalization, conditioning, or addition procedure on the normal random vector and then consider the hidden truncation process. This will lead to exactly the same distribution, as shown in Domínguez-Molina, González-Farías, and Gupta (2003). The same argument applies when we obtain the joint distribution of independent CSN random variables. Using the conditioning approach of Domínguez-Molina, González-Farías, and Gupta (2003), we provide a simple derivation of the distribution function of a CSN random vector which proves to be useful in the study of dependence structures via copulas (Nelsen 2006).

The next theorem is due to Domínguez-Molina, González-Farías, and Gupta (2007).

Theorem 11.2. *The distribution function of a CSN random vector* $\mathbf{y}$*, with parameters* μ*,* Σ*,* $\mathbf{D}$*,* ν*,* Δ *is given by*

$$Fp,q(\mathbf{y}_0; \mu, \Sigma, \mathbf{D}, \nu, \Delta) = C\Phi_{p+q}\left[\begin{pmatrix}\mathbf{y}_0\\ \mathbf{0}\end{pmatrix}; \begin{pmatrix}\mu\\ \nu\end{pmatrix}, \begin{pmatrix}\Sigma & -\Sigma\mathbf{D}'\\ -\mathbf{D}\Sigma & \Delta + \mathbf{D}\Sigma\mathbf{D}'\end{pmatrix}\right]$$

where C *is as given in (11.2).*

PROOF: By definition, $Fp,q(\mathbf{y}_0; \mu, \Sigma, \mathbf{D}, \nu, \Delta) = P(\mathbf{y} \leq \mathbf{y}_0)$. Now, from the extension of the Copas and Li model given in González-Farías, Domínguez-Molina, and Gupta (2004a), we obtain that

$$\begin{aligned} P(\mathbf{y} \le \mathbf{y}_0) &= P(\mathbf{w}_0 \le \mathbf{y}_0 | \mathbf{z} \ge \mathbf{0}) \\ &= \frac{P(\mathbf{w}_0 \le \mathbf{y}_0, \mathbf{z} \ge \mathbf{0})}{P(\mathbf{z} \ge \mathbf{0})} \\ &= \frac{P(\mathbf{w}_0 \le \mathbf{y}_0, -\mathbf{z} \ge \mathbf{0})}{P(-\mathbf{z} \ge \mathbf{0})} \\ &= CP(\mathbf{w}_0 \le \mathbf{y}_0, -\mathbf{z} \ge \mathbf{0}) . \end{aligned}$$

The result follows on noting that

$$\begin{pmatrix} \mathbf{w}_0 \\ -\mathbf{z} \end{pmatrix} \sim \Phi_{p+q}\left[\begin{pmatrix} \mu \\ \nu \end{pmatrix}, \begin{pmatrix} \Sigma & -\Sigma\mathbf{D}' \\ -\mathbf{D}\Sigma & \Delta + \mathbf{D}\Sigma\mathbf{D}' \end{pmatrix} \right] . \quad \blacksquare$$

Although the conditioning argument provides a means with which to derive elegant proofs for certain results, it cannot be used, for instance, in the calculation of moments. For the latter, the representation in terms of sums is far more useful. Hence, we will use the marginal representation given in Theorem 11.1 when considering the application of the matrix variate extension of the CSN distribution to the stochastic frontier analysis in Sect. 11.4.

11.2 Matrix Variate Skew Normal Distribution

In this section we introduce the matrix variate generalization of the CSN distribution.

First, we define the $p \times m$ random matrix of observations as

$$\mathbf{X} = \begin{pmatrix} x_{11} & \cdots & x_{p1} \\ \vdots & \ddots & \vdots \\ x_{p1} & \cdots & x_{pn} \end{pmatrix} = (\mathbf{x}_1, \ldots, \mathbf{x}_m) ,$$

where $\mathbf{x}_i$ $(p \times 1)$, $i = 1, \ldots, m$ is the ith column of $\mathbf{X}$. Here, $\mathbf{x}_1, \ldots, \mathbf{x}_m$ can be thought as a sample of size m from a p-dimensional population, but it is not necessary to assume that $\mathbf{x}_, \ldots, \mathbf{x}_m$ are independent. The random matrix $\mathbf{X}$ is said to have a matrix variate normal distribution with mean matrix $\mathbf{M}$ $(p \times m)$ and covariance matrix Ω $(pm \times pm)$ if $vec(\mathbf{X}') \sim N_{pm}(vec(\mathbf{M}'), \Omega)$. We denote the probability density function and the cumulative distribution function of $\mathbf{X}$ as

$$\phi_{p,m}(\mathbf{X}; \mathbf{M}, \Omega) = \phi_{pm}(vec(\mathbf{X}); vec(\mathbf{M}), \Omega)$$

and

$$\Phi_{p,m}(\mathbf{X}; \mathbf{M}, \Omega) = \Phi_{pm}(vec(\mathbf{X}); vec(\mathbf{M}), \Omega) .$$

Using the above and the material on the CSN distribution presented in the preceding two sections, we are now in the position to define its matrix variate extension.

Definition 11.2. A random matrix $\mathbf{Y}$ $(p \times m)$ is said to have a matrix variate closed skew normal (MVCSN) distribution with parameters $\mathbf{M}$ $(p \times m)$, $\mathbf{S}$ $(pm \times pm)$, $\mathbf{B}$ $(qn \times pm)$, $\mathbf{L}$ $(q \times n)$, and $\mathbf{Q}$ $(qn \times qn)$, with $\mathbf{S} > \mathbf{0}$ and $\mathbf{Q} > \mathbf{0}$, if

$$vec(\mathbf{Y}') \sim CSN_{pm;qn}[vec(\mathbf{M}'), \mathbf{S}, \mathbf{B}, vec(\mathbf{L}'), \mathbf{Q}].$$

We use the notation

$$\mathbf{Y} \sim CSN_{p,m;q,n}(\mathbf{M}, \mathbf{S}, \mathbf{B}, \mathbf{L}, \mathbf{Q}) \tag{11.4}$$

to denote the fact. In most cases, the matrices $\mathbf{S}$ and $\mathbf{B}$ will have specific structures. Properties for the parametrization (11.4) are obtained immediately from González-Farías, Domínguez-Molina, and Gupta (2004a).

When working with random matrices it is important to bear in mind how the random matrix, $\mathbf{Y}$, is assembled. Here we consider the situation in which $\mathbf{Y} = (\mathbf{y}_1, \ldots, \mathbf{y}_n)$ is a sample of independent and identically distributed random vectors with $\mathbf{y}_i \sim CSN_{p,q}(\mu, \Sigma, \mathbf{D}, \nu, \Delta)$ random vectors. Due to Corollary 2.4.1 of González-Farías, Domínguez-Molina, and Gupta (2004b), we know that the distribution of $vec(\mathbf{Y}) = (\mathbf{y}_1', \ldots, \mathbf{y}_n')'$ is

$$CSN_{pn,qn}(\mathbf{1}_n \otimes \mu, \mathbf{I}_n \otimes \Sigma, \mathbf{I}_n \otimes \mathbf{D}, \mathbf{1}_n \otimes \nu, \mathbf{I}_n \otimes \Delta),$$

and hence

$$\mathbf{Y}' \sim CSN_{p,n;q,n}(\mathbf{1}_n' \otimes \mu, \mathbf{I}_n \otimes \Sigma, \mathbf{I}_n \otimes \mathbf{D}, \mathbf{1}_n' \otimes \nu, \mathbf{I}_n \otimes \Delta).$$

Thus, assuming iid columns for $\mathbf{Y}$ we obtain the distribution of $\mathbf{Y}'$, not that of $\mathbf{Y}$ as we might have hoped for. In order to obtain the distribution of $\mathbf{Y}$, we first consider the distribution of the transpose of a MVCSN matrix. We start by defining the commutation matrix which transforms $vec(\mathbf{A})$ into $vec(\mathbf{A}')$. The commutation matrix, K_{mp} $(mp \times mp)$ is defined as $K_{mp} = \sum_{i=1}^{m} \sum_{j=1}^{p} (\mathbf{H}_{ij} \otimes \mathbf{H}_{ij}')$, where the (i,j)th element of $\mathbf{H}_{ij}$ $(m \times p)$ is 1 and all its other elements are 0. Then, if

$$\mathbf{X} \sim CSN_{p,m;q,n}(\mathbf{M}, \mathbf{S}, \mathbf{B}, \mathbf{L}, \mathbf{Q}),$$

the distribution of $\mathbf{X}'$ can be obtained from the fact that $vec(\mathbf{X}) = \mathbf{K}_{mp} vec(\mathbf{X}')$. Using Theorem 1 of González-Farías, Domínguez-Molina, and Gupta (2004b) and Theorem 1.2.22 of Gupta and Nagar (2000), we then obtain that

$$\mathbf{X}' \sim CSN_{m,p;n,q}(\mathbf{M}', \mathbf{K}_{mp}\mathbf{S}\mathbf{K}_{pm}, \mathbf{B}\mathbf{K}_{pm}, \mathbf{L}, \mathbf{Q}).$$

Moreover, if $\mathbf{S} = \Psi \otimes \Sigma$ with $\Sigma\,(p \times p) > \mathbf{0}$ and $\Psi\,(m \times m) > \mathbf{0}$, then

$$\mathbf{X}' \sim CSN_{m,p;n,q}(\mathbf{M}', \Psi \otimes \Sigma, \mathbf{B}\mathbf{K}_{pm}, \mathbf{L}, \mathbf{Q}).$$

This follows because, from Eqs. (1.2.3) and (1.2.5) of Gupta and Nagar (2000), $\mathbf{K}_{mp}^{-1} = \mathbf{K}_{pm}$ and $\mathbf{K}_{pm}(\Psi \otimes \Sigma)\mathbf{K}_{mp} = \Psi \otimes \Sigma$. Finally, returning to the distribution of $\mathbf{Y} = (\mathbf{y}_1, \ldots, \mathbf{y}_n)$, we can use the above results to obtain

$$\mathbf{Y} \sim CSN_{p,n;q,n}(\mathbf{1}_n \otimes \mu', \Sigma \otimes \mathbf{I}_n, \mathbf{I}_n \otimes \mathbf{D}, (\mathbf{1}_n' \otimes \nu)\mathbf{K}_{pn}, \mathbf{I}_n \otimes \Delta).$$

Alternatively, a matrix variate CSN distribution can be obtained using an extension of the hidden truncation argument of Copas and Li (1997). This construction, which may be more natural in many experimental settings, proceeds as follows.

Define the independent normal random matrices $\mathbf{U}_1 \sim N_{p,m}(\mathbf{0}, \mathbf{S})$ and $\mathbf{U}_2 \sim N_{q,n}(\mathbf{0}, \mathbf{Q})$, where, as previously, $\mathbf{S}$ is $mp \times mp$ and $\mathbf{Q}$ is $nq \times nq$. Now, consider the matrices $\mathbf{W} = \mathbf{M} + \mathbf{U}_1$ and $\mathbf{Z} = -\mathbf{L} + \mathbf{D}\mathbf{U}_1\mathbf{E}' + \mathbf{U}_2$, where $\mathbf{D}$ is $q \times p$, $\mathbf{E}$ is $n \times m$ and, as before, $\mathbf{M}$ is $p \times m$ and $\mathbf{L}$ is $q \times n$. Then the joint distribution of $\mathbf{W}$ and $\mathbf{Z}$ is

$$\begin{pmatrix} \mathbf{W} \\ \mathbf{Z} \end{pmatrix} \sim N_{qn+pm}\left[\begin{pmatrix} \mathbf{M} \\ -\mathbf{L} \end{pmatrix}, \Omega\right],$$

where

$$\Omega = \begin{pmatrix} \mathbf{S} & \mathbf{S}(\mathbf{D}' \otimes \mathbf{E}') \\ (\mathbf{D} \otimes \mathbf{E})\mathbf{S} & \mathbf{Q} + (\mathbf{D} \otimes \mathbf{E})\mathbf{S}(\mathbf{D}' \otimes \mathbf{E}') \end{pmatrix}.$$

Now, if $\mathbf{Y} \approx \mathbf{W} | \{\mathbf{Z} \geq \mathbf{0}\}$ we obtain that

$$f(\mathbf{Y}) = K\phi_{p,m}(\mathbf{Y}; \mathbf{M}, \mathbf{S})\Phi_{q,n}[\mathbf{E}(\mathbf{Y} - \mathbf{M})\mathbf{D}'; \mathbf{L}, \mathbf{Q}],$$

where $K^{-1} = \Phi_{q,n}[\mathbf{0}; \mathbf{L}, \mathbf{Q} + (\mathbf{D} \otimes \mathbf{E})\mathbf{S}(\mathbf{D}' \otimes \mathbf{E}')]$. Hence,

$$\mathbf{Y} \sim CSN_{p,m;q,n}(\mathbf{M}, \mathbf{S}, \mathbf{D} \otimes \mathbf{E}, \mathbf{L}, \mathbf{Q}),$$

which is a particular case of (11.4).

11.2.1 Basic Properties

Here we present certain basic properties of the MVCSN distribution. We consider the distribution of linear transformations of MVCSN matrices and then give the distributions moment generating function.

First, we consider a closure property for linear transformations of MVCSN matrices of the form $\mathbf{W} = \mathbf{A}_1\mathbf{Y}\mathbf{A}_2$. This kind of transformation admits contrasts among rows as well as among columns which, for the usual setting of random matrices, would allow contrasts among individuals and among attributes.

Theorem 11.3. *Consider* $\mathbf{Y} \sim CSN_{p,m;q,n}(\mathbf{M},\mathbf{S},\mathbf{B},\mathbf{L},\mathbf{Q})$ *and let* $\mathbf{A}_1$ $(n_1 \times p)$ *and* $\mathbf{A}_2$ $(m \times n_2)$ *be matrices such that* $\mathbf{A} = \mathbf{A}_1 \otimes \mathbf{A}_2'$ *has full row rank. If* $\mathbf{W} = \mathbf{A}_1\mathbf{Y}\mathbf{A}_2$ *then* $\mathbf{W} \sim CSN_{n_1,n_2;q,n}(\mathbf{M}_\mathbf{A},\mathbf{S}_\mathbf{A},\mathbf{B}_\mathbf{A},\mathbf{L},\mathbf{Q}_\mathbf{A})$, *where* $\mathbf{M}_\mathbf{A} = \mathbf{A}_1\mathbf{M}\mathbf{A}_2$, $\mathbf{S}_\mathbf{A} = \mathbf{A}\mathbf{S}\mathbf{A}'$, $\mathbf{B}_\mathbf{A} = \mathbf{B}\mathbf{S}\mathbf{A}'\mathbf{S}_\mathbf{A}^{-1}$, *and* $\mathbf{Q}_\mathbf{A} = \mathbf{Q} + \mathbf{B}\mathbf{S}\mathbf{B}' - \mathbf{B}\mathbf{S}\mathbf{A}'\mathbf{S}_\mathbf{A}^{-1}\mathbf{A}\mathbf{S}\mathbf{B}'$.

PROOF: Using Theorem 1.2.22 of Gupta and Nagar (2000) we obtain that $vec(\mathbf{W}') = (\mathbf{A}_1 \otimes A_2')vec(\mathbf{Y}')$. The result then follows from Theorem 1 of González-Farías, Domínguez-Molina, and Gupta (2004b). ■

Prior to presenting the moment generating function of the MVCSN distribution, we need to introduce some additional notation. We consider the partitioned matrices $\mathbf{B} = (\mathbf{B}_1', \ldots, \mathbf{B}_q')'$ and $\mathbf{S} = (\mathbf{S}_1', \ldots, \mathbf{S}_m')'$, where $\mathbf{B}_i$ is $n \times mp$, $i = 1, \ldots, q$ and $\mathbf{S}_j$ is $p \times mp$, $j = 1, \ldots, m$. Let $\mathbf{T}$ $(p \times m)$ be an arbitrary matrix,

$$\tilde{\mathbf{T}} = [\mathbf{B}_1\mathbf{S}vec(\mathbf{T}'), \ldots, \mathbf{B}_q\mathbf{S}vec(\mathbf{T}')] \quad \text{and} \quad \tilde{\mathbf{S}} = [\mathbf{S}_1 vec(\mathbf{T}'), \ldots, \mathbf{S}_m vec(\mathbf{T}')].$$

Theorem 11.4. *Let* $\mathbf{Y} \sim CSN_{p,m;q,n}(\mathbf{M},\mathbf{S},\mathbf{B},\mathbf{L},\mathbf{Q})$. *Then the moment generating function of* $\mathbf{Y}$ *is given by*

$$M_\mathbf{Y}(\mathbf{T}) = Eetr(\mathbf{Y}'\mathbf{T}) = \frac{\Phi_{q,n}(\tilde{\mathbf{T}};\mathbf{L}',\mathbf{Q}+\mathbf{B}\mathbf{S}\mathbf{B}')}{\Phi_{q,n}(\mathbf{0};\mathbf{L}',\mathbf{Q}+\mathbf{B}\mathbf{S}\mathbf{B}')} etr\left(\mathbf{M}'\mathbf{T} + \frac{1}{2}\tilde{\mathbf{S}}'\mathbf{T}\right). \tag{11.5}$$

PROOF: Due to the fact that $tr(\mathbf{Y}'\mathbf{T}) = (vec(\mathbf{T}'))'vec(\mathbf{Y}')$, and also that $vec(\mathbf{Y}') \sim CSN_{pm;qn}(vec(\mathbf{M}'),\mathbf{S},\mathbf{B},vec(\mathbf{L}'),\mathbf{Q})$, we obtain from (11.3) that

$$\begin{aligned} E(etr(\mathbf{Y}'\mathbf{T})) = & \frac{\Phi_{nq}(\mathbf{B}\mathbf{S}vec(\mathbf{T}');vec(\mathbf{L}'),\mathbf{Q}+\mathbf{B}\mathbf{S}\mathbf{B}')}{\Phi_{nq}(\mathbf{0};vec(\mathbf{L}'),\mathbf{Q}+\mathbf{B}\mathbf{S}\mathbf{B}')} \\ & \times \exp\left[(vec(\mathbf{T}'))'vec(\mathbf{M}') + \frac{1}{2}(vec(\mathbf{T}'))'\mathbf{S}vec(\mathbf{T}')\right]. \end{aligned} \tag{11.6}$$

Now, by noting that $\mathbf{B}\mathbf{S}vec(\mathbf{T}') = vec(\tilde{\mathbf{T}})$ and $\mathbf{S}vec(\mathbf{T}') = vec(\tilde{\mathbf{S}})$, we obtain

$$(vec(\mathbf{T}'))'\mathbf{S}vec(\mathbf{T}') = tr(\tilde{\mathbf{S}}'\mathbf{T}).$$

Finally, (11.5) results by making use of these results, together with the definition of $\Phi_{q,n}()$, in (11.6). ■

The moment generating function for the MVCSN distribution with the parametrization $\mathbf{S} = \Sigma \otimes \Psi$ and $\mathbf{B} = \mathbf{D} \otimes \mathbf{E}$, where Σ $(p \times p)$ and Ψ $(m \times m)$ are positive definite and $\mathbf{D}$ $(n \times p)$ and $\mathbf{E}$ $(q \times m)$ are arbitrary matrices, is given by the following corollary.

Corollary 11.1. *Let* $\mathbf{Y} \sim CSN_{p,m;q,n}(\mathbf{M}, \Sigma \otimes \Psi, \mathbf{D} \otimes \mathbf{E}, \mathbf{L}, \mathbf{Q})$. *Then the moment generating function of* $\mathbf{Y}$ *is given by*

$$M_{\mathbf{Y}}(\mathbf{T}) = \frac{\Phi_{q,n}(\mathbf{E}\Psi\mathbf{T}'\Sigma\mathbf{D}'; \mathbf{L}, \mathbf{Q} + (\mathbf{D}\Sigma\mathbf{D}') \otimes (\mathbf{E}\Psi\mathbf{E}'))}{\Phi_{q,n}(\mathbf{0}; \mathbf{L}, \mathbf{Q} + (\mathbf{D}\Sigma\mathbf{D}') \otimes (\mathbf{E}\Psi\mathbf{E}'))} etr\left(\mathbf{M}'\mathbf{T} + \frac{1}{2}\mathbf{T}'\Sigma\mathbf{T}\Psi\right).$$

11.3 Quadratic Forms of the Matrix Variate Skew Normal Distributions

As is well known, the distributional properties of quadratic forms of normal variables play a key role in classical inference. Certain results for quadratic forms of skew normal variates have appeared recently in the literature. Azzalini and Capitanio (1999, Sect. 3.3), discuss the independence of quadratic forms and present a theorem which is similar to the Fisher-Cochran theorem given in Rao (1973, Sect. 3b.4). Loperfido (2001) considers quadratic forms for skew normal random vectors. Genton, He, and Liu (2001) derive the moments of skew normal random vectors and their quadratic forms, and consider applications in time series analysis and spatial statistics. Finally, Wang, Boyer, and Genton (2004) establish an equivalence between the chi-square and generalized skew normal distributions. They also show how properties of the chi-square distribution extend to the univariate and multivariate skew normal distributions. In what follows, we present three results related to the quadratic forms of MVCSN matrices. As will become evident, these results draw heavily on the work of Domínguez-Molina, González-Farías, and Gupta (2003) on quadratic forms of CSN variates.

Theorem 11.5. *Let* $\mathbf{A}$ $(r \times m)$, $\mathbf{B}$ $(p \times p)$, $\mathbf{C}$ $(m \times s)$, $r \leq m$, $s \leq m$, *and*

$$\mathbf{Y} \sim CSN_{p,m;q,n}(\mathbf{0}, \Sigma \otimes \Psi, \mathbf{D} \otimes \mathbf{E}, \mathbf{L}, \mathbf{Q}).$$

Then the moment generating function of $\mathbf{Z} = \mathbf{A}\mathbf{Y}'\mathbf{B}\mathbf{Y}\mathbf{C}$ *is*

$$M_{\mathbf{Z}}(\mathbf{T}) = \frac{\Phi_{q,n}[\mathbf{0}; \mathbf{L}, \mathbf{Q} + (\mathbf{D} \otimes \mathbf{E})\Theta(\mathbf{D}' \otimes \mathbf{E}')]}{\Phi_{q,n}[\mathbf{0}; \mathbf{L}, \mathbf{Q} + (\mathbf{D}\Sigma\mathbf{D}') \otimes (\mathbf{E}\Psi\mathbf{E}')]} |\mathbf{I}_{mp} - 2(\Sigma\mathbf{B}) \otimes (\Psi\mathbf{C}\mathbf{T}'\mathbf{A})|^{-\frac{1}{2}}, \tag{11.7}$$

where $\Theta = [\mathbf{I}_{mp} - 2(\mathbf{B}\Sigma) \otimes (\mathbf{C}\mathbf{T}'\mathbf{A}\Psi)]^{-1}$.

PROOF: From (1.2.6) of Gupta and Nagar (2000), we obtain that

$$tr(\mathbf{A}\mathbf{Y}'\mathbf{B}\mathbf{Y}\mathbf{C}\mathbf{T}) = (vec(\mathbf{Y}'))'(\mathbf{B} \otimes (\mathbf{C}\mathbf{T}'\mathbf{A}))vec(\mathbf{Y}').$$

The result then follows from Proposition 13 of Domínguez-Molina, González-Farías, and Gupta (2003). ■

Corollary 11.2. *Let* $\mathbf{Y} \sim CSN_{p,m;1,1}(\mathbf{0}, \Sigma \otimes \Psi, \mathbf{D} \otimes \mathbf{E}, 0, \mathbf{v})$, $\mathbf{A} = \mathbf{C} = \mathbf{I}_m$, *then* $\mathbf{Y}'\Sigma^{-1}\mathbf{Y}$ *has a Wishart distribution with parameters m, p, and* Ψ, *that is* $\mathbf{Y}'\Sigma^{-1}\mathbf{Y} \sim W_m(p, \Psi)$.

PROOF: Using the specified values of the parameters of the distribution of Y in (11.7), we obtain that

$$M_{\mathbf{Z}}(\mathbf{T}) = \frac{\Phi_1[0; 0, \mathbf{v} + (\mathbf{D} \otimes \mathbf{E})\Theta(\mathbf{D}' \otimes \mathbf{E}')]}{\Phi_1[0; 0, \mathbf{v} + (\mathbf{D}\Sigma\mathbf{D}') \otimes (\mathbf{E}\Psi\mathbf{E}')]} |\mathbf{I}_{mp} - 2\mathbf{I}_p \otimes (\Psi\mathbf{T}')|^{-\frac{1}{2}},$$

which simplifies to $M_{\mathbf{Z}}(\mathbf{T}) = |\mathbf{I}_m - 2\Psi\mathbf{T}\prime|^{-\frac{p}{2}}$. ■

Note that, as a direct consequence of Corollary 11.2, if $\mathbf{y}$ has a $CSN_{p,1}(\mathbf{0}, \Sigma, \delta, 0, 1)$ distribution then $\mathbf{y}\mathbf{y}' \sim W_p(1, \Sigma)$.

Corollary 11.3. *Let* $\mathbf{Y} \sim CSN_{p,1;p,n}(\mathbf{0}, \Sigma, \Gamma \otimes \mathbf{E}, \mathbf{0}, \mathbf{Q})$, *where* Γ *is part of the spectral decomposition of* Σ, $\Sigma = \Gamma\Lambda\Gamma'$ *and* $\mathbf{Q}$ *is diagonal. Then* $\mathbf{Y}'\Sigma^{-1}\mathbf{Y} \sim \chi^2_p$.

PROOF: Given that T is a real number, we deduce that

$$\Theta = [\mathbf{I}_p - 2(\Sigma^{-1}\Sigma) \otimes T]^{-1} = [\mathbf{I}_p - 2\mathbf{I}_p \otimes T]^{-1} = [\mathbf{I}_p - 2\mathbf{I}_p T]^{-1} = (1 - 2T)^{-1}\mathbf{I}_p.$$

Now,

$$\begin{aligned} M_Z(T) &= \frac{\Phi_{np}[\mathbf{0}; \mathbf{0}, \mathbf{Q} + (\mathbf{D} \otimes \mathbf{E})(1 - 2T\Psi)^{-1}\mathbf{I}_p(\mathbf{D}' \otimes \mathbf{E}')]}{\Phi_{np}[\mathbf{0}; \mathbf{0}, \mathbf{Q} + (\mathbf{D}\Sigma\mathbf{D}') \otimes (\mathbf{E}\mathbf{E}')]} |\mathbf{I}_p - 2(\Sigma\Sigma^{-1}) \otimes T|^{-\frac{1}{2}} \\ &= \frac{\Phi_{np}[\mathbf{0}; \mathbf{0}, \mathbf{Q} + (1 - 2T)^{-1}((\mathbf{D}\mathbf{D}') \otimes (\mathbf{E}\mathbf{E}'))]}{\Phi_{np}[\mathbf{0}; \mathbf{0}, \mathbf{Q} + (\mathbf{D}\Sigma\mathbf{D}') \otimes (\mathbf{E}\mathbf{E}')]} |\mathbf{I}_p - 2\mathbf{I}_p T|^{-\frac{1}{2}} \\ &= (1 - 2T)^{-\frac{p}{2}}. \end{aligned}$$

■

11.4 A Multivariate Stochastic Frontier Model

In this section we extend the relationship between the closed skew normal distribution and the stochastic frontier models to the matrix case using a similar approach to that used by Domínguez-Molina, González-Farías, and Ramos-Quiroga (2004) for the vector case.

In what follows, we will use the notation $\mathbf{U} \sim N^{\mathbf{C}}_{m,n}(\mathbf{M}, \mathbf{S})$ to denote that $\mathbf{U}$ is a $N_{m,n}(\mathbf{M}, \mathbf{S})$ random matrix truncated below at $\mathbf{C}$. That is, the truncation is of the type $\mathbf{U} \geq \mathbf{C}$, where $\mathbf{W} \geq \mathbf{C}$ means $W_{ij} \geq C_{ij}$, $i = 1, \ldots, m$, $j = 1, \ldots, n$. Note that $\mathbf{U} \geq \mathbf{C} \Rightarrow vec(\mathbf{U}') \geq vec(\mathbf{C}')$.

Consider, now, production data on p firms at time t. We assume a stochastic frontier model for time t of the form $\mathbf{y}_t = (\mathbf{X_t}; \boldsymbol{\beta}_\mathbf{t}) + \boldsymbol{\varepsilon}_\mathbf{t}$, where

$(\mathbf{X_t};\beta_t)=(\mathbf{f}(\mathbf{x_{1t}};\beta_t),\ldots,\mathbf{f}(\mathbf{x_{pt}};\beta_t))'$, $\mathbf{X}_t = (\mathbf{x}_{1t},\ldots,\mathbf{x}_{pt})'$ is a known matrix of covariates, β_t is unknown, $\varepsilon_t = (\varepsilon_{1t},\ldots,\varepsilon_{pt})'$ is a random vector of compound errors and $\varepsilon_t = \mathbf{v}_t + \mathbf{G}\mathbf{u}_t$, with $\mathbf{v}_t = (\mathbf{v}_{1t},\ldots,\mathbf{v}_{pt})'$, $\mathbf{u}_t = (\mathbf{u}_{1t},\ldots,\mathbf{u}_{pt})'$ and $\mathbf{G}$ is a $p\times q$ weighting matrix. We use $\mathbf{Y}$ to denote the $p\times m$ matrix of the value added observations for the p firms at times $t = 1,\ldots,m$, i.e.,

$$\mathbf{Y} = \begin{pmatrix} y_{11} & \cdots & x_{1m} \\ \vdots & \ddots & \vdots \\ y_{p1} & \cdots & x_{pm} \end{pmatrix} = (\mathbf{y}_1,\ldots,\mathbf{y}_m).$$

A joint model for such production data can be written as

$$\mathbf{Y} = \mathbf{F} + \Theta, \tag{11.8}$$

where $\mathbf{F} = (f(\mathbf{y}_1,\beta_1),\ldots,f(\mathbf{y}_m,\beta_m))$, $\Theta = \mathbf{V}+\mathbf{G}\mathbf{U}$, $\mathbf{V} = (\mathbf{v}_1,\ldots,\mathbf{v}_m)$, and $\mathbf{U} = (\mathbf{u}_1,\ldots,\mathbf{u}_m)$.

We choose, in fact, to consider a slightly more general model for the compound errors, namely,

$$\Theta = \mathbf{V} + \mathbf{D}\mathbf{U}\mathbf{E}',$$

where $\mathbf{V}\sim N_{p,m}(\mathbf{0},\mathbf{S})$, $\mathbf{U}\sim N^{\mathbf{C}}_{q,m}(\mathbf{L},\mathbf{Q})$, $\mathbf{D}$ $(p\times q)$, $\mathbf{E}$ $(m\times m)$, and $\mathbf{V}$ is independent of $\mathbf{U}$. By pre-multiplying the matrix of technical inefficiencies, $\mathbf{U}$, by $\mathbf{D}$ we can incorporate common inefficiencies within groups of similar companies. Similarly, by post-multiplying $\mathbf{U}$ by $\mathbf{E}'$, time related inefficiency effects can be allowed for. Note that the matrix $\mathbf{V}$ is no longer constrained to merely reflect measurement error. Indeed, depending on the structure of the variance matrix $\mathbf{S}$, it can also incorporate random effects such as random intercepts and time-induced correlations among the columns of $\mathbf{Y}$. Given that $vec(\Theta') = vec(\mathbf{V}') + (\mathbf{D}\otimes\mathbf{E})vec(\mathbf{U}')$, we obtain from Domínguez-Molina, González-Farías, and Ramos-Quiroga (2004) that the density of the compound error $\Theta = \mathbf{V}+\mathbf{D}\mathbf{U}\mathbf{E}'$ is

$$\begin{aligned} g(\Theta) &= \Phi^{-1}_{q,m}(\mathbf{0};\mathbf{C}-\mathbf{L},\mathbf{Q})\phi_{p,m}(\Theta;\mathbf{D}\mathbf{L}\mathbf{E}',\theta) \\ &\quad\times \Phi_{mq}\{\mathbf{Q}(\mathbf{D}'\otimes\mathbf{E}')\theta^{-1}[vec(\Theta-\mathbf{D}\mathbf{L}\mathbf{E}')];vec(\mathbf{C}-\mathbf{L}),\Upsilon\}, \end{aligned}$$

where $\theta = \mathbf{S}+(\mathbf{D}\otimes\mathbf{E})\mathbf{Q}(\mathbf{D}'\otimes\mathbf{E}')$ and $\Upsilon = \mathbf{Q}-\mathbf{Q}(\mathbf{D}'\otimes\mathbf{E}')\theta^{-1}(\mathbf{D}\otimes\mathbf{E})\mathbf{Q}$. Thus, Θ has a matrix variate closed skew normal distribution. Specifically,

$$\Theta \sim CSN_{p,m;q,m}(\mathbf{D}\mathbf{L}\mathbf{E}',\theta,\mathbf{Q}(\mathbf{D}'\otimes\mathbf{E}')\theta^{-1},\mathbf{C}-\mathbf{L},\Upsilon).$$

Model (11.8), with the compound error structure $\Theta = \mathbf{V}+\mathbf{D}\mathbf{U}\mathbf{E}'$, includes the following submodels as special cases:

- *Model I (Homoscedastic and uncorrelated errors).* $\mathbf{D}$ an arbitrary $p \times q$ matrix, $\mathbf{E} = \mathbf{I}_m$, $\mathbf{S} = \mathbf{I}_m \otimes \Sigma$, and $\mathbf{Q} = \mathbf{I}_m \otimes \Delta$, where Σ $(p \times p)$ and Δ $(q \times q)$ are covariance matrices.
- *Model II (Heteroscedastic and uncorrelated errors).* $\mathbf{D}$ an arbitrary $p \times q$ matrix, $\mathbf{E} = \mathbf{I}_m$, $\mathbf{S}$ and $\mathbf{Q}$ block diagonal matrices of the form $\mathbf{S} = \oplus_{i=1}^m \Sigma_i$, and $\mathbf{Q} = \oplus_{i=1}^m \Delta_i$, with $\oplus$ denoting the matrix direct sum operator (see Horn and Johnson 1985, p. 24). The result of $A \oplus B$ is a block diagonal matrix. Here, Σ_i $(p \times p)$ and Δ_i $(m \times m)$ are covariance matrices, $i = 1, \ldots, m$.
- *Model III (Correlated errors).* If any of the matrices $\mathbf{E}$, $\mathbf{S}$, or $\mathbf{Q}$ are non block diagonal.

11.5 Global Minimum Variance Portfolio Under the Skew Normality

In this section, we study the impact of skewness on the performance of the global minimum variance portfolio. The GMVP plays an important role in the Markowitz's mean-variance analysis. This portfolio lies on the vertex of the efficient frontier which is a parabola in the mean-variance space (see e.g., Merton 1972; Bodnar and Schmid 2008b, 2009). It is also a unique portfolio, whose weights are independent of the mean vector μ of the asset returns. Because of this property, the estimator of the GMVP weights does not suffer from the error in the means which is much larger than the error in the variances and covariances (see, e.g. Merton 1980; Best and Grauer 1991). It makes the GMVP portfolio attractive for practitioners as well as for researchers in the financial sector (see, e.g., Jagannathan and Ma 2003; Bodnar and Schmid 2008a).

Let $\mathbf{w}$ denote the weights of the portfolio. Then, the weights of the GMVP are obtained by minimizing the portfolio variance $\mathbf{w}'\mathbf{V}\mathbf{w}$ under the constraint $\mathbf{w}'\mathbf{1}_p = 1$. The solution is given by

$$\mathbf{w}_{GMV} = \frac{\mathbf{V}^{-1}\mathbf{1}_p}{\mathbf{1}_p'\mathbf{V}^{-1}\mathbf{1}_p}. \tag{11.9}$$

Because $\mathbf{V}$ is an unknown parameter of the asset return distribution, the vector of the GMVP weights cannot be calculated in practice. The investor estimates $\mathbf{V}$ by $\hat{\mathbf{V}}$ and then plugs $\hat{\mathbf{V}}$ in (11.9) instead of $\mathbf{V}$. We consider the sample estimator of the covariance matrix given by

$$\hat{\mathbf{V}} = \frac{1}{n-1}\sum_{j=1}^{n}(\mathbf{x}_j - \hat{\mu})(\mathbf{x}_j - \hat{\mu})' = \frac{1}{n-1}\mathbf{X}'\mathbf{M}\mathbf{X} \quad \text{with} \quad \hat{\mu} = \frac{1}{n}\sum_{j=1}^{n}\mathbf{x}_j. \tag{11.10}$$

Here, $\mathbf{M} = \mathbf{I}_n - \frac{1}{n}\mathbf{1}_n\mathbf{1}_n'$ is a symmetric idempotent matrix such that $\mathbf{M} = \mathbf{M}'$ and $\mathbf{M}\mathbf{M} = \mathbf{M}$.

The weights of the global minimum variance portfolio are estimated by

$$\hat{\mathbf{w}}_{GMV} = \frac{\hat{\mathbf{V}}^{-1}\mathbf{1}_p}{\mathbf{1}_p'\hat{\mathbf{V}}^{-1}\mathbf{1}_p}. \tag{11.11}$$

We consider a linear combination of the GMVP weights. Let $\mathbf{l}_i \in I\!R^p$, $i = 1,\ldots,q$, $1 \le q \le p-1$, and $\mathbf{L}' = (\mathbf{l}_1,\ldots,\mathbf{l}_q)$. We are interested in

$$\mathbf{w}_{L;q} = \mathbf{L}\mathbf{w}_{GMV} = \frac{\mathbf{L}\mathbf{V}^{-1}\mathbf{1}_p}{\mathbf{1}_p'\mathbf{V}^{-1}\mathbf{1}_p} = \left(\frac{\mathbf{l}_1'\mathbf{V}^{-1}\mathbf{1}_p}{\mathbf{1}_p'\mathbf{V}^{-1}\mathbf{1}_p},\ldots,\frac{\mathbf{l}_q'\mathbf{V}^{-1}\mathbf{1}_p}{\mathbf{1}_p'\mathbf{V}^{-1}\mathbf{1}_p}\right)'. \tag{11.12}$$

Using the estimator (11.10), we get

$$\hat{\mathbf{w}}_{L;q} = \mathbf{L}\hat{\mathbf{w}}_{GMV} = \left(\frac{\mathbf{l}_1'\hat{\mathbf{V}}^{-1}\mathbf{1}_p}{\mathbf{1}_p'\hat{\mathbf{V}}^{-1}\mathbf{1}_p},\ldots,\frac{\mathbf{l}_q'\hat{\mathbf{V}}^{-1}\mathbf{1}_p}{\mathbf{1}_p'\hat{\mathbf{V}}^{-1}\mathbf{1}_p}\right)'. \tag{11.13}$$

In order to derive the distributional properties of $\hat{\mathbf{w}}_{L;q}$ we need the following result.

Theorem 11.6. *Let $\mathbf{X} \sim CSN_{n,p;1,1}(\mathbf{1}_n \otimes \mu', \mathbf{I}_n \otimes \Sigma, \mathbf{D}' \otimes \mathbf{E}', 0, \nu)$ with $n > p$. Then $(n-1)\hat{\mathbf{V}} \sim W_p(n-1,\Sigma)$ (p-dimensional Wishart distribution with $n-1$ degrees of freedom and the covariance matrix Σ).*

PROOF: From Corollary 11.2 of Domínguez-Molina, González-Farías, and Gupta (2007) the moment generating function of $(n-1)\hat{\mathbf{V}} = \mathbf{X}'\mathbf{M}\mathbf{X}$ is given by

$$\begin{aligned} M_{(n-1)\hat{\mathbf{V}}}(\mathbf{T}) &= \frac{\Phi(0;0,\nu + (\mathbf{D}'\otimes\mathbf{E}')(\mathbf{I}_{np} - 2(\mathbf{M}\Sigma)\otimes\mathbf{T}'\Sigma)^{-1}(\mathbf{D}\otimes\mathbf{E}))}{\Phi(0;0,\nu + (\mathbf{D}'\mathbf{D})\otimes(\mathbf{E}'\Sigma\mathbf{E}))} \\ &\quad \times |\mathbf{I}_{np} - 2(\mathbf{M}\otimes\Sigma\mathbf{T}')|^{-1/2} \\ &= |\mathbf{I}_{np} - 2(\mathbf{M}\otimes\Sigma\mathbf{T}')|^{-1/2}. \end{aligned}$$

The last equality is independent of $\mathbf{E}$ and $\mathbf{D}$. Hence, the expression of the moment generating function is equal to the moment generating function of the sample covariance matrix under the assumption of normality, i.e. if $\mathbf{D} = \mathbf{0}_n$ or $\mathbf{E} = \mathbf{0}_p$. Because under normality the sample covariance matrix is Wishart with $n-1$ degrees of freedom, the proposition follows.

Using Theorem 11.6 and the proof of Theorem 1 of Bodnar and Schmid (2008a), we obtain the distribution of $\hat{\mathbf{w}}_{L;q}$ when the matrix of the asset returns follows a matrix variate closed skew normal distribution.

Theorem 11.7. *Let $\mathbf{X} \sim CSN_{n,p;1,1}(\mathbf{1}_n \otimes \mu', \mathbf{I}_n \otimes \Sigma, \mathbf{D}' \otimes \mathbf{E}', 0, \nu)$ with $n > p > q \ge 1$. Let $\tilde{\mathbf{M}}' = (\mathbf{L}', \mathbf{1}_p)$ and $rk(\tilde{\mathbf{M}}) = q+1$. Then*

$$\hat{\mathbf{w}}_{L;q} \sim t_q(n-p+1,\, \tilde{\mathbf{w}}_{L;q},\, \frac{1}{n-p+1}\frac{\mathbf{LRL}'}{\mathbf{1}_p'\Sigma^{-1}\mathbf{1}_p}),$$

where

$$\tilde{\mathbf{w}}_{L;q} = \frac{\mathbf{L}\Sigma^{-1}\mathbf{1}_p}{\mathbf{1}_p'\Sigma^{-1}\mathbf{1}_p} \quad \textit{and} \quad \mathbf{R} = \Sigma^{-1} - \Sigma^{-1}\mathbf{1}_p\mathbf{1}_p'\Sigma^{-1}/\mathbf{1}_p'\Sigma^{-1}\mathbf{1}_p \tag{11.14}$$

As a consequence of Theorem 11.7, we obtain some interesting results. First, because the parameters of the multivariate t-distribution, namely the scale vector and the dispersion matrix, are only functions of Σ, the distribution of $\hat{\mathbf{w}}_{L;q}$ is independent of $\mathbf{D}$ and $\mathbf{E}$. On the other hand, the distribution of the asset returns does depend on $\mathbf{D}$ and $\mathbf{E}$. Since the covariance function is a function of $\mathbf{D}$ and $\mathbf{E}$, we obtain that the true vector of the GMVP weights is given by

$$\begin{aligned}\mathbf{w}_{GMV} &= \frac{\mathbf{V}^{-1}\mathbf{1}_p}{\mathbf{1}_p'\mathbf{V}^{-1}\mathbf{1}_p} = \frac{(\Sigma - \frac{2}{\pi}\frac{d_j^2}{\nu+\mathbf{D}'\mathbf{D}\mathbf{E}'\Sigma\mathbf{E}}\mathbf{E}\mathbf{E}')^{-1}\mathbf{1}_p}{\mathbf{1}_p'(\Sigma - \frac{2}{\pi}\frac{d_j^2}{\nu+\mathbf{D}'\mathbf{D}\mathbf{E}'\Sigma\mathbf{E}}\mathbf{E}\mathbf{E}')^{-1}\mathbf{1}_p} \\ &= \frac{(\Sigma^{-1} + 2d_j^2\frac{\Sigma^{-1}\mathbf{E}\mathbf{E}'\Sigma^{-1}}{\pi(\nu+\mathbf{D}'\mathbf{D}\mathbf{E}'\Sigma\mathbf{E})-2d_j^2\mathbf{E}'\Sigma^{-1}\mathbf{E}})\mathbf{1}_p}{\mathbf{1}_p'(\Sigma^{-1} + 2d_j^2\frac{\Sigma^{-1}\mathbf{E}\mathbf{E}'\Sigma^{-1}}{\pi(\nu+\mathbf{D}'\mathbf{D}\mathbf{E}'\Sigma\mathbf{E})-2d_j^2\mathbf{E}'\Sigma^{-1}\mathbf{E}})\mathbf{1}_p}\end{aligned} \tag{11.15}$$

and, hence, $\hat{\mathbf{w}}_{L;q}$ is a biased estimator of $\mathbf{w}_{L;q}$ if $\mathbf{D} \neq \mathbf{0}_n$ and $\mathbf{E} \neq \mathbf{0}_p$. Moreover, the GMVP weights cannot be estimated by replacing the matrix $\mathbf{V}$ by the sample covariance matrix $\hat{\mathbf{V}}$. In Sect. 11.5.1, we present an alternative method of estimating $\mathbf{V}$.

For providing further investigation on how large is the impact of skewness on the performance of the global minimum variance portfolio, a test for linear combination of the GMVP weights is applied. We consider the general linear hypothesis which is given by

$$H_0 : \mathbf{L}\mathbf{w}_{GMV} = \mathbf{w}_0 \qquad \text{against} \qquad H_1 : \mathbf{L}\mathbf{w}_{GMV} \neq \mathbf{w}_0 \,. \tag{11.16}$$

This means that the investor is interested in knowing whether the weights of the GMVP fulfill q linear restrictions or not. This is a very general testing problem and it includes many important special cases (cf. Greene 2003, pp. 95–96).

To test (11.16) Bodnar and Schmid (2008a) derived the following test statistic

$$T = \frac{n-k}{q}\left(\mathbf{1}_p'\hat{\mathbf{V}}^{-1}\mathbf{1}_p\right)\left(\hat{\mathbf{w}}_{L;q} - \mathbf{w}_0\right)'\left(\mathbf{L}\hat{\mathbf{R}}\mathbf{L}'\right)^{-1}\left(\hat{\mathbf{w}}_{L;q} - \mathbf{w}_0\right). \tag{11.17}$$

Although this quantity is very similar to the F statistic for testing a linear hypothesis within the linear regression model, its distribution is different than in the case of a linear model. Consequently, the well-known results, obtained for the linear regression model, cannot be applied directly.

Theorem 11.8 is taken from Bodnar and Gupta (2013).

Theorem 11.8. *Let* $\mathbf{X} \sim CSN_{n,p;1,1}(\mathbf{1}_n \otimes \mu', \mathbf{I}_n \otimes \Sigma, \mathbf{D}' \otimes \mathbf{E}', 0, \nu)$ *with* $n > p > q \geq 1$. *Let* $\tilde{\mathbf{M}}' = (\mathbf{L}', \mathbf{1}_p)$ *and* $rk(\tilde{\mathbf{M}}) = q+1$. *The density of* T *is given by*

$$f_T(x) = f_{q,n-p}(x)\,(1+\lambda)^{-(n-p+q)/2} \times {}_2F_1\left(\frac{n-p+q}{2}, \frac{n-p+q}{2}, \frac{q}{2}; \frac{qx}{n-p+qx}\frac{\lambda}{1+\lambda}\right) \tag{11.18}$$

with $\lambda = \mathbf{1}_p'\Sigma^{-1}\mathbf{1}_p(\mathbf{w}_0 - \tilde{\mathbf{w}}_{L;q})'(\mathbf{LRL}')^{-1}(\mathbf{w}_0 - \tilde{\mathbf{w}}_{L;q})$.

The proof of Theorem 11.8 follows from Theorem 11.6 and the proof of Theorem 2a of Bodnar and Schmid (2008a). If $\mathbf{E} = \mathbf{0}_p$ or $\mathbf{D} = \mathbf{0}_n$ the null hypothesis is rejected if $T > F_{q,n-p;1-\alpha}$, where $F_{q,n-p;1-\alpha}$ stands for the $1-\alpha$ quantile of the central F-distribution with q and $n-p$ degrees of freedom. However, if $\mathbf{E} \neq \mathbf{0}_p$ and $\mathbf{D} \neq \mathbf{0}_n$ the decision rule $T > F_{q,n-p;1-\alpha}$ might reject the null hypothesis with the probability larger than α although the vector of the target GMVP weights $\mathbf{w}_0$ is correctly specified.

11.5.1 Model Estimation

Since the dimension of the vector $\mathbf{D}$ is n and we deal with the sample of size n, in the following it is assumed that the skewness is the same within the sample of each asset, i.e. $\mathbf{D} = \mathbf{1}_n$. In this section, we study on the estimation procedure for the suggested model of Sect. 11.2. For this aim, we modify the estimation method for the closed skew normal distributions considered by Flecher, Naveaua, and Allard (2009). Namely, the parameters of the model are estimated by the method of the weighted moments. In Theorem 11.9, we derive an expression for these moments, which is used later.

Theorem 11.9. *Let* $\mathbf{Y} \sim CSN_{p,1;1,1}(\mu, \Sigma, \mathbf{E}', 0, \nu)$ *with* $p \geq 1$. *Let* $h(\mathbf{y})$ *be a real valued function such that* $E\left(h(\mathbf{Y})exp\left(c\mathbf{Y}'\Sigma^{-1}\mathbf{Y}\right)\right)$ *exists. Then*

$$E\left((1-2c)^{p/2}h(\mathbf{Y})exp\left(c\mathbf{Y}'\Sigma^{-1}\mathbf{Y}\right)\right) = exp\left(\frac{c}{1-2c}\mu'\Sigma^{-1}\mu\right)E(h(\mathbf{Y}^*)) \tag{11.19}$$

for $c < 1/2$ *where* $\mathbf{Y}^* \sim CSN_{p,1;1,1}(\frac{1}{1-2c}\mu, \frac{1}{1-2c}\Sigma, \mathbf{E}', 0, \nu)$.

PROOF: We have

$$E\left((1-2c)^{p/2}h(\mathbf{Y})exp\left(c\mathbf{Y}'\Sigma^{-1}\mathbf{Y}\right)\right)$$

$$= 2\int_{\mathbf{y}}(1-2c)^{p/2}h(\mathbf{y})exp\left(c\mathbf{y}'\Sigma^{-1}\mathbf{y}\right)\phi_p(\mathbf{y};\mu,\Sigma)\Phi(\mathbf{E}'(\mathbf{y}-\mu)/\sqrt{v})d\mathbf{y}$$

$$= 2exp\left(\frac{c}{1-2c}\mu'\Sigma^{-1}\mu\right)\int_{\mathbf{y}}h(\mathbf{y})\phi_p(\mathbf{y};\frac{1}{1-2c}\mu,\frac{1}{1-2c}\Sigma)\Phi(\mathbf{E}'(\mathbf{y}-\mu)/\sqrt{v})d\mathbf{y}.$$

The theorem is proved.

Next, we present the estimation procedure. The scale matrix Σ is estimated from the fact that $(n-1)\hat{\mathbf{V}} \sim W_p(n-1,\Sigma)$, where $\hat{\mathbf{V}}$ is defi ned in (11.10). Note that it is not our aim to derive an estimation procedure for all of the model parameters. In Sect. 11.6, we calculate the probability of type I error rate for the test (11.16) in the case of real data and for this reason we need only the estimators of the scale matrix Σ and of the product $\tilde{\mathbf{E}} = (v+n\mathbf{E}'\Sigma\mathbf{E})^{-1/2}\mathbf{E}$. Moreover, since in the moment identities such a product is present, it seems that the estimation of v could be difficult. Note that

$$E(\mathbf{x}_j) = \mu + \frac{\sqrt{2}}{\sqrt{\pi}}(v+n\mathbf{E}'\Sigma\mathbf{E})^{-1/2}\mathbf{E}$$

$$E((1-2c)^{p/2}\mathbf{x}_j exp\left(c\mathbf{x}_j'\Sigma^{-1}\mathbf{x}_j\right))$$

$$= exp\left(\frac{c}{1-2c}\mu'\Sigma^{-1}\mu\right)\left(\frac{1}{1-2c}\mu + \frac{\sqrt{2}}{\sqrt{\pi}}(v+\frac{n}{1-2c}\mathbf{E}'\Sigma\mathbf{E})^{-1/2}\mathbf{E}\right)$$

$$E((1-2c)^{p/2}exp\left(c\mathbf{x}_j'\Sigma^{-1}\mathbf{x}_j\right)) = exp\left(\frac{c}{1-2c}\mu'\Sigma^{-1}\mu\right).$$

Equating the theoretical moments with the sample moments we obtain

$$\hat{\Sigma} = \hat{\mathbf{V}} \tag{11.20}$$

$$\bar{\mathbf{x}} = \hat{\mu} + \frac{\sqrt{2}}{\sqrt{\pi}}(\hat{v}+n\hat{\mathbf{E}}'\hat{\Sigma}\hat{\mathbf{E}})^{-1/2}\hat{\mathbf{E}} \tag{11.21}$$

$$\bar{\mathbf{x}}_\phi = exp\left(\frac{c}{1-2c}\hat{\mu}'\hat{\Sigma}^{-1}\hat{\mu}\right)\left(\frac{1}{1-2c}\hat{\mu}+\frac{\sqrt{2}}{\sqrt{\pi}}(\hat{v}+\frac{n}{1-2c}\hat{\mathbf{E}}'\hat{\Sigma}\hat{\mathbf{E}})^{-1/2}\hat{\mathbf{E}}\right) \tag{11.22}$$

$$\bar{\phi} = exp\left(\frac{c}{1-2c}\hat{\mu}'\hat{\Sigma}^{-1}\hat{\mu}\right), \tag{11.23}$$

where

$$\bar{\mathbf{x}} = \hat{r} = \frac{1}{n}\sum_{i=1}^{n}\mathbf{x}_i,$$

$$\bar{\mathbf{x}}_\phi = \frac{(1-2c)^{p/2}}{n}\sum_{i=1}^{n}\mathbf{x}_i exp\left(c\mathbf{x}_i'\hat{\Sigma}^{-1}\mathbf{x}_i\right),$$

$$\bar{\phi} = \frac{(1-2c)^{p/2}}{n}\sum_{i=1}^{n} exp\left(c\mathbf{x}_i'\hat{\Sigma}^{-1}\mathbf{x}_i\right).$$

Now let $\hat{a} = \sqrt{\hat{v}+n\hat{\mathbf{E}}'\hat{\Sigma}\hat{\mathbf{E}}}$ and $\hat{b} = \sqrt{\hat{v}+\frac{n}{1-2c}\hat{\mathbf{E}}'\hat{\Sigma}\hat{\mathbf{E}}}$, then from (11.21), (11.22), and (11.23), we obtain

$$\begin{cases} \hat{a}\bar{\mathbf{x}} - \frac{\hat{b}}{\bar{\phi}}\bar{\mathbf{x}}_\phi = (\hat{a}-\frac{1}{1-2c}\hat{b})\hat{\mu} \\ \frac{1}{1-2c}\bar{\mathbf{x}} - \frac{1}{\bar{\phi}}\bar{\mathbf{x}}_\phi = \frac{\sqrt{2}}{\sqrt{\pi}}(\frac{1}{\hat{a}}-\frac{1}{\hat{b}})\hat{\mathbf{E}}. \end{cases}$$

Application of (11.23) and the identity $\hat{\mathbf{E}}'\hat{\Sigma}\hat{\mathbf{E}} = \frac{1-2c}{2nc}(\hat{b}^2-\hat{a}^2)$ leads to

$$\begin{cases} (\hat{a}\bar{\mathbf{x}} - \frac{\hat{b}}{\bar{\phi}}\bar{\mathbf{x}}_\phi)'\hat{\Sigma}^{-1}(\hat{a}\bar{\mathbf{x}} - \frac{\hat{b}}{\bar{\phi}}\bar{\mathbf{x}}_\phi) = \frac{1-2c}{c}(\hat{a}-\frac{1}{1-2c}\hat{b})^2 ln(\bar{\phi}) \\ (\frac{1}{1-2c}\bar{\mathbf{x}} - \frac{1}{\bar{\phi}}\bar{\mathbf{x}}_\phi)'\hat{\Sigma}(\frac{1}{1-2c}\bar{\mathbf{x}} - \frac{1}{\bar{\phi}}\bar{\mathbf{x}}_\phi) = \frac{(1-2c)}{\pi nc}(\frac{1}{\hat{a}}-\frac{1}{\hat{b}})^2(\hat{b}^2-\hat{a}^2). \end{cases}$$

Further let $\tilde{a} = a/b$ and $\tilde{c} = \frac{1}{1-2c}$. Then,

$$(\hat{\tilde{a}}\bar{\mathbf{x}} - \frac{1}{\bar{\phi}}\bar{\mathbf{x}}_\phi)'\hat{\Sigma}^{-1}(\hat{\tilde{a}}\bar{\mathbf{x}} - \frac{1}{\bar{\phi}}\bar{\mathbf{x}}_\phi) - \frac{2}{\tilde{c}-1}(\hat{\tilde{a}}-\tilde{c})^2 ln(\bar{\phi}) = 0 \quad (11.24)$$

$$(\tilde{c}\bar{\mathbf{x}} - \frac{1}{\bar{\phi}}\bar{\mathbf{x}}_\phi)'\hat{\Sigma}(\tilde{c}\bar{\mathbf{x}} - \frac{1}{\bar{\phi}}\bar{\mathbf{x}}_\phi) + \frac{2}{(\tilde{c}-1)\pi n}(1-\hat{\tilde{a}})^2(1-\frac{1}{\hat{\tilde{a}}^2}) = 0. \quad (11.25)$$

The identity (11.24) is a quadratic equation with respect to $\hat{\tilde{a}}$. Solving it we get

$$\hat{\tilde{a}} = \frac{\frac{1}{\bar{\phi}}\bar{\mathbf{x}}\hat{\Sigma}^{-1}\bar{\mathbf{x}}_\phi - \frac{2\tilde{c}}{\tilde{c}-1}ln(\bar{\phi})}{\bar{\mathbf{x}}\hat{\Sigma}^{-1}\bar{\mathbf{x}} - \frac{2}{\tilde{c}-1}ln(\bar{\phi})} \quad (11.26)$$

$$\pm \frac{\sqrt{(\frac{1}{\bar{\phi}}\bar{\mathbf{x}}\hat{\Sigma}^{-1}\bar{\mathbf{x}}_\phi - \frac{2\tilde{c}}{\tilde{c}-1}ln(\bar{\phi}))^2 - (\bar{\mathbf{x}}\hat{\Sigma}^{-1}\bar{\mathbf{x}} - \frac{2}{\tilde{c}-1}ln(\bar{\phi}))((\frac{1}{\bar{\phi}})^2\bar{\mathbf{x}}_\phi\hat{\Sigma}^{-1}\bar{\mathbf{x}}_\phi - \frac{2\tilde{c}^2}{\tilde{c}-1}ln(\bar{\phi}))}}{\bar{\mathbf{x}}\hat{\Sigma}^{-1}\bar{\mathbf{x}} - \frac{2}{\tilde{c}-1}ln(\bar{\phi})}.$$

Substituting from (11.26) into (11.25), we obtain an identity which depends only on one parameter $\tilde{c}$. It is solved by applying the regula falsi (see, e.g., Conte and de Boor 1981). Then the estimators of Σ, μ, and $\tilde{\mathbf{E}} = (v+n\mathbf{E}'\Sigma\mathbf{E})^{-1/2}\mathbf{E}$ are calculated yielding

$$\hat{\Sigma} = \mathbf{V}, \quad (11.27)$$

$$\hat{\mu} = \frac{1}{\hat{\tilde{a}}-\tilde{c}}(\hat{\tilde{a}}\bar{\mathbf{x}} - \frac{1}{\bar{\phi}}\bar{\mathbf{x}}_\phi), \quad (11.28)$$

$$\hat{\hat{\mathbf{E}}} = \frac{\sqrt{\pi}}{\sqrt{2}} \frac{1}{1-\hat{\hat{a}}} (\tilde{c}\bar{\mathbf{x}} - \frac{1}{\bar{\phi}} \bar{\mathbf{x}}_{\phi}) . \tag{11.29}$$

The suggested estimator (11.28) is not surprising. It is noted that μ is no longer the mean vector of the closed skew normal distribution and, consequently, the sample mean does not provide a good fit in general case. Note that an improved estimator of the covariance matrix can be obtained by

$$\hat{\hat{\mathbf{V}}} = \hat{\mathbf{V}} - \frac{2}{\pi} \hat{\hat{\mathbf{E}}} \hat{\hat{\mathbf{E}}}' = \hat{\mathbf{V}} - \frac{1}{(1-\hat{\hat{a}})^2} (\tilde{c}\bar{\mathbf{x}} - \frac{1}{\bar{\phi}} \bar{\mathbf{x}}_{\phi})(\tilde{c}\bar{\mathbf{x}} - \frac{1}{\bar{\phi}} \bar{\mathbf{x}}_{\phi})' . \tag{11.30}$$

11.5.2 Goodness-of-Fit Test

When a model is fitted to real data, one would like to know how good it can describe the observable dynamics of data and if it can be applied at all. In statistics such questions are usually treated with goodness-of-fit tests. Although the skew normal distributions have been already successfully applied to real data (cf. Adcock 2005; Harvey, Leichty, Leichty, and Muller 2010; Framstad 2011), the problem of testing their validity is not dealt in detail. This is the aim of the present section, namely we derive a goodness-of-fit test for the matrix variate closed skew normal distribution.

Several goodness-of-fit tests for the univariate skew normal distributions are discussed in the literature (see e.g., Gupta and Chen 2001; Mateu-Figueras, Puig, and Pewsey 2007; Meintanis 2007, 2010; Cabras and Castellanos 2009; Pérez Rodríguez and Villaseñor Alva 2010). It seems that the first goodness-of-fit test is suggested by Gupta and Chen (2001) who applied the Kolmogorov-Smirnov test statistic and the Pearson's χ^2 test. Mateu-Figueras, Puig, and Pewsey (2007) extended these results to the case with estimated parameters, while Cabras and Castellanos (2009) presented the Bayesian goodness-of-fit test for the skew normal model. Further approaches can be found by Meintanis (2007, 2010) as well as by Pérez Rodríguez and Villaseñor Alva (2010).

On the other hand, there is only one paper that deals with testing the hypothesis of skew normality in the multivariate case (cf. Meintanis and Hlávka 2010), and no result for the matrix variate skew normal distribution is available. Meintanis and Hlávka (2010) suggested the application of the moment generating function for testing the multivariate skew normality and dealt the bivariate case in detail. The situation is much more complicated in the matrix variate case. Here, the decision about the goodness of model is based only on a single observation. In this section, we extend the approach of Meintanis and Hlávka (2010) to the matrix variate case. In the derivation, we use the analytical expression of the moment generating function of the matrix variate closed skew normal distribution.

Let $\mathbf{X} \sim CSN_{n,p;1,1}(\mathbf{1}_n \otimes \mu', \mathbf{I}_n \otimes \Sigma, \mathbf{D}' \otimes \mathbf{E}', 0, \nu)$. Then the moment generating function of $\mathbf{X}$ is obtained from the fact that $vec(\mathbf{X}') \sim CSN_{np;1}(vec(\mathbf{1}_n' \otimes \mu), \Sigma \otimes \mathbf{I}_n, \mathbf{E}' \otimes \mathbf{D}', 0, \nu)$ and it is expressed as (cf. Domínguez-Molina, González-Farías, and Gupta 2007, Proposition 3.2)

$$M_X(\mathbf{T}) = E(etr(\mathbf{X}'\mathbf{T})) = 2\Phi\left((vec(\mathbf{T}'))'vec(\Sigma\mathbf{E}\mathbf{D}');0,\nu+\mathbf{D}'\mathbf{D}\mathbf{E}'\Sigma\mathbf{E}\right) \tag{11.31}$$

$$\times\, exp\left((vec(\mathbf{T}'))'vec(\mathbf{1}_n'\otimes\mu)+\frac{1}{2}(vec(\mathbf{T}'))'(\Sigma\otimes\mathbf{I}_n)vec(\mathbf{T}')\right). \tag{11.32}$$

The application of the rules of matrix differentiation (see Harville 1997, Chap. 15) leads to

$$\begin{aligned}\frac{\partial(M_X(\mathbf{T}))}{\partial(vec(\mathbf{T}'))} &= M_X(\mathbf{T})\left(vec(\mathbf{1}_n'\otimes\mu)+(\Sigma\otimes\mathbf{I}_n)vec(\mathbf{T}')\right) \\ &+ exp\left((vec(\mathbf{T}'))'vec(\mathbf{1}_n'\otimes\mu)+\frac{1}{2}(vec(\mathbf{T}'))'(\Sigma\otimes\mathbf{I}_n)vec(\mathbf{T}')\right) \\ &\times 2(\nu+\mathbf{D}'\mathbf{D}\mathbf{E}'\Sigma\mathbf{E})^{-1/2} \\ &\times \phi\left((vec(\mathbf{T}'))'vec(\Sigma\mathbf{E}\mathbf{D}');0,\nu+\mathbf{D}'\mathbf{D}\mathbf{E}'\Sigma\mathbf{E}\right)vec(\Sigma\mathbf{E}\mathbf{D}').\end{aligned} \tag{11.33}$$

Let $\mathbf{a}$ be an arbitrary vector which is orthogonal to $vec(\Sigma\mathbf{E}\mathbf{D}')$ and has a norm equal to one, i.e.

$$\mathbf{a}'vec(\Sigma\mathbf{E}\mathbf{D}') = 0 \quad \text{and} \quad \mathbf{a}'\mathbf{a} = 1. \tag{11.34}$$

Then from (11.33) we get

$$\mathbf{a}'\frac{\partial(M_X(\mathbf{T}))}{\partial(vec(\mathbf{T}'))} - M_X(\mathbf{T})\mathbf{a}'\left(vec(\mathbf{1}_n'\otimes\mu)+(\Sigma\otimes\mathbf{I}_n)vec(\mathbf{T}')\right) = 0. \tag{11.35}$$

Equation (11.35) is used for the derivation of the test statistic defined by

$$T_{n,w} = np\int_{-\infty}^{\infty}\hat{D}_n^2 w(\mathbf{T})d\mathbf{T}, \tag{11.36}$$

where $w(\mathbf{T})$ is a suitable weight function and $\hat{D}_n^2$ is the empirical counterpart of D_n^2 expressed as

$$D_n^2 = \left(\mathbf{a}'\frac{\partial(M_X(\mathbf{T}))}{\partial(vec(\mathbf{T}'))} - M_X(\mathbf{T})\mathbf{a}'\left(vec(\mathbf{1}_n'\otimes\mu)+(\Sigma\otimes\mathbf{I}_n)vec(\mathbf{T}')\right)\right)^2, \tag{11.37}$$

which is obtained by substituting for the moment generating function $M_X(\mathbf{T})$ by the empirical one given by

$$\hat{M}_X(\mathbf{T}) = etr(\mathbf{X}'\mathbf{T}) = exp\left((vec(T'))'vec(\mathbf{X}')\right). \tag{11.38}$$

If the assumption of the matrix variate closed skew normal distribution is valid for the data analyzed then the values of the statistic $T_{n,w}$ should be close to zero. On the other hand, a large value of $T_{n,w}$ is a signal for the rejection of the null hypothesis.

Next, we rewrite (11.37). The aim is to obtain an analytical expression that is easy to use in practice. Following Meintanis and Hlávka (2010) we let

$$w(\mathbf{T}) = \prod_{i=1}^{n}\prod_{j=1}^{p} \frac{exp(-t_{ij}^2)}{\sqrt{\pi}} = exp\left(-(vec(\mathbf{T}'))'vec(\mathbf{T}')\right)/\pi^{np/2}. \tag{11.39}$$

Inserting (11.39) into (11.37) and using the identity

$$\frac{\partial \hat{M}_X(\mathbf{T})}{\partial vec(\mathbf{T}')} = exp\left((vec(\mathbf{T}'))'vec(\mathbf{X}')\right) vec(\mathbf{X}') \tag{11.40}$$

we get

$$\begin{aligned} T_{n,w} &= \frac{np}{\pi^{np/2}} \int_{-\infty}^{\infty} \left(\mathbf{a}'vec(\mathbf{X}') - \mathbf{a}'\left(vec(\mathbf{1}_n' \otimes \mu) + (\Sigma \otimes \mathbf{I}_n)vec(\mathbf{T}')\right)\right)^2 \\ &\times exp\left(2(vec(\mathbf{T}'))'vec(\mathbf{X}')\right) exp\left(-(vec(\mathbf{T}'))'vec(\mathbf{T}')\right) d\mathbf{T} \\ &= \frac{np}{\pi^{np/2}} exp\left(-(vec(\mathbf{X}'))'vec(\mathbf{X}')\right) \\ &\times \int_{-\infty}^{\infty} \left(\mathbf{a}'(vec(\mathbf{X}') - vec(\mathbf{1}_n' \otimes \mu)) - \mathbf{a}'(\Sigma \otimes \mathbf{I}_n)vec(\mathbf{T}')\right)^2 \\ &\times exp\left(-(vec(\mathbf{T}') - vec(\mathbf{X}'))'(vec(\mathbf{T}') - vec(\mathbf{X}'))\right) d\mathbf{T} \end{aligned}$$

The last integral is simply

$$\pi^{np/2} E\left(\left(\mathbf{a}'(vec(\mathbf{X}') - vec(\mathbf{1}_n' \otimes \mu)) - \mathbf{a}'(\Sigma \otimes \mathbf{I}_n)vec(\mathbf{Y}')\right)^2\right),$$

where $vec(\mathbf{Y}') \sim N_{np}\left(vec(\mathbf{X}'), \frac{1}{2}\mathbf{I}_{np}\right)$. Hence,

$$\begin{aligned} T_{n,w} &= npexp\left(-(vec(\mathbf{X}'))'vec(\mathbf{X}')\right) \\ &\times E\left(\left(\mathbf{a}'(vec(\mathbf{X}') - vec(\mathbf{1}_n' \otimes \mu)) - \mathbf{a}'(\Sigma \otimes \mathbf{I}_n)vec(\mathbf{Y}')\right)^2\right) \\ &= npexp\left(-(vec(\mathbf{X}'))'vec(\mathbf{X}')\right) \\ &\times \left((\mathbf{a}'(vec(\mathbf{X}') - vec(\mathbf{1}_n' \otimes \mu)) - \mathbf{a}'(\Sigma \otimes \mathbf{I}_n)vec(\mathbf{X}')\right)^2 \\ &+ \mathbf{a}'(\Sigma^2 \otimes \mathbf{I}_n)\mathbf{a}/2\right). \end{aligned}$$

Hence, for testing the null hypothesis of the matrix variate closed skew normal distribution, i.e.,

$$H_0 : \mathbf{X} \sim CSN_{n,p;1,1}(\mathbf{1}_n \otimes \mu', \mathbf{I}_n \otimes \Sigma, \mathbf{D}' \otimes \mathbf{E}', 0, \nu) \tag{11.41}$$

against

$$H_1 : \mathbf{X} \not\sim CSN_{n,p;1,1}(\mathbf{1}_n \otimes \mu', \mathbf{I}_n \otimes \Sigma, \mathbf{D}' \otimes \mathbf{E}', 0, \nu)$$

the following test statistic is derived

$$\begin{aligned} T_{n,w} &= npexp\left(-(vec(\mathbf{X}'))'vec(\mathbf{X}')\right) \\ &\times \left((\mathbf{a}'(vec(\mathbf{X}') - vec(\mathbf{1}_n' \otimes \mu)) - \mathbf{a}'(\Sigma \otimes \mathbf{I}_n)vec(\mathbf{X}'))^2 + \mathbf{a}'(\Sigma^2 \otimes \mathbf{I}_n)\mathbf{a}/2\right). \end{aligned} \tag{11.42}$$

It is noted that the distribution of the test statistic $T_{n,w}$ is not trivial under both the null and the alternative hypothesis. The situation is even more complicated when the parameter uncertainty is taken into account, i.e. if the parameters μ, Σ, and $\mathbf{E}$ are replaced by the corresponding estimators derived in Sect. 11.5.1.

For the application of the suggested testing procedure in a practical situation we use parametric bootstrap as follows:

(1) In the first stage a sample of size N from the matrix variate closed skew normal distribution is generated with the corresponding parameters. For the simulation of the sample we make use of the relationship between the matrix variate closed skew normal distribution and the corresponding multivariate one given by (cf. Domínguez-Molina, González-Farías, and Gupta 2007)

$$\text{If} \quad \mathbf{X} \sim CSN_{n,p;1,1}(\mathbf{1}_n \otimes \mu', \mathbf{I}_n \otimes \Sigma, \mathbf{D}' \otimes \mathbf{E}', 0, \nu)$$

$$\text{then} \quad vec(\mathbf{X}') \sim CSN_{np;1,1}(\mathbf{1}_n' \otimes \mu, \Sigma \otimes \mathbf{I}_n, \mathbf{E}' \otimes \mathbf{D}', 0, \nu).$$

Then the stochastic representation of the multivariate closed skew normal distribution, namely (see, e.g., Domínguez-Molina, González-Farías, and Gupta 2007, Proposition 2.1)

$$\begin{aligned} vec(\mathbf{X}') &\approx \mathbf{1}_n' \otimes \mu + (\Sigma^{-1} \otimes \mathbf{I}_n + (\mathbf{E} \otimes \mathbf{D})(\mathbf{E}' \otimes \mathbf{D}')/\nu)^{-1/2} vec(\mathbf{Z}') \\ &+ (\Sigma \otimes \mathbf{I}_n)(\mathbf{E} \otimes \mathbf{D})(\nu + \mathbf{D}'\mathbf{D}\mathbf{E}'\Sigma\mathbf{E})^{-1/2}|z_0| \\ &= \mathbf{1}_n' \otimes \mu + (\Sigma^{-1} \otimes \mathbf{I}_n + (\tilde{\mathbf{E}} \otimes \mathbf{D})(\tilde{\mathbf{E}}' \otimes \mathbf{D}')(1 + \tilde{\nu}))^{-1/2} vec(\mathbf{Z}') \\ &+ (\Sigma \otimes \mathbf{I}_n)(\tilde{\mathbf{E}} \otimes \mathbf{D})|z_0|, \end{aligned}$$

where $\tilde{\nu} = \mathbf{D}'\mathbf{D}\mathbf{E}'\Sigma\mathbf{E}/\nu$, $vec(\mathbf{Z}') \sim N_{np}(\mathbf{0}, \mathbf{I}_{np})$, $z_0 \sim \mathcal{N}(0,1)$ and $vec(\mathbf{Z}')$, z_0 are independent, is applied. In Sect. 11.5.1 we derive estimators for μ, Σ, and $\tilde{\mathbf{E}}$, but it is pointed out that it is not easy to estimate ν or/and $\tilde{\nu}$. Thus, several values of $\tilde{\nu}$ are considered in the first stage of the bootstrap procedure, and for each value of $\tilde{\nu}$ a sample of size N is generated.

Table 11.1 The estimated parameters of the skew normal distribution ($\tilde{c} = 1.0287$ and $\hat{\tilde{a}} = 0.9560254$)

$\bar{\mathbf{x}}$	$\hat{\mu}$	$\hat{\tilde{\mathbf{E}}}$	$\hat{\mathbf{V}}$				
−0.003192	−0.012447	0.0191695	0.0032143	0.0018796	0.0015245	0.0007313	0.0013129
−0.001172	−0.011944	0.0223120	0.0018796	0.0022471	0.0014159	0.0006440	0.0009754
0.001057	−0.011135	0.0252522	0.0015245	0.0014159	0.0023945	0.0005513	0.0012291
0.0003413	−0.008289	0.0178766	0.0007313	0.0006440	0.0005513	0.0011300	0.0006633
−0.000847	−0.016930	0.0333126	0.0013129	0.0009754	0.0012291	0.0006633	0.0022025

(2) For each element of the generated sample the value of the test statistic $T_{n,w;k}(\tilde{v})$, $k = 1, \ldots, N$ is calculated.
(3) Finally, the sequences $T_{n,w;k}(\tilde{v})$ are used for determining the critical values of the test for each $\tilde{v}$.
(4) The p-values can alternatively be computed in the third stage. It is performed by including the value of the test statistic calculated from the data into the sequence $T_{n,w;k}(\tilde{v})$ for each $\tilde{v}$ and determining its position in the corresponding ordered sequence $T_{n,w}^{(k)}(\tilde{v})$.

11.5.3 Application to Some Stocks in the Dow Jones Index

In order to get a better understanding for the results presented in Sect. 11.5 we consider an example of real data in this section. Further results of applications of skew distributions in portfolio theory were considered by Athayde and Flôres (2004), Patton (2004), Mencía and Sentana (2009), Adcock (2010). We make use of weekly data of five stocks which are included in the Dow Jones index. The choice of data is motivated by the paper of Jondeau and Rockinger (2006) who showed that the skewness in weekly stock returns of the USA companies can significantly influence the portfolio selection on this market. The assets in the study are the stocks of the Boeing Co, Disney (Walt) Co, Hewlett-Packard, Altria Group INC, and Microsoft Corp. The data is taken for the period from the 3rd of January 2007 to the 5th of October 2009 and it includes $n = 145$ observations for each stock.

In Table 11.1, the estimators of the parameters of the skew normal distribution described in Sect. 11.5.1 are presented. We observe that $\bar{\mathbf{x}}$ provides a poor estimator of the location vector μ. The improved estimator $\hat{\mu}$ consists of much smaller elements which are more than 20 times smaller than the corresponding components of $\bar{\mathbf{x}}$. As a result, the application of the vector $\bar{\mathbf{x}}$ leads to misleading results, i.e. the investor expects more (or less) from the holding asset. Consequently, the decision about buying, holding, or selling the asset could be wrong when $\bar{\mathbf{x}}$ is used for estimating of μ.

In Fig. 11.1a–e, the histograms for the data considered are plotted. The densities presented in the figures are calculated by using the kernel density estimation with

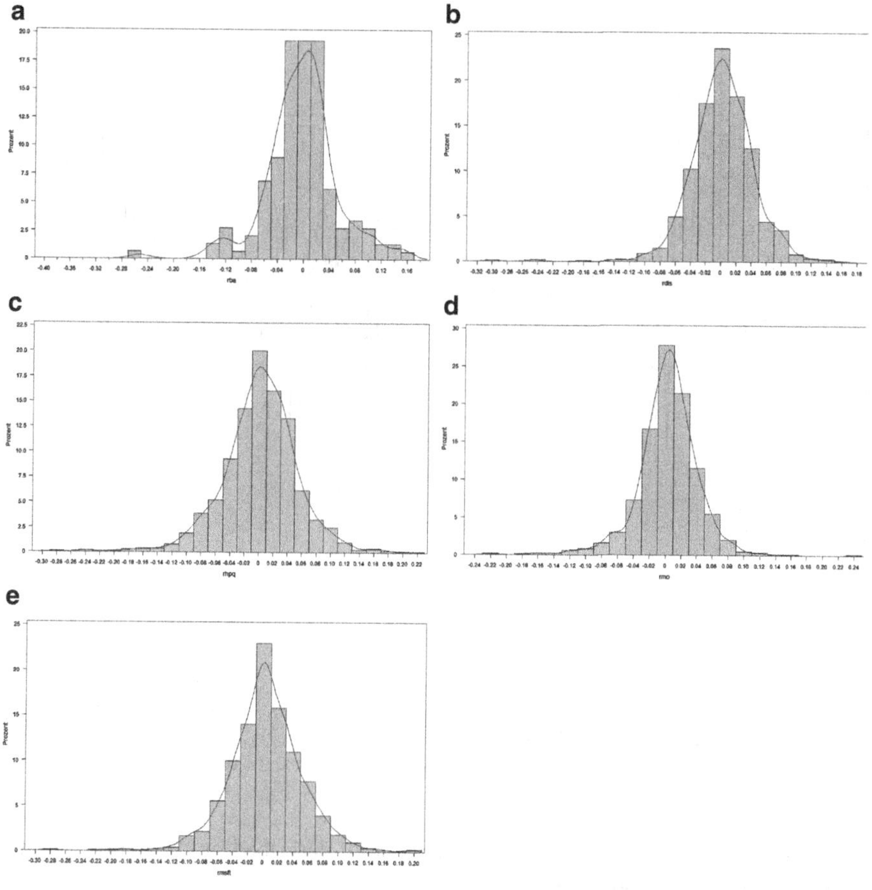

Fig. 11.1 Histograms and the kernel density estimators for the weekly returns of the **(a)** Boeing Co, **(b)** Disney (Walt) Co, **(c)** Hewlett-Packard, **(d)** Altria group INC, and **(e)** Microsoft Corp

the normal kernel. We observe that the densities of the Boeing Co and Altria Group INC returns are skewed to the left, while the densities of the Disney (Walt) Co, Hewlett-Packard, and Microsoft Corp returns are skewed to the right.

Stronger results are presented in Fig. 11.2. Here, we apply the goodness-of-fit test of Sect. 11.5.2 to the data considered with the estimated parameters given in Table 11.1 for $\tilde{\nu} \in [0.1, 10]$. Although, in Table 11.1 the estimator of $\tilde{\mathbf{E}}$ instead of $\mathbf{E}$ is present it can be used since we need only the direction of the vector $\mathbf{E}$ for determining $\mathbf{a}$ which is orthogonal to $vec(\Sigma\mathbf{E}\mathbf{D}') = (w_1, \mathbf{w}_2')'$. The vector $\mathbf{a} = (a_1, \mathbf{a}_2')'/(a_1^2 + \mathbf{a}_2'\mathbf{a}_2)$ is calculated with $\mathbf{a_2} = \mathbf{w}_2$ and $a_1 = -\mathbf{w}_2'\mathbf{w}_2/w_1$.

In Fig. 11.2, we plot the p-values of the test as a function of $\tilde{\nu}$. The calculated value of the test statistic is 0.0018504. We observe that the p-values are large for all of the considered values of $\tilde{\nu}$. They are always larger than 0.948. This result

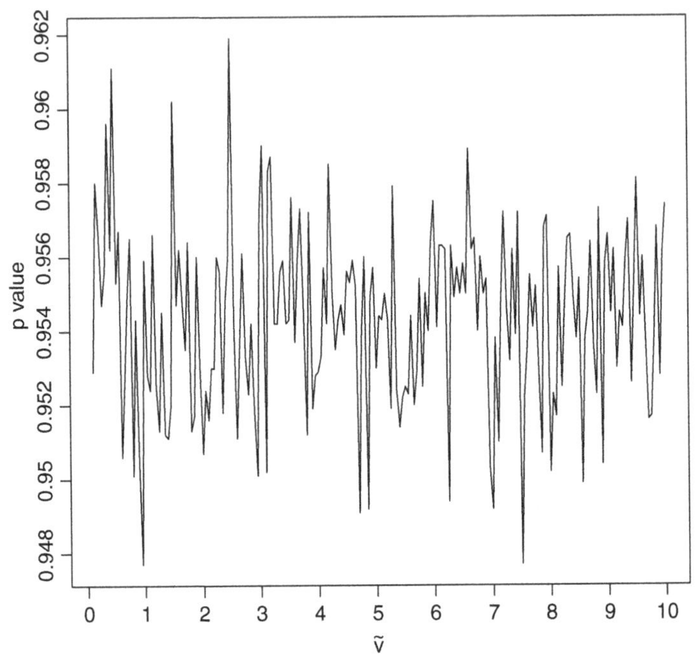

Fig. 11.2 The p-values of the goodness-of-fit test for the matrix variate closed skew normal distribution as a function of $\tilde{v} = \mathbf{D}'\mathbf{D}\mathbf{E}'\Sigma\mathbf{E}/v$ applied to the weekly returns of the Boeing Co, Disney (Walt) Co, Hewlett-Packard, Altria group INC, and Microsoft Corp

confirms that the weekly asset returns can be well described by the matrix variate closed skew normal distribution.

Next, the main results of this section are presented. We apply the results of Theorem 11.8 for studying the impact of skewness on the inference procedures about the GMVP weights. For each estimated weight of the portfolio we test if this weight corresponds to the true value calculated from (11.15), i.c thc following hypothesis is tested

$$H_{0,i} : w_{GMV,i} = w_{0,i} = \tilde{w}_{GMV,i} \qquad \text{against} \qquad H_{1,i} : w_{GMV,i} \neq w_{0,i} = \tilde{w}_{GMV,i} \tag{11.43}$$

for $i = 1,2,3,4,5$. For testing $H_{0,i}$ the following statistic is considered (c.f. Theorem 2)

$$T_i = (n-p)\frac{\left(\mathbf{1}'_p\hat{\mathbf{V}}^{-1}\mathbf{1}_p\right)^2(\hat{w}_{GMV,i} - \tilde{w}_{GMV,i})^2}{\mathbf{1}'_p\hat{\mathbf{V}}^{-1}\mathbf{1}_p;\hat{v}_{ii}^{(-)} - (\Sigma_{j=p}^{k}\hat{v}_{ij}^{(-)})^2}$$

with $\hat{\mathbf{V}}^{-1} = (\hat{v}_{ij}^{(-)})$.

Table 11.2 The estimated and true weights of the global minimum variance portfolio are given in the first two columns ($p = 5$ and $n = 145$). The estimated values of λ for the individual test on the GMVP weight of each asset and the corresponding estimated probability of the type I error are presented in the third column and the fourth column of the table ($\alpha = 0.05$) respectively

$\hat{\mathbf{w}}_{GMV}$	$\tilde{\mathbf{w}}_{GMV}$	$\hat{\lambda}$	$\hat{P}(T > F_{q,n-p;1-\alpha} \mid \mathbf{w}_0)$
−0.047986	−0.103944	0.0061412	0.152132
0.1379042	0.1618053	0.0007404	0.061879
0.1278034	0.1280838	0.00000014	0.050002
0.6523426	0.5494764	0.0195875	0.378095
0.1299362	0.2645785	0.0334062	0.575389

The results are given in Table 11.2. In the first column we present the estimators of the GMVP weights obtained from (11.11) with $\hat{\mathbf{V}}$ as in (11.10). The vector of true weights is obtained from (11.15) by substituting the unknown parameters $\mathbf{V}$ and $\tilde{\mathbf{E}}$ with the corresponding estimators (11.27) and (11.29). We observe that the elements of the vectors $\hat{\mathbf{w}}_{GMV}$ and $\tilde{\mathbf{w}}_{GMV}$ are different. The largest deviation is present for the Microsoft Corp asset while the Altria Group INC asset is ranked second. This result shows that the skewness influences the covariance matrix of the asset returns and as a result it has a significant impact on the portfolio weights. Moreover, the results obtained about the distributional properties of the GMVP weights are in line with the statistical theory. The sample mean and the sample covariance matrix are hardly interpretable if the distribution of data is not symmetric.

In the third column of Table 11.2 we present the estimators of λ for each test (11.43). The largest value of $\hat{\lambda}$ is obtained for the Microsoft Corp asset and it is equal to 0.0334062. Finally, in the last column of Table 11.2 the estimated probabilities of the type I error for each test are given. In the case of the Altria Group INC asset and the Microsoft Corp asset, the estimated probabilities are larger than 37 % with the significance level of 5 %. Hence, because of the skewness, which is ignored when the normal distribution is fitted to the asset returns, the probability of rejection of the null hypothesis is more than seven times larger than the significance level of test.

11.6 General Class of the Skew Elliptically Contoured Distributions

Definition 11.3. A random matrix $\mathbf{Y}$ $(p \times m)$ is said to have a matrix variate extended skew elliptical (MVESE) distribution with pdf generator h and parameters $\mathbf{M}$ $(p \times m)$, $\mathbf{S}$ $(mp \times mp)$, $\mathbf{B}$ $(nq \times mp)$, $\mathbf{L}$ $(q \times n)$, $\mathbf{Q}$ $(nq \times nq)$, where $\mathbf{S} > \mathbf{0}$ and $\mathbf{Q} > \mathbf{0}$, if

$$vec(Y') \sim ESE_{pm,nq}[vec(\mathbf{M}'), \mathbf{S}, \mathbf{B}, vec(\mathbf{L}'), \mathbf{Q}, h].$$

Here, ESE denotes the extended skew-elliptical distribution as given in González-Farías, Domínguez-Molina, and Gupta (2004a). We denote this relation by $\mathbf{Y} \sim ESE_{p,m;q,n}(\mathbf{M}, \mathbf{S}, \mathbf{B}, \mathbf{L}, \mathbf{Q}, h)$.

Using similar arguments to those employed in Sect. 11.2, it is also possible to derive the matrix variate skew-elliptical distribution for which the parameter matrix $\mathbf{B} = \mathbf{D} \otimes \mathbf{E}$, where $\mathbf{D}$ $(n \times p)$ and $\mathbf{E}$ $(q \times m)$ are arbitrary matrices.

As partial cases of this general family of distributions in the next two subsections we consider the skew t-distribution and the skew Cauchy distribution which are investigated by Ramos-Quiroga (2004) in detail.

11.6.1 Multivariate Skew t-Distribution

The multivariate skew normal distribution has been studied by Gupta and Huang (2002), Gupta and Kollo (2003), Azzalini and Dalla Valle (1996), Azzalini (2005), Arellano-Valle and Azzalini (2006), Pewsey (2000), and its applications are given in Azzalini and Capitanio (1999). This class of distributions includes the normal distribution and has some properties like the normal and yet is skew. It is useful in studying robustness. Following Gupta and Kollo (2003), the random vector $\mathbf{x}$ $(p \times 1)$ is said to have a multivariate skew normal distribution if it is continuous and its density function is given by

$$f_{\mathbf{x}}(\mathbf{x}) = 2\phi_p(\mathbf{x};\Sigma)\Phi(\alpha'\mathbf{x}), \ \mathbf{x} \in I\!R^p, \tag{11.44}$$

where $\Sigma > 0$, $\alpha \in I\!R^p$. It is denoted by $\mathbf{x} \sim SN_p(\Sigma, \alpha)$, to mean that the random vector $\mathbf{x}$ has p-variate skew normal density (11.44). The moment generating function of $\mathbf{x}$ is

$$M_{\mathbf{x}}(\mathbf{t}) = 2e^{\frac{1}{2}\mathbf{t}'\Sigma\mathbf{t}}\Phi\left(\frac{\alpha'\Sigma\mathbf{t}}{(1+\alpha'\Sigma\alpha)^{\frac{1}{2}}}\right), \ \mathbf{t} \in I\!R^p. \tag{11.45}$$

This distribution family is not in the elliptically contoured family (see Gupta and Nagar 2000).

The mean vector and the covariance matrix of $\mathbf{x}$ are given by

$$\mu_{\mathbf{x}} = E(\mathbf{x}) = \sqrt{\frac{2}{\pi}}\,\delta$$

$$\operatorname{cov}(\mathbf{x}) = \Sigma - \mu_x\mu_x'$$

where $\delta = (1+\alpha'\Sigma\alpha)^{-\frac{1}{2}}\alpha$. Note that the mean vector μ_x given by Azzalini and Capitanio (1999) is in error. Gupta and Kollo (2003) have further defined the

$SN_p(\mu, \Sigma, \alpha)$, family where μ is the location parameter as $\mathbf{y} = \mathbf{x} + \mu \sim SN_p(\mu, \Sigma, \alpha)$. Azzalini and Dalla Valle (1996) defined $SN_p(\Omega, \alpha)$ where Ω is the correlation matrix.

It may be noted that in the multivariate case very few models are available for dealing with non-normal data, especially so for modeling the skewness. The univariate skew distributions have been studied by Gupta, Chang, and Huang (2002). Next, we define the general multivariate skew t-distributions.

Definition 11.4. Let $\mathbf{x} = (x_1, \dots, x_p)' \sim SN_p(\Sigma, \alpha)$ and $W \sim \chi^2_\nu$ independent of $\mathbf{x}$. Then the joint distribution of $y_j = x_j/\sqrt{W/\nu},\ j = 1, \dots, p$ is defined as the multivariate skew t-distribution with ν degrees of freedom.

We denote it as $\mathbf{y} = (y_1, \dots, y_p)' \sim SMT_\nu(\alpha)$. This may be called multivariate skew t with common denominator as in Cornish (1954), Dunnett and Sobel (1954), and Laurent (1955). Branco and Dey (2001) derive similar result by considering t-distribution as a special case of scale mixture of normal distribution.

Now we can find the join density of $\mathbf{y}$. The joint p.d.f. of $y_1, y_2, \dots, y_p$ is given by

$$f_\nu(\mathbf{y}; \alpha) = \frac{2}{(2\pi\nu)^{\frac{p}{2}} |\Sigma|^{\frac{1}{2}} 2^{\frac{\nu}{2}} \Gamma\left(\frac{\nu}{2}\right)} \int_0^\infty w^{\frac{\nu+p}{2}-1} e^{-(1+\nu^{-1}\mathbf{y}'\Sigma^{-1}\mathbf{y})\frac{w}{2}} \Phi\left(\sqrt{\frac{w}{\nu}}\alpha'\mathbf{y}\right) d\mathbf{y} \tag{11.46}$$

To evaluate the integral in (11.46) we need the following result.

Lemma 11.1. *Let $U \sim \chi^2_k$, then*

$$E_U \Phi(a\sqrt{U} + b) = F'_{T_{k,b^2}}(a\sqrt{k}) \tag{11.47}$$

where $\Phi(\cdot)$ is the c.d.f. of standard normal distribution and $F'_{T_{k,\delta^2}}(\cdot)$ is the c.d.f. of non-central t-distribution with k degrees of freedom and non-centrality parameter $\delta^2 = b^2$.

PROOF: Let $Z \sim N(0,1)$, then

$$\begin{aligned} E_U \Phi(a\sqrt{U} + b) &= E_U P[Z < a\sqrt{U} + b|U] \\ &= E_U P\left[\frac{Z-b}{\sqrt{U/k}} < a\sqrt{k}|U\right] \\ &= P[T'_k \le a\sqrt{k}] \\ &= F'_{T_{k,b^2}}(a\sqrt{k}). \end{aligned}$$

■

Now using the above lemma we can write (11.46) as

$$f_\nu(\mathbf{y};\alpha) = \frac{2(1+\nu^{-1}\mathbf{y}'\Sigma^{-1}\mathbf{y})^{-\frac{\nu+p}{2}}\Gamma\left(\frac{\nu+p}{2}\right)}{(\pi\nu)^{\frac{p}{2}}|\Sigma|^{\frac{1}{2}}\Gamma\left(\frac{\nu}{2}\right)} \cdot E_U \Phi\left(\frac{\alpha'\mathbf{y}}{(\nu+\mathbf{y}'\Sigma^{-1}\mathbf{y})^{\frac{1}{2}}}\sqrt{U}\right)$$

where $U \sim \chi^2_{\nu+p}$. Hence

$$f_\nu(\mathbf{y};\alpha) = \frac{2(1+\nu^{-1}\mathbf{y}\Sigma^{-1}\mathbf{y})^{-\frac{\nu+p}{2}}\Gamma\left(\frac{\nu+p}{2}\right)}{(\pi\nu)^{\frac{p}{2}}|\Sigma|^{\frac{1}{2}}\Gamma\left(\frac{\nu}{2}\right)} F_{T_{\nu+p}}\left(\frac{\alpha'\mathbf{y}}{(\nu+\mathbf{y}'\Sigma^{-1}\mathbf{y})^{\frac{1}{2}}}\sqrt{\nu+p}\right) \tag{11.48}$$

where $F_{T_k}(\cdot)$ is the c.d.f. of central t-distribution with k degrees of freedom. Or equivalently $\mathbf{Y} \sim SMT_\nu(\alpha)$ if its p.d.f. is given by

$$f_\nu(\mathbf{y};\alpha) = 2f_{T_\nu}(\mathbf{y})F_{T_{\nu+p}}\left(\frac{\alpha'\mathbf{y}}{(\nu+\mathbf{y}'\Sigma^{-1}\mathbf{y})^{\frac{1}{2}}}\sqrt{\nu+p}\right), \quad \mathbf{y} \in \mathbb{R}^p. \tag{11.49}$$

From the definition and the density function (11.49) of $SMT_\nu(\alpha)$, the following properties can be easily seen.

(i) $SMT_\nu(0) \equiv MT_\nu-$ multivariate t-distribution with ν degrees of freedom, where usually Σ is replaced by the corresponding correlation matrix R.
(ii) $\lim_{\nu\to\infty} f_\nu(\mathbf{y},\alpha) = 2\phi_p(\mathbf{y};\Sigma)\Phi(\alpha'\mathbf{y})$ i.e. the multivariate skew t-distribution tends to the multivariate skew normal distribution as $\nu \to \infty$.
(iii) Note that $Y_j^2 = \frac{X_j^2}{W/\nu} \sim F(1,\nu) \equiv t_\nu^2$ since $X_j^2 \sim \chi_1^2$. Here $F(m,n)$ denotes the Snedecor's F-distribution with degrees of freedom m and n. Furthermore, the joint distribution of $(Y_1^2,\ldots,Y_p^2)$ is multivariate F-distribution with parameters $1,1,\ldots,1,\nu+p$ (see Finney 1941).
(iv) The quadratic form $\mathbf{y}'\Sigma^{-1}\mathbf{y} \sim pF(p,\nu)$, since

$$\mathbf{y}'\Sigma^{-1}\mathbf{y} = \frac{\mathbf{x}'\Sigma^{-1}\mathbf{x}}{W/\nu},$$

and $\mathbf{x}'\Sigma^{-1}\mathbf{x} \sim \chi_p^2$ (see Gupta and Kollo 2003; Azzalini and Dalla Valle 1996). Note that this distribution does not depend on α. This therefore extends the multivariate normal theory to the multivariate skew normal or the multivariate t- to multivariate skew t-distribution.

11.6.1.1 Expected Values

Let $\mathbf{y} \sim SMT_\nu(\alpha)$. Since $\mathbf{y} = \nu^{1/2}W^{-1}\mathbf{x}$ and the mean vector and the covariance matrix of $\mathbf{x} \sim SN_p(\Sigma,\alpha)$ are given by Gupta and Kollo (2003) and Azzalini and Dalla Valle (1996), we get

$$E(\mathbf{y}) = \sqrt{\frac{2\nu}{\pi}}\frac{\Sigma\alpha}{(\nu-2)(1+\alpha'\Sigma\alpha)^{\frac{1}{2}}}, \quad \nu > 2 \tag{11.50}$$

$$\text{cov}(\mathbf{y}) = \frac{\nu}{(\nu-2)(\nu-4)}\Sigma\left[I - \frac{2(\nu-4)\alpha\alpha'\Sigma}{\pi(\nu-2)(1+\alpha'\Sigma\alpha)}\right], \quad \nu > 4. \tag{11.51}$$

For $\alpha = \mathbf{0}$, from (11.50) and (11.51) we get the moments of multivariate t-distribution. Further, since $x_j = x_j/\sqrt{\frac{W}{\nu}}$, $j = 1,\dots,p$, and W and x_j are independent, the product moments of $y_1, y_2, \dots, y_p$ are easily found as

$$\begin{aligned}
\mu'_{r_1,r_2,\dots,r_p} &= E\left[\prod_{j=1}^{p} y_j^{r_j}\right] \\
&= \nu^{\frac{1}{2}\Sigma_1^p r_j} E\left(W^{-\Sigma_1^p r_j}\right) E\left(\prod_1^p x_j^{r_j}\right), \\
&= \left(\frac{\sqrt{\nu}}{2}\right)^{\frac{1}{2}\Sigma_1^p r_j} \frac{\Gamma\left(\frac{\nu}{2} - \Sigma_1^p r_j\right)}{\Gamma\left(\frac{\nu}{2}\right)} E\left(\prod_1^p x_j^{r_j}\right), \quad \sum_1^p r_j < \frac{1}{2}\nu.
\end{aligned} \tag{11.52}$$

If $x_1, \dots, x_p$ are mutually independent, then

$$\mu'_{r_1,\dots,r_p} = \left(\frac{\nu}{4}\right)^{\frac{1}{2}\sum_1^p r_j} \frac{\Gamma\left(\frac{\nu}{2} - \prod_1^p r_j\right)}{\Gamma\left(\frac{\nu}{2}\right)} \left\{\prod_{j=1}^{p} E(X_j^{r_j})\right\}. \tag{11.53}$$

The marginal distribution of x_j is univariate skew normal and its moments can be computed easily.

11.6.2 Multivariate Skew Cauchy Distribution

The special case of the multivariate skew t-distribution for $\Sigma = I$ is

$$\begin{aligned}
f_\nu(\mathbf{y}, \alpha) &= \frac{2\Gamma\left(\frac{\nu+p}{2}\right)}{(\pi\nu)^{\frac{p}{2}}\Gamma\left(\frac{\nu}{2}\right)}\left(1 + \nu^{-1}\sum_1^p y_j^2\right)^{-\frac{\nu+p}{2}} \\
&\times F_{T_{\nu+p}}\left(\frac{\alpha'\mathbf{y}}{\left(\nu + \sum_1^p y_j^2\right)^{\frac{1}{2}}}\sqrt{\nu+p}\right), \quad \mathbf{y} \in I\!R^p,
\end{aligned} \tag{11.54}$$

and may be called standard multivariate skew t-distribution. For $v = 1$, this distribution is defined as the multivariate skew Cauchy distribution with density

$$f(\mathbf{y};\alpha) = \frac{2\Gamma\left(\frac{p+1}{2}\right)}{(\pi)^{\frac{p+1}{2}}}\left(1+\sum_1^p y_j^2\right)^{-\frac{p+1}{2}} F_{T_{p+1}}\left(\frac{\alpha'\mathbf{y}\sqrt{p+1}}{\left(1+\sum_1^p y_j^2\right)^{\frac{1}{2}}}\right), \quad \mathbf{y} \in I\!R^p. \tag{11.55}$$

Definition 11.5. The random vector $\mathbf{y}$ is said to have multivariate skew Cauchy distribution if its density function is given by (11.55).

It will be denoted by $\mathbf{y} \sim SMC(\alpha)$. From the density (11.55) it is seen that the multivariate skew-Cauchy distribution does not belong to the class of spherical distributions (see Gupta and Nagar 2000) whereas the multivariate Cauchy does.

It may be noted that the $SMC(0)$, which is

$$f(\mathbf{y};0) = \frac{\Gamma\left(\frac{p+1}{2}\right)}{(\pi)^{\frac{p+1}{2}}}\left(1+\sum_1^p y_j^2\right)^{-\frac{p+1}{2}}, \quad -\infty < y_j < \infty \tag{11.56}$$

gives the multivariate Cauchy density. Then (11.56), for $p = 1$, gives the standard Cauchy $C(0,1)$, density. However the univariate skew Cauchy, obtained from (11.55) for $p = 1$ is

$$f(y;\alpha) = \frac{2}{\pi(1+y^2)}F_{T_2}\left(\sqrt{2}\alpha\frac{y}{\sqrt{1+y^2}}\right), \quad -\infty < y < \infty. \tag{11.57}$$

This definition of univariate skew Cauchy distribution is not unique (see Gupta, Chang, and Huang 2002).

References

Abramowitz, M., Stegun, I.A.: Handbook of Mathematical Functions with Formulas, Graphs, and Mathematical Tables. Dover Publications, Inc., New York (1965)

Adcock, C.J.: Exploiting skewness to build an optimal hedge fund with a currency overlay. European Journal of Finance **11**, 445–462 (2005)

Adcock, C.J.: Asset pricing and portfolio selection based on the multivariate extended skew-Student-t distribution. Annals of Operation Research **176**, 221–234 (2010)

Aigner, D., Lovell, C.A.K., Schmidt, P.: Formulation and estimation of stochastic frontier production function models. Journal of Econometrics **6**, 21–37 (1977)

Andersen, T.G., Bollerslev, T., Diebold, F.X., Ebens, H.: The distribution of stock return volatility. Journal of Financial Economics **61**, 43–76 (2001)

Andersen, T.G., Bollerslev, T., Diebold, F.X.: Parametric and nonparametric measurements of volatility. In: Aït-Sahalia Y., Hansen L.P. (eds.): Handbook of financial econometrics. North-Holland, Amsterdam (2005)

Anderson, T.W.: An Introduction to Multivariate Statistical Analysis, 3rd Edition. Wiley, New York (2003)

Anderson, T.W., Fang, K.T.: Inference in multivariate elliptically contoured distributions based on maximum likelihood. Technical Report 1. Department of Statistics, Stanford University, California (1982a) (Reprinted in Fang and Anderson (1990))

Anderson, T.W., Fang, K.T.: On the theory of multivariate elliptically contoured distributions and their applications. Technical Report 54. Department of Statistics, Stanford University, California (1982b) (Reprinted in Fang and Anderson (1990))

Anderson, T.W. and Fang, K.T.: Cochran's theorem for elliptically contoured distributions. Sankhya, Ser. A **49**, 305–315 (1987)

Anderson, T.W., Fang, K.T., Hsu, H.: Maximum likelihood estimates and likelihood-ratio criteria for multivariate elliptically contoured distributions. The Canadian Journal of Statistics **14**, 55–59 (1986)

Arellano-Valle, R.B., Azzalini, A.: On the unification of families of skew-normal distributions. Scandinavian Journal of Statistics **33**, 561–574 (2006)

Athayde, G.M., Flôres, R.G.: Finding a maximum skewness portfolio- a general solution to three-moments portfolio choice. Journal of Economic Dynamics and Control **28**, 1335–1352 (2004)

Azzalini, A.: The skew-normal distribution and related multivariate families. Scandinavian Journal of Statistics **32**, 159–188 (2005)

Azzalini, A., Capitanio, A.: Statistical applications of the multivariate skew normal distribution. Journal of the Royal Statistal Society B **61**, 579–602 (1999)

Azzalini, A., Dalla Valle, A.: The multivariate skew normal distribution. Biometrika **83**, 715–726 (1996)

A.K. Gupta et al., *Elliptically Contoured Models in Statistics and Portfolio Theory*, DOI 10.1007/978-1-4614-8154-6, © Springer Science+Business Media New York 2013

Barberis, N.: Investing for the long run when returns are predictable. The Journal of Finance **55**, 225–264 (1999)

Baringhaus, L.: Testing for spherical symmetry of a multivariate distributions. The Annals of Statistics **19**, 899–917 (1991)

Basak, G.K., Jagannathan, R., Ma, T.: Estimation the Risk in Sample Efficient Portfolios. Working Paper, Northwestern University (2005)

Bentler, P.M., Berkane, M.: Developments in the elliptical theory generalization of normal multivariate analysis. American Statistical Association, Proceedings of Social Statistics Section, 291–295 (1985)

Beran, R.: Testing ellipsoidal symmetry of a multivariate density. The Annals of Statistics **7**, 150–162 (1979)

Berk, J.B.: Necessary conditions for the CAPM. Journal of Economic Theory **73**, 245–257 (1997)

Berkane, M., Bentler, P.M.: Moments of elliptically distributed random variates. Statistics and Probability Letters **4**, 333–335 (1986a)

Berkane, M., Bentler, P.M.: Characterizing parameters of elliptical distributions. American Statistical Association, Proceedings of Social Statistics Section, 278–279 (1986b)

Best, M.J., Grauer, R.R.: On the sensitivity of mean-variance-efficient portfolios to changes in asset means: some analytical and computational results. Review of Financial Studies **4**, 315–342 (1991)

Billingsley, P.: Probability and Measure. Wiley, New York (1979)

Blattberg, R.C., Gonedes, N.J.: A comparison of the stable and student distributions as statistical models for stock prices, Journal of Business **47**, 244–280 (1974)

Bodnar, O.: Sequential procedures for monitoring covariances of asset returns. In: Gregoriou, G.N. (ed.) Advances in Risk Management, pp. 241–264. Palgrave, London, (2007)

Bodnar, O.: Sequential surveillance of the tangency portfolio weights. International Journal of Theoretical and Applied Finance **12**, 797–810 (2009)

Bodnar, O., Bodnar, T.: On the unbiased estimator of the efficient frontier. International Journal of Theoretical and Applied Finance **13**, 1065–1073 (2010)

Bodnar, T., Gupta, A.K.: Construction and Inferences of the Efficient Frontier in Elliptical Models. Journal of the Japan Statistical Society **39**, 193–207 (2009)

Bodnar, T., Gupta, A.K.: Robustness of the Inference Procedures for the Global Minimum Variance Portfolio Weights in a Skew Normal Model. To appear in European Journal of Finance. DOI: 10.1080/1351847X.2012.696073 (2013)

Bodnar, T., Schmid, W.: The distribution of the sample variance of the global minimum variance portfolio in elliptical models, Statistics **41**, 65–75 (2007)

Bodnar, T., Schmid, W.: A test for the weights of the global minimum variance portfolio in an elliptical model. Metrika **67**, 127–143 (2008a)

Bodnar, T., Schmid, W.: Estimation of optimal portfolio compositions for gaussian returns, Statistics & Decisions **26**, 179–201 (2008b)

Bodnar, T., Schmid, W.: Econometrical analysis of the sample efficient frontier, European Journal of Finance **15**, 317–335 (2009)

Bollerslev, T.: Generalized autoregressive conditional heteroscedasticity, Journal of Econometrics **31**, 307–327 (1986)

Branco, M.D., Dey, D.K.: A general class of multivariate skew-elliptical distributions. Journal of Multivariate Analysis **79**, 99–113 (2001)

Britten-Jones, M.: The sampling error in estimates of mean-variance efficient portfolio weights. Journal of Finance **54**, 655–671 (1999)

Browne, M.W.: Asymptotically distribution-free methods for the analysis of covariance structures. British Journal of Mathematical and Statistical Pschology **37**, 62–83 (1984)

Cabras, S., Castellanos, M.E.: Default Bayesian goodness-of-fit tests for the skew-normal model. Journal of Applied Statistics **36**, 223–232 (2009)

Cacoullos, T., Koutras, M.: Quadratic forms in spherical random variables: generalized noncentral χ^2 distribution. Naval Research Logistics Quarterly **41**, 447–461 (1984)

Cacoullos, T., Koutras, M.: Minimum-distance discrimination for spherical distributions. In: Matusita, K. (ed) Statistical Theory and Data Analysis, pp. 91–102. Elsevier Science Publishers B.V., North Holland (1985)

Cambanis, S., Huang, S., Simons, G.: On the theory of elliptically contoured distributions. Journal of Multivariate Analysis **11**, 368–385 (1981)

Cellier, D., Fourdrinier, D., Robert, C.: Robust shrinkage estimators of the location parameter for elliptically symmetric distributions. Journal of Multivariate Analysis **29**, 39–52 (1989)

Chamberlain, G.A.: A characterization of the distributions that imply mean-variance utility functions. Journal of Economic Theory **29**, 185–201 (1983)

Chan, L.K.C., Karceski, J., Lakonishok, J.: On portfolio optimization: forecasting and choosing the risk model. The Review of Financial Studies **12**, 937–974 (1999)

Chen, J.T., Gupta, A.K., Troskie, C.G.: The distribution of stock returns when the market is up. Communications in Statistics: Theory and Methods **32**, 1541–1558 (2003)

Chmielewski, M.A.: Invariant scale matrix hypothesis tests under elliptical symmetry. Journal of Multivariate Analysis **10**, 343–350 (1980)

Chmielewski, M.A.: Elliptically symmetric distributions: a review and bibliography. International Statistical Review **49**, 67–74 (1981)

Chopra, V.K., Ziemba, W.T.: The effect of errors in means, variances and covariances on optimal portfolio choice. The Journal of Portfolio Management **Winter 1993**, 6–11 (1993)

Chu, K.C.: Estimation and decision for linear systems with elliptical random processes. IEEE Transactions on Automatic Control **18**, 499–505 (1973)

Cléroux, R., Ducharme, G.R.: Vector correlation for elliptical distributions. Communications in Statistics–Theory and Methods **18**, 1441–1454 (1989)

Cochrane, J.H.: Portfolio advice for a multifactor world. NBER working paper 7170 (1999)

Coelli, T.J., Prasada Rao, D.S., O'Donnell, C.J., Battese, G.: An Introduction to Efficiency and Productivity Analysis. Springer Science+Business Media, New York (2005)

Constandinidis, G.M., Malliaris, A.G.: Portfolio theory. In: Jarrow, R., Maksimovic, V., Ziemba, W.T.:(eds.) Handbooks in Operations Research and Management Science, Vol. 9, pp 1–30. North-Holland, Amsterdam (1995).

Conte, S.D., de Boor, C.: Elementary Numerical Analysis. McGraw-Hill, London (1981)

Copas, J.B., Li, H.G.: Inference for non-random samples. Journal of the Royal Statistical Society B **59**, 55–95 (1997)

Cornish, E.A., The multivariate t-distribution associated with a set of normal sample deviates. Australian Journal of Physics **7**, 531–542 (1954)

Dawid, A.P.: Spherical matrix distributions and a multivariate model. Journal of the Royal Statistical Society Ser. B **39**, 254–261 (1977)

Dickey, J.M.: Matrix variate generalizations of the multivariate t distribution and the inverted multivariate t distribution. Annals of Mathematical Statistics **38**, 511–518 (1967)

Domínguez-Molina, J.A., González-Farías, G., Gupta, A.K.: The multivariate closed skew normal distribution. Technical Report No. 03–12, Department of Mathematics and Statistics, Bowling Green State University, Bowling Green, OH (2003)

Domínguez-Molina, J.A., González-Farías, G., Gupta, A.K.: A Matrix Variate Closed Skew-Normal Distribution with Applications to Stochastic Frontier Analysis. Communications in Statistics: Theory and Methods 36, 1691–1703 (2007)

Domínguez-Molina, J.A., González-Farías, G., Ramos-Quiroga, R.: Skew normality in stochastic frontier analysis. In: Genton, M.G. (ed.) Skew-Elliptical Distributions and Their Applications: A Journey Beyond Normality, pp. 223–241. Chapman and Hall/CRC, Boca Raton, FL (2004)

Dunnett, C.W., Sobel, M.A.: Bivariate generalization of student's t-distribution, with tables for certain special cases. Biometrika **41**, 153–159 (1954)

Eaton, M.L.: Multivariate Statistical Analysis. Institut for Mathematisk Statistik, Kobenhavns Universitet, Kobenhavn (1972)

Engle, R.F.: Autoregressive conditional heteroscedasticity with estimates of the variance of U.K. inflation. Econometrica **50**, 987–1008 (1982)

Engle, R.F.: Dynamic conditional correlation – a simple class of multivariate GARCH models. Journal of Business and Economic Statistics **20**, 339–350 (2002)
Fama, E.F.: The behavior of stock market prices. Journal of Business **38**, 34–105 (1965)
Fama, E.F.: Foundations of finance. Basic Books, New York (1976)
Fan, J., Fang, K.T.: Minimax estimators and Stein's two-stage estimators of location parameters. Chinese Journal of Applied Probability and Statistics **2**, 103–114 (1985a) (Reprinted in Fang and Anderson (1990))
Fan, J., Fang, K.T.: Inadmissibility of sample mean and sample regression coefficients for elliptically contoured distributions. Northeastern Mathematical Journal **1**, 68–81 (1985b) (Reprinted in Fang and Anderson (1990))
Fang, K.T., Anderson, T.W.: Statistical Inference in Elliptically Contoured and Related Distributions. Allerton Press Inc., New York (1990)
Fang, K.T., Chen, H.F.: Relationships among classes of spherical matrix distributions. Acta Mathematicae Sinica (English Series) **1(2)**, 138–148 (1984) (Reprinted in Fang and Anderson (1990))
Fang, K.T., Kotz, S., Ng, K.W.: Symmetric Multivariate and Related Distributions. Chapman and Hall, London, New York (1990)
Fang, K.T., Wu, Y.: Distribution of quadratic forms and Cochran's theorem. Mathematics in Economics **1**, 29–48 (1984)
Fang, K.T., Zhang, Y.T.: Generalized Multivariate Analysis. Springer-Verlag, New York (1990)
Feller, W.: An Introduction to Probabilit Theoryy and Its Applications, vol. I, 2nd Edition. Wiley, New York (1957)
Finney, D.J.: The joint distribution of variance ratios based on a common error mean square. Annals of Eugenics **11**, 136–140 (1941)
Flecher, C., Naveaua, P., Allard, D.: Estimating the closed skew-normal distribution parameters using weighted moments. Statistics and Probability Letters **79**, 1977–1984 (2009)
Fleming, J., Kirby, C., Ostdiek, B.: The economic value of volatility timing. The Journal of Finance **56**, 329–352 (2001)
Framstad, N.C.: Portfolio separation properties of the skew-elliptical distributions, with generalizations. Statistics and Probability Letters **81**, 1862–1866 (2011)
Genton, M.G.: Skew-Elliptical Distributions and Their Applications: A Journey Beyond Normality. Boca Raton, FL: Chapman and Hall/CRC (2004)
Genton, M.G., He, L., Liu, X.: Moments of skew-normal random vectors and their quadratic forms. Statistics and Probability Letter **51**, 319–325 (2001)
Gibbons, M.: Multivariate tests of financial models: a new approach. Journal of Financial Economics **10**, 3–27 (1982)
Gibbons, M.R., Ross, S.A., Shanken, J.: A test of the efficiency of a given portfolio. Econometrica **57**, 1121–1152 (1989)
González-Farías, G., Domínguez-Molina, J.A., Gupta, A.K.: The closed skew normal distribution. In: Genton, M.G. (ed.) Skew-Elliptical Distributions and Their Applications: A Journey Beyond Normality, pp. 25–42. Chapman and Hall/CRC, Boca Raton, FL (2004a)
González-Farías, G., Domínguez-Molina, J.A., Gupta, A.K.: Additive properties of skew normal random vectors. Journal of Statistical Planning and Inference **126**, 521–534 (2004b)
Graybill, F.A.: Introduction to Matrices with Applications to Statistics. Wadsworth Publishing Company, Belmont, California (1969)
Greene, W.H.: Econometric Analysis. Pearson/Prentice Hall, New Jersey (2003)
Grübel, R., Rocke, D.M.: On the cumulants of affine equivariant estimators in elliptical families. Department of Mathematics, Imperial College, London. Unpublished (1989)
Gupta, A.K.: Noncentral distribution of Wilks' statistic in MANOVA. Annals of Mathematical Statistics **42**, 1254–1261 (1971)
Gupta, A.K.: On a stochastic inequality for the Wilks' statistic. Annals of the Institute of Statistical Mathematics **27**, 341–348 (1975)
Gupta, A.K.: On the sphericity test criterion in the multivariate Gaussian distribution. Australian Journal of Statistics **19**, 202–205 (1977)

Gupta, A.K.: Multivariate skew t-distribution. Statistics **37**, 359–363 (2004)

Gupta, A.K., Chang, F.C., Huang, W.J: Some skew-symmetric models. Random Operators and Stochastic Equations **10**, 133–140 (2002)

Gupta, A.K., Chattopadhyay, A.K., Krishnaiah, P.R.: Asymptotic distributions of the determinants of some random matrices. Communications in Statistics **4**, 33–47 (1975)

Gupta, A.K., Chen, J.T.: Goodness-of-fit tests for the skew normal distribution. Communications in Statistics-Simulation and Computation 30, 907–930 (2001)

Gupta, A.K., Huang, W.J.: Quadratic forms in skew normal variates. Journal of Mathematical Analysis and Applications **273**, 558–564 (2002)

Gupta, A.K., Javier, W. R. Nonnull distribution of the determinant of B-statistic in multivariate analysis. South African Journal of Statistics **20**, 87–102 (1986)

Gupta, A.K., Kollo, T.: Density expansions based on multivariate skew normal distribution. Sankhya **65**, 821–835 (2003)

Gupta, A.K., Nagar, D.K.: Likelihood ratio test for multisample sphericity. In: Gupta, A.K. (ed) Advances in Multivariate Statistical Analysis, pp. 111–139. Reidel Publishing Company, Dordretch (1987)

Gupta, A.K., Nagar, D.K.: Asymptotic expansion of the nonnull distribution of likelihood ratio statistic for testing multisample sphericity. Communications in Statistics–Theory and Methods **17**, 3145–3156 (1988)

Gupta, A.K., Nagar, D.K.: Matrix Variate Distributions. Chapman and Hall/CRC, Boca Raton (2000)

Gupta, A.K., Tang, J.: Distribution of likelihood ratio statistic for tetsing equality of covariance matrices of multivariate Gaussian models. Biometrika **71**, 555–559 (1984)

Gupta, A.K., Tang, J.: On a general distribution theory for a class of likelihood ratio criteria. Australian Journal of Statistics **30**, 359–366 (1988)

Gupta, A.K., Varga, T.: Characterization of joint density by conditional densities. Communications in Statistics–Theory and Methods **19**, 4643–4652 (1990)

Gupta, A.K., Varga, T.: Rank of a quadratic form in an elliptically contoured matrix random variable. Statistics and Probability Letters **12**, 131–134 (1991)

Gupta, A.K., Varga, T.: Characterization of matrix variate normal distributions. Journal of Multivariate Analysis **41**, 80–88 (1992)

Gupta, A.K., Varga, T.: Characterization of matrix variate normality through conditional distributions. Mathematical Methods of Statistics **3**, 163–170 (1994a)

Gupta, A.K., Varga, T.: A new class of matrix variate elliptically contoured distributions. Journal of Italian Statistical Society **3**, 255–270 (1994b)

Gupta, A.K., Varga, T.: Moments and other expected values for matrix variate elliptically contoured distributions. Statistica **54**, 361–373 (1994c)

Gupta, A.K., Varga, T.: Some applications of the stochastic representation of elliptically contoured distribution. Random Operators and Stochastic Equations **2**, 1–11 (1994d)

Gupta, A.K., Varga, T.: Normal mixture representation of matrix variate elliptically contoured distributions. Sankhya, Ser. A **51**, 68–78 (1995a)

Gupta, A.K., Varga, T.: Some inference problems for matrix variate elliptically contoured distributions. Statistics **26**, 219–229 (1995b)

Gupta, A.K., Varga, T.: Characterization of matrix variate elliptically contoured distributions. In: Johnson, N.L. and Balakrishnan, N. (eds.) Advances in the Theory and Practice of Statistics: A volume in honor of S. Kotz, pp. 455–467. Wiley, New York, (1997)

Harvey, C.R., Leichty, J.C., Leichty, M.W., Muller, P.: Portfolio selection with higher moments. Quantitative Finance **10**, 469–485 (2010)

Harville, D.A.: Matrix Algebra from a Statistician's Perspective. Springer-Verlag, New York (1997)

Hayakawa, T.: Normalizing and variance stabilizing transformations of multivariate statistics under an elliptical population. Annals of the Institute of Statistical Mathematics **39A**, 299–306 (1987)

Heathcote, C.R., Cheng, B., Rachev, S.T.: Testing multivariate symmetry. Journal of Multivariate Analysis **54**, 91–112 (1995)

Hocking, R.R.: The Analysis of Linear Models. Brooks and Cole Publishing Company, Belmont, California (1985)

Hodgson, D.J., Linton, O., Vorkink, K.: Testing the capital asset pricing model efficiency under elliptical symmetry: a semiparametric approach. Journal of Applied Econometrics **17**, 617–639 (2002)

Horn, R.A., Johnson, C.R.: Matrix Analysis. Cambridge University Press, Cambridge (1985)

Hsu, H.: Generalized T^2-test for multivariate elliptically contoured distributions. Technical Report 14. Department of Statistics, Stanford University, California(1985a) (Reprinted in Fang and Anderson (1990))

Hsu, H.: Invariant tests for multivariate elliptically contoured distributions. Technical Report 14. Department of Statistics, Stanford University, California (1985b) (Reprinted in Fang and Anderson (1990))

Jagannathan, R., Ma, T.: Risk reduction in large portfolio: why imposing wrong constraints helps. Journal of Finance **58**, 1651–1683 (2003)

Jajuga, K.: Elliptically symmetric distributions and their application to classification and regression. Proceedings of the Second International Tampere Conference in Statistics. Department of Mathematical Sciences, University of Tampere, 491–498 (1987)

Javier, W.R., Gupta, A.K.: On matrix variate t-distribution. Communications in Statistics–Theory and Methods **14**, 1413–1425 (1985a)

Javier, W.R., Gupta, A.K.: On generalized matrix variate beta distributions. Statistics **16**, 549–558 (1985b)

Jobson, J.D.: Confidence regions for the mean-variance efficient set: an alternative approach to estimation risk. Review of Quantitative Finance and Accounting **1**, 235–257 (1991)

Jobson, J.D., Korkie, B.: Estimation of Markowitz efficient portfolios. Journal of the American Statistical Association **75**, 544–554 (1980)

Jobson, J.D., Korkie, B.: A performance interpretation of multivariate tests of asset set intersection, spanning, and mean-variance efficiency. Journal of Financial and Quantitative Analysis **24**, s185–204 (1989)

Johnson, M.E.: Multivariate Statistical Simulation. Wiley, New York (1987)

Johnson, N.L., Kotz, S.: Distributions in Statistics: Continuous Multivariate Distributions. Wiley, New York (1972)

Johnson, N.L., Kotz, S., Balakrishnan, N.: Continuous Univariate Distributions, vol.2. Wiley, New York (1995)

Jondeau, E., Rockinger, M.: Optimal portfolio allocation under higher moments. European Financial Management **12**, 29–55 (2006)

Kan, R., Smith, D.R.: The distribution of the sample minimum-variance frontier. Management Science **54**, 1364–1380 (2008)

Kandel, S.: Likelihood ratio statistics of mean-variance efficiency without a riskless asset. Journal of Financial Economics **13**, 575–592 (1984)

Kariya, T.A.: robustness property of Hotelling's T^2-test. The Annals of Statistics **9(1)**, 211–214(1981)

Karlin, S.: Decision theory for Polya-type distributions. Case of two actions, I. Proceedings of the Third Berkeley Symposium on Mathematical Statistics and Probability **1**, pp. 115–129. University of California Press, Berkeley (1956)

Kelker, D.: Distribution theory of spherical distributions and a location-scale parameter generalization. Sankhya, Ser. A **32**, 419–430 (1970)

Khatri, C.G.: A note on an inequality for elliptically contoured distributions. Gujarat Statistical Review **7**, 17–21 (1980)

Khatri, C.G.: Quadratic forms and null robustness for elliptical distributions. Proceedings of the Second International Tampere Conference in Statistics. Department of Mathematical Sciences, University of Tampere, 177–203 (1987)

Khatri, C.G.: Some asymptotic inferential problems connected with elliptical distributions. Journal of Multivariate Analysis **27**, 319–333 (1988)

Khatri, C.G., Mukerjee, R.: Characterization of normality within the class of elliptical contoured distributions. Statistics and Probability Letters **5**, 187–190 (1987)
Khattree, R., Peddada, S.D.: A short note on Pitman nearness for elliptically symmetric estimators. Journal of Statistical Planning and Inference **16**, 257–260 (1987)
Kingman, J.F.C.: On random sequences with spherical symmetry. Biometrika **59**, 494–494 (1972)
Konishi, S., Gupta, A.K.: Testing the equality of several intraclass correation coefficients. Journal of Statistical Planning and Inference **19**, 93–105 (1989)
Krishnaiah, P.R., Lin, J.: Complex elliptically symmetric distributions. Communications in Statistics–Theory and Methods **15**, 3693–3718 (1986)
Kroll, Y., Levy, H., Markowitz, H.M.: Mean-variance versus direct utility maximization. Journal of Finance **39**, 47–61 (1984)
Kuritsyn, Y.G.: On the least-squares method for elliptically contoured distributions. Theory of Probability and its Applications **31**, 738–740 (1986)
Lam, Y.-M.: Confidence limits for non-centrality parameters of noncentral chi-squared and F distributions. ASA Proceedings of the Statistical Computing Section, 441–443 (1987)
Laurent, A.G.: Distribution d-echantillon et de caractéristiques d'echantillions quand la population de référence est Laplace-Caussienne de parametèters inconnus. Journal de la Société de Statistique de Paris **96**, 262–296 (1955)
Lehmann, E.L.: Testing Statistical Hypothesis. Wiley, New York (1959)
Loperfido, N.: Quadratic forms of skew-normal random vectors. Statistics and Probability Letter **54**, 381–387 (2001)
Lukacs, E.: Characterizations of populations by properties of suitable statistics. Proceedings of the Third Berkeley Symposium on Mathematical Statistics and Probability **2**, University of California Press, Los Angeles and Berkeley (1956)
MacKinley, A.C., Pastor, L.: Asset pricing models: implications for expected returns and portfolio selection. The Review of Financial Studies **13**, 883–916 (2000)
Magnus, J.R., Neudecker, H.: Matrix Differential Calculus with Applications in Statistics and Econometrics. Wiley, New York (1988)
Manzotti, A., Perez, F.J., Quiroz, A.J., A statistic for testing the null hypothesis of elliptical symmetry. Journal of Multivariate Analysis **81**, 274–285 (2002)
Marcinkiewicz, J.: Sur les fonctions indépendantes III. Fundamenta Mathematicae **31**, 86–102 (1938)
Markowitz, H.M.: Portfolio selection. The Journal of Finance **7**, 77–91 (1952)
Markowitz, H.M.: Foundations of portfolio theory. The Journal of Finance **7**, 469–477 (1991)
Mateu-Figueras, G., Puig, P., Pewsey, A.: Goodness-of-fit tests for the skew-normal distribution when the parameters are estimated from the data. Communications in Statistics–Theory and Methods **36**, 1735–1755 (2007)
Marshall, A., Olkin, I.: Inequalities: Theory of Majorization and Its Applications, Academic Press, New York (1979)
Meeusen, W., van Den Broeck, J.: Efficiency estimation from Cobb-Douglas production functions with composed error. International Economic Review **18**, 435–444 (1977)
Meintanis, S.G.: A Kolmogorov-Smirnov type test for skew normal distributions based on the empirical moment generating function. Journal of Statistical Planning and Inference **137**, 2681–2688 (2007)
Meintanis, S.G.: Testing skew normality via the moment generating function. Mathematical Methods of Statistics **19**, 64–72 (2010)
Meintanis S.G., Hlávka, Z.: Goodness-of-fit tests for bivariate and multivariate skew-normal distributions. Scandinavian Journal of Statistics **37**, 701–714 (2010)
Mencía, J., Sentana, E.: Multivariate location-scale mixtures of normals and mean-variance-skewness portfolio allocation. Journal of Econometrics **153**, 105–121 (2009)
Merton, R.C.: An analytical derivation of the efficient frontier. Journal of Financial and Quantitative Analysis **7**, 1851–1872 (1972)
Merton, R.C.: On estimating the expected return on the market: an exploratory investigation. Journal of Financial Economics **8**, 323–361 (1980)

Mitchell, A.F.S.: The information matrix, skewness tensor and L-connections for the general multivariate elliptic distributions. Annals of the Institute of Statistical Mathematics **41**, 289–304 (1989)

Mitchell, A.F.S., Krzanowski, W.J.: The Mahalanobis distance and elliptic distributions. Biometrika **72(2)**, 464–467 (1985)

Mittnik, S., Rachev, S.T.: Modelling asset returns with alternative stable distributions, Econometric Reviews, **12**, 261–330 (1993)

Mood, A.M., Graybill, F.A., Boes, D.C.: Introduction to the Theory of Statistics. McGraw-Hill, New York (1974)

Muirhead, R.J.: Aspects of Multivariate Statistical Theory. Wiley, New York (1982)

Nagao, H.: On some test criteria for covariance matrix. Annals of Statistics **1**, 700–709 (1973)

Nagar, D.K., Gupta, A.K.: Nonnull distribution of LRC for testing $H_0 : \mu = \mathbf{0}, \Sigma = \sigma^2\mathbf{I}$ in multivariate normal distribution. Statistica **46**, 291–296 (1986)

Nel, H.M.: On distributions and moments associated with matrix normal distributions. Technical Report 24, Department of Mathematical Statistics, University of the Orange Free State (1977)

Nelsen, R.B.: An Introduction to Copulas. 2nd ed. Springer Science+Business Media, New York (2006)

Nelson, D.: Conditional heteroscedasticity in stock returns: a new approach. Econometrica **59**, 347–370 (1991)

Okamoto, M.: Distinctness of the eigenvalues of a quadratic form in a multivariate sample. The Annals of Statistics **1**, 763–765 (1973)

Okhrin, Y., Schmid, W.: Distributional properties of portfolio weights. Journal of Econometrics **134**, 235–256 (2006)

Olkin, I., Rubin, H.: Multivariate beta distributions and independence properties of the Wishart distribution. Annals of Mathematical Statistics **35**, 261–269 (1964)

Osborne, M.F.M.: Brownian motion in the stock market. Operation Research **7**, 145–173 (1959)

Owen, J., Rabinovitch, R.: On the class of elliptical distributions and their applications to the theory of portfolio choice. Journal of Finance **38**, 745–752 (1983)

Patton, A.J.: On the out-of-sample importance of skewness and asymmetric dependence for asset allocation. Journal of Financial Econometrics **2**, 130–168 (2004)

Pewsey, A.: Problems of inference for Azzalini's skew-normal distribution. Journal of Applied Statistics **27**, 859–870 (2000)

Pérez Rodríguez, P., Villaseñor Alva, J.A.: On testing the skew normal hypothesis. Journal of Statistical Planning and Inference **140**, 3148–3159 (2010)

Pillai, K.C.S., Gupta, A.K.: Exact distribution of Wilks' likelihood ratio criterion. Biometrika **56**, 109–118 (1969)

Press, S.J.: Applied Multivariate Analysis. Holt, Rinehart, and Winston Incorporated, New York (1972)

Quan, H.: Some optimal properties of testing hypotheses of elliptically contoured distributions. Acta Mathematicae Applicatae Sinica **3:1**, 1–14 (1987) (Reprinted in Fang and Anderson (1990))

Quan, H., Fang, K.T.: Unbiasedness of the parameter tests for generalized multivariate distributions. Acta Mathematicae Applicatae Sinica **10**, 215–234 (1987) (Reprinted in Fang and Anderson (1990))

Rachev, S.T., Mittnik, S.: Stable Paretian Models in Finance. Wiley, New York (2000)

Rao, C.R.: Linear Statistical Inference and Its Applications, 2nd Edition. Wiley, New York (1973)

Rao, C.R., Mitra, S.K.: Generalized Inverse of Matrices and its Applications. Wiley, New York (1971)

Rao, C.R., Toutenburg, H.: Linear Models. Springer, New York, Berlin, Heidelberg (1995)

Richards, D.: Hyperspherical models, fractional derivatives and exponential distributions on matrix spaces. Sankhya, Ser. A **46**, 155–165 (1984)

Sampson, A.R.: Positive dependence properties of elliptically symmetric distributions. Journal of Multivariate Analysis **13**, 375–381 (1983)

Samuelson, P.A.: The fundamental approximation theorem of portfolio analysis in terms of means, variances, and higher moments. Review of Economical Studies **36**, 537–542 (1970)
Schoenberg, I.J.: Metric spaces and completely monotone functions. Annals of Mathematics **39**, 811–841 (1938)
Shanken, J.: Multivariate test of the zero-beta CAPM. Journal of Financial Economics **14**, 327–348 (1985)
Siegel, A.F., Woodgate, A.: Performance of portfolios optimized with estimation error. Management Science **53**, 1005–10015 (2007)
Siotani, M., Hayakawa, T., Fujikoshi, Y.: Modern Multivariate Statistical Analysis: A Graduate Course and Handbook. American Sciences Press, Inc., Columbus, Ohio (1985)
Smith, M.D.: On the expectation of a ratio of quadratic forms in normal variables. Journal of Multivariate Analysis **31**, 244–257 (1989)
Stambaugh, R.F.: On the exclusion of assets from tests of the two parameter model: a sensitivity analysis. Journal of Financial Economics **10**, 237–268 (1982)
Stiglitz, J.E.: Discussion: mutual funds, capital structure, and economic efficiency. In: Bhattacharya, S., Constantinides, G.M. (eds.) Theory and valuation. Rowman & Littlefield Publishers, New York (1989)
Stoyan, D.: Comparison Methods for Queues and Other Stochastic Models. Wiley, New York (1983)
Sutradhar, B.C.: Testing linear hypothesis with t-error variable. Sankhya Ser. B **50**, 175–180 (1988)
Sutradhar, B.C., Ali, M.M.: A generalization of the Wishart distribution for the elliptical model and its moments for the multivariate *t* model. Journal of Multivariate Analysis **29**, 155–162 (1989)
Tang, J., Gupta, A.K.: On the distribution of the product of independent beta random variables. Statistics and Probability Letters **2**, 165–168 (1984)
Tang, J., Gupta, A.K.: Exact distribution of certain general test statistics in multivariate analysis. Australian Journal of Statistics **28**, 107–114 (1986)
Tang, J., Gupta, A.K.: On the type-B integral equation and the distribution of Wilks' statistic for testing independence of several groups of variables. Statistics **18**, 379–387 (1987)
Timm, N.H.: Multivariate Analysis with Applications in Education and Psychology. Wadsworth Publishing Company, Inc., Belmont, California (1975)
Wang, J., Boyer, J., Genton, M.G.: A note on the equivalence between chi-square and generalized skew-normal distributions. Statistics and Probability Letter **66**, 395–398 (2004)
Zellner, A.: Bayesian and non-Bayesian analysis of the regression model with multivariate student-t error terms. Journal of the American Statistical Association **71**, 400–405 (1976)
Zhou, G.: Asset-pricing tests under alternative distributions. The Journal of Finance **48**, 1927–1942 (1993)
Zhu, L.X., Neuhaus, G.: Conditional tests for elliptical symmetry. Journal of Multivariate Analysis **84**, 284–298 (2003)

Author Index

A.K. Gupta et al., *Elliptically Contoured Models in Statistics and Portfolio Theory*, DOI 10.1007/978-1-4614-8154-6, © Springer Science+Business Media New York 2013

Subject Index

A.K. Gupta et al., *Elliptically Contoured Models in Statistics and Portfolio Theory*,
DOI 10.1007/978-1-4614-8154-6, © Springer Science+Business Media New York 2013

Printed in the United States
By Bookmasters